Encyclopedia of
ELECTRONIC CIRCUITS

Volume 3

Rudolf F. Graf

TAB Books
Division of McGraw-Hill, Inc.
New York   San Francisco   Washington, D.C.   Auckland   Bogotá
Caracas   Lisbon   London   Madrid   Mexico City   Milan
Montreal   New Delhi   San Juan   Singapore
Sydney   Tokyo   Toronto

FIRST EDITION
SIXTH PRINTING

© 1991 by **Rudolf F. Graf**.
Published by TAB Books.
TAB Books is a division of McGraw-Hill, Inc.

Library of Congress Cataloging-in-Publication Data

(Revised for Vol. 3)
Graf, Rudolf F.
    Encyclopedia of electronic circuits.

    Includes bibliographical references and indexes.
      1. Electronic circuits—Encyclopedias.   I. Title.
TK867.G66   1985   621.381′5   84-26772
ISBN 0-8306-0938-5 (v. 1)
ISBN 0-8306-1938-0 (pbk. : v. 1)
ISBN 0-8306-3138-0 (v. 2)
ISBN 0-8306-3138-0 (pbk. : v. 2)
ISBN 0-8306-7348-2 (v. 3)
ISBN 0-8306-3348-0 (pbk. : v. 3)

Acquisitions Editor: Roland S. Phelps
Technical Editor: Andrew Yoder
Director of Production: Katherine G. Brown

Contents

To Sheryl Melissa,
a budding scholar
From Popsi

Preface

Volume III of *The Encyclopedia of Electronic Circuits* adds about 1,000 new circuits to the ready-to-use files that were established by the publication of volumes I and II of this set of circuits encyclopedias.

These three volumes now offer an invaluable storehouse of about 3,000 carefully arranged and categorized, easy-to-access circuits. Volume IV is scheduled for publication in 1992.

Once again it gives me great pleasure to extend my gratitude to William Sheets for his comments and contributions, and to Mrs. Stella Dillon for her virtuoso performance on the word processor.

1

Active Antennas

The sources of the following circuits are contained in the Sources section beginning on page 782. The figure number contained in the box of each circuit correlates to the sources entry in the Sources section.

Active Antenna
Active Antenna

ACTIVE ANTENNA

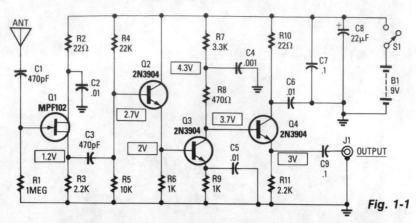

Fig. 1-1

Antennas that are much shorter than 1/4 wavelength present a very small and highly relative impedance that is dependent on the received frequency. It is difficult to match impedances over a decade of frequency coverage. Instead, input stage Q1 is an FET source-follower. A high-impedance input successfully bridges antenna characteristics at any frequency.

Transistor Q2 is used as an emitter-follower to provide a high-impedance load for Q1, but more importantly, it provides a low-drive impedance for common-emitter amplifier Q3, which provides all of the amplifier's voltage gain. Transistor Q4 transforms Q3's moderate output impedance into low impedance, thereby providing sufficient drive for a receiver's 50-Ω, antenna-input impedance.

ACTIVE ANTENNA

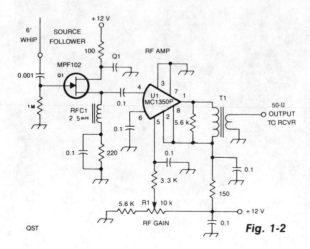

All capacitors in this circuit are disc ceramic. Fixed-value resistors are 1/4- or 1/2-W carbon. R1 controls the gain of U1. RFC1 is a miniature 2.5-mH rf choke. T1 has 30 primary turns of #28 enamel wire on an Amidon FT50 – 43 ferrite toroid core, and the secondary has four turns of #28 wire.

Fig. 1-2

2

Alarm and Security Circuits

The sources of the following circuits are contained in the Sources section beginning on page 782. The figure number contained in the box of each circuit correlates to the sources entry in the Sources section.

Auto Alarm
One-Chip Burglar Alarm
Semiconductor Fail-Safe Alarm
Single-IC Auto Alarm
Burglar Alarm
Burglar Alarm

AUTO ALARM

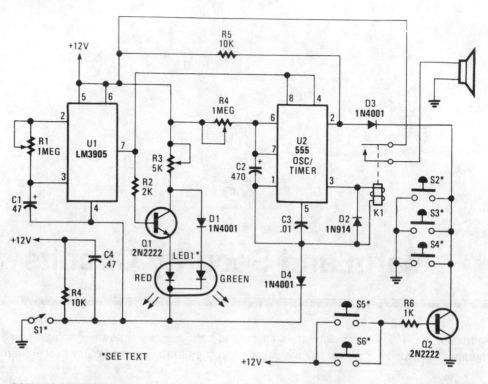

POPULAR ELECTRONICS

Fig. 2-1

In operation, the alarm circuit allows a 0–47 second time delay, as determined by the R1/C1 combination, after the switch is armed to allow the vehicle's motion sensor to settle down. This allows you time to get a bag of groceries out of the trunk and not have the hassle of juggling the groceries and the key switch at once.

During the time delay, half of LED1, which is actually a single, bi-colored, three-legged common cathode device, lights green. At the same time, pins 8 and 4 of U2 (a 555 oscillator/timer) are held low by U1 (a 3905 oscillator/timer), causing the alarm to remain silent. Once the delay is over, LED1 turns red, indicating that the circuit is armed.

At that point, a ground at pin 2 of U2 forces pin 3 of U2 high, closing the contacts of K1 and sounding the siren for a time duration determined by R4 and C2. Once the time has elapsed, pin 3 is pulled low, K1 opens, and the circuit is again ready to go. The circuit can be manually reset by the simple expedient of opening and closing the key switch. Potentiometer R3 controls the LED's illumination intensity. Diode D1 ensures that the green segment of LED1 is fully extinguished when Q1 is turned on – which turns the LED to red. Resistors R4 and R5 must be connected to the +V bus, not to pin 7 of U1, otherwise U2 will mysteriously trigger itself each time the initial delay ends.

ONE-CHIP BURGLAR ALARM

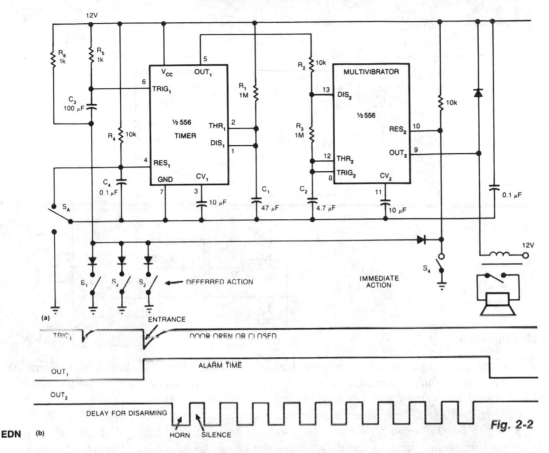

Fig. 2-2

The single-chip, burglar-alarm circuit shown uses a dual 556 timer, draws 10 mA of standby current, and generates a pulsing alarm signal that conserves battery energy. Once activated, the alarm will remain on, independent of the subsequent state of any of the sensors. The sensors support both deferred and immediate-action modes. You can attach this circuit to your car's internal lighting circuitry using a single wire and a relay. To arm this circuit, you open your car door and close switch SA. The switch discharges capacitor C4 and holds the timer (one half of the 556 IC) in a reset state to prevent false triggering while you're arming the circuit. When you close your car door, the circuit enters a standby mode. If the door is then reopened, the sensors apply a negative-going pulse to trigger 1. Output 1 then increases, enabling the alarm for 1.1R1C1 seconds. Output 1's high state triggers the multivibrator, the other half of the 556, which begins to cycle after a delay equal to 1.1 (R2 + R3) C2 seconds. As long as the timer's output stays high, the multivibrator will continue to cycle, turning the horn off and on at 3.3-second intervals. During the interval between time that the timer's output increases and the time that the multivibrator's output decreases, you can disarm this circuit using switch SA. To prevent false triggering caused by switch contacts, at S1, S2, and S3, that may bounce when closing the door, make the R6C3 time constant as large as possible. In addition, capacitors C1 and C2 should be tantalum types and should exhibit leakage of less than 1 μA at room temperature.

SEMICONDUCTOR FAIL-SAFE ALARM

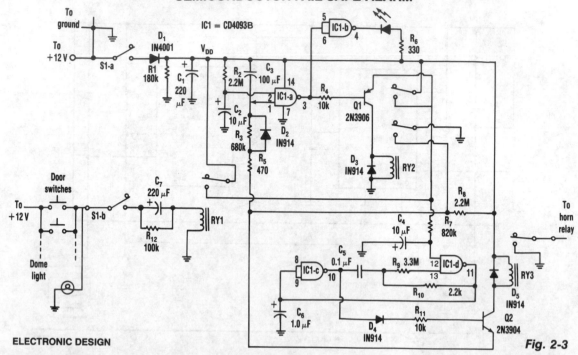

ELECTRONIC DESIGN

Fig. 2-3

False alarms produced by semiconductor failure are impossible with this burglar-alarm circuit equipped with relays. What's more, the circuit is virtually immune to false triggering. With a standby current of less than 0.1 mA, the circuit offers all the features an alarm needs: entry and exit delays, a timed alarm period, and automatic reset after an intrusion.

One CMOS CD4093B quad NAND gate, IC1, supplies both logic and analog timing functions with the aid of Schmitt-trigger switching action. Relays make the circuit fail-safe in the alarm-active mode, even when the semiconductors fail. The relays are 12-V, with coil resistances of 250 Ω or more.

Closing switch S1 initiates circuit operation. Capacitor C2 begins charging through resistor R2 and arming indicator LED1 lights. When pin 2 of IC1a reaches its switching point, its output decreases, extinguishing LED1 and indicating that the exit delay has ended. That output also drives the base of Q1 low, so that if the emitter circuit completes to the V_{DD} line, Q1 conducts. The circuit is now armed, and current drain drops to less than 0.1 mA.

When the vehicle is entered, relay RY1 contacts close momentarily, completing the emitter circuit of Q1 and causing the RY2 contacts to close. Charging C4 through R7 determines the entry-delay period. If the system isn't turned off by opening S1 during this period, the oscillator circuit of IC1c and IC1d activates, and a rapid on/off horn-honking cycle kicks on with the aid of Q2 and RY3.

The alarm cycle ends after about a minute, when C2 charges through R3 to the threshold voltage of IC1a at pin 1. This voltage resets the timing circuit, readying it for another entry/alarm cycle.

RY1 is connected for vehicles that use door switches connected to +12 V. For vehicles that use grounding door switches, the bottom of the RY1 coil should connect to +12 V instead of ground. In the latter case, the polarity of C7 should be reversed.

For home use, the R3C3 time constant should be increased to give a longer alarm.

SINGLE-IC AUTO ALARM

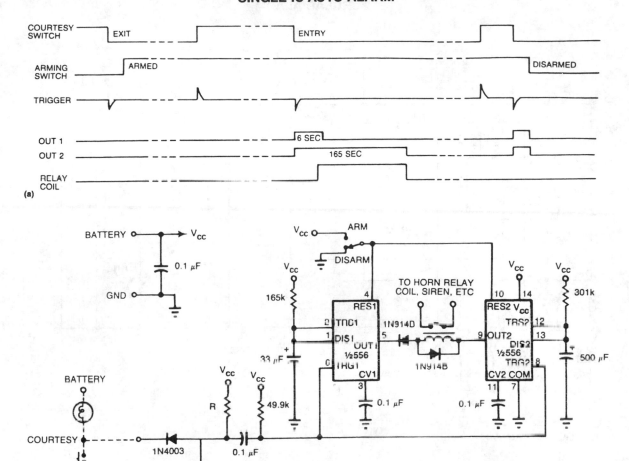

(a)

(b)

EDN

Fig. 2-4

See (a) for the timing information for the alarm circuit in (b). When leaving your vehicle, flip the arming switch and close the door to arm the device. Subsequent opening of an entrance triggers both timers. After the expiration of the entry delay timer, the alarm sounds for a time determined by the second timer. The value of R should be less than 1 KΩ. If you use an incandescent lamp instead of a resistor, you get an extra function—an open-entrance indicator. By keeping the resistance low, you avoid false tripping should water collect under the hood. If your door switch connects the courtesy light to 12 V rather than ground, use a single transistor as an inverter at the input.

BURGLAR ALARM

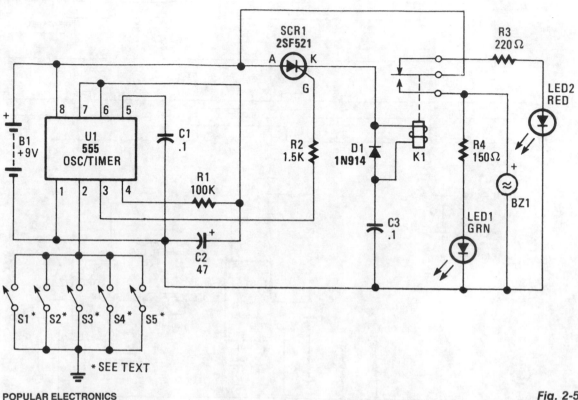

POPULAR ELECTRONICS

Fig. 2-5

The heart of the circuit is a 555 oscillator/timer, U1, configured for monostable operation. The output of U1 at pin 3 is tied to the gate of SCR1. As long as S1–S5, which are connected to the trigger input of U1, are open, the circuit remains in the ready state, and does not trigger SCR1 into conduction. Because the relay is not energized, battery current is routed through the relay's normally-closed terminal and through current-limiting resistor R3 to LED2, causing it to light.

However, when one of the switches (S1–S5) is closed, grounding U1 pin 2, the output of U1 at pin 3 increases, activating SCR1. That energizes the relay, pulling the wiper of K1 to the normally-open terminal, causing LED1 to light and BZ1 to sound.

The duration of the output is determined by the RC time-constant circuit, formed by R1 and C1. Resistor R2 regulates the output of U1 to a safe value for the gate of SCR1. Switches S1–S5 are to doors, windows, etc. A switch can be connected in series with B1 to activate and deactivate the alarm circuit when it's not needed.

BURGLAR ALARM

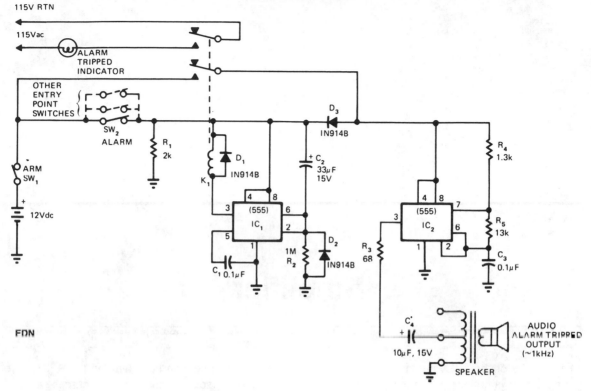

Fig. 2-6

This circuit cannot be shut off for 10 to 60 seconds—even if the trip condition is immediately removed. It draws no standby power from the battery and is self-resetting.

3

Amplifiers

The sources of the following circuits are contained in the Sources section beginning on page 782. The figure number contained in the box of each circuit correlates to the sources entry in the Sources section.

Input/Output Buffer Amplifier for
    Analog Multiplexers
Absolute-Value Norton Amplifier
Intrinsically Safe Protected Op Amp
±15-V Chopper Amplifier
Composite Amplifier
Cascaded Amplifier
Inverting Amplifier
Noninverting Amplifier
Differential Amplifier
Active Clamp-Limiting Amplifier
Wide-Band AGC Amplifier

Polarity-Reversing Low-Power Amplifier
Summing Amplifier
Ac-Coupled Dynamic Amplifier
Forward-Current Booster
Dc-Stabilized Fast Amplifier
Write Amplifier
Low-Noise Photodiode Amplifier
Voltage-Follower Amplifier for
    Signal-Supply Operation
Current-Shunt Amplifier
Constant-Bandwidth Amplifier

INPUT/OUTPUT BUFFER AMPLIFIER FOR ANALOG MULTIPLEXERS

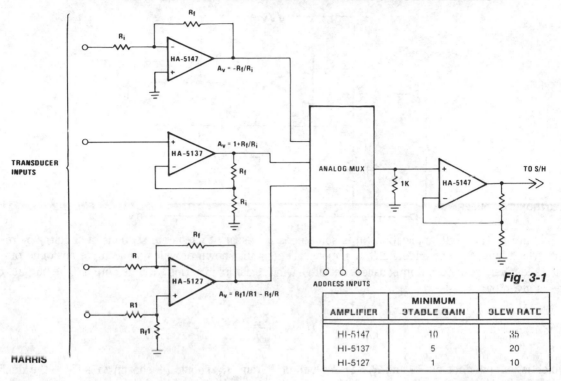

Fig. 3-1

AMPLIFIER	MINIMUM STABLE GAIN	SLEW RATE
HI-5147	10	35
HI-5137	5	20
HI-5127	1	10

The precision input characteristics of the HA-5147 help simplify system *error budgets*, while its speed and drive capabilities provide fast charging of the multiplexer's output capacitance. This speed eliminates an increased multiplexer acquisition time, which can be induced by more limited amplifiers. The HA-5147 accurately transfers information to the next stage while effectively reducing any loading effects on the multiplexer's output.

ABSOLUTE VALUE NORTON AMPLIFIER

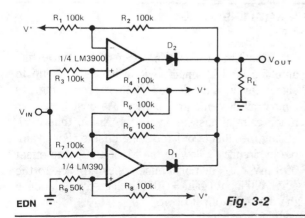

Fig. 3-2

The noninverting amplifier has a gain of R2/R3 (1 in this case) and produces a voltage of V_{out} during a positive excursion of V_{in} with respect to ground. The inverting amplifier accommodates the negative excursions of V_{in}; its gain is given by $-$R6/R7, which equals -1 to maintain symmetry with the noninverting amplifier. R9 provides adjustment for the symmetry, supply variations, and offsets. Even though the circuit operates on a single supply, V_{in} can go negative to the same extent that it goes positive.

11

INTRINSICALLY SAFE PROTECTED OP AMP

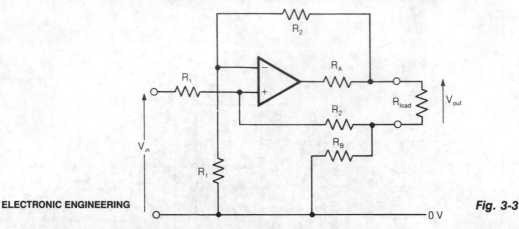

ELECTRONIC ENGINEERING

Fig. 3-3

In intrinsically safe applications, it is sometimes necessary to separate sections of circuitry by resistors which limit current under fault conditions. The circuit shown provides an accurate analogue output with effectively zero output impedance, despite having resistors in series with the output. The output voltage is given by:

$$V_{\text{out}} = \frac{V_{\text{in}}\, R_2}{R_1}$$

which is independent of R_A and R_B. The values of R_A and R_B should be chosen to achieve the desired current limiting, but note that a proportion of the voltage given at the op-amp output will be dropped across these resistors. This limits the output swing at the load to approximately:

$$\frac{V_S\, R_{\text{load}}}{R_A + R_B + R_{\text{load}}}$$

where: V_S = voltage swing at the op-amp output. Any type of op amp would be suitable.

± 15-V CHOPPER AMPLIFIER

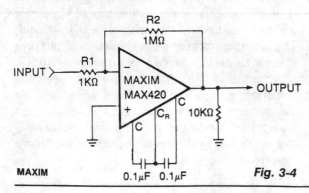

MAXIM

Fig. 3-4

This simple circuit is a gain-of-1000 inverting amplifier. It will amplify submillivolt signals up to signal levels suitable for further processing. In almost all system applications, it is best to use as much gain as possible in the MAX420, thus minimizing the effects of later-stage offsets. For example, if circuitry following the MAX420 has an offset of 5 mV, the additional offset referred back to the MAX420 input (gain = 1000) will be 5 μV, doubling the system's offset error.

COMPOSITE AMPLIFIER

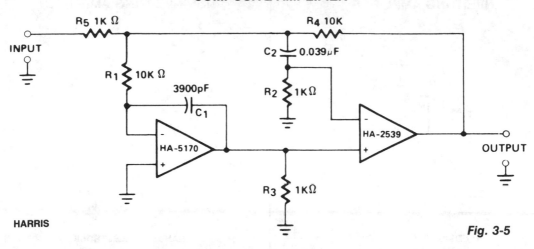

HARRIS

Fig. 3-5

A composite configuration greatly reduces dc errors without compromising the high-speed, wideband characteristics of HA-2539. The HA-2540 could also be used, but with slightly lower speeds and bandwidth response.

The HA-2539 amplifies signals above 40 kHz which are fed forward via C2; R2 and R5 set the voltage gain at -10. The slew rate of this circuit was measured at 350 V/μs. Settling time to a 0.1% level for a 10-V output step is under 150 ns and the gain bandwidth product is 300 MHz.

The HA-5170 amplifies signals below 40 kHz, as set by C1 and R1, and controls the dc input characteristics such as offset voltage, drift, and bias currents of the composite amplifier. Therefore, it has an offset voltage of 100 μV, drift of 2 μV/°C, and bias currents in the 20-pA range. The offset voltage can be externally nulled by connecting a 20-KΩ pot to pins 1 and 5, with the wiper tied to the negative supply. The dc gains of the HA-5170 and HA-2539 are cascaded; this means that the dc gain of the composite amplifier is well over 160 dB.

The excellent ac and dc performance of this composite amplifier is complemented by its low noise performance, 0.5-μV rms from 0.1 Hz to 100 Hz. It is very useful in high-speed data acquisition systems.

CASCADED AMPLIFIER

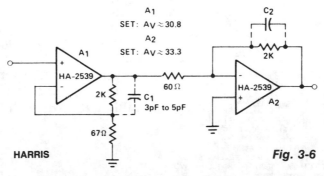

HARRIS

Fig. 3-6

Cascaded amplifier sections are used to extend bandwidth and increase gain. Using two HA-2539 devices, this circuit is capable of 60-dB gain at 20 MHz.

INVERTING AMPLIFIER

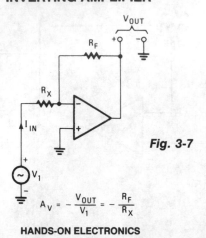

Fig. 3-7

$$A_V = -\frac{V_{OUT}}{V_1} = -\frac{R_F}{R_X}$$

HANDS-ON ELECTRONICS

NONINVERTING AMPLIFIER

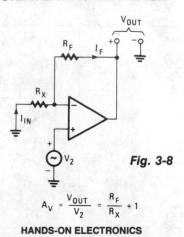

Fig. 3-8

$$A_V = \frac{V_{OUT}}{V_2} = \frac{R_F}{R_X} + 1$$

HANDS-ON ELECTRONICS

DIFFERENTIAL AMPLIFIER

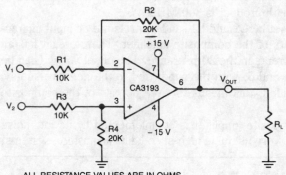

ALL RESISTANCE VALUES ARE IN OHMS

$$V_{OUT} = V_2\left(\frac{R4}{R3 + R4}\right)\left(\frac{R1 + R2}{R1}\right) - V_1\left(\frac{R2}{R1}\right)$$

IF R4 = R2, R3 = R1 AND $\frac{R2}{R1} = \frac{R4}{R3}$

THEN $V_{OUT} = (V_2 - V_1)\left(\frac{R2}{R1}\right)$

FOR VALUES ABOVE $V_{OUT} = (V_2 - V_1)$

IF A_V IS TO BE MADE 1 AND IF R1 = R3 = R4 = R
WITH R2 = 0.999R (0.1% MISMATCH IN R2)

THEN $V_{OCM} = 0.0005\ V_{IN}$ OR CMRR = 66 dB
THUS, THE CMRR OF THIS CIRCUIT IS LIMITED BY
THE MATCHING OR MISMATCHING OF THIS NETWORK
RATHER THAN THE AMPLIFER

This differential amplifier uses a CA3193 BiMOS op amp. This classical, differential input-to-signal-ended output converter when used with low-resistance signal source will maintain level of CMRR, if $R_1 = R_3 + R_4$.

GE/RCA

Fig. 3-9

14

ACTIVE CLAMP-LIMITING AMPLIFIER

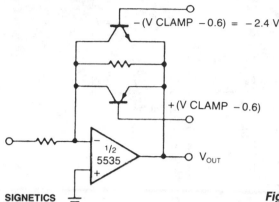

The modified inverting amplifier uses an active clamp to limit the output swing with precision. Allowance must be made for the V_{BE} of the transistors. The swing is limited by the base-emitter breakdown of the transistors. A simple circuit uses two back-to-back zener diodes across the feedback resistor, but tends to give less precise limiting and cannot be easily controlled.

SIGNETICS

Fig. 3-10

WIDE-BAND AGC AMPLIFIER

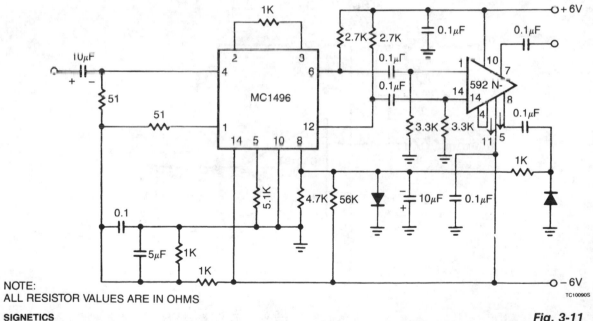

NOTE:
ALL RESISTOR VALUES ARE IN OHMS

SIGNETICS

TC10090S

Fig. 3-11

   The NE592 is connected in conjunction with a MC1496 balanced modulator to form an excellent automatic gain control system. The signal is fed to the signal input of the MC1496 and rc-coupled to the NE592. Unbalancing the carrier input of the MC1496 causes the signal to pass through unattenuated. Rectifying and filtering one of the NE592 outputs produces a dc signal which is proportional to the ac signal amplitude. After filtering, this control signal is applied to the MC1496, causing its gain to change.

POLARITY-REVERSING LOW-POWER AMPLIFIER

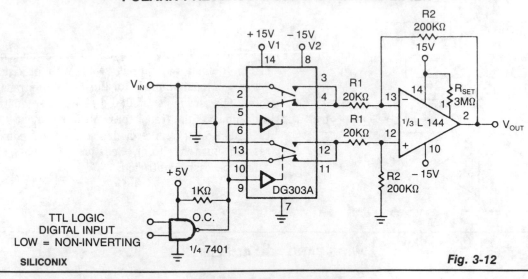

Fig. 3-12

SUMMING AMPLIFIER

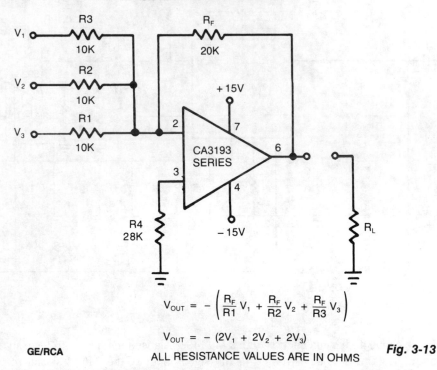

$$V_{OUT} = - \left(\frac{R_F}{R1} V_1 + \frac{R_F}{R2} V_2 + \frac{R_F}{R3} V_3 \right)$$

$$V_{OUT} = - (2V_1 + 2V_2 + 2V_3)$$

GE/RCA

ALL RESISTANCE VALUES ARE IN OHMS

Fig. 3-13

This circuit uses a CA3193 BiMOS op amp. Because input noise of the amplifier is increased by $R_F/R1//R2//R3$, and the gain that a single input will amplify is the gain of only one of the input channels ($R_F/R1$), for good noise performance, use the smallest number of inputs.

AC-COUPLED DYNAMIC AMPLIFIER

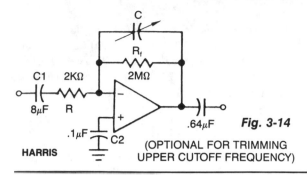

Fig. 3-14

(OPTIONAL FOR TRIMMING
UPPER CUTOFF FREQUENCY)

HARRIS

This circuit acts as a bandpass filter with gain and would be most useful for biomedical instrumentation. Low-frequency cutoff is set at 10 Hz while the high-frequency breakpoint is given by the open-loop rolloff characteristic of the HA-5141/42/44. In this case, the A_{VCL} = 60 dB where the rolloff occurs at approximately 300 Hz. This corner frequency may be trimmed by inserting a capacitor in parallel with R_f.

FORWARD-CURRENT BOOSTER

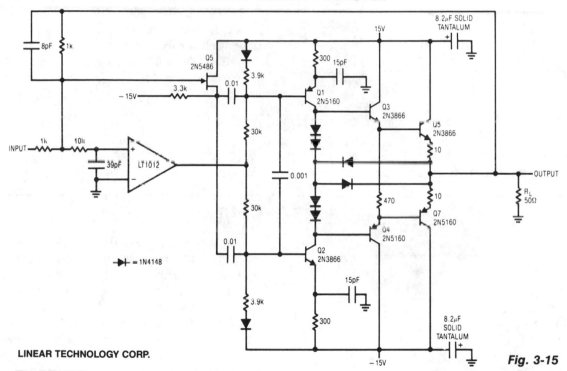

LINEAR TECHNOLOGY CORP.

Fig. 3-15

The LT1012 corrects dc errors in the booster stage, and does not set high-frequency signals. Fast signals are fed directly to the stage via Q5 and the 0.01-μF coupling capacitors. Dc and low-frequency signals drive the stage via the op-amp's output. The output stage consists of current sources, Q1 and Q2, driving the Q3 – Q5 and Q4 – Q7 complementary emitter follows. The diode network at the output steers drive away from the transistor bases when output current exceeds 250 mA, providing fast short-circuit protection. The circuit's high frequency summing node is the junction of the 1-K and 10-K resistors at the LT1012. The 10 K/39 pF pair filters high frequencies, permitting accurate dc summation at the LT1012's positive input. This current-boosted amplifier has a slew rate in excess of 1000 V/μs, a full power bandwidth of 7.5 MHz and a 3-dB point of 14 MHz.

DC-STABILIZED FAST AMPLIFIER

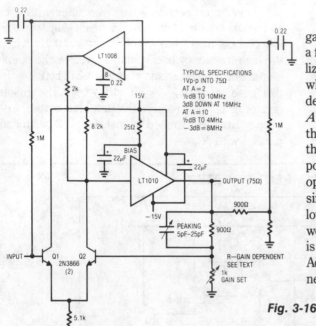

TYPICAL SPECIFICATIONS
1Vp-p INTO 75Ω
AT A = 2
½ dB TO 10MHz
3dB DOWN AT 16MHz
AT A = 10
½ dB TO 4MHz
−3dB = 8MHz

This amplifier functions over a wide range of gains, typically 1 – 10. It combines the LT1010 and a fast discrete stage with an LT1008 based dc stabilizing loop. Q1 and Q2 form a differential stage which single-ends into the LT1010. The circuit delivers 1 V pk-pk into a typical 75-Ω video load. At $A = 2$, the gain is within 0.5 dB to 10 MHz with the −3-dB point occurring at 16 MHz. At $A = 10$, the gain is flat (±0.5 dB to 4 MHz) with a −3-dB point at 8 MHz. The peaking adjustment should be optimized under loaded output conditions. This is a simple stage for fast applications where relatively low output swing is required. Its 1 V pk-pk output works nicely for video circuits. A possible problem is the relatively high bias current, typically 10 μA. Additional swing is possible, but more circuitry is needed.

Fig. 3-16

LINEAR TECHNOLOGY CORP.

WRITE AMPLIFIER

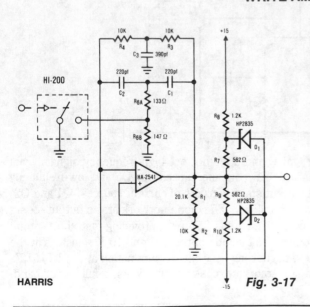

The proliferation of industrial and computerized equipment containing programmable memory has increased the need for reliable recording media. The magnetic tape medium is presently one of the most widely used methods. The primary component of any magnetic recording mechanism is the write mechanism. The concept of the write generator is very basic. The digital input causes both a change in the output amplitude, as well as a change in frequency. This type of operation is accomplished by altering the value of a resistor in the standard twin-tee oscillator. A HI-201 analog switch was used to facilitate the switching action. The effect of the external components on the feedback network requires R6A and R6B to be much smaller than would normally have been expected when using the twin-tee feedback scheme.

HARRIS   *Fig. 3-17*

LOW-NOISE PHOTODIODE AMPLIFIER

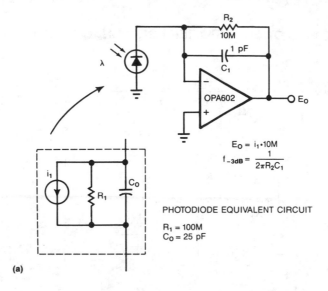

$$E_O = i_1 \cdot 10M$$
$$f_{-3dB} = \frac{1}{2\pi R_2 C_1}$$

PHOTODIODE EQUIVALENT CIRCUIT

$R_1 = 100M$
$C_O = 25$ pF

(a)

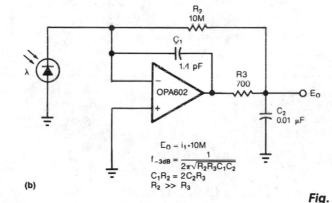

$$E_O = i_1 \cdot 10M$$
$$f_{-3dB} = \frac{1}{2\pi \sqrt{R_2 R_3 C_1 C_2}}$$
$$C_1 R_2 = 2 C_2 R_3$$
$$R_2 \gg R_3$$

(b)

EDN

Fig. 3-18

Adding two passive components to a standard photodiode amplifier reduces noise. Without the modification, the shunt capacitance of the photodiode reacting with the relatively large feedback resistor of the transimpedance (current-to-voltage) amplifier, creates excessive noise gain.

The improved circuit, Fig. 3-18b, adds a second pole, formed by R3 and C2. The modifications reduce noise by a factor of 3. Because the pole is within the feedback loop, the amplifier maintains its low output impedance. If you place the pole outside the feedback loop, you have to add an additional buffer, which would increase noise and dc error.

The signal bandwidth of both circuits is 16 kHz. In the standard circuit (Fig. 3-18a), the 1-pF stray capacitance in the feedback loop forms a single 16-kHz pole. The improved circuit has the same bandwidth as the first, but exhibits a 2-pole response.

VOLTAGE-FOLLOWER AMPLIFIER FOR SIGNAL-SUPPLY OPERATION

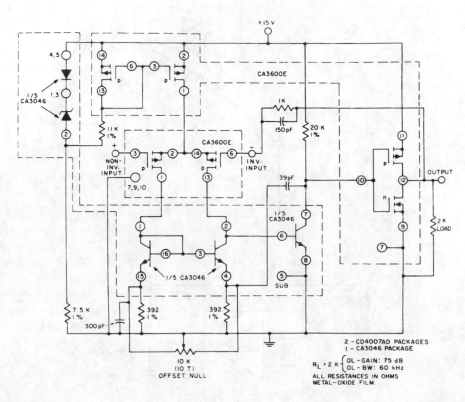

Fig. 3-19

This unity-gain follower amplifier has a CMOS p-channel input, an npn second-gain stage, and a CMOS inverter output. The IC building blocks are two CA3600E's (CMOS transistor pairs) and a CA3046 npn transistor array. A zener-regulated leg provides bias for a 400-μA p-channel source, feeding the input stage, which is terminated in an npn current mirror. The amplifier voltage-offset is nulled with the 10-KΩ balance potentiometer. The second-stage current level is established by the 20-KΩ load, and is selected to approximately the first-stage current level, to assure similar positive and negative slew rates. The CMOS inverter portion forms the final output stage and is terminated in a 2-KΩ load, a typical value used with monolithic op amps. Voltage gain is affected by the choice of load resistance value. The output stage of this amplifier is easily driven to within 1 mV of the negative supply voltage.

CURRENT-SHUNT AMPLIFIER

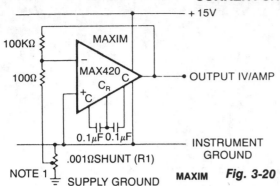

This circuit measures the power-supply current of a circuit without really having a current-shunt resistor: R1 is only 3 cm of #20 gauge copper wire. A length of the power distribution wiring can be used for R1. The MAX420's CMVR includes its own negative power supply; therefore, it can both be powered by and measure current in the ground line.

Fig. 3-20

CONSTANT-BANDWIDTH AMPLIFIER

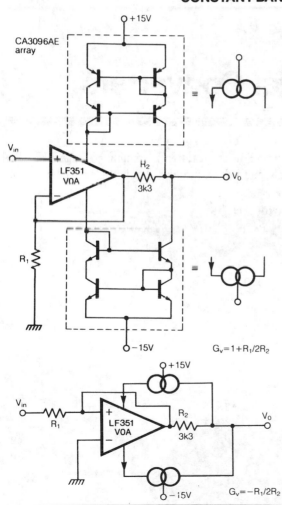

$G_v = 1 + R_1/2R_2$

$G_v = -R_1/2R_2$

The traditional restriction of constant gain-bandwidth products for a voltage amplifier can be overcome by employing feedback around a current amplifier. Two current mirrors, constructed from transistors in a CA3096AE array, effectively turn the LF351 op amp into a current amplifier. Feedback is then applied by using R2 and R1, turning the whole circuit into a feedback voltage amplifier with a noninverting gain of G of $1 + R_1/2R_2$.

Using the values shown, a constant bandwidth of 3.5 MHz is obtained for all voltage gains up to and beyond 100 at 10 V pk-pk output, equivalent to a gain-bandwidth product of 350 MHz from an op amp with an advertised unity gain-bandwidth of 10 MHz. An inverting gain configuration is also possible (see Fig. 2) where $G = R_1/2R_2$. Slewing rates are significantly improved by this approach; even a 741 can manage 100 V μs under these conditions since its output is a virtual earth. However, because the new configurations use current feedback to achieve bandwidth independence, an output buffer should be added for circuits where a significant output current is required.

Fig. 3-21

ELECTRONIC ENGINEERING

21

4

Analog-to-Digital Converters

The sources of the following circuits are contained in the Sources section beginning on page 782. The figure number contained in the box of each circuit correlates to the sources entry in the Sources section.

Switched-Capacitor ADC
Tracking ADC
ADC
Half-Flash ADC

SWITCHED-CAPACITOR ADC

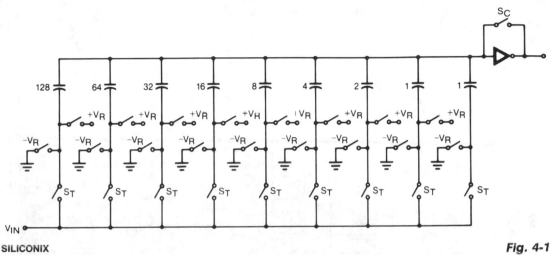

SILICONIX

Fig. 4-1

The CMOS comparator in the successive-approximation system determines each bit by examining the charge on a series of binary-weighted capacitors. In the first phase of the conversion process, the analog input is sampled by closing switch *SC* and all *ST* switches, and by simultaneously charging all the capacitors to the input voltage.

In the next phase of the conversion process, all *ST* and *SC* switches are opened and the comparator begins identifying bits by identifying the charge on each capacitor relative to the reference voltage. In the switching sequence, all 8 capacitors are examined separately until all 8 bits are identified, and then the charge-convert sequence is repeated. In the first step of the conversion phase, the comparator looks at the first capacitor (binary weight = 128). One pole of the capacitor is switched to the reference voltage, and the equivalent poles of all the other capacitors on the ladder are switched to ground. If the voltage at the summing node is greater than the trip point of the comparator—approximately 1/2 the reference voltage, a bit is placed in the output register, and the 128-weight capacitor is switched to ground. If the voltage at the summing node is less than the trip point of the comparator, this 128-weight capacitor remains connected to the reference input through the remainder of the capacitor-sampling (bit-counting) process. The process is repeated for the 64-weight capacitor, the 32-weight capacitor, and so forth down the line, until all bits are tested. With each step of the capacitor-sampling process, the initial charge is redistributed among the capacitors. The conversion process is successive-approximation, but relies on charge shifting rather than a successive-approximation register—and reference d/a—to count and weigh the bits from MSB to LSB.

TRACKING ADC

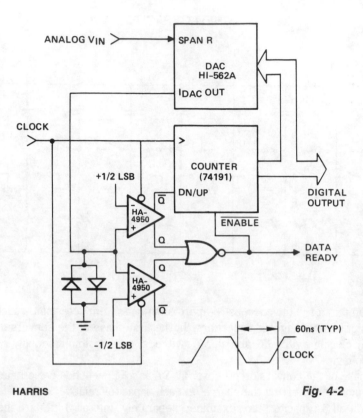

HARRIS

Fig. 4-2

The analog input is fed into the span resistor of a DAC. The analog input voltage range is selectable in the same way as the output voltage range of the DAC. The net current flow through the ladder termination resistance; i.e., 2 KΩ for HI-562A; produces an error voltage at the DAC output. This error voltage is compared with $1/2$ LSB by a comparator. When the error voltage is within $\pm 1/2$ LSB range, the Q output of the comparators are both low, which stops the counter and gives a data ready signal to indicate that the digital output is correct. If the error exceeds the $\pm 1/2$ LSB range, the counter is enabled and driven in an up or down direction depending on the polarity of the error voltage.

The digital output changes state only when there is a significant change in the analog input. When monitoring a slowly varying input, it is necessary to read the digital output only after a change has taken place. The data ready signal could be used to trigger a flip-flop to indicate the condition and reset it after readout. The main disadvantage of the tracking ADC is the time required to initially acquire a signal; for a 12-bit ADC, it could be up to 4096 clock periods. The input signal usually must be filtered so that its rate of change does not exceed the tracking range of the ADC—1 LSB per clock period.

ADC

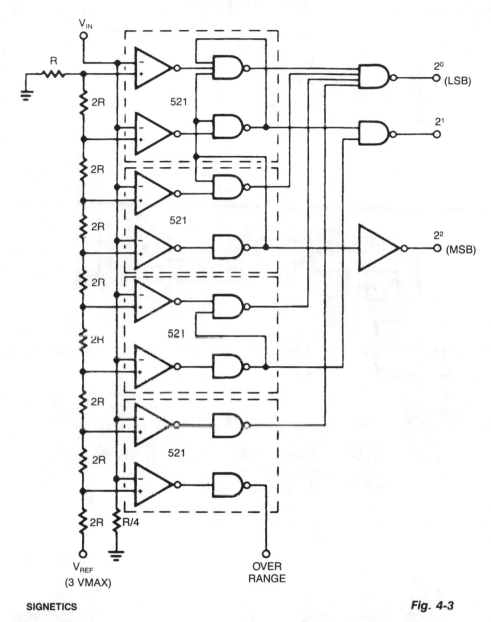

Fig. 4-3

Conversion speed of this design is the sum of the delay through the comparator and the decoding gates. Reference voltages for each bit are developed from a precision resistor ladder network. Values of R and $2R$ are chosen so that the threshold is $1/2$ of the least significant bit. This assures maximum accuracy of $\pm 1/2$ bit. The individual strobe line and duality features of the NE521 greatly reduced the cost and complexity of the design.

HALF-FLASH ADC

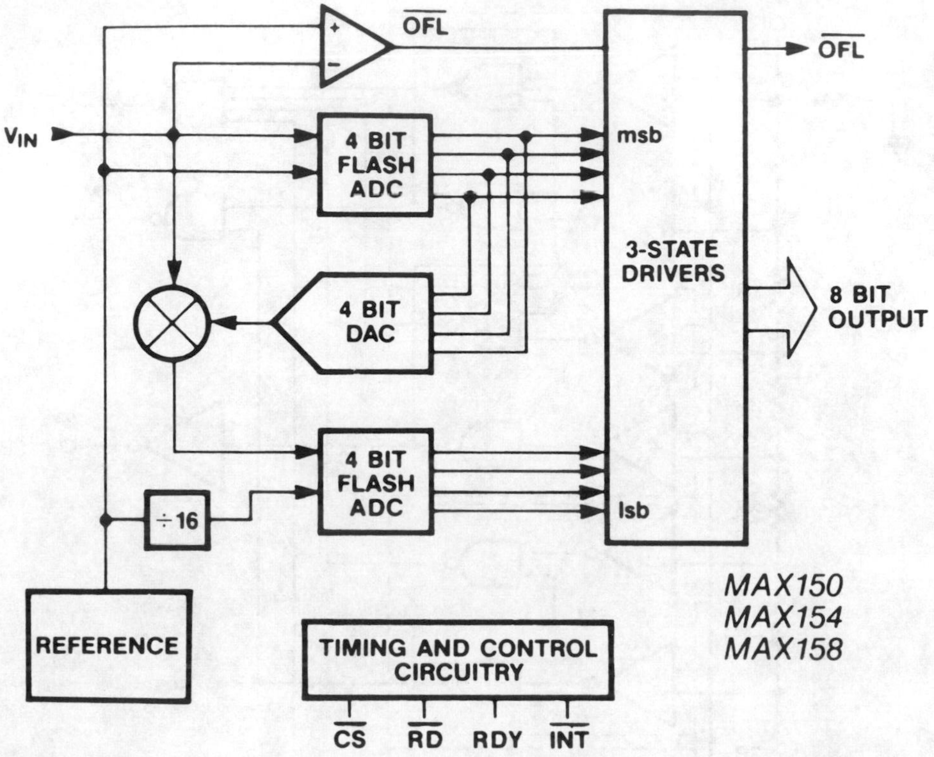

MAXIM

MAX150
MAX154
MAX158

Fig. 4-4

An a/d conversion technique which combines some of the speed advantages of flash conversion with the circuitry savings of successive approximation is termed *half-flash*. In an 8-bit, half-flash converter, two 4-bit flash a/d sections are combined. The upper flash a/d compares the input signal to the reference and generates the upper 4 data bits. This data goes to an internal DAC, whose output is subtracted from the analog input. Then, the difference can be measured by the second flash a/d, which provides the lower 4 data bits.

5

Annunciator

The sources of the following circuits are contained in the Sources section beginning on page 782. The figure number contained in the box of each circuit correlates to the sources entry in the Sources section.

Transformerless Tone Annunciator

TRANSFORMERLESS TONE ANNUNCIATOR

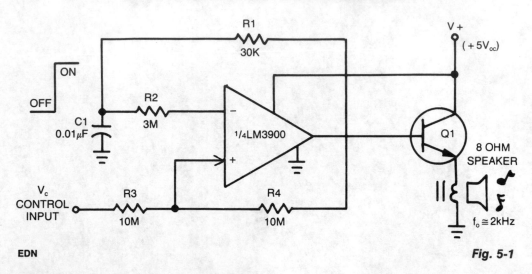

EDN

Fig. 5-1

This circuit does not require an output transformer or an output coupling capacitor; the annunciator can easily be turned on or off by a control input voltage driving a 10-MΩ input resistor, R3. For a smaller acoustic output, replace output transistor, Q1, with a 100-Ω resistor, while also raising the voice coil impedance to 100 Ω, to prevent loading of the IC.

6

Attenuators

The sources of the following circuits are contained in the Sources section beginning on page 782. The figure number contained in the box of each circuit correlates to the sources entry in the Sources section.

Digitally Programmable Attenuator
Programmable Attenuator
Voltage-Controlled Attenuator

DIGITALLY PROGRAMMABLE ATTENUATOR

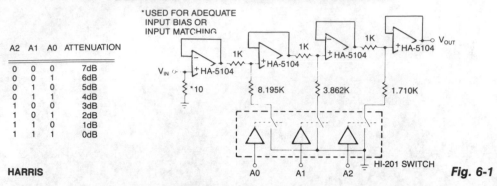

A2	A1	A0	ATTENUATION
0	0	0	7dB
0	0	1	6dB
0	1	0	5dB
0	1	1	4dB
1	0	0	3dB
1	0	1	2dB
1	1	0	1dB
1	1	1	0dB

HARRIS

Fig. 6-1

The first stage is a simple buffer used to isolate the signal source from the attenuator stages to follow. Each of the subsequent stages is preceded by a voltage divider formed by two resistors and CMOS switch. Provided that the CMOS switch for each stage is closed, the drive signal will be attenuated according to the basic voltage divider relationship at each stage. In the event a switch is open, nearly all of the signal strength will be passed to the next stage through the 1-KΩ resistor. The amplifiers act as buffers for divider networks and reduce the interaction between stages. Eight levels of attenuation are possible with the circuit as illustrated, but more stages could be added. Each divider network must be closely matched to the resistor ratios shown or the level of attenuation will not match the levels in the logic chart.

PROGRAMMABLE ATTENUATOR

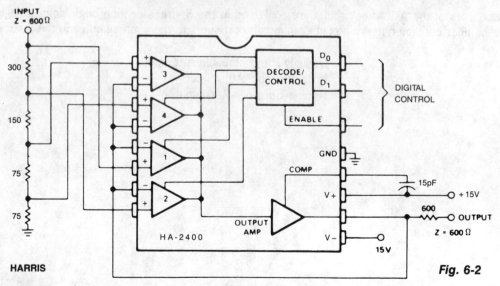

HARRIS

Fig. 6-2

This circuit performs the function of dividing the input signal by a selected constant (1, 2, 4, 8, etc.). While T, Z, or L sections could be used in the input attenuator, this is not necessary since the amplifier loading is negligible and a constant input impedance is maintained. The circuit is thus much simpler and more accurate than the usual method of constructing a constant impedance ladder, and switching sections in and out with analog switches. Two identical circuits can be used to attenuate a balanced line.

VOLTAGE-CONTROLLED ATTENUATOR

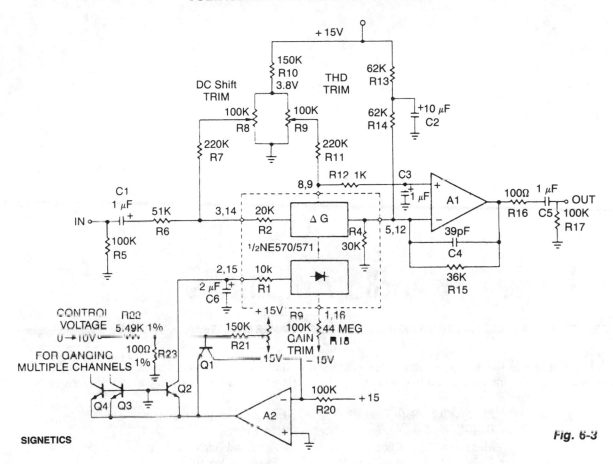

SIGNETICS

Fig. 6-3

This typical circuit uses an external op amp for better performance and an exponential converter to get a control characteristic of −6 dB/V. Trim networks are shown to null out distortion and dc shift, and to fine trim the gain to 0 dB with 0 V of control voltage.

7

Audio Amplifiers

The sources of the following circuits are contained in the Sources section beginning on page 782. The figure number contained in the box of each circuit correlates to the sources entry in the Sources section.

AGC with Squelch Control
Gain-Controlled Stereo Amplifier
Microphone Amplifier
Audio Circuit Bridge Load Drive
20-dB Audio Booster
Micro-Size Amplifier
Audio Amplifier
Line-Operated Amplifier

Magnetic Phono Preamplifier
RIAA Preamplifier
Professional Audio NAB Tape
   Playback Preamplifier
Mini-Stereo
Speaker Amplifier for Hand-Held
   Transceivers
TV Audio Amplifier

AGC WITH SQUELCH CONTROL

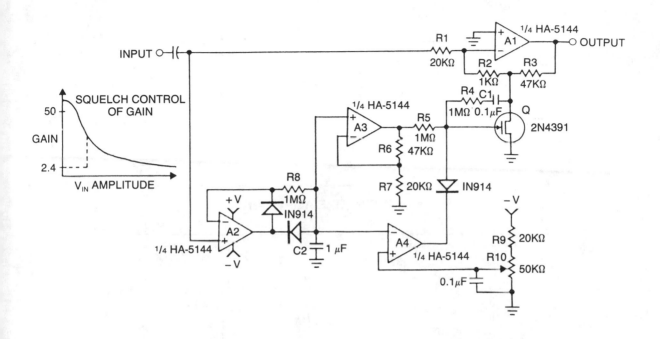

HARRIS

Fig. 7-1

Automatic gain control is a very useful feature in a number of audio amplifier circuits: tape recorders, telephone speaker phones, communication systems and PA systems. This circuit consists of a HA-5144 quad op amp and a FET transistor used as a voltage-controlled resistor to implement an AGC circuit with squelch control. The squelch function helps eliminate noise in communications systems when no signal is present and allows remote hands-free operation of tape recorder systems. Amplifier A1 is placed in an inverting-gain *T* configuration in order to provide a fairly wide gain range and a small signal level across the FET. The small signal level and the addition of resistors R5 and R6 help reduce nonlinearities and distortion. Amplifier A2 acts as a negative peak detector to keep track of signal amplitude. Amplifier A3 can be used to amplify this peak signal if the cutoff voltage of the FET is higher than desired. Amplifier A4 acts as a comparator in the squelch control section of the circuit. When the signal level falls below the voltage set by R10, the gate of the FET is pulled low—turning it off completely—and reducing the gain to 2.4. The output A4 can also be used as a control signal in applications, such as a hands-free tape recorder system.

GAIN-CONTROLLED STEREO AMPLIFIER

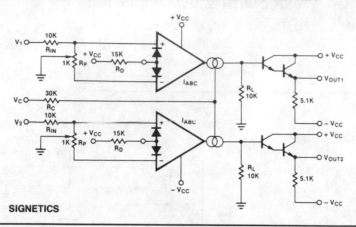

SIGNETICS

Fig. 7-2

MICROPHONE AMPLIFIER

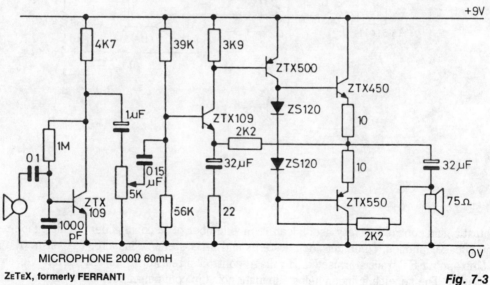

ZᴇTᴇX, formerly FERRANTI

Fig. 7-3

This circuit features the ZTX450/ZTX550 transistors in a push-pull output stage. The following readings were taken at maximum volume:

Input: 0.4 mV rms
Output: 1.8 V rms
Voltage gain: 4500
Max. output before distortion: 2.25 V rms − supply current = 3.5 mA
Zero output-supply current: 3.5 mA
Wattage: 0.034 W
Frequency response: 250 Hz to 28 kHz

AUDIO CIRCUIT BRIDGE LOAD DRIVE

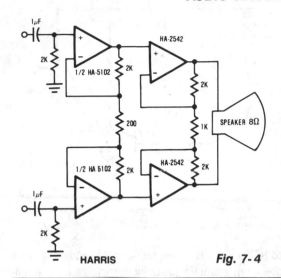

This circuit shows a method which increases the power capability of a drive system for audio speakers. Two HA-2542s are used to operate on half cycles only, which greatly increases their power handling capability. Bridging the speaker, as shown, makes 200 mA of output current available to drive the load. The HA-5102 is used as an ac-coupled, low noise preamplifier, which drives the bridge circuit.

HARRIS   *Fig. 7-4*

20-dB AUDIO BOOSTER

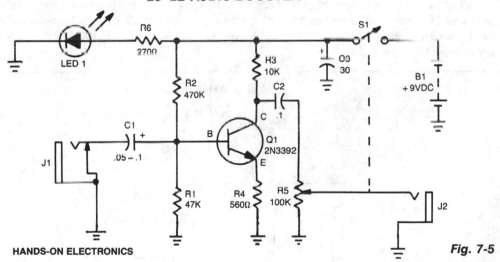

HANDS-ON ELECTRONICS   *Fig. 7-5*

The amplifier's gain is nominally 20 dB. Its frequency response is determined primarily by the value of just a few components—primarily C1 and R1. The values in the schematic diagram provide a response of ±3.0 dB from about 120 to over 20,000 Hz. Actually, the frequency response is flat from about 170 to well over 20,000 Hz; it's the low end that deviates from a flat frequency response. The low end's rolloff is primarily a function of capacitor C1, since R1's resistive value is fixed. If C1's value is changed to 0.1 μF, the low end's corner frequency—the frequency at which the low end rolloff starts—is reduced to about 70 Hz. If you need an even deeper low end rolloff, change C1 to a 1.0-μF capacitor. If it's an electrolytic type, make certain that it's installed into the circuit with the correct polarity – with the positive terminal connected to Q1's base terminal.

MICRO-SIZED AMPLIFIER

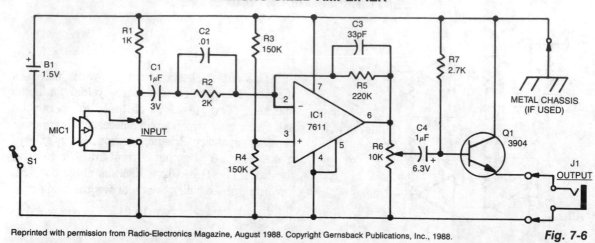

Fig. 7-6

Sound detected by electret microphone MIC1 is fed to IC1's input through resistor R2, and capacitors C1 and C2. Resistors R2 and R5 determine the overall stage gain, while C2 partially determines the amplifier's frequency response. To ensure proper operation, use a single-ended power supply. R3 and R4 simulate a null condition equal to half the power supply's voltage at IC1's noninverting input. The output of IC1 is transferred to emitter-follower amplifier Q1 via volume control R6. The high-Z-in/low-Z-out characteristic of the emitter-follower matches the moderately high-impedance output of IC1 to a low-impedance headphone load.

AUDIO AMPLIFIER

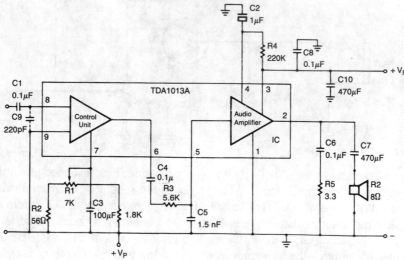

SIGNETICS

Fig. 7-7

C9 is necessary to filter-out rf input interferences. R3 in combination with C5 is used to limit the af frequency bandwidth. The 470-μF power supply decoupling capacitor is C10.

LINE-OPERATED AMPLIFIER

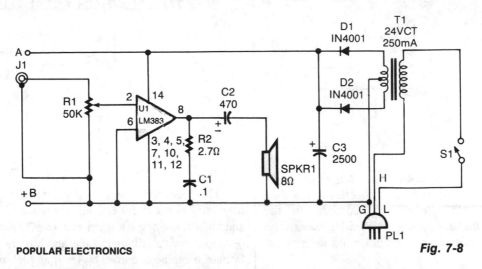

Fig. 7-8

T1 isolates the unit from the line, and has a 24-V, center-tapped secondary. The output of the transformer is rectified by diodes D1 and D2 and filtered by capacitor C3 to provide 15 to 18 Vdc. The LM383 has built-in protection against speaker shorts.

MAGNETIC PHONO PREAMPLIFIER

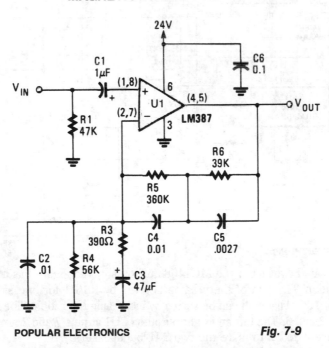

Fig. 7-9

RIAA PREAMPLIFIER

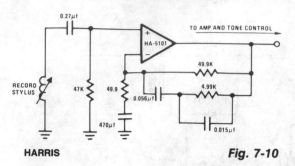

HARRIS   **Fig. 7-10**

The circuit essentially provides low-frequency boost below 318 Hz and high-frequency attenuation above 3150 Hz. Recent modifications to the response standard include a 31.5-Hz peak gain region to reduce dc-oriented distortion from external vibration.

PROFESSIONAL AUDIO NAB TAPE PLAYBACK PREAMPLIFIER

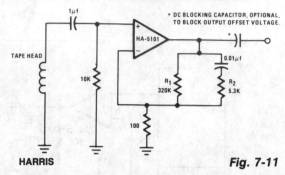

HARRIS   **Fig. 7-11**

The preamplifier is configured to provide low-frequency boost to 50 Hz, flat response to 3 kHz, and high-frequency attenuation above 3 kHz. Compensation for variations in tape and tape head performance can be achieved by trimming R1 and R2.

MINI-STEREO

HANDS-ON ELECTRONICS   **Fig. 7-12**

This circuit is built around two chips: the MC1458 dual op amp, configured as a preamplifier, and the LM378 dual 4-watt amplifier. The gain of the preamp is given by R3/R1 for one side and R4/R2 for the other side, which is about 100. That gain can be varied by increasing the ratios. The left and right channel inputs are applied to pins 2 and 6. The left and right outputs of U1 at pins 7 and 2 are coupled through C5/R10 and C3/R6, respectively, to U2 to drive the two 8-Ω loudspeakers.

SPEAKER AMPLIFIER FOR HAND-HELD TRANSCEIVERS

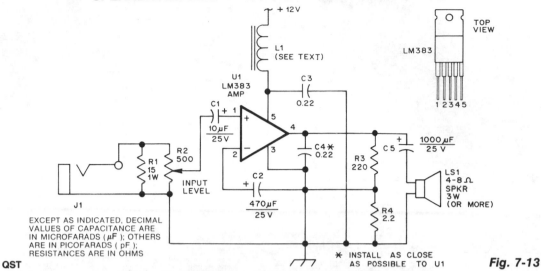

Fig. 7-13

The LM383 is an audio-power amplifier that is capable of producing up to 8 W of audio output. R1 is essentially a load resistor for the hand-held transceiver's audio output. R2 can be composed of two fixed resistors in a 10:1 divider arrangement, but using a potentiometer makes it easy to set the amplifier's maximum gain. When powered from a vehicle's electrical system, the amplifier's +12 V power source requires filter L1 to eliminate alternator whine. The LM383 can be mounted directly on the heatsink because the mounting tab is at ground potential.

TV AUDIO AMPLIFIER

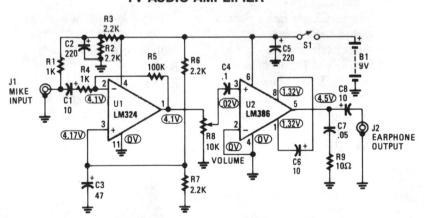

Fig. 7-14

The amplifier picks up the TV's audio output signal and amplifies it to drive a set of earphones for private listening. It is built around an LM324 quad op amp and an LM386 low-power audio amplifier. The circuit uses an inexpensive electret microphone element as the pick-up and a set of earphones as the output device.

8

Automotive Circuits

The sources of the following circuits are contained in the Sources section beginning on page 782. The figure number contained in the box of each circuit correlates to the sources entry in the Sources section.

Automobile Ignition Substitute
Courtesy Light Delay Switch
Lights-On Reminder
Automobile Locater
Read-Head Preamplifier
Delayed Extra Brake Light
Digital Tach/Dwell Meter
Automobile Air Conditioner
   Smart Clutch
Door Ajar Monitor

Tachometer with Set Point
Automobile Voltage Regulator
Directional Signals Monitor
Automatic Headlight Delay
Back-Up Beeper
Electronic Car Horn
Courtesy Light Extender
Flashing Third Brake Light
Headlight Alarm

AUTOMOBILE IGNITION SUBSTITUTE

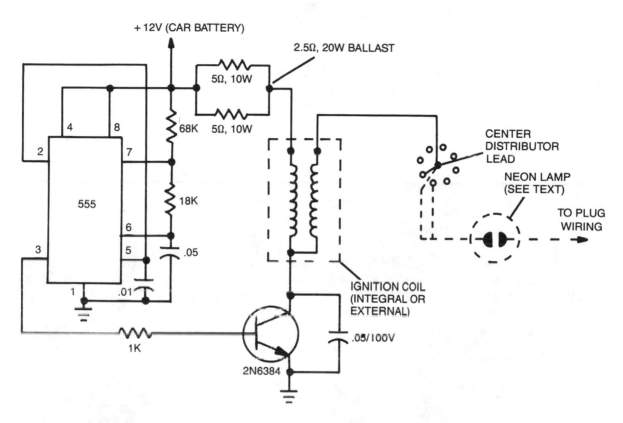

GERNSBACK PUBLICATIONS INC.

Fig. 8-1

The ignition substitute provides a constant power source for the ignition coil. Its frequency, 0.5 – 1.0 kHz, is that used by an 8-cylinder engine with an idling speed of 650 RPM, and the unit provides a rapid spark at a 17% duty cycle, while staying within the power dissipation limits of the components.

The circuit consists of a 555 timer IC configured as an astable free-running multivibrator that is used to drive a high-current npn transistor, such as a 2N6384. The transistor should be heavily heatsinked because it might be drawing several amps over quite a long period of time.

The coil ballast can be from 0.68 to 6.5 Ω, depending on what's available. The 2.5-Ω, 20-W ballast shown works well. All the other resistors can be either 1/4- or 1/2-W devices, and the capacitor, between pins 1 and 5 of the 555, can range from 0.01 to 0.05 μF. Do not omit the 100-V, 0.05-μF capacitor across the transistor; it prevents voltage spikes from damaging the device.

Although designed for an 8-cylinder engine, this device can be used with other types. In addition, a neon bulb can be added to the circuit to verify the presence of a spark.

COURTESY LIGHT DELAY SWITCH

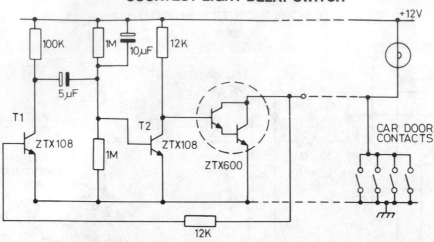

ZᴇTᴇX, formerly FERRANTI

Fig. 8-2

This circuit holds on the internal light for approximately one minute after the car doors are closed. When the door contacts open, a + VE pulse is applied to the base of T1. This transistor turns on, turning off T2 and charging the 10-μF capacitor. T3 turns on, holding on the internal light. The capacitor takes one minute to discharge when the circuit reverts to its original state.

LIGHTS-ON REMINDER

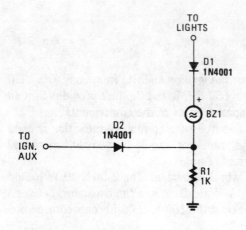

With both the ignition and the car lights on, piezo transducer BZ1 draws no current and remains silent. With only the ignition on, diode D1 is reverse-biased and so prevents current flow through BZ1.

However, when the lights are on and the ignition is off, the transducer becomes energized and sounds to alert you to turn the lights off. With the ignition off and the lights on, D2 is reverse-biased, preventing current from flowing to the ignition. Resistor R1 prevents a short circuit when the ignition is on.

POPULAR ELECTRONICS   *Fig. 8-3*

AUTOMOBILE LOCATER

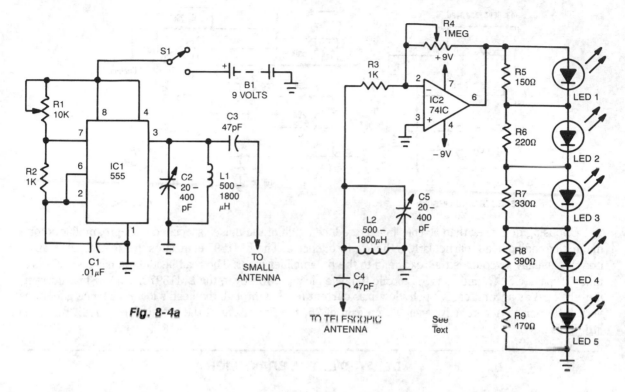

Fig. 8-4a

Fig. 8-4b

GERNSBACK PUBLICATIONS INC.

This locater is made up of two parts. The first is an rf oscillator, whose circuit is shown in Fig. 8-4a. The second is a sensitive receiver shown in Fig. 8-4b. The heart of the oscillator is a 555 timer IC. Tank circuit C2 and L1 is used to tune the transmitter. The antenna is coupled to the transmitter through C3. A telescopic antenna or a length of hookup wire will work quite well. At the receiver, the incoming signal is tuned by C5 and L2 before being passed on to the 741 IC. The five LEDs are used to indicate signal strength, they light up in order (1 to 5) as the signal gets stronger.

After the devices are built, the receiver and transmitter will need to be tuned. Tune the transmitter until all of the receiver's LEDs light. Separate the receiver and the transmitter—the farther apart they are the better—and adjust R4 until you get a maximum strength reading only when the receiver's antenna is pointed directly at the transmitter. Place the transmitter on the dashboard and completely extend the antenna. To find your car, just extend the telescope antenna to its full length and hold it parallel to the ground. Point the antenna to your far left, then swing it to your far right. Do that until you find in which direction the strongest signal lies, as indicated by the LEDs. The antenna will be pointing at your car.

READ-HEAD PREAMPLIFIER

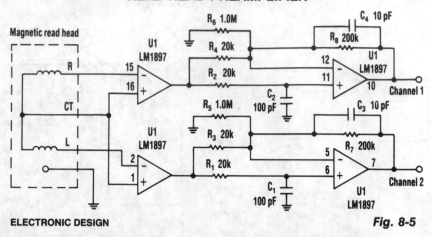

ELECTRONIC DESIGN

Fig. 8-5

Choosing dc rather than ac coupling can reduce much of the noise associated with preamplifiers for a magnetic reading head, particularly in the low frequencies. The LM1897 eliminates the need for the capacitor that usually ac couples the read head to the preamplifier input. The read head itself has a small resistance, typically 50 Ω, and so is less prone to noise pickup. Moreover, the LM1897 has a low-bias current; merely 2 μA as a worst case. Such a low-bias current flowing through the head's low resistance generates very little noise. Accordingly, even with a gain of 25, the first stage of the preamplifier circuit produces little noise.

DELAYED EXTRA BRAKE LIGHT

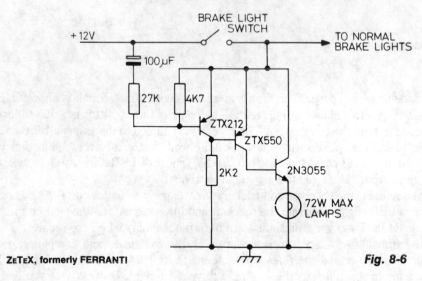

ZᴇTᴇX, formerly FERRANTI

Fig. 8-6

Operating the brake pedal of the car brings on the normal brake lights and then, after a delay, the extra lights are turned on. A bimetal strip in series with the lights would make them flash.

DIGITAL TACH/DWELL METER

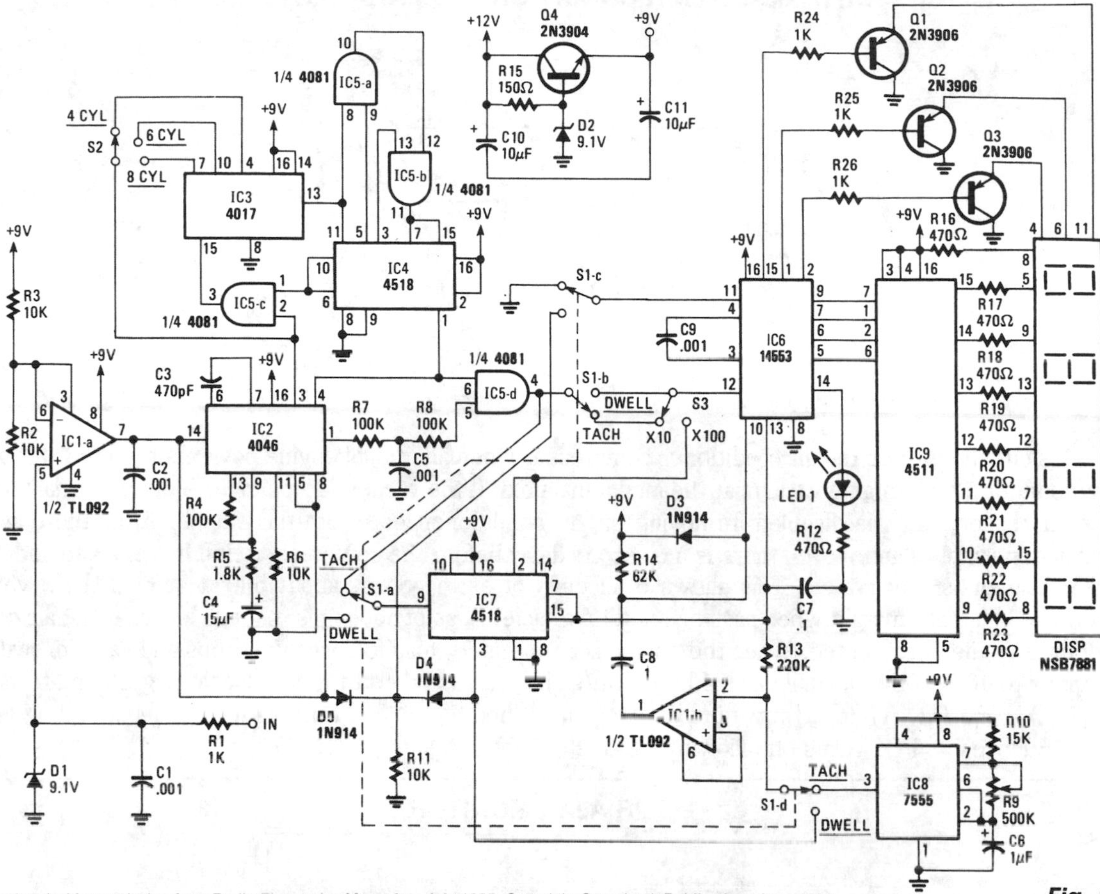

Fig. 8-7

The heart of the circuit is IC2, a 4046 micropower phase-locked loop (PLL). The incoming signals are fed to the PLL after being buffered by IC1a and its associated components. The frequency of the incoming signal is multiplied by either 90, 60, or 45, depending on the setting of the cylinder select switch, S2. That switch selects the proper output from counters IC3 and IC4, which are set to divide the output frequency of the PLL by those amounts, and then send the divided output back to the comparator to the PLL to keep it locked on to the input signal. The phase pulses output at pin 4 of IC2, then go through an AND gate IC5d—which only passes the signals if the PLL is locked on to an input signal, preventing stray readings—and then to the input of IC6. When in the tach mode, IC6 counts the number of pulses present at pin 12, during the timing interval generated by IC8 and the associated circuitry of IC1b. Because of the varied multiplication rate for the different cylinder selections—90, 60, and 45 for 4, 6, and 8 cylinders, respectively—the time interval is always constant at 1/3 of a second. The time interval is adjusted with R9, a 500-KΩ potentiometer; it is the only adjustment in the circuit.

In the high-tach (TACH 1 or ×100) range of 0–9990 rpm, the output of IC2 is routed by switches S1a and S3 through IC7, a divide-by-ten counter, which increases the count range tenfold. In the low-tach (TACH 2 or ×10) range of 0–999 rpm, the counter is bypassed.

AUTOMOBILE AIR CONDITIONER SMART CLUTCH

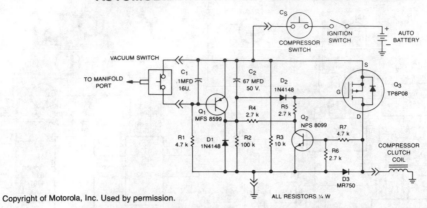

ALL RESISTORS ¼ W

Fig. 8-8

This circuit disables the air conditioner compressor when additional engine power is required. It does so by monitoring the engine vacuum at the intake manifold. If the vacuum drops to 40% of its normal level, the compressor clutch is disabled, removing the air conditioner load from the engine. After the engine returns to normal vacuum level, there is a 6 second delay before the compressor clutch is enabled and the air conditioner is reactivated. This allows 6 seconds of extra power, about 500 ft at 60 MPH, which increases the safety margin when passing another vehicle. Loss of cooling is minimal because the air conditioner fan is not interrupted. When the engine is accelerated, manifold vacuum drops and vacuum switch VS opens to 40% of the normal manifold pressure. This causes Q1 to turn on, discharging C2 and turning off Q3 via diode D2. When Q3 turns off, so does Q2. When the engine reaches its normal operating vacuum, VS closes and Q1 turns off, allowing C2 to charge for 6 seconds until Q3 turns on again.

DOOR AJAR MONITOR

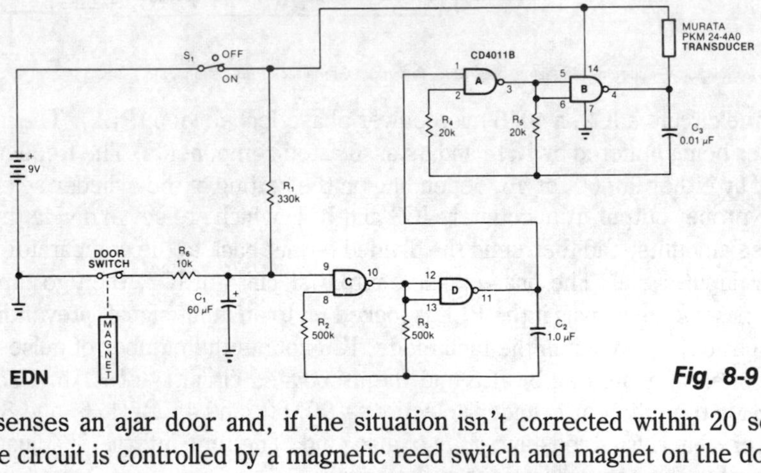

EDN

Fig. 8-9

The monitor senses an ajar door and, if the situation isn't corrected within 20 seconds, sounds a beeping alarm. The circuit is controlled by a magnetic reed switch and magnet on the door. With the door closed, the switch is closed and the alarm is disarmed. Opening the door opens switch, C1 starts charging up through R1. Approximately 20 seconds later, the voltage at pin 9 is high enough to turn on the oscillator formed from C, D, R2, R3, and C2. That pulses the piezoelectric transducer's 3-kHz oscillator. For lower standby drain on the battery, change R1 to 66 MΩ and C1 to 1 mF (film).

TACHOMETER WITH SET POINT

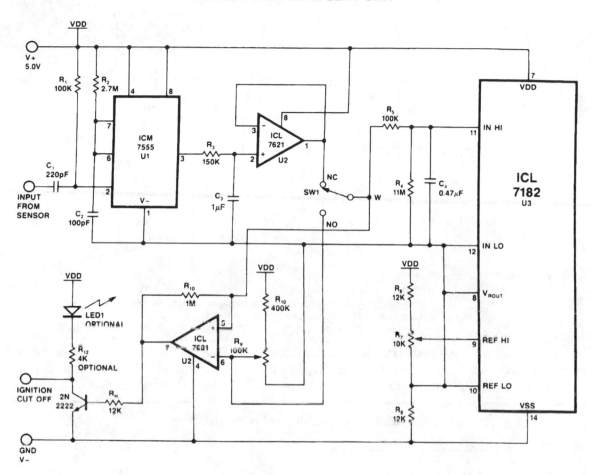

$$\frac{V_O}{V_{IN}} = \frac{1/RC}{S + 1/R_3C_3}$$

$$fc = \frac{1}{2\pi R_3C_3}$$

SW1 Momentary
   Switch SPST

V_{IN} = 264 mV @ 5000 RPM
4 Stroke V8

RPM	Hz	Period
600	10	100 ms
1000	16.7	60 ms
5000	83	12 ms
10,000	166.7	6 ms

No. of Cylinders	Events Per Cycle	Strokes Per Cycle
1	0.5	4
4	2	4
6	3	4
8	4	4

INTERSIL

Fig. 8-10

AUTOMOBILE VOLTAGE REGULATOR

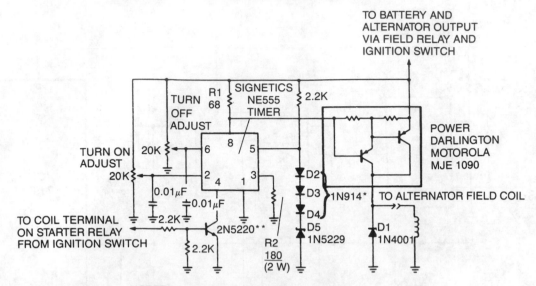

NOTES:
* Can be any general purpose Silicon diode or 1N4157 device.
** Can be any general purpose Silicon transistor.
All resistor values are in ohms.

SIGNETICS

Fig. 8-11

A monolithic 555-type timer is the heart of this simple automobile voltage regulator. When the timer is off so that its output at pin 3 is low, the power Darlington transistor pair is off. If battery voltage becomes too low, less than 14.4 V in this case, the timer turns on and the Darlington pair conducts.

DIRECTIONAL SIGNALS MONITOR

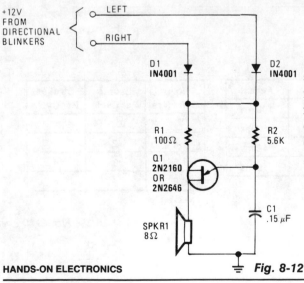

A unijunction transistor audio oscillator drives a small speaker. The oscillator's frequency is determined by resistor R2 and capacitor C2. The operating voltage is supplied from the car's turn-signal circuit(s) through D1 and D2. The diodes conduct current from the blinker circuit that is energized, and prevent stray current flow to the other blinker circuit.

HANDS-ON ELECTRONICS   **Fig. 8-12**

AUTOMATIC HEADLIGHT DELAY

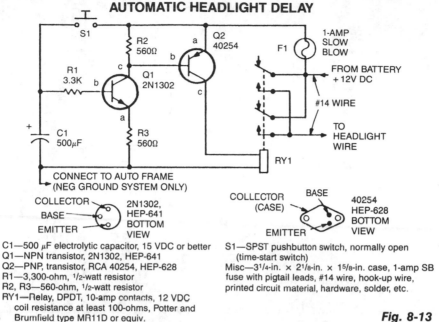

C1—500 µF electrolytic capacitor, 15 VDC or better
Q1—NPN transistor, 2N1302, HEP-641
Q2—PNP, transistor, RCA 40254, HEP-628
R1—3,300-ohm, 1/2-watt resistor
R2, R3—560-ohm, 1/2-watt resistor
RY1—Relay, DPDT, 10-amp contacts, 12 VDC
coil resistance at least 100-ohms, Potter and
Brumfield type MR11D or equiv.

S1—SPST pushbutton switch, normally open
(time-start switch)
Misc—3 1/4-in. × 2 1/8-in. × 1 5/8-in. case, 1-amp SB
fuse with pigtail leads, #14 wire, hook-up wire,
printed circuit material, hardware, solder, etc.

TAB BOOKS

Fig. 8-13

When the driver depresses pushbutton switch S1, timing capacitor C1 charges to 12 V and turns on transistor Q1, which drives power transistor Q2 into conduction. This, in turn, energizes the relay which has its contacts connected in parallel with the headlight switch. The relay will stay energized until C1 discharges to the Q1 turn-off level. The lights-on period is determined by the value of C1, R1, and the characteristics of transistor Q1. With values chosen on the schematic, about 60 light-on seconds are provided.

BACK-UP BEEPER

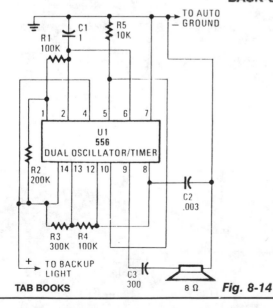

TAB BOOKS   **Fig. 8-14**

Put the car in reverse and the circuit provides a loud, audible beep at the rate of about one per second (1 Hz). Half of U1, a 556 dual oscillator/timer, is used as a slow-pulse oscillator with a rate of about 1 Hz. Components R2, R1, and C1 form the long time constant. You can calculate on time by $t = .7 (R1 + R2) C1$ or 1.15 seconds. The off time is shorter than the on time, at .77 second. Enabling pin 4 (reset) is held high to keep the oscillator free-running when voltage is applied to pin 14. The output at pin 5 is coupled to pin 10 of U1 enabling oscillator 2. Oscillator 2 of U1 produces an audio output of about 1 kHz, as determined by C2, R3, and R4. Pin 10 (reset) of oscillator 2 is connected to the pin 5 output of oscillator 1. So when pin 5 becomes positive, the oscillator beeps a short pulsed tone of 1 kHz.

ELECTRONIC CAR HORN

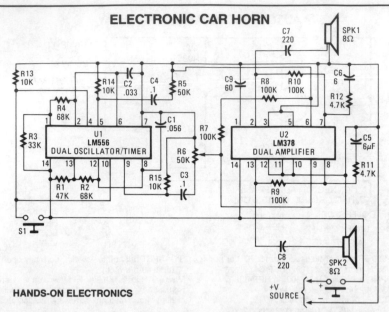

HANDS-ON ELECTRONICS

Fig. 8-15

An LM556 dual oscillator/timer, U1, configured as a two-tone oscillator drives U2, a dual 4-watt amplifier. One of the oscillators, pins 1 to 6, contained in U1 produces the upper frequency signal of about 200 Hz, while the second oscillator, pins 8 to 13, provides the lower frequency signal of about 140 Hz. Increase or decrease the frequencies by changing the values of C2 and C3. U1's outputs, pins 9 and 5, are connected to separate potentiometers to provide control over volume and balance. Each half of U2 produces 4 W of audio that is delivered to two 8-Ω loudspeakers via capacitors C7 and C8.

COURTESY LIGHT EXTENDER

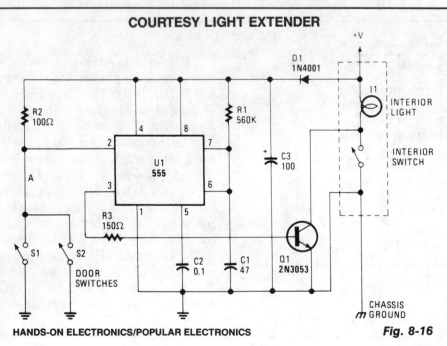

HANDS-ON ELECTRONICS/POPULAR ELECTRONICS

Fig. 8-16

COURTESY LIGHT EXTENDER (*Cont.*)

The circuit keeps the courtesy light on for 30 seconds after you close the door. The lead from the door switch is removed and connected to the 555 circuit. The 555 is arranged in a monostable mode, and is triggered by the door switches. The output drives Q1, which is connected across the interior light switch. The interior light is turned on for 30 seconds after the door is opened. If the door(s) are held open for longer than 30 seconds, it will not reset until after the doors are closed. In that case, the lights go out immediately.

FLASHING THIRD BRAKE LIGHT

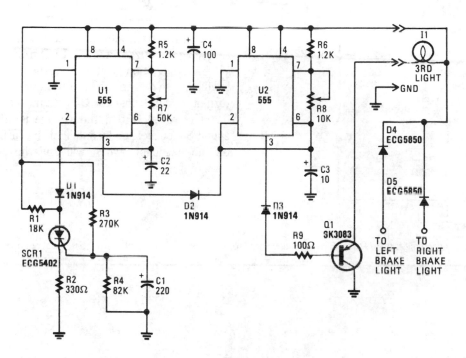

POPULAR ELECTRONICS/HANDS-ON ELECTRONICS

Fig. 8-17

When power is first applied, three things happen: light-driving transistor Q1 is switched on due to a low output from U2 pin 3; timer U1 begins its timing cycle, with the output, pin 3, becoming high, inhibiting U2's trigger, pin 2, via D2; and charge current begins to move through R3 and R4 to C1.

When U1's output becomes low, the inhibiting bias on U2 pin 2 is removed, so U2 begins to oscillate, flashing the third light via Q1, at a rate determined by R8, R6, and C3. That oscillation continues until the gate-threshold voltage of SCR1 is reached, causing it to fire and pull U1's trigger, pin 2, low.

With its trigger low, U1's output is forced high, disabling U2's triggering. With triggering inhibited, U2's output switches to a low state, which makes Q1 conduct, turning on I1 until the brakes are released. Of course, removing power from the circuit resets SCR1, but the rc network consisting of R4 and C1 will not discharge immediately and will trigger SCR1 earlier. So, frequent brake use means fewer flashes.

HEADLIGHT ALARM

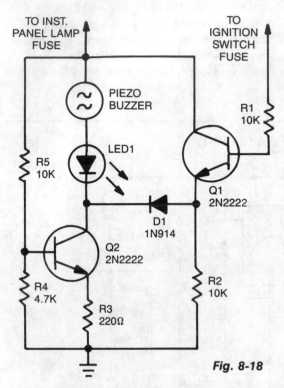

TO INST.
PANEL LAMP
FUSE

PIEZO
BUZZER

LED1

R5
10K

Q2
2N2222

R4
4.7K

R3
220Ω

TO
IGNITION
SWITCH
FUSE

R1
10K

Q1
2N2222

D1
1N914

R2
10K

The base of Q1 is connected to the car's ignition circuit. One side of the piezoelectric buzzer is connected to the instrument-panel light fuse. When the headlights are off, no current reaches the buzzer, and therefore nothing happens. What happens when the headlights are on depends on the state of the ignition switch. When the ignition switch is on, transistors Q1 and Q2 are biased on, removing the buzzer and the LED from the circuit. When the ignition switch is turned off, but the headlight switch remains on; transistor Q1 is turned off, but transistor Q2 continues to be biased on. The result is that the voltage is sufficient to sound the buzzer loudly and light the LED. Turning off the headlight switch will end the commotion quickly.

Fig. 8-18

Reprinted with permission from Radio-Electronics Magazine, April 1987. Copyright Gernsback Publications, Inc., 1987.

9

Battery Chargers

The sources of the following circuits are contained in the Sources section beginning on page 782. The figure number contained in the box of each circuit correlates to the sources entry in the Sources section.

PUT Battery Charger

Lead/Acid Battery Charger

Lead/Acid Low Battery Detector

Universal Battery Charger

UJT Battery Charger

Portable NiCad Battery Charger

Universal Battery Charger

Low-Battery Warning

PUT BATTERY CHARGER

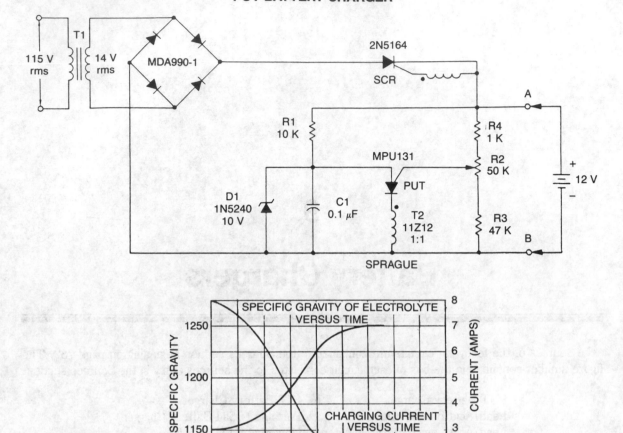

Copyright of Motorola, Inc. Used by permission.

Fig. 9-1

A short-circuit-proof battery charger will provide an average charging current of about 8 A to a 12-V lead/acid storage battery. The charger circuit has an additional advantage; it will not function nor will it be damaged by improperly connecting the battery to the circuit. With 115 V at the input, the circuit commences to function when the battery is properly attached. The battery provides the current to charge the timing capacitor C1 used in the PUT relaxation oscillator. When C1 charges to the peak point voltage of the PUT, the PUT fires turning the SCR on, which in turn applies charging current to the battery. As the battery charges, the battery voltage increases slightly which increases the peak point voltage of the PUT. This means that C1 has to charge to a slightly higher voltage to fire the PUT. The voltage on C1 increases until the zener voltage of D1 is reached, which clamps the voltage on C1, and thus prevents the PUT oscillator from oscillating and charging ceases. The maximum battery voltage is set by potentiometer R2 which sets the peak point firing voltage of the PUT. In the circuit shown, the charging voltage can be set from 10 V to 14 V—the lower limit being set by D1 and the upper limit by T1.

LEAD/ACID BATTERY CHARGER

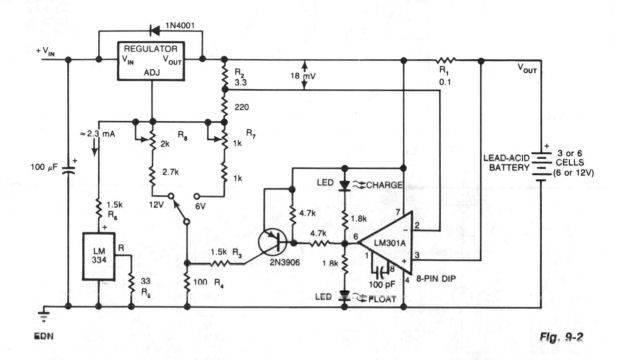

EDN

Fig. 9-2

This circuit furnishes an initial voltage of 2.5 V per cell at 25°C to rapidly charge a battery. The charging current decreases as the battery charges, and when the current drops to 180 mA, the charging circuit reduces the output voltage to 2.35 V per cell, leaving the battery in a fully charged state. This lower voltage prevents the battery from overcharging, which would shorten its life.

The LM301A compares the voltage drop across R1 with an 18 mV reference set by R2. The comparator's output controls the voltage regulator, forcing it to produce the lower float voltage when the battery-charging current, passing through R1, drops below 180 mA. The 150 mV difference between the charge and float voltages is set by the ratio of R3 to R4. The LEDs show the state of the circuit.

Temperature compensation helps prevent overcharging, particularly when a battery undergoes wide temperature changes while being charged. The LM334 temperature sensor should be placed near or on the battery to decrease the charging voltage by 4 mV/°C for each cell. Because batteries need more temperature compensation at lower temperatures, change R5 to 30 Ω for a tc of −5 mV/°C per cell if application will see temperatures below −20°C.

The charger's input voltage must be filtered dc that is at least 3 V higher than the maximum required output voltage: approximately 2.5 V per cell. Choose a regulator for the maximum current needed: LM371 for 2 A, LM350 for 4 A, or LM338 for 8 A. At 25°C and with no output load, adjust R7 for a V_{OUT} of 7.05 V, and adjust R8 for a V_{OUT} of 14.1 V.

LEAD/ACID LOW-BATTERY DETECTOR

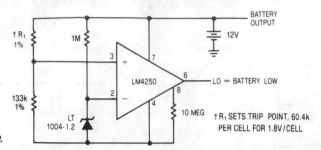

LINEAR TECHNOLOGY CORP.

$\dagger R_1$ SETS TRIP POINT, 60.4k
PER CELL FOR 1.8V/CELL

Fig. 9-3

UNIVERSAL BATTERY CHARGER

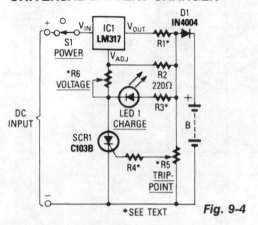

*SEE TEXT   *Fig. 9-4*

Reprinted with permission from Radio-Electronics Magazine, July 1986.
Copyright Gernsback Publications, Inc. 1986.

When power is applied to the circuit, SCR1 is off, so there is no bias-current path to ground; thus, LM317 acts as a current regulator. The LM317 is connected to the battery through steering diode D1, limiting resistor R1, and bias resistor R2. The steering diode prevents the battery from discharging through the LED and the SCR when power is removed from the circuit. As the battery charges, the voltage across trip-point potentiometer R5 rises, and at some point, turns on the SCR. Then, current from the regulator can flow to ground, so the regulator now functions in the voltage mode. When the SCR turns on, it also provides LED1 with a path to ground through R3. So, when LED1 is on, the circuit is in the voltage-regulating mode; when LED1 is off, the circuit is in the current-regulating mode.

UJT BATTERY CHARGER

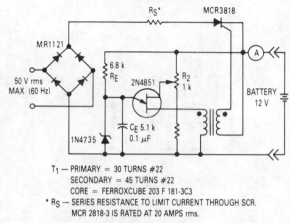

T_1 — PRIMARY = 30 TURNS #22
  SECONDARY = 45 TURNS #22
  CORE = FERROXCUBE 203 F 181-3C3
* R_S — SERIES RESISTANCE TO LIMIT CURRENT THROUGH SCR.
  MCR 2818-3 IS RATED AT 20 AMPS rms.

Copyright of Motorola, Inc. Used by permission.   *Fig. 9-5*

This circuit will not work unless the battery to be charged is connected with proper polarity. The battery voltage controls the charger and when the battery is fully charged, the charger will not supply current to the battery. The battery charging the current is obtained through the SCR when it is triggered into the conducting state by the UJT relaxation oscillator. The oscillator is only activated when the battery voltage is low. V_{B2B1} of the UJT is derived from the voltage of the battery to be charged, and since $V_P = V_D = V_{B2B1}$; the higher V_{B2B1}, the higher V_P. When V_P exceeds the breakdown voltage of the zener diode Z1, the UJT will cease to fire and the SCR will not conduct. This indicates that the battery has attained its desired charge as set by R2.

PORTABLE NICAD BATTERY CHARGER

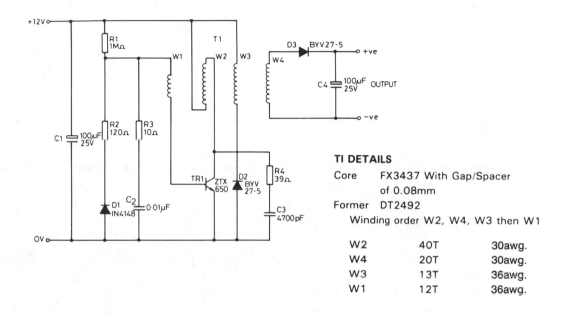

TI DETAILS

Core	FX3437 With Gap/Spacer	
	of 0.08mm	
Former	DT2492	

Winding order W2, W4, W3 then W1

W2	40T	30awg.
W4	20T	30awg.
W3	13T	36awg.
W1	12T	36awg.

ZᴇTᴇX, formerly FERRANTI

Fig. 9-6

This circuit was designed to charge NiCad battery packs in the range of 4.8 to 15.6 V from a convenient remote power source, such as an automobile battery. When power is first applied to the circuit, a small bias current supplied by R1 via winding W1, starts to turn on the transistor TR1. This forces a voltage across W2 and the positive feedback given by the coupling of W1 and W2 causes the transistor to turn hard on, applying the full supply across W2. The base drive voltage induced across W1 makes the junction between R1 and R2 become negative with respect to the 0-V supply, forward-biasing diode D1 to provide the necessary base current to hold TR1 on.

With the transistor on, a magnetizing current builds up in W2, which eventually saturates the ferrite core of transformer T1. This results in a sudden increase on the collector current flowing through TR1, causing its collector-emitter voltage to rise, and thus reducing the voltage across W2. The current flowing in W2 forces the collector voltage of the TR1 to swing positive until restricted by transformer output loading. Rc network R4 and C3 limits the turn off transient TR1. R3 and C2 maintain the loop gain of the circuit when diode D1 is not conducting.

UNIVERSAL BATTERY CHARGER

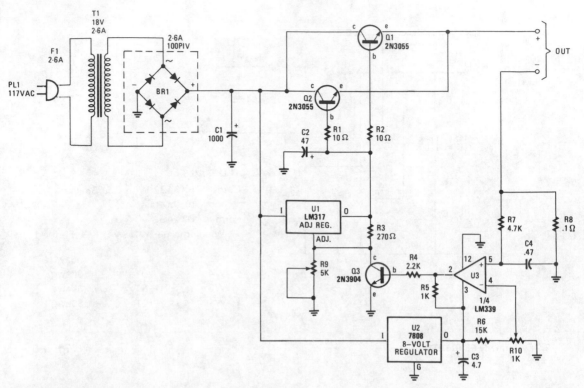

Fig. 9-7

The charger's output voltage is adjustable and regulated, and has an adjustable constant-current charging circuit that makes it easy to use with most NiCad batteries. The charger can charge a single cell or a number of series-connected cells up to a maximum of 18 V.

Power transistors Q1 and Q2 are connected as series regulators to control the battery charger's output voltage and charge-current rate. An LM317 adjustable voltage regulator supplies the drive signal to the bases of power transistors Q1 and Q2. Potentiometer R9 sets the output-voltage level. A current-sampling resistor, R8 (a 0.1-Ω, 5-W unit), is connected between the negative output lead and circuit ground. For each amp of charging that flows through R8, a 100 mV output is developed across it. The voltage developed across R8 is fed to one input of comparator U3. The other input of the comparator is connected to variable resistor R10.

As the charging voltage across the battery begins to drop, the current through R8 decreases. Then the voltage feeding pin 5 of U3 decreases, and the comparator output follows, turning Q3 back off, which completes the signal's circular path to regulate the battery's charging current.

The charging current can be set by adjusting R10 for the desired current. The circuit's output voltage is set by R9.

LOW-BATTERY WARNING

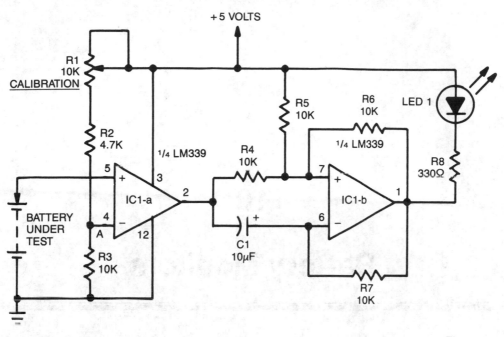

Fig. 9-8

A voltage divider consisting of R1, R2, and R3 is used to set the input reference voltage below which the batteries are to be replaced. That reference voltage, at point A, is varied by R1. With the voltage divider shown in Fig. 9-7, a range of 2 to 3.5 V is possible. When the battery voltage drops below that at point A, the output of IC1a, 1/4 of a LM339 quad comparator, switches from high to low. That triggers IC1b, which is configured as an astable multivibrator. Feedback resistors R6 and R7, coupled with capacitor C1, determine the time constant of the multivibrator. The output from IC1b is connected to LED1 through dropping resistor R8. With the circuit values as shown, the LED will flash at a rate of 3 Hz. Although this circuit was designed specifically to monitor RAM back-up batteries, it can of course be modified for use in just about any application where the condition of a battery must be found.

10

Battery Monitors

The sources of the following circuits are contained in the Sources section beginning on page 782. The figure number contained in the box of each circuit correlates to the sources entry in the Sources section.

QUICKLY DEACTIVATING BATTERY SENSOR

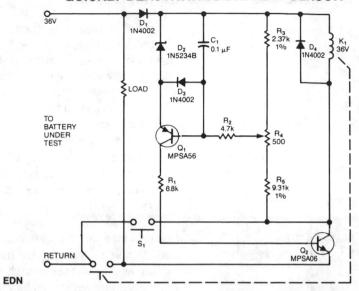

EDN

Fig. 10-1

The sensing circuit rapidly disconnects the battery voltage and load whenever the voltage drops below a preset threshold. One-way operation prevents the circuit from reconnecting the load if the voltage should then rise above the threshold. C1 ensures that the circuit doesn't activate while making connections to the battery; if you accidentally reverse these connections, D1 will block the turn on the relay.

After you connect the battery, nothing happens until you depress pushbutton switch S1, which allows relay K1 to energize. When you release S1, the relay remains on only if the battery voltage is above the minimum level. You preset this threshold—to 31.5 V when testing 36-V batteries, for example—using R4. Q1 begins to turn off as the battery voltage drops. Once the threshold level is reached, Q2 also begins to turn off, and its rising collector voltage provides positive feedback to the base of Q1, accelerating the turn off. When Q2 turns off, the relay drops out, disconnecting the battery from its load.

AUTOMATIC SHUTOFF FOR BATTERY-POWERED PROJECTS

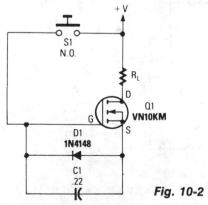

Fig. 10-2

When S1 is depressed, C1 begins to charge to the supply voltage. That places a forward bias on the gate of Q1 turning it on and supplying current to load resistor RL. When the charge on C1 leaks off, the transistor shuts off, cutting off current to the load. That load could be anything from a transistor radio to a child's toy. Transistor Q1, available from Radio Shack as part No. 276–2070, is rated at 0.5 A at 60 Vdc. With a supply voltage of 9 Vdc and with C1 rated at 0.22 µF, a delay of about one minute is produced; with C1 rated at 10 µF, the delay is about an hour.

POPULAR ELECTRONICS / HANDS-ON ELECTRONICS

NICAD-BATTERY PROTECTION CIRCUIT

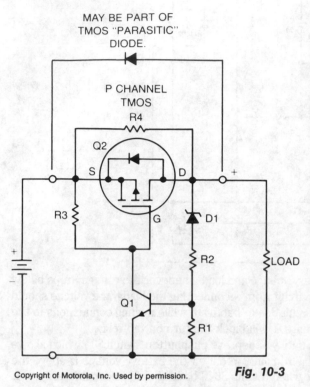

MAY BE PART OF TMOS "PARASITIC" DIODE.

P CHANNEL TMOS

Copyright of Motorola, Inc. Used by permission.

Fig. 10-3

If a NiCad battery is discharged to a point at which the lowest capacity cell becomes fully discharged and reverses polarity, that cell will usually short internally and become unusable. To prevent this type of damage, this circuit detects a one-cell drop of 1.25 V and turns the load off before cell reversal can occur.

Low-current zener or other voltage sensor D1 and resistors R1 and R2 establish a reference level for transistor Q1. These resistors bias the zener to a few microamperes above its "knee." Therefore, if battery voltage falls more than 1.25 V, Q1 turns off, turning off Q2, and disconnecting the load. After the load is disconnected, if the battery returns to nominal voltage, the high value of resistor shunting Q2 provides enough output voltage to reset the voltage sensor and turn Q2 back on. If desirable, shunt diode D2 or the parasitic diode of the TMOS device, if suitable, allows the battery to be charged from the load terminals.

The protection circuit presents a shunt current of only 10 mA at nominal battery voltage, which is low relative to the internal leakage of the batteries.

9-V BATTERY LIFE EXTENDER

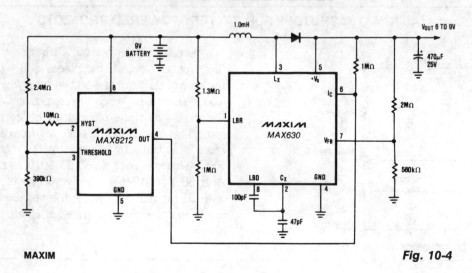

MAXIM

Fig. 10-4

9-V BATTERY LIFE EXTENDER (*CONT.*)

Circuit provides a minimum of 7 V until the 9-V battery voltage falls to less than 2 V. When the battery voltage is above 7 V, the MAX630's IC pin is low, putting it into the shutdown mode which draws only 10 nA. When the battery voltage falls to 7 V, the MAX8212 voltage detector's output increases. The MAX630 then maintains the output voltage at 7 V. The low battery detector (LBD) is used to decrease the oscillator frequency when the battery voltage falls to 3 V, thereby increasing the output current capability of the circuit. Note that this circuit, with or without the MAX8212, can be used to provide 5 V from 4 alkaline cells. The initial voltage is approximately 6 V, and the output is maintained at 5 V even when the battery voltage falls to less than 2 V.

AUTO BATTERY ALTERNATOR MONITOR	LOW-BATTERY DETECTOR

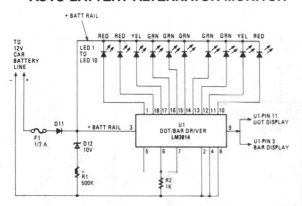

HANDS-ON ELECTRONICS   *Fig. 10-5*

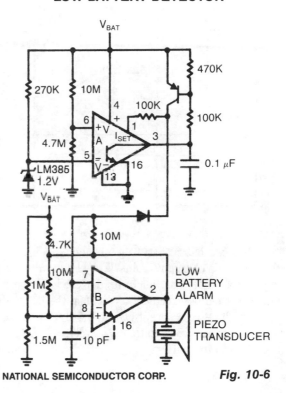

NATIONAL SEMICONDUCTOR CORP.   *Fig. 10-6*

Most of the circuitry is contained in the LM3914 dot/bar-graph driver IC chip. In addition to the comparator circuitry within the package, it also contains a stable reference supply and the drivers for the LEDs. Resistor R2 acts as the current limiter for all the LEDs. Resistor R2 may be varied for LED brightness.

The unit will illuminate one LED for each voltage condition encountered in the charging system. This system is called a dot-graph display; it is achieved by wiring the mode control at pin 9 to pin 11 on U1. It is possible to wire the monitor so that each lamp will be illuminated up to the maximum voltage on the line at that moment. The latter is referred to as a bar-graph display. By connecting pin 9 to pin 3 on U1, the bar-graph mode will be enabled.

Comparator A detects when the supply voltage drops to 4 V and enables comparator B to drive a piezoelectric alarm.

I_S: 6 V at 45 μA
I_S: 3.8 V at 1 μA
f: 3 kHz

NICAD-BATTERY ANALYZER

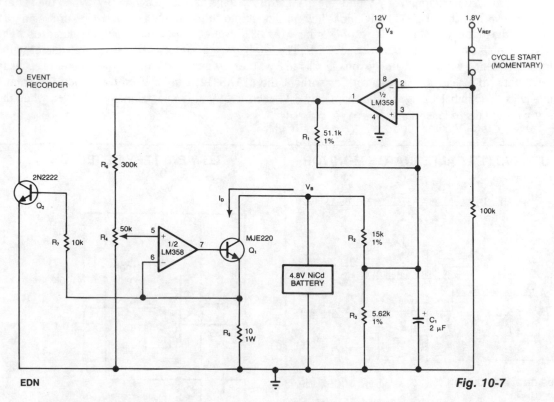

EDN

Fig. 10-7

Because NiCad batteries maintain a constant output voltage, it is difficult to determine how much of the battery's charge remains. The circuit provides a way of determining the capacity of a battery by draining it at a preset current to its depleted voltage of 1 V/cell. Measure the discharge time of the cells and perform a simple calculation to obtain the battery's capacity.

Set the drain current (I_D) to $0.5C$ (C = battery capacity in mA/hr) by selecting an appropriate value for R4. Choose R5 such that: $I_D \times R5 = 1$ V. V_{REF} is set so the comparator turns off the drain current and timer when the battery reaches its depleted voltage, V_B (usually 1 V/cell). You calculate V_{REF} as follows:

$$V_{REF} = \frac{R_3 \left[R_2(V_S - 1.3) + R_1 V_B \right]}{R_1 R_2 + R_2 R_3 + R_1 R_3}$$

With the battery in place, activate the circuit by grounding V_{REF} with the momentary switch. The battery drains at I_D until it reaches V_B, turning off the drain circuit and the timer. Hysteresis keeps the circuit from restarting. Determine the battery's capacity using the following equation:

$$C_{(mAhr)} = \text{Time of Cycle} \times I_D$$

The circuit shown tests 4.8 V, 180 mA/hr batteries. I_D is 100 mA and V_B is 4 V.

LOW-BATTERY PROTECTOR

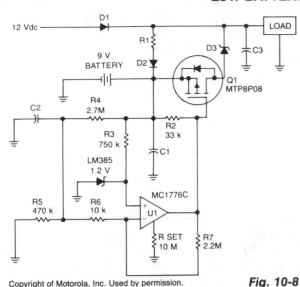

To prevent battery damage due to over-discharge, a low-voltage detector and switch should be included in the design of the battery backup circuit. The detector circuit should consume extremely low current. The switch should exhibit a low-voltage drop and be easy to control.

R1 and D2 provide a trickle charge for the battery. Chosen for its low forward voltage drop, Schottky diode D3 prevents forward polarization of the diode incorporated in Q1. When the battery voltage is above approximately 8 V, the output of U1 is low and Q1 is turned on. If the battery voltage falls below 8 V, the output of U1 increases and turns off Q1.

Fig. 10-8

LOW-BATTERY WARNING/DISCONNECTOR

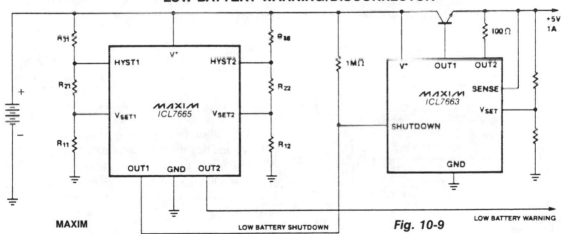

Fig. 10-9

NiCad batteries are excellent rechargeable power sources for portable equipment, but care must be taken to ensure that the batteries are not damaged by overdischarge. Specifically, a NiCad battery should not be discharged to the point where the polarity of the cell is reversed and is reverse-charged by the higher-capacity cells. This reverse charging will dramatically reduce the life of a NiCad battery. This circuit both prevents reverse charging and also gives a low-battery warning. A typical low-battery warning voltage is 1 V per cell. Since a NiCad, 9-V battery is ordinarily made up of six cells with a nominal voltage of 7.2 V, a low-battery warning of 6 V is appropriate, with a small hysteresis of 100 mV. To prevent overdischarge of a battery, the load should be disconnected when the battery voltage is $1 \text{ V} \times (N-1)$, where N = number of cells. In this case, the low-battery load disconnect should occur at 5 V. Since the battery voltage will rise when the load is disconnected, 800 mV of hysteresis is used to prevent repeated on-off cycling.

BATTERY CAPACITY TESTER

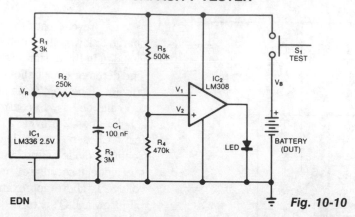

EDN   *Fig. 10-10*

The test circuit gives an indication of the capacity remaining in a battery. By noting the time in seconds that the LED remains on after you depress the test switch S1. The circuit has proven reliable in testing NiCad-, carbon-, and alkaline-type batteries. Closing S1 activates the circuit by applying voltage from the battery under test. Voltage V_1 jumps to a value $V_0 = V_R R_3/(R_2 + R_3)$ when the switch closes and then increases with a time constant $T = C_1 (R_2 + R_3)$. The divider R4/R5 fixes V_2. The reference circuit IC1 sets V_R to approximately 2.5 V. The op amp's output remains high (LED on) until V_1 rises to the level of V_2, when the LED turns off. Calculate the on-time t_{ON} as follows:

$$t_{ON} = T \ln \frac{V_R - V_0}{V_R - V_2}$$

BATTERY SPLITTER

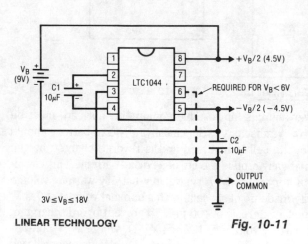

LINEAR TECHNOLOGY   *Fig. 10-11*

A common need in many systems is to obtain positive and negative supplies from a single battery. Where current requirements are small, the circuit shown is a simple solution. It provides symmetrical ± output voltages, both equal to one half the input voltage. The output voltages are referenced to pin 3, output common. If the input voltage between pin 8 and pin 5 exceeds 6 V, pin 6 should also be connected to pin 3, as shown by the dashed line. Higher current requirements are served by an LT1010 buffer. The splitter circuit can source or sink up to ±150 mA with only 5 mA quiescent current. The output capacitor, C2, can be made as large as necessary to absorb current transients. An input capacitor is also used on the buffer to avoid high frequency instability that can be caused by high source impedance.

ELECTRIC VEHICLE BATTERY SAVER

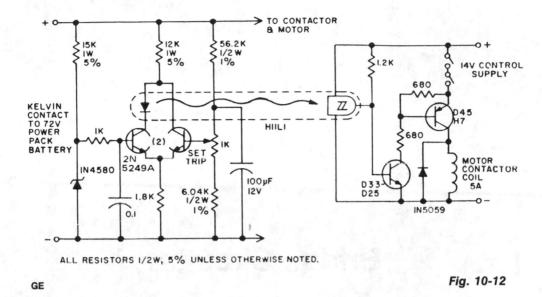

ALL RESISTORS 1/2W, 5% UNLESS OTHERWISE NOTED.

GE

Fig. 10-12

The battery life and operating cost of an electric vehicle is severely affected by overdischarge of the battery. This circuit provides both warning and shutdown. An electronic switch is placed in series with the propulsion motor contactor coil. Three modes of operation are possible:

- When the propulsion power pack voltage is above the 63-V trip point, the electronic switch has no effect on operation
- When the propulsion power pack no load voltage is below 63 V, power will not be supplied to the propulsion motor since the electronic switch will prevent contactor operation
- When the propulsion power pack loaded voltage drops below 63 V, the contactor will close and open because of the electronic switch. The *bucking operation* indicates to the operator need to charge the batteries

11

Bridge Circuits

The sources of the following circuits are contained in the Sources section beginning on page 782. The figure number contained in the box of each circuit correlates to the sources entry in the Sources section.

Auto-Zeroing Scale
Accurate Null/Variable Gain Circuit
Remote Sensor Loop Transmitter
Bridge Transducer Amplifier
Strain Gauge Signal Conditioner with
   Bridge Excitation

AUTO-ZEROING SCALE

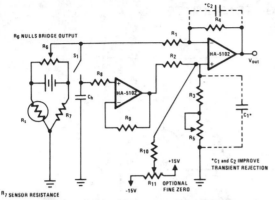

Fig. 11-1

HARRIS

Electronic scales have come into wide use and the HA-510X, as a very low noise device, can improve such designs. This circuit uses a stain-gauge sensing element as part of a resistive Wien-bridge. An auto-zero circuit is also incorporated into this design by including a sample-and-hold network.

The bridge signal drives the inverting input of a differentially configured HA-5102. The noninverting input is driven by the other half of the HA-5102 used as a buffer for the holding capacitor, *CH*. This second amplifier and its capacitor *CH* form the sampling circuit used for automatic output zeroing. The 20-KΩ resistor between the holding capacitor *CH* and the input terminal, reduces the drain from the bias currents. A second resistor *RG* is used in the feedback loop to balance the effect of R8. If R7 is approximately equal to the resistance of the strain gauge, the input signal from the bridge can be roughly nulled with R6. With very close matching of the ratio R4/R1 to R3/R2, the output offset can be nulled by closing S1. This will charge *CH* and provide a 0-V difference to the inputs of the second amplifier, which results in a 0-V output. In this manner, the output of the strain gauge can be indirectly zeroed. R10 and potentiometer R11 provide an additional mechanism for fine tuning V_{OUT}, but they can also increase offset voltage away from the zero point. C1 and C2 reduce the circuit's susceptibility to noise and transients.

ACCURATE NULL/VARIABLE GAIN CIRCUIT

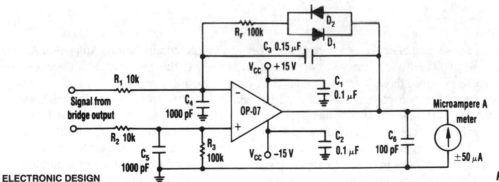

ELECTRONIC DESIGN

Fig. 11-2

The circuit can use any general-purpose, low-offset, low-drift op amp, such as the OP-07. The differential signal from the bridge feeds an amplifier that drives an ordinary, rugged ±50-µA meter. Near the null point, however, the drastically reduced signal level from the bridge requires very high gain to achieve a high null resolution. To provide the variable-gain feature, the op amp's feedback path needs a dynamic resistance that increases as the input signal drops. Two common signal diodes, D1 and D2, in an antiparallel configuration in the feedback path supply function for all positive and negative inputs. To stabilize the op amp circuit at high gain, capacitors C3, C5, and C6 reduce response to high frequencies; capacitors C1 and C2 bypass the amplifier's power supplies.

REMOTE SENSOR LOOP TRANSMITTER

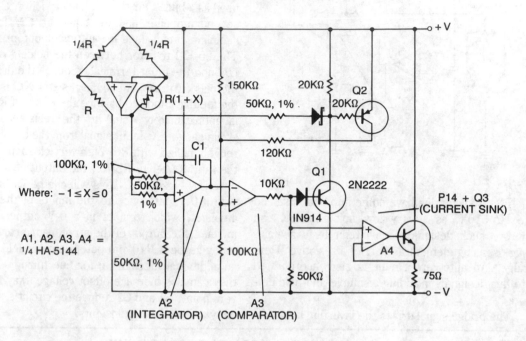

HARRIS

Fig. 11-3

This circuit shows amplifier A1 as a sensor amplifier in a bridge configuration. Amplifiers A2 and A3 are configured as a voltage to frequency converter and A4 is used as the transmitter. This entire sensor/transmitter can be powered directly from a 4 to 20 mA current loop.

The bridge configuration produces a linear output with respect to the changes in resistance of the sensor. The voltage at the output of A1 causes the integrator output A2 to ramp down until it crosses the comparator threshold voltage of A3. A3 turns on Q1 and Q2. A1 causes the output of A2 to ramp up at a rate nearly equal to its negative slope, while Q2 provides hysteresis for the comparator. In addition, Q1 and Q2 help eliminate changes in power supply loop voltage. Amplifier A4 and Q3 are configured as a constant current sink which turns on when the comparator current increases. The resulting increase in loop current transmits the frequency of the V/F converter back to the control circuitry.

BRIDGE TRANSDUCER AMPLIFIER

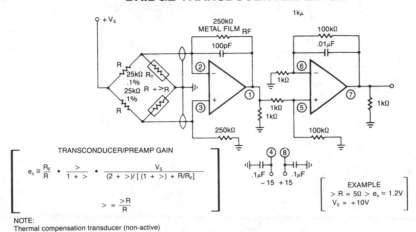

SIGNETICS

Fig. 11-4

In applications involving strain gauges, accelerometers, and thermal sensors, a bridge transducer is often used. Frequently, the sensor elements are high resistance units requiring equally high bridge resistance for good sensitivity. This type of circuit then demands an amplifier with high input impedance, low bias current and low drift. The circuit shown represents a possible solution to these general requirements.

STRAIN GAUGE SIGNAL CONDITIONER WITH BRIDGE EXCITATION

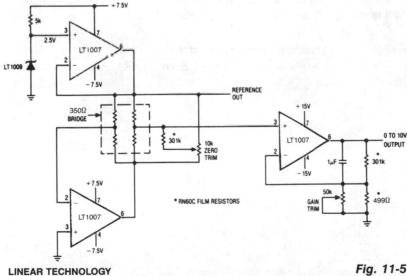

LINEAR TECHNOLOGY

Fig. 11-5

The LT1007 is capable of providing excitation current directly to bias the 350-Ω bridge at 5 V. With only 5 V across the bridge, as opposed to the usual 10 V, total power dissipation and bridge warm-up drift is reduced. The bridge output signal is halved, but the LT1007 can amplify the reduced signal accurately.

12

Burst Generators

The sources of the following circuits are contained in the Sources section beginning on page 782. The figure number contained in the box of each circuit correlates to the sources entry in the Sources section.

Portable Rf Burst Generator
Tone Burst Generator for European Repeaters

PORTABLE RF BURST GENERATOR

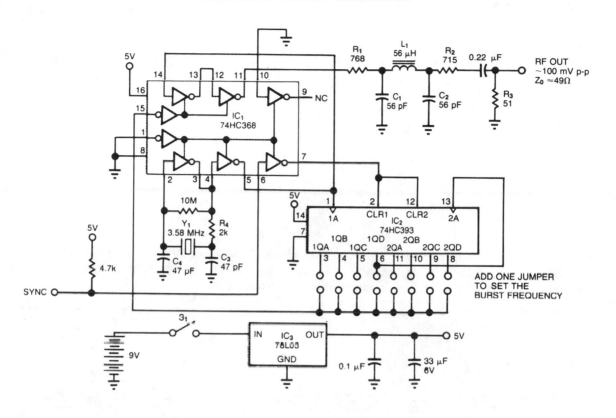

Fig. 12-1

The circuit generates low-level rf bursts having frequencies as high as 10 MHz, thus permitting field testing of high frequency receivers. A jumper-selectable binary fraction ($1/2$ to $1/256$ of the Y1 crystal frequency gates the output rf signal. Output amplitude (open circuit) is approximately 100 mV; output impedance is approximately 49 Ω. The rf source is a clock oscillator based on a 3.58-MHz, color-burst crystal and two inverting buffers. The oscillator drives two cascaded 4-bit binary counters, IC2, and the sync signal resets the counters with a logic-high pulse—logic low at the counters. Select the desired output frequency by adding a jumper to one of the counter's eight output lines, which provides an enable signal for the two 3-state output buffers. The square-wave output at IC1, pin 11, is attenuated by R1, R2, and R3 to fix the output resistance at approximately 49 Ω. Resistor R3 is the only critical component; for clean gating, isolate it from the rest of the circuit.

TONE BURST GENERATOR FOR EUROPEAN REPEATERS

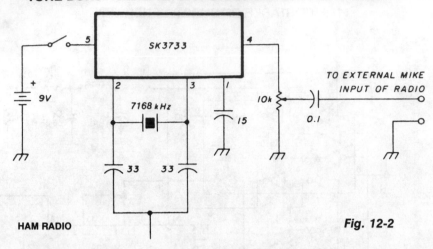

HAM RADIO

Fig. 12-2

Most European repeaters must be brought up with a 1750-Hz tone. The SK3733 (also known as an ECG1197) IC contains a crystal oscillator and is divided by −256, 1024, 2048, and 4096. A 7168-kHz crystal is used; the divide-by-4096 output produces a 1750-Hz signal.

13

Capacitance Meters

The sources of the following circuits are contained in the Sources section beginning on page 782. The figure number contained in the box of each circuit correlates to the sources entry in the Sources section.

3½-Digit A/D Capacitance Meter
Capacitance Meter

3¹/₂-DIGIT A/D CAPACITANCE METER

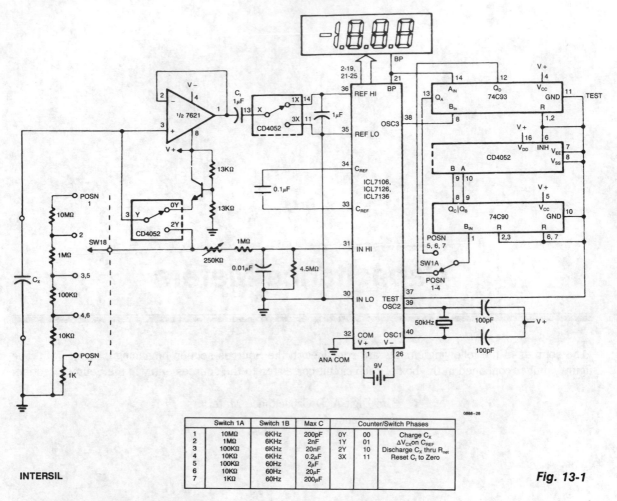

	Switch 1A	Switch 1B	Max C	Counter/Switch Phases			
1	10MΩ	6KHz	200pF	0Y	00	Charge C_X	
2	1MΩ	6KHz	2nF	1Y	01	ΔV_{CX} on C_{REF}	
3	100KΩ	6KHz	20nF	2Y	10	Discharge C_X thru R_{net}	
4	10KΩ	6KHz	0.2µF	3X	11	Reset C_t to Zero	
5	100KΩ	60Hz	2µF				
6	10KΩ	60Hz	20µF				
7	1KΩ	60Hz	200µF				

INTERSIL

Fig. 13-1

The circuit charges and discharges a capacitor at a crystal-controlled rate, and stores on a sample-and-difference amplifier the change in voltage achieved. The current that flows during the discharge cycle is averaged, and ratiometrically measured in the a/d using the voltage change as a reference. Range switching is done by changing the cycle rate and current metering resistor. The cycle rate is synchronized with the conversion rate of the a/d by using the externally divided internal oscillator and the internally divided back plane signals. For convenience in timing, the switching cycle takes 5 counter states, although only four switch configurations are used. Capacitances up to 200 µF can be measured, and the resolution on the lowest range is down to 0.1 pF.

The zero integrator time can be set initially at ¹/₃ to ¹/₂, the minimum auto-zero time, but if an optimum adjustment is required, look at the comparator output with a scope under worst-case overload conditions. The output of the delay timer should stay low until after the comparator has come off the rail, and is in the linear region (usually fairly noisy).

CAPACITANCE METER

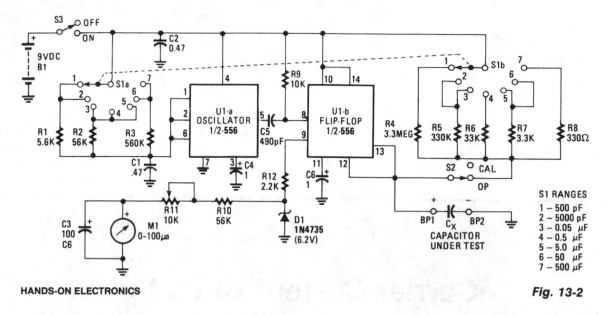

HANDS-ON ELECTRONICS

Fig. 13-2

U1a is an oscillator and U1b the measurement part of the circuit. It converts unknown capacity into a pulse-width modulated signal the same way an automotive dwell meter works. The meter is linear so the fraction or percentage of time that the output is high is directly proportional to the unknown capacitance (*CX* in the schematic). Meter M1 reads the average voltage of those pulses since its mechanical frequency response is low compared to the oscillator frequency of U1a.

14

Carrier-Current Circuits

The sources of the following circuits are contained in the Sources section beginning on page 782 . The figure number contained in the box of each circuit correlates to the sources entry in the Sources section.

Carrier-Current Audio Transmitter
Carrier-Current FM Receiver
Carrier-Current AM Receiver
Power-Line Modem

CARRIER-CURRENT AUDIO TRANSMITTER

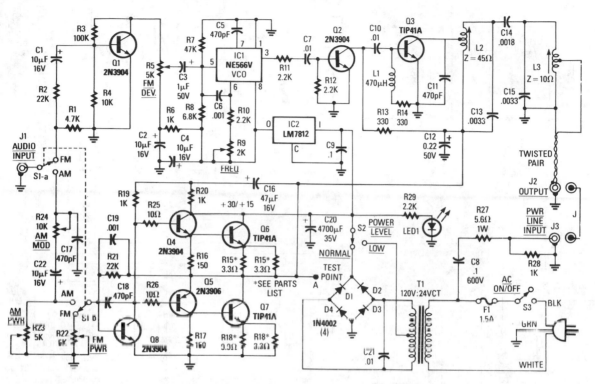

Reprinted with permission from Radio-Electronics Magazine, January 1989. Copyright Gernsback Publications, Inc., 1989.

Fig. 14-1

The decision to use either AM, narrowband FM (less than 15 kHz), or wideband FM (greater than 30 kHz) depends on the application. For the transmission of music, FM is better because it has greater noise immunity. For speech or other noncritical applications, AM may be satisfactory. Our transmitter permits either mode by switch selection.

Audio is fed from S1a to either the FM or AM circuitry. Starting with the FM section, amplifier Q1 accepts an audio signal in the 10 Hz to 20 kHz range of about 0.5 V peak-to-peak. The audio gain is adjusted via R5 to provide up to 60 kHz deviation of voltage-controlled oscillator IC1 which is set to nominally 280 kHz. IC1 and Q1 are supplied with a regulated 12 V from IC2. A square-wave signal from IC1 pin 3 drives Q2, and Q2 drives the output amplifier Q3. A coupling network is used to match the nominal 45-Ω output impedance of Q3 to the 10-Ω ac line impedance.

In the AM mode, audio is coupled to Q8 via R24 and then amplified again by transistors Q4 to Q7. The normally stable dc voltage at test point A is thereby varied at an audio rate. Because Q2 and Q3 obtain their dc V_{CC} from test point A, the VCO carrier input to Q2 is amplitude modulated by the varying V_{CC} amplitude. That produces an amplitude-modulated output from the transmitter. Careful setting of carrier level R23 and audio level R24 provides up to 100% modulation. The kit is available from North Country Radio, P.O. Box 53, Wykagyl Station, NY 10804.

CARRIER-CURRENT FM RECEIVER

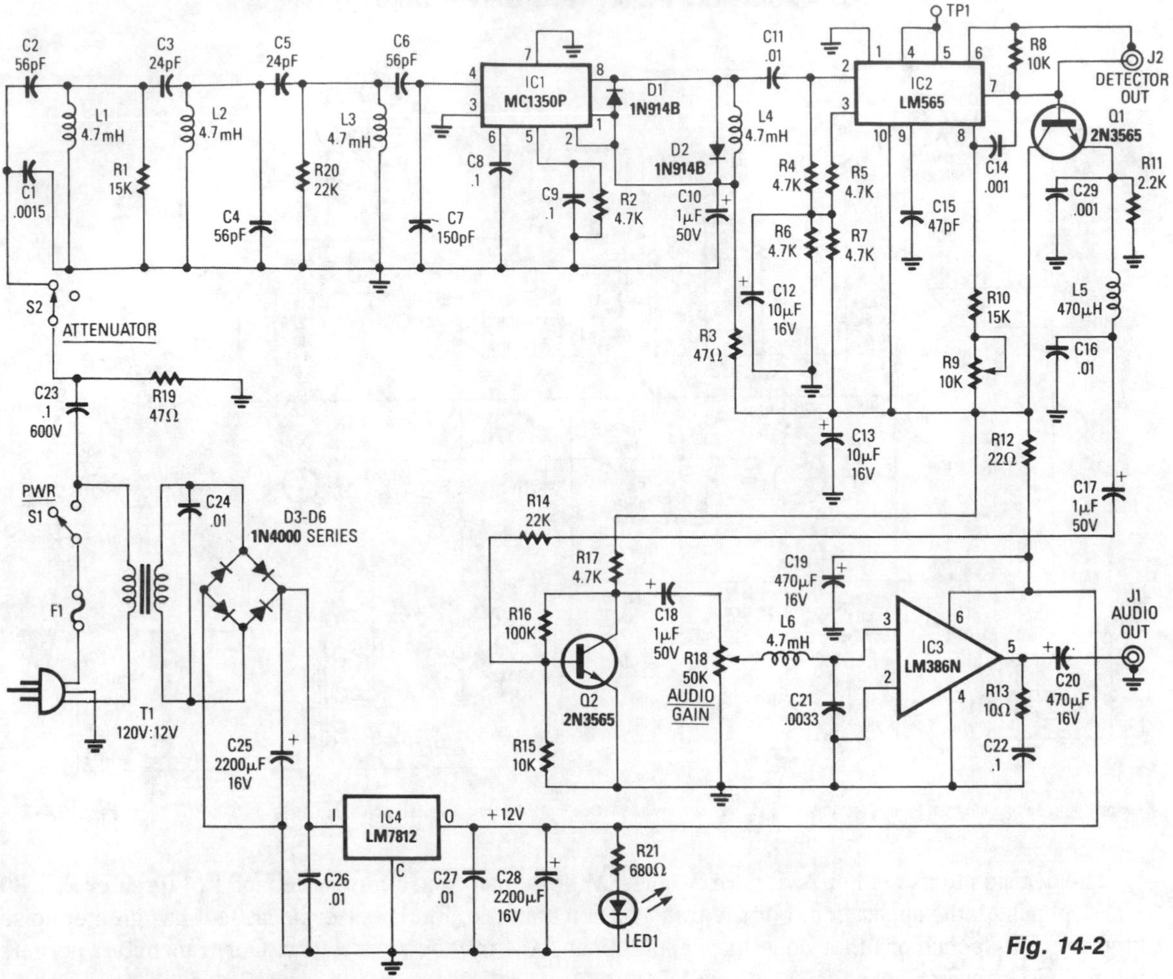

Fig. 14-2

Reprinted with permission from Radio-Electronics Magazine, February 1989. Copyright Gernsback Publications, Inc., 1989.

Input signals from the power line are coupled through C23 and R19 to the input filter network. C23 must be rated at 600 volts. Switch S2 is used as an attenuator. Components C2 through C7, L1 through L3, R1, and R20 form a triple-tuned bandpass filter having a passband from 220–340 kHz. Signals from the filter are fed to an MC1350P gain block IC, which is used as a tuned rf amplifier.

IC2, the LM565 PLL, is used as an FM demodulator. Pins 8 and 9 are connected to an internal VCO and components R9, R10, and C15 set the VCO's free running frequency. The VCO signal and the input signal from pin 2 are compared in the phase detector. The output from the phase detector is internally amplified, and then appears at pin 7. The output at pin 7 is a replica of the original modulation on the FM input signal to the receiver; the output at pin 7 is therefore the recovered audio. C17 and R14 couple audio to the base of Q2, which, in conjunction with R15, R16, R17, and C18, form an audio amplifier that brings the recovered audio up to around 1 V peak-to-peak. The signal is then fed into an LM386N audio amplifier, which can deliver up to 1/2 W of audio, coupled via C20, to any standard 8-Ω external speaker. The kit is available from North Country Radio, P.O. Box 53, Wykagyl Station, NY 10804.

CARRIER-CURRENT AM RECEIVER

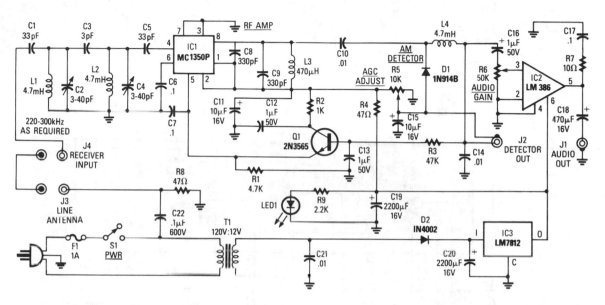

Fig. 14-3

The AM Tuned Radio Frequency (TRF) receiver, has a sensitivity of about 1 mV at the input for an audio output of 1/2 W. Capacitor C22 couples audio signals from the power line to the PC board—it must be rated at 600 Vdc. R8 will cause F1 to blow, if C22 shorts. The signal from C22 goes to a tuned network (C1 through C5, L1, and L2) that has a 20-kHz bandwidth, which allows only the desired signal to pass through.

IC1 is a *gain block* i-f chip that has AGC capability and approximately 60 dB of gain. Components C8, C9, and L3, which are placed across the output of IC1, are broadly resonant around 280 kHz. C10 couples rf to detector-diode D1, which is used as an envelope detector.

The detector output is taken from C14, which sets the upper frequency limit at about 10 kHz or so. By reducing the value of C14, high frequency response can be obtained. The detector output is connected to an external jack. Audio components are fed to audio-gain control R6, through C16 to IC2, an audio amplifier. C18 couples up to 1/2 watt of audio to an external speaker. The kit is available from North Country Radio, P.O. Box 53, Wykagyl Station, NY 10804.

POWER-LINE MODEM

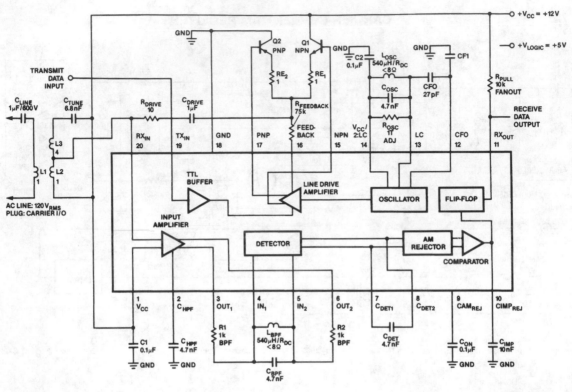

SIGNETICS

Fig. 14-4

In the 100-kHz application from left to right, the coupling network feeds into the receiver section on the bottom of the chip. (The external components are summarized later.) The receive data output is pulled up via R_{PULL} = 10 KΩ. A minimum current of 10 mA sets the voltage drop across R_{PULL}. Another voltage supply, V_{LOGIC}, is shown if the user wants to have the output sent at TTL levels. Across the top is the transmitter section; going from right to left, the oscillator network, the class AB output stage (note feedback resistor $R_{FEEDBACK}$) and the drive section. The LC values on the oscillator network should match those on the bandpass filter in the receiver. The drive stage feeds into the coupling network and back into the receive section. This enables the on-chip collision detection with listen-while-talking capability. This effect can be cancelled, although the transmitter will still be connected to the receiver.

15

Clock Circuits

The sources of the following circuits are contained in the Sources section beginning on page 782. The figure number contained in the box of each circuit correlates to the sources entry in the Sources section.

Digital Clock with Alarm
Oscillator/Clock Generator
Single Op Amp Clock
Wide-Frequency TTL Clock

DIGITAL CLOCK WITH ALARM

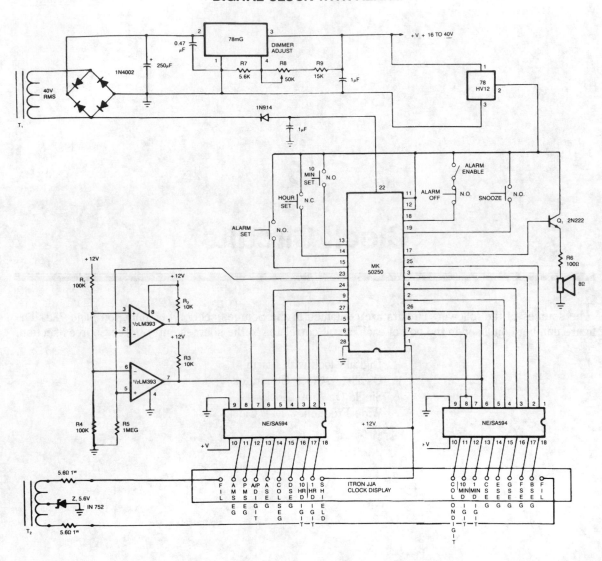

SIGNETICS

Fig. 15-1

OSCILLATOR/CLOCK GENERATOR

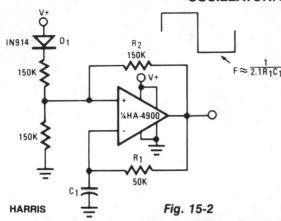

$$F \approx \frac{1}{2.1R_1C_1}$$

HARRIS   **Fig. 15-2**

This self-starting fixed-frequency oscillator circuit gives excellent frequency stability. R1 and C1 comprise the frequency-determining network, while R2 provides the regenerative feedback. Diode D1 enhances the stability by compensating for the difference between V_{OH} and V_{SUPPLY}. In applications where a precision clock generator up to 100 kHz is required, such as in automatic test equipment, C1 might be replaced by a crystal.

SINGLE OP AMP CLOCK

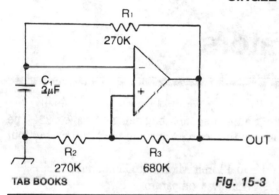

TAB BOOKS   **Fig. 15-3**

Capacitor C1 is charged through timing resistor R1 when the clock output is high. When C1 reaches the upper threshold voltage, the output signal decreases, and then C1 discharges through R1 until its voltage reaches the lower threshold point. When this happens, the output increases again and the cycle repeats itself. Using the parts values shown results in a frequency of 1 Hz. The output frequency can be adjusted by trimming the value of R1 slightly.

WIDE-FREQUENCY TTL CLOCK

EDN

Fig. 15-4

This free-running TTL square-wave oscillator has a variable frequency output over a 20:1 range or better through use of four of the six inverters in an SN7404 chip and the additional components shown. Frequency of oscillation is determined by the capacitor and the settings of potentiometers R2 and R4; the first pot controls width T1 and the second controls width T2 of the square-wave output. These adjustments are not completely independent.

16

Comparators

The sources of the following circuits are contained in the Sources section beginning on page 782. The figure number contained in the box of each circuit correlates to the sources entry in the Sources section.

Window Comparator
Microvolt Comparator with Hysteresis
Comparator/Latch
Frequency-Detecting Comparator
Precision Comparator with Balanced Outputs
    and Variable Offset

Dual Limit Microvolt Comparator
Window Comparator
Four-Channel Comparator

WINDOW COMPARATOR

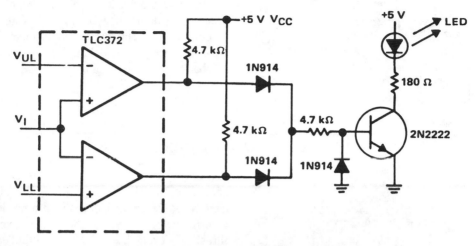

Window Comparator with LED Indicator

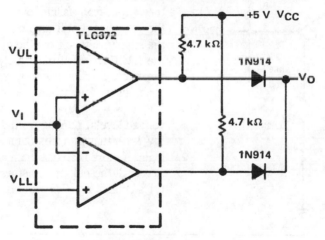

Basic Window Comparator   *Fig. 16-1*

A window detector is a specialized comparator circuit designed to detect the presence of a voltage between two prescribed limits; that is, within a voltage *window*. This circuit is implemented by logically combining the outputs of two single-ended comparators by the IN914 diodes. When the input voltage is between the upper limit, V_{UL}, and the lower limit, V_{LL}, the output voltage is zero; otherwise it equals a logic high level. The output of this circuit can be used to drive a logic gate, LED driver, or relay driver circuit. The circuit shown in Fig. 16-1 shows a 2N2222 npn transistor being driven by the window comparator. When the input voltage to the window comparator is outside the range set by the V_{UL} and V_{LL} inputs, the output changes to positive, which turns on the transistor and lights the LED indicator.

MICROVOLT COMPARATOR WITH HYSTERESIS

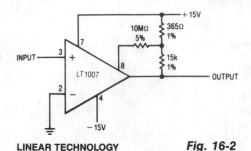

Fig. 16-2

Positive feedback to one of the nulling terminals creates approximately 5 μV of hysteresis. The output can sink 16 mA; the input offset voltage is typically changed less than 5 μV because of the feedback.

COMPARATOR/LATCH

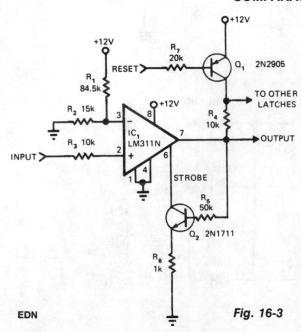

Fig. 16-3

The primary advantage of this circuit, when compared to other comparators, is its ability to latch after the input has reached a predetermined threshold level. When the input exceeds the threshold level, the LM311N output increases. This transition enables the strobe input, preventing the output from falling low. A high-level voltage on the reset input will turn off Q1, thereby removing the supply voltage from the open collector output of the LM311N. With no supply to the strobe input, the latch condition is removed and the output is again allowed to follow the input excursions. The LM311N will operate with a wide variety of supply voltage levels, ranging from dual ± 15 V to a single 5 V level that provides compatibility with digital IC logic. If more than one latch is used with a common reset, all the pull-up resistors may be connected to Q1's collector.

FREQUENCY-DETECTING COMPARATOR

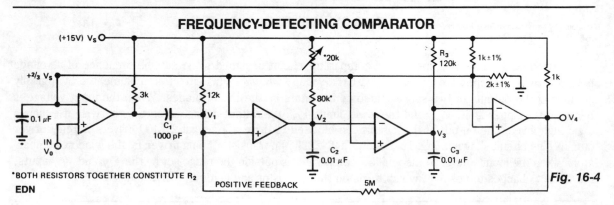

*BOTH RESISTORS TOGETHER CONSTITUTE R$_2$

Fig. 16-4

FREQUENCY-DETECTING COMPARATOR (*CONT.*)

A quad comparator forms the basis of a frequency detector that is faster and less expensive than more complex versions designed around frequency-to-voltage converter chips. Positive feedback through a 5-MΩ resistor allows the circuit to resolve changes as small as two percent; the output responds to those changes in about one cycle. When the input frequency is high, V_2 is pulled low; it's never allowed to exceed $2/3$ V. When the input frequency is lower than the limit, V_2 exceeds $2/3$ V once each cycle, but V_3 is held below that limit. The trip frequency is defined by $F = 1/(1.1R_2C_2)$. R2 can be adjusted to permit trimming of the trip point, but the value of R3 must remain larger than R2.

PRECISION COMPARATOR WITH BALANCED INPUTS AND VARIABLE OFFSET

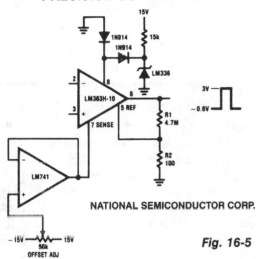

$$t_{PD} \approx 15 \ \mu s \ \text{at 1-mV overdrive}$$

$$V_{OUT} = V_2 + 0.6 \text{ V}$$

$$\text{Hysteresis} = \frac{V_{OUT}}{R \ (R_1 + R_2)} = 2 \text{ mV}$$

$$\text{Offset} = V_{SENSE}/G$$
$$\pm \ 1.3 \text{ range}$$

NATIONAL SEMICONDUCTOR CORP.

Fig. 16-5

DUAL LIMIT MICROVOLT COMPARATOR

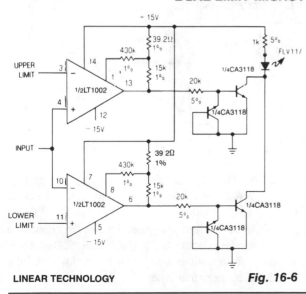

When the upper or lower limit is exceeded, the LED lights up. Positive feedback to one of the nulling terminals creates 5 to 20 μV of hysteresis on both amplifiers. This feedback changes to offset voltage of the LT1002 by less than 5 μV. Therefore, the basic accuracy of the comparator is limited only by the low offset voltage of the LT1002.

LINEAR TECHNOLOGY

Fig. 16-6

WINDOW COMPARATOR

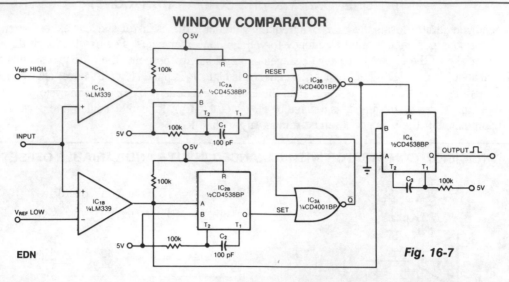

EDN

Fig. 16-7

This window comparator generates an output pulse for each event that occurs within a specified window. That is, each output pulse signifies an input voltage pulse or level change that exceeds V_{REFLOW}, but not $V_{REFHIGH}$. The monostable multivibrators, IC2A and IC2B, produce a 10-μs pulse at their Q output in response to a rising edge at their A input. Comparator IC1B produces a rising edge when the input exceeds V_{REFLOW}, and comparator IC2A produces a rising edge when the input exceeds $V_{REFHIGH}$. The NOR gates, IC3A and IC3B, form a bistable latch whose Q output, when low, disables IC4. IC4, unless disabled, produces output pulses in response to falling edges at the IC1B comparator output. You set the width of these pulses by selecting the value of C3. The circuit can handle an input waveform containing 0 to 2 V amplitudes and 10-Hz to 10-kHz frequency components.

FOUR-CHANNEL COMPARATOR

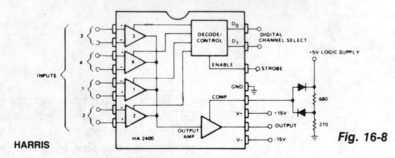

HARRIS

Fig. 16-8

When operated as an open loop without compensation, the HA-2400 becomes a comparator with four selectable input channels. The clamping network at the compensation pin limits the output voltage to allow DTL or TTL digital circuits to be driven with a fanout of up to ten loads.

The circuit can be used to compare several signals against each other or against fixed references; or a single signal can be compared against several references. A window comparator, which assures that a signal is within a voltage range, can be formed by monitoring the output polarity, while rapidly switching between two channels with different reference inputs and the same signal input.

17

Compressor/Expander Circuits

The sources of the following circuits are contained in the Sources section beginning on page 782. The figure number contained in the box of each circuit correlates to the sources entry in the Sources section.

Low-Voltage Compander
Hi-Fi Compressor with Pre-Emphasis
Variable Slope Compressor/Expander
Hi-Fi Expander with De-Emphasis

LOW-VOLTAGE COMPANDER

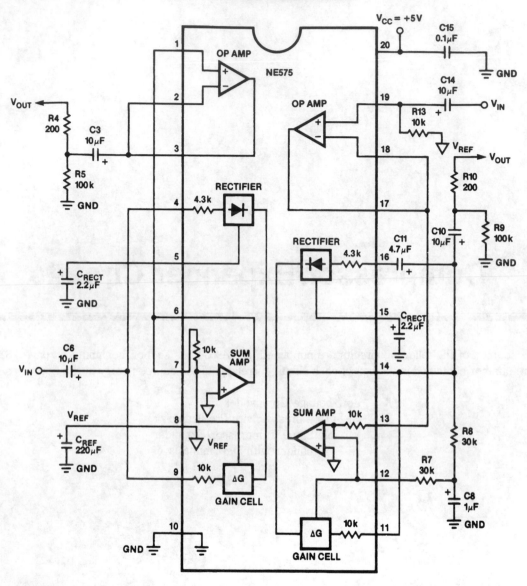

NOTE:
Left channel in expander mode; right channel in compressor mode.
For additional information, call the factory.

SIGNETICS

Fig. 17-1

The NE575 is a dual-gain control circuit designed for low voltage applications. The NE575's channel 1 is an expander, while channel 2 can be configured either for expander, compressor, or automatic level controller (ALC) applications.

HI-FI COMPRESSOR WITH PRE-EMPHASIS

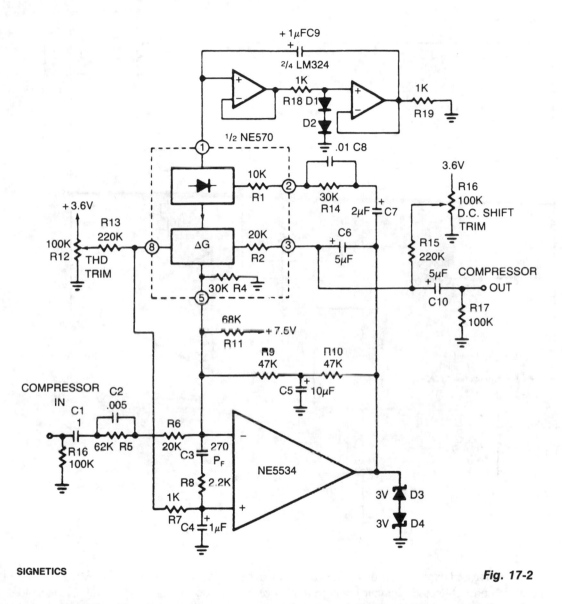

SIGNETICS

Fig. 17-2

The compressor contains a high-frequency, pre-emphasis circuit (C2, R5, and C8, R14), which helps solve this problem. Matching de-emphasis on the expander is required. More complex designs could make the pre-emphasis variable.

VARIABLE SLOPE COMPRESSOR/EXPANDER

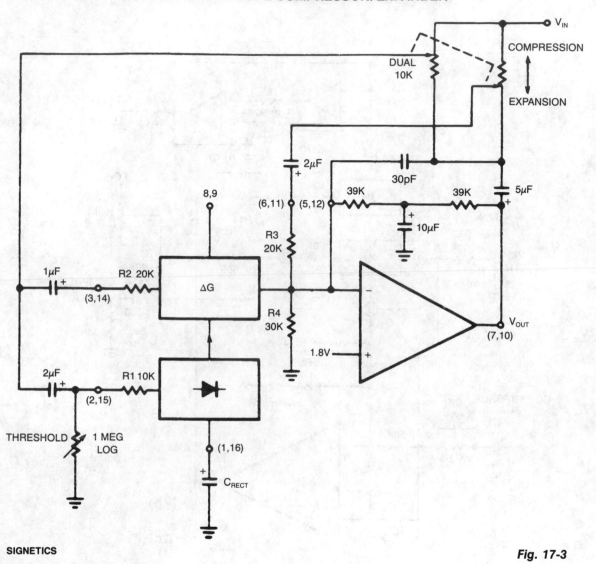

SIGNETICS

Fig. 17-3

Compression and expansion ratios other than 2:1 can be achieved by the circuit shown. Rotation of the dual potentiometer causes the circuit hook-up to change from a basic compressor to a basic expander. In the center of rotation, the circuit is 1:1, has neither compression nor expansion. The (input) output transfer characteristic is thus continuously variable from 2:1 compression to 1:2 expansion. If a fixed compression or expansion ratio is desired, proper selection of fixed resistors can be used instead of the potentiometer. The optional threshold resistor will make the compression or expansion ratio deviate towards 1:1 at low levels. A wide variety of (input) output characteristics can be created with this circuit.

HI-FI EXPANDER WITH DE-EMPHASIS

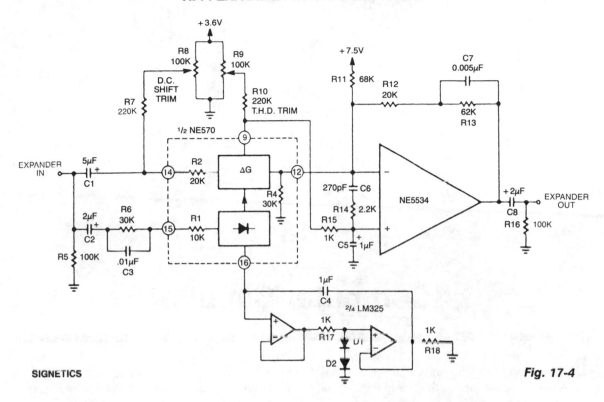

SIGNETICS

Fig. 17-4

The expander to complement the compressor is shown. An external op amp is used for high slew rate. Both the compressor and expander have unity gain levels of 0 dB. Trim networks are shown for distortion (THD) and dc shift. The distortion trim should be done first, with an input of 0 dB at 10 kHz. The dc shift should be adjusted for minimum envelope bounce with tone bursts. When applied to consumer tape recorders, the subjective performance of this system is excellent.

18

Computer Circuits

The sources of the following circuits are contained in the Sources section beginning on page 782. The figure number contained in the box of each circuit correlates to the sources entry in the Sources section.

Automatic RS-232 Dataselector
Interface to 680X, 650X, and 8080 Families
RGB Blue Box
5V-Powered EEPROM Pulse Generator
One-of-Eight Channel Transmission System
Microprocessor-Controlled Analog Signal
   Attenuator
Multiple Input Detector
RS-232-to-CMOS Line Receiver

RS-232C LED Circuit
Spare Flip-Flop Inverter
Coprocessor Socket Debugger
20-MHz-to-NuBus Clock Phase Lock
XOR Gate Up/Down Converter
Eight-Digit Microprocessor Display
Logic Line Monitor
Long Delay Line for Logic Signals

AUTOMATIC RS-232 DATASELECTOR

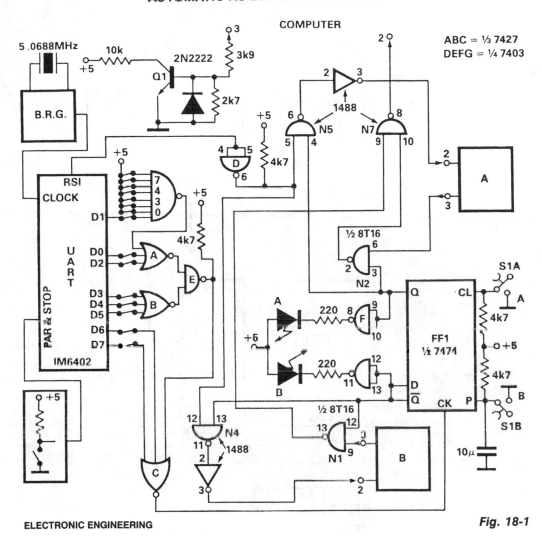

ELECTRONIC ENGINEERING

Fig. 18-1

With this dataselector, only one RS-232 port is used to connect two RS-232 devices (i.e., printer, plotter, etc.) with a mini- or microcomputer. The operation is very simple. Power on will reset FFI (Q_{FFI} = Low), which enables gates N1, N5, and N7. Now communication between computer and device B is possible. Detection of the switch command, i.e., Control B character = CHR$(2), selectable with wire-wrap pins, on the parallel outputs of the UART (IM 6402 or equivalent) will set: Q_{FFI} = High. Gates N2, N5, and N7 are open, so device A is connected with the computer until Control B character is detected again.

Transistor Q1 converts RS-232 levels to TTL levels while two LEDs indicate whether device A or B is linked. The baud-rate generator provides the 16× clock needed for the UART. Any baud rate ranging from 50 to 19200 can be selected. Manual control of the selector is available with toggleswitch S1.

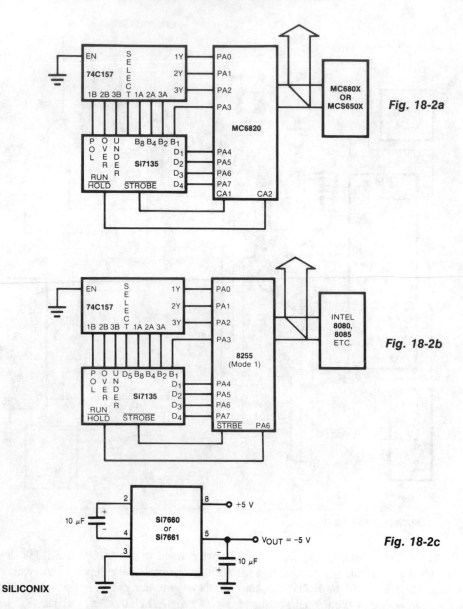

Fig. 18-2a

Fig. 18-2b

Fig. 18-2c

SILICONIX

Circuits to interface the Si7135 directly with two popular microprocessors are shown in Figs. 18-2a and b. The 8080/8048 and the MC6800 families with 8-bit words need to have polarity, overrange, and underrange multiplexed onto the digit 5 word. In each case, the microprocessor can instruct the ADC when to begin a measurement and when to hold this measurement. The Si7135 is designed to work from ±5 V supplies. However, if a negative supply is not available, it can be generated using 2 capacitors, and an inexpensive Si7660 or Si7661 IC, as shown in Fig. 18-2c.

RGB BLUE BOX

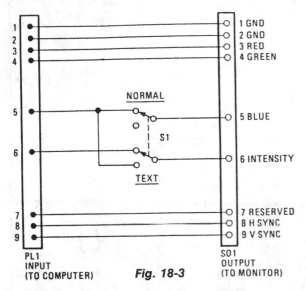

PL1
INPUT
(TO COMPUTER)

Fig. 18-3

SO1
OUTPUT
(TO MONITOR)

HANDS-ON-ELECTRONICS/POPULAR ELECTRONICS

The RGB blue box turns your PC's RGB-monitor screen blue at the flip of a switch. That is, it makes your computer display bright white text on a blue background, instead of the usual low-intensity white on black. The RGB blue box connects between your IBM PC color graphics adapter, or equivalent, and your RGB color monitor. By flipping a switch, you choose between two modes. One mode passes the signal from the PC to the monitor unaltered; the other transforms it to make text more readable. The monitor has four TTL-level inputs—red, green, blue, and intensity—and it interprets disconnected wires as on. That's why the screen turns white if you disconnect the monitor from the computer, and blue if you disconnect only the blue line. Instead of just discarding the blue signal, the blue box reroutes it to the intensity input. As a result, most of the text colors come out intensified.

5V-POWERED EEPROM PULSE GENERATOR

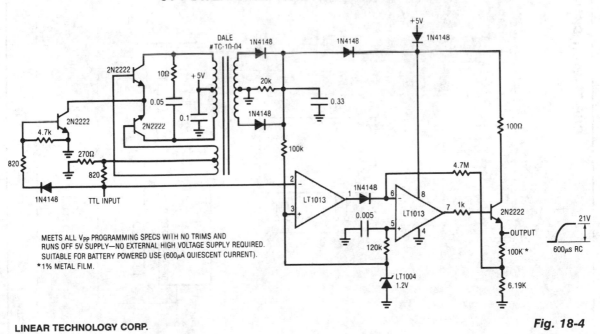

MEETS ALL V$_{pp}$ PROGRAMMING SPECS WITH NO TRIMS AND
RUNS OFF 5V SUPPLY—NO EXTERNAL HIGH VOLTAGE SUPPLY REQUIRED.
SUITABLE FOR BATTERY POWERED USE (600μA QUIESCENT CURRENT).
*1% METAL FILM.

LINEAR TECHNOLOGY CORP.

Fig. 18-4

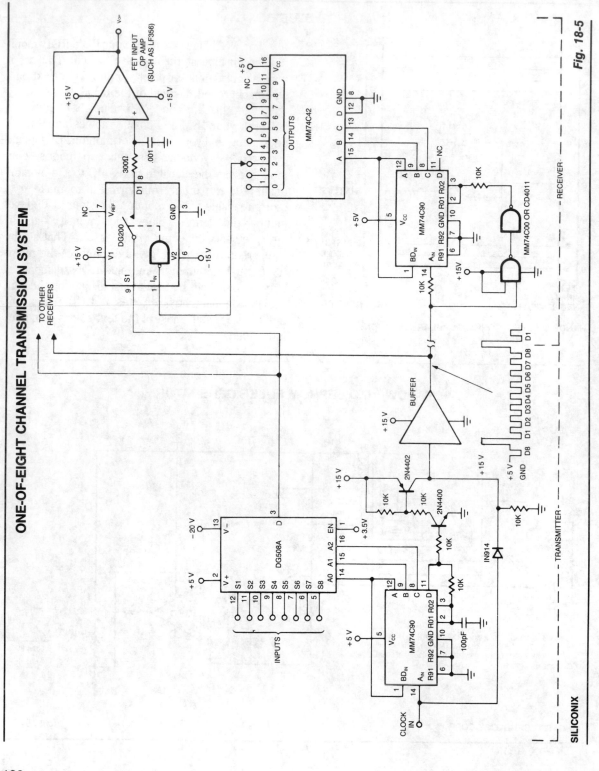

ONE-OF-EIGHT CHANNEL TRANSMISSION SYSTEM

Fig. 18-5

SILICONIX

100

MICROPROCESSOR-CONTROLLED ANALOG SIGNAL ATTENUATOR

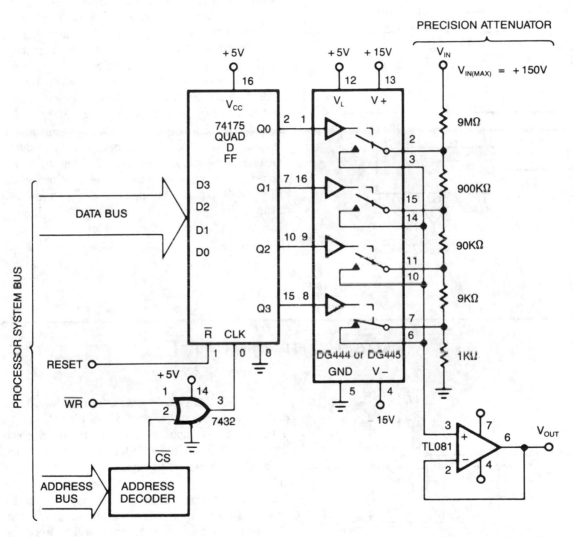

SILICONIX

Fig. 18-6

MULTIPLE INPUT DETECTOR

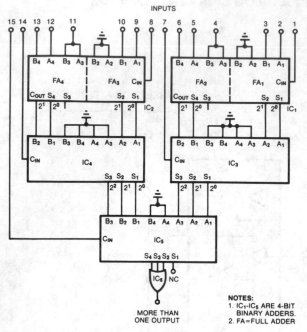

This circuit determines whether more than one input in a group of digital inputs is active. It provides a digital measure of the number of active inputs, and it allows you to establish a threshold for majority-decision applications. That is, whether the number of active inputs is more than, less than, or equal to a value between 1 and 15. You can monitor more inputs by cascading the adders.

Each binary adder, IC1 and IC2, forms two full adders (FAs). Each FA monitors three input lines and generates a 2-bit output representing the number of inputs active. IC3 and IC4, by summing the outputs of two FAs plus an input line, individually measure how many in a group of seven inputs are active. Similarly, by monitoring the 3-bit outputs of IC3 and IC4 plus one input, IC5 measures how many in the group of 15 are active. The OR gate, IC6, simply indicates whether more than one input is active.

NOTES:
1. IC_1-IC_5 ARE 4-BIT BINARY ADDERS.
2. FA=FULL ADDER

EDN

Fig. 18-7

RS-232-TO-CMOS LINE RECEIVER

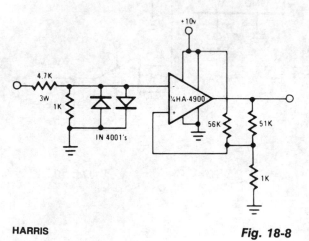

This RS-232 type line receiver to drive CMOS logic uses a Schmitt-trigger feedback network to give about 1-V input hysteresis for added noise immunity. A possible problem in an interface which connects two pieces of equipment, each plugged into a different ac receptacle, is that the power line voltage might appear at the receiver input when the interface connection is made or broken. The two diodes and a 3-W input resistor will protect the inputs under these conditions.

HARRIS

Fig. 18-8

RS-232C LED CIRCUIT

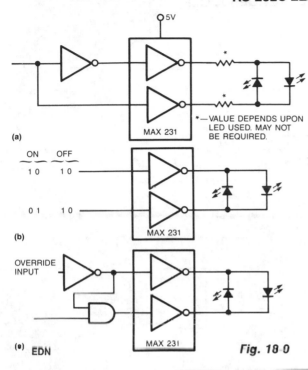

(a)

(b)

(s) EDN

*—VALUE DEPENDS UPON LED USED. MAY NOT BE REQUIRED.

Fig. 18-0

Use a pair of Maxim's 5V-powered MAX231 RS-232C transmitters as drivers to obtain a 2-color LED. The transmitters require only a single-ended, 5-V input to generate ± 10 V internally. Their outputs are short-circuit-proof and can supply as much as 10 mA—enough to drive most LEDs. Depending on which LED you select, their current-limiting feature might also eliminate the need for external series resistors. Using the simple circuits, you can implement a variety of functions.

SPARE FLIP-FLOP INVERTER

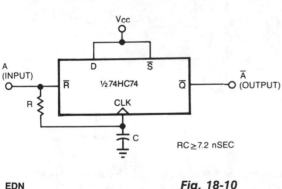

$RC \geq 7.2$ nSEC

EDN

Fig. 18-10

The circuit uses one-half of a dual D flip-flop as an inverter. When the input decreases, the flip-flop resets, and its Q output increases. When the input increases, the reset line is released and Q gets clocked low. The rc delay between applying the input signal to the flip-flop's reset input and its clock input enables clocking the flip-flop on the input's positive edge. A 74HC74 dual D flip-flop, for example, requires a minimum recovery time of 5 ns after releasing the reset input before strobing its clock input. Therefore, speccing rc at greater than 7.5 ns provides adequate margin. The slight slowing of the clock edge presents no problem, because the clock input's maximum allowable rise time is a much longer 500 ns. To prevent skewing of the output's symmetry, limit the maximum input frequency to less than 10 MHz.

COPROCESSOR SOCKET DEBUGGER

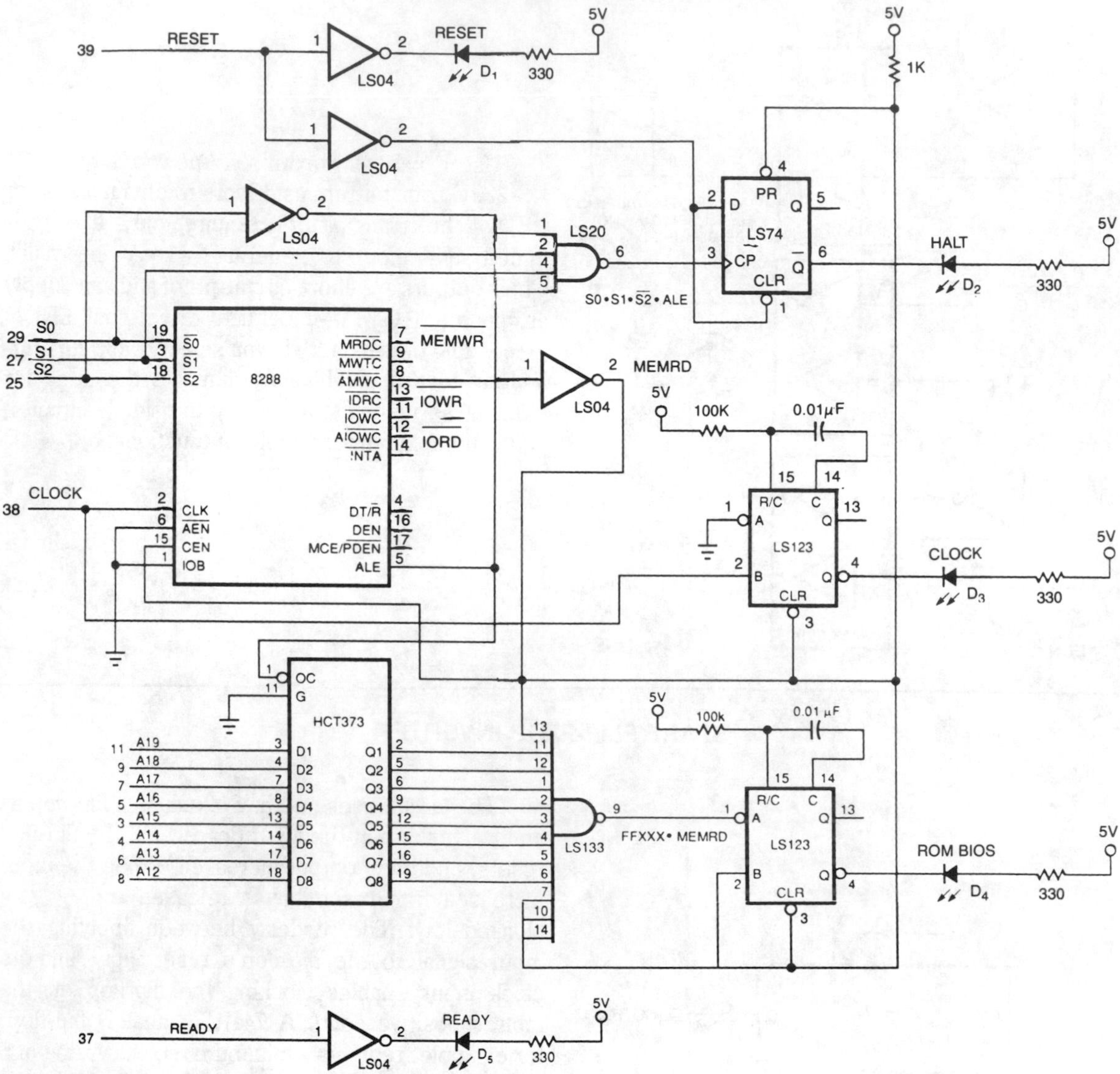

Fig. 18-11

The IBM PC debugger plugs into the PC's math-coprocessor socket. The 8288 bus controller, IC1, regenerates control signals from the processor's status signals, S0, S1, and S2. Reset LED D1 lights if reset is active and holding the processor. Clock status LED D3 indicates that the processor is receiving a toggling clock signal. The address-decode logic detects when the processor is doing a jump-on-reset to the PC's BIOS ROM's power-on; self-test then detects a fatal error and halts the processor.

20-MHz-TO-NUBUS CLOCK PHASE LOCK

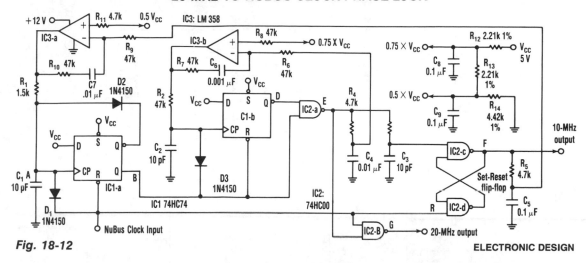

Fig. 18-12

ELECTRONIC DESIGN

The 20-MHz clock phase-locks to Apple's Mac II 10-MHz NuBus clock. It uses a simple, inexpensive CMOS circuitry to generate 10- and 20-MHz square waves. The output duty cycle settings are insensitive to V_{CC} variations. The input to the circuit is a NuBus clock signal with specifications that call for a 75 percent duty cycle at 10 MHz—a square wave that's high for 75 ns and low for the remaining 25 ns. To generate the 20-MHz signal, the circuit produces a 25-ns negative-going pulse, delayed 50 ns from the falling edge of the 10-MHz NuBus clock input at point E. NORing that pulse with the NuBus clock produces the 20-MHz clock at point G. Finally, applying the 25-ns pulse to the set input of a set-reset input, results in a 10-MHz square wave at F.

XOR GATE UP/DOWN COUNTER

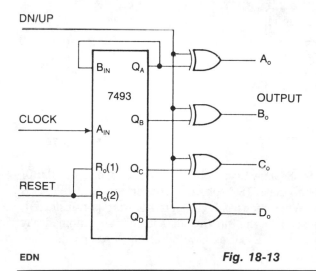

DN/UP

7493

CLOCK

RESET

OUTPUT

EDN

Fig. 18-13

One can transform an ordinary binary counter, such as a 7493, into an up/down counter with mode control by adding XOR gates 7486 to the counter's outputs. The circuit counts up when the DN/UP line is low and down when the DN/UP line is high.

To use the 7493 counter to count out its maximum count length of 0–15, connect the Q_A output to the B_{IN} input and apply clock pulses to the A_{IN} input. The reset input, when high, inhibits the count inputs and simultaneously returns the outputs A_o through D_o to low in the up-count mode or 15 in the down-count mode. For normal counting, the reset input must be low. One can easily cascade this counter by feeding the Q_D line to the clock input of a succeeding counter.

EIGHT-DIGIT MICROPROCESSOR DISPLAY

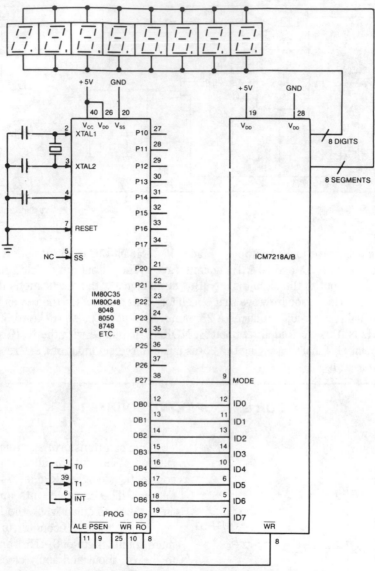

Fig. 18-14

Display interface uses the ICM7218A/B with an 8048 family microcontroller. The 8-bit data bus (DB0/DB7 – ID0/ID7) transfers control and data information to the 7218 display interface on successive WRITE pulses. The mode input pins on the microcontroller. When mode is high, a control word is transferred; when mode is low, data is transferred. Sequential locations in the 8-byte static memory are automatically loaded on each successive WRITE pulse. After eight WRITE pulses have occurred, further pulses are ignored until a new control word is transferred.

LONG DELAY LINE FOR LOGIC SIGNALS

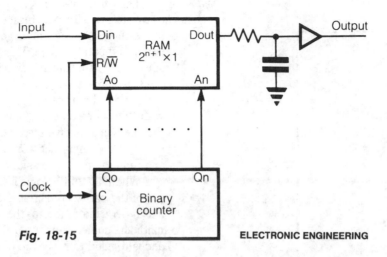

Fig. 18-15

ELECTRONIC ENGINEERING

An extremely long delay of logic signal can be accomplished with this circuit. The logic signals to be delayed are applied to the D_{IN} of RAM. Address lines $Ao, A1, \ldots An$ are connected to outputs $Qo, Q1, \ldots Qn$ of a binary counter. Clock input of counter and $R/\overline{W}$ input of RAM are joined together. However, it is sometimes necessary to put an inverter between those inputs, depending on the RAM and counter employed in line. In the first half of clock interval, content on outputs of counter is increased for 1 and content of chosen memory cell is read; in the second half of the clock interval, new content from D_{IN} in the same memory cell is written. When full cycles of counting reaches the same memory cell, again we can read, in the first half of the clock interval of the following counting cycles, the chosen content. Delay time is:

$$Td = 2^{n+1} \cdot tcl$$

If clock frequency is not a multiple of input signal frequency, distortion of input signal is proportioned to the clock period. But if the clock frequency is a multiple of input signal frequency, there is no distortion. If we use RAM organized according to $2^{n+1} \times 4$ with separated data inputs and data outputs, we can have four parallel long delay lines. The resistor, capacitor, and buffer on D_{OUT} of RAM are used to save output signal in writing time, when output of RAM becomes high impedance.

LOGIC LINE MONITOR

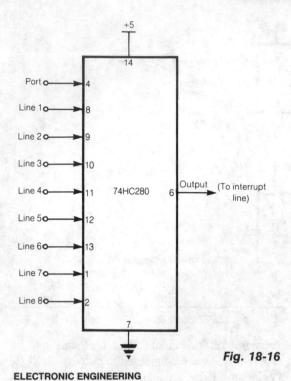

This circuit requires only one CMOS IC, which is available in a 14-pin surface-mount package. The figure shows the logic lines going to a standard 9-bit parity comparator chip. This device is conventionally used in data transmission and recording applications to provide a means of error-detection by comparing the received eight- or nine-bit words with their corresponding parity bits. If the sum of the one's in a received word is odd but the odd-parity bit is low, then that word is known to be in error and requires retransmission.

When one of the logic lines decreases, the output of the parity comparator decreases, generating a *wake-up* interrupt to the microprocessor. The ninth line comes from a port on the microprocessor and is toggled to reset the output signal high again, ready for the next logic change.

Fig. 18-16

ELECTRONIC ENGINEERING

19

Converters

The sources of the following circuits are contained in the Sources section beginning on page 782. The figure number contained in the box of each circuit correlates to the sources entry in the Sources section.

10 Hz-TO-10kHz V/F CONVERTER

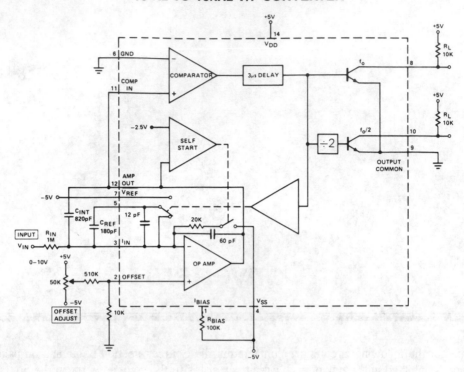

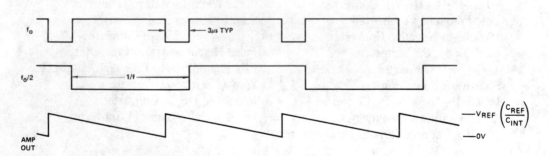

Notes:
1. To adjust f_{min}, set V_{IN} = 10 mV and adjust the 50 k offset for 10 Hz out.
2. To adjust f_{max}, set V_{IN} = 10 V and adjust R_{IN} or V_{REF} for 10 kHz out.
3. To increase f_{OUT} MAX to 100 kHz change C_{REF} to 27 pF and C_{INT} to 75 pF.
4. For high performance applications use high stability components for R_{IN}, C_{REF}, V_{REF} (metal film resistors and glass film capacitors). Also separate the output ground (Pin 9) from the input ground (Pin 6).

Fig. 19-1

LOW-FREQUENCY CONVERTER

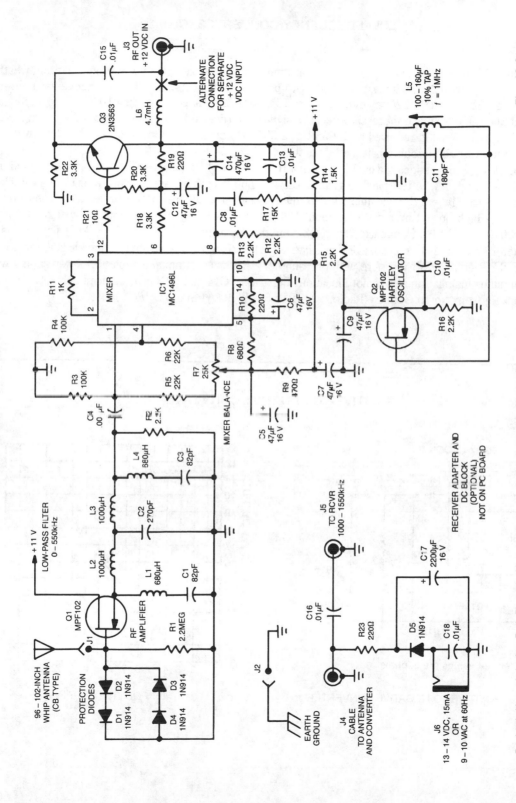

Reprinted with permission from Radio-Electronics Magazine, September 1989. Copyright Gernsback Publications, Inc., 1989.

Fig. 19-2

LOW-FREQUENCY CONVERTER (*Cont.*)

Among the signals below 550 kHz are maritime mobile, distress, radio beacons, aircraft weather, European Longwave-AM broadcast, and point-to-point communications. The low-frequency converter converts the 10 to 500 kHz LW range to a 1010 to 1550 kHz MW range, by adding 1000 kHz to all received signals. Radio calibration is unnecessary because signals are received at the AM-radio's dial setting, plus 1 MHz; a 100-kHz signal is received at 1100 kHz, a 335-kHz signal at 1335 kHz, etc. The low-frequency signals are fed to IC1, a doubly-balanced mixer.

Transistor Q2 and associated circuitry form a Hartley 1000-kHz local oscillator, which is coupled from Q2's drain, through C8, to IC1 pin 8. Signals in the $10 - 550$ kHz range are converted to $1010 - 1550$ kHz. The mixer heterodynes the incoming low-frequency signal and local-oscillator signal. Transistor Q3 reduces IC1's high-output impedance to about 100 Ω to match most receiver inputs. Capacitor C15 couples the $1010 - 1550$ kHz frequencies from Q3's emitter to output jack J3, while blocking any dc bias.

Inductor L6 couples the dc voltage that's carried in the rf signal cable from the rcvr/dc adaptor. The dc voltage and rf signals don't interfere with one another; that saves running a separate power-supply wire, which simplifies installation at a remote location. Capacitors C14 and C13 provide dc supply filtering. The kit is available from North Country Radio, P.O. Box 53, Wykagyl Station, NY 10804.

POSITIVE-TO-NEGATIVE CONVERTER

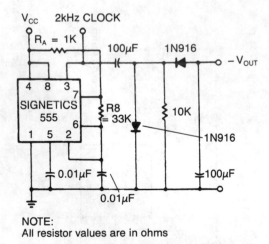

NOTE:
All resistor values are in ohms

(a) POSITIVE-TO-NEGATIVE CONVERTER

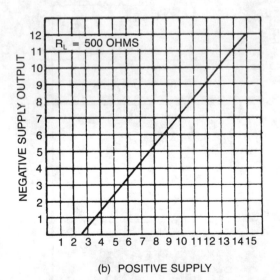

(b) POSITIVE SUPPLY

POSITIVE-TO-NEGATIVE CONVERTER (*Cont.*)

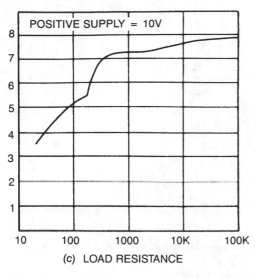

The transformerless dc-dc converter derives a negative supply voltage from a positive. As a bonus, the circuit also generates a clock signal. The negative output voltage tracks the dc-input voltage linearity (a), but its magnitude is about 3 V lower. Application of a 500-Ω load, (b), causes 10% change from the no-load value.

SIGNETICS   *Fig. 19-3*

BUCK/BOOST CONVERTER

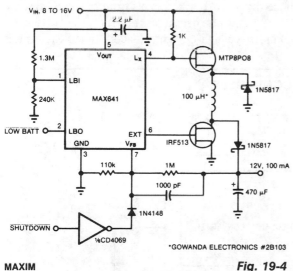

This converter can accommodate wide input-voltage swings, such as the 8 to 15-V swing typical of a 12-V sealed lead/acid battery. The low battery output indicates when input voltage drops below 8 V. Pulling shutdown turns off the circuit.

MAXIM   *Fig. 19-4*

4 – 18 MHz CONVERTER

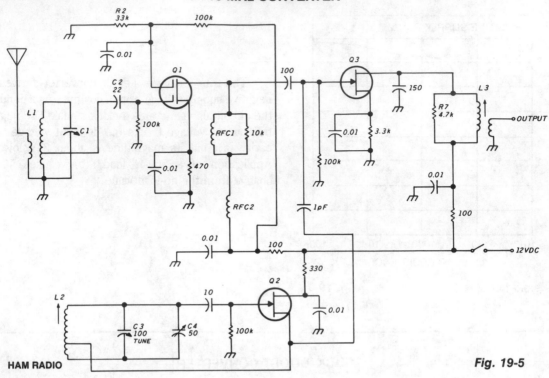

Fig. 19-5

HAM RADIO

The unit consists of rf amplifier Q1, local oscillator Q2, and mixer Q3. The two bands are covered without a bandswitch by using an i-f or 3.5 MHz. The oscillator range is 7.5 to 14.5 MHz. Incoming signals from 4 to 11 MHz are mixed with the oscillator to produce the 3.5-MHz i-f. Signals from 11 to 18 MHz are mixed with the oscillator to also produce an i-f of 3.5 MHz. At any one oscillator frequency, the two incoming signals are 7 MHz apart. Rf amplifier input C1/L1 comprises a high-Q, lightly loaded, tuned circuit; this is essential for good band separation.

SHORTWAVE CONVERTER

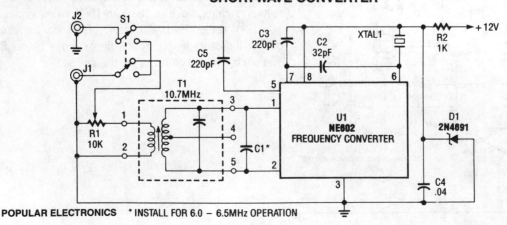

POPULAR ELECTRONICS   * INSTALL FOR 6.0 – 6.5MHz OPERATION

Fig. 19-6

SHORTWAVE CONVERTER (*Cont.*)

The NE602, U1, contains oscillator and mixer stages. The mixer combines the oscillator signal with the input rf signal to produce signals whose frequencies are the sum and difference of the input frequencies. For example, a 7.5-MHz signal is picked up by the antenna and mixes with the 8.5-MHz oscillator frequency. The difference between those two signals is 1 MHz—right in the center of your AM dial. Transformer T1 is a 10.7-MHz i-f transformer.

ISOLATED +15 V DC-DC CONVERTER

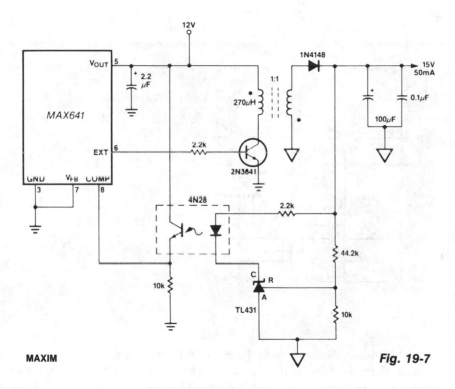

MAXIM

Fig. 19-7

In this circuit, a TL431 shunt regulator is used to sense the output voltage. The TL431 drives the LED of a 4N28 optocoupler which provides feedback to the MAX641 while maintaining isolation between the input, +12 V, and the output, +15 V. In this circuit, the +15 V output is fully regulated with respect to both line and load changes.

VOLTAGE RATIO-TO-FREQUENCY CONVERTER

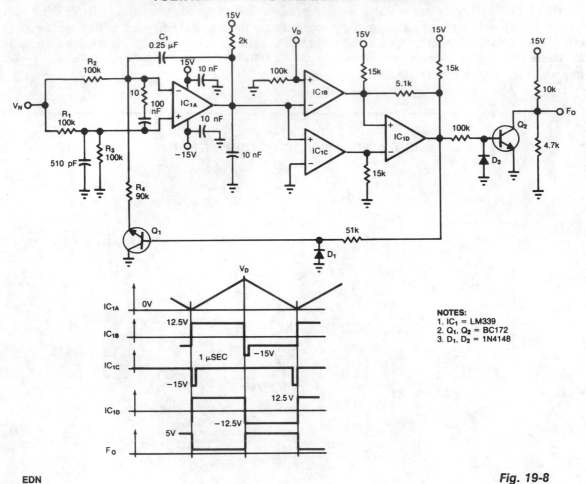

EDN

Fig. 19-8

NOTES:
1. IC_1 = LM339
2. Q_1, Q_2 = BC172
3. D_1, D_2 = 1N4148

The circuit accepts two positive-voltage inputs V_N and V_D and provides a TTL-compatible output pulse train whose repetition rate is proportional to the ratio V_N/V_D. Full-scale output frequency is about 100 Hz, and linearity error is below 0.5 percent. The output F_o equals KV_N/V_D, where $K = 1/(4R_2C_1)$ provided $R_1 = R_3$. Op amp IC1A alternately integrates $V_N/2$ and $-V_N/2$, producing a sawtooth output that ramps between the V_D level and ground. When transistor Q1 is on, for example, IC1A integrates $-V_N/2$ until its output equals V_D. At that time, the IC1B comparator switches low, causing IC1D's bistable output to go low, which turns off Q1. IC1A's output then ramps in the negative direction. When the output reaches 0 V, the IC1C comparator switches, Q1 turns on, and the cycle repeats. Transistor Q2 converts the IC1D output to TTL-compatible output logic levels. Setting V_D to 1.00 V yields a linear voltage-to-frequency converter ($F_o = KV_N$), and setting V_N to 1.00 V yields a reciprocal voltage-to-frequency converter ($F_o = KV_D$).

50-MHz THERMAL RMS-TO-DC CONVERTER

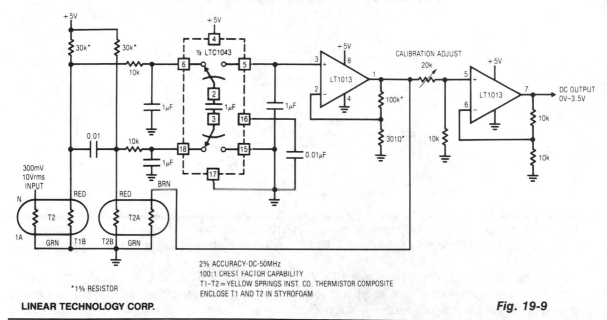

2% ACCURACY-DC-50MHz
100:1 CREST FACTOR CAPABILITY
T1–T2 = YELLOW SPRINGS INST. CO. THERMISTOR COMPOSITE
ENCLOSE T1 AND T2 IN STYROFOAM

*1% RESISTOR

LINEAR TECHNOLOGY CORP.

Fig. 19-9

PULSE WIDTH-TO-VOLTAGE CONVERTER

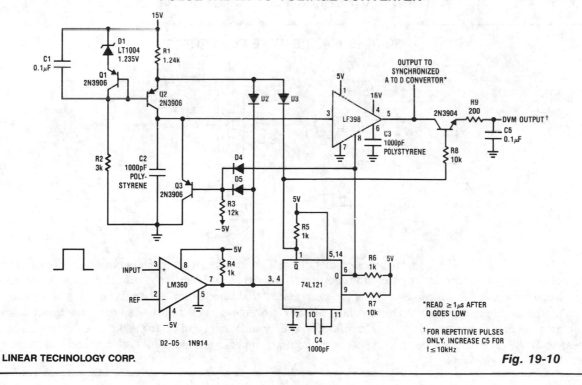

*READ ≥ 1μs AFTER
Q GOES LOW

† FOR REPETITIVE PULSES
ONLY. INCREASE C5 FOR
f ≤ 10kHz

D2–D5 1N914

LINEAR TECHNOLOGY CORP.

Fig. 19-10

117

STEP UP/DOWN DC-DC CONVERTER

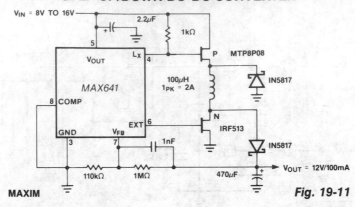

MAXIM

Fig. 19-11

Positive output step-up and step-down dc-dc converters have a common limitation in that neither can handle input voltages that are both greater than or less than the output. For example, when converting a 12-V sealed lead/acid battery to a regulated + 12 V output, the battery voltage might vary from a high of 15 V down to 10 V.

By using a MAX641 to drive separate P- and N-channel MOSFETs, both ends of the inductor are switched to allow noninverting buck/boost operation. A second advantage of the circuit over most boost-only designs is that the output goes to 0 V when shutdown is activated. Inefficiency is a drawback because two MOSFETs and two diodes increase the losses in the charge and discharge path of the inductor. The circuit delivers + 12 V at 100 mA at 70 percent efficiency with an 8-V input.

SQUARE-TO-SINE WAVE CONVERTER

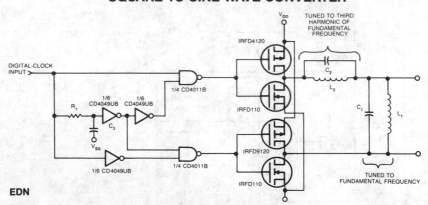

EDN

Fig. 19-12

Two pairs of MOSFETs form a bridge that alternately switches current in opposite directions. Two parallel-resonant LC circuits complete the converter. The L1/C1 combination is resonant at the fundamental frequency; the L2/C2 combination is resonant at the clock frequency's third harmonic and acts as a trap. T1 and C3 ensure that both halves of the MOSFET bridge are never on at the same time by providing a common delay to the gate drive of each half. Select the values of R1 and C3 to yield a time constant that's less than 5% of the clock's period. You can add an output amplifier for additional buffering and conditioning of the circuit's sine-wave output.

PULSE HEIGHT-TO-WIDTH CONVERTER

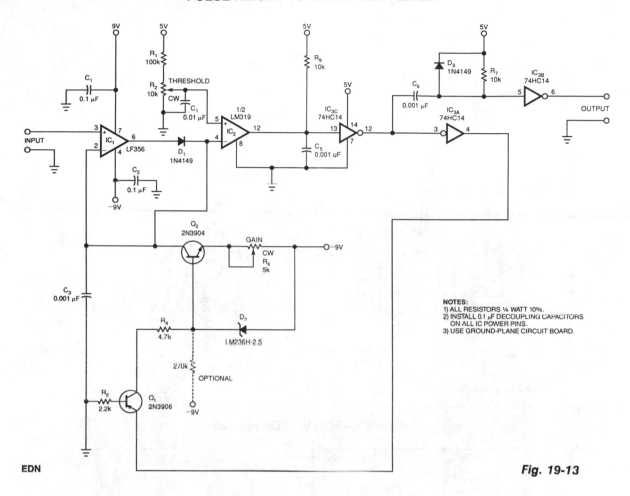

EDN

Fig. 19-13

The output-pulse width from the circuit is a linear function of the input pulse's height. You can set the circuit's input threshold to discriminate against low-level pulses, while fixed components limit the circuit's maximum output-pulse width.

With a 270-KΩ resistor connected from the −9 V supply to the base lead of Q2, this circuit can handle input pulses separated by 20 μs for correct operation. The turn-off time of zener diode D2 sets the lower limit for the input-pulse repetition rate.

IC1, D1, and C3 detect the peak of the input pulse. The comparator IC2, triggers at your preset threshold. The RC delay network, R9 and C5, hold off inverter IC3's changing state until the completion of peak detection. After IC3A changes state, Q1 turns on and then turns on Q2, a constant-current source.

Constant-current source Q1 then discharges C3, the peak-detecting capacitor. When C3 has discharged below IC2's threshold, IC2's output decreases, as do pins 6 and 4 of IC3. The output-pulse width is a function of this discharge time, which you can adjust with R6. C6 and R7 control the maximum output-pulse width, which is 8 μs max.

PIN PHOTODIODE-TO-FREQUENCY CONVERTER

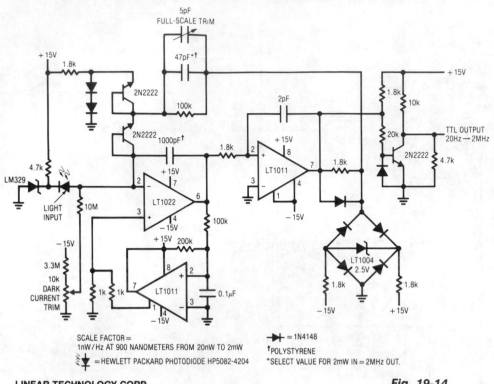

SCALE FACTOR =
1nW/Hz AT 900 NANOMETERS FROM 20nW TO 2mW

➤| = 1N4148

†POLYSTYRENE

*SELECT VALUE FOR 2mW IN = 2MHz OUT.

= HEWLETT PACKARD PHOTODIODE HP5082-4204

LINEAR TECHNOLOGY CORP.

Fig. 19-14

ZERO I_B ERROR V/I CONVERTER

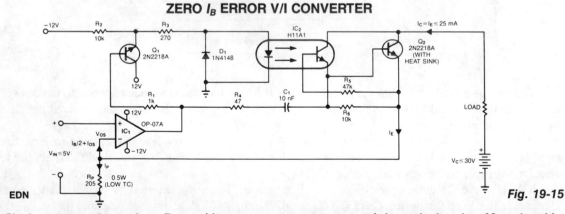

EDN

Fig. 19-15

Single programming resistor R_p provides an output-current range of about six decades. Note that this resistor's TC is also a potential source of error; it dissipates 125 mW when V_{IN} = 5 V. The maximum deviation is typically 50 nA or 0.0002% of full scale. This voltage-controlled current source uses optocoupler IC2 to eliminate an error found in more conventional circuits and which is caused by the output transistor's base current.

REGULATED DC-DC CONVERTER

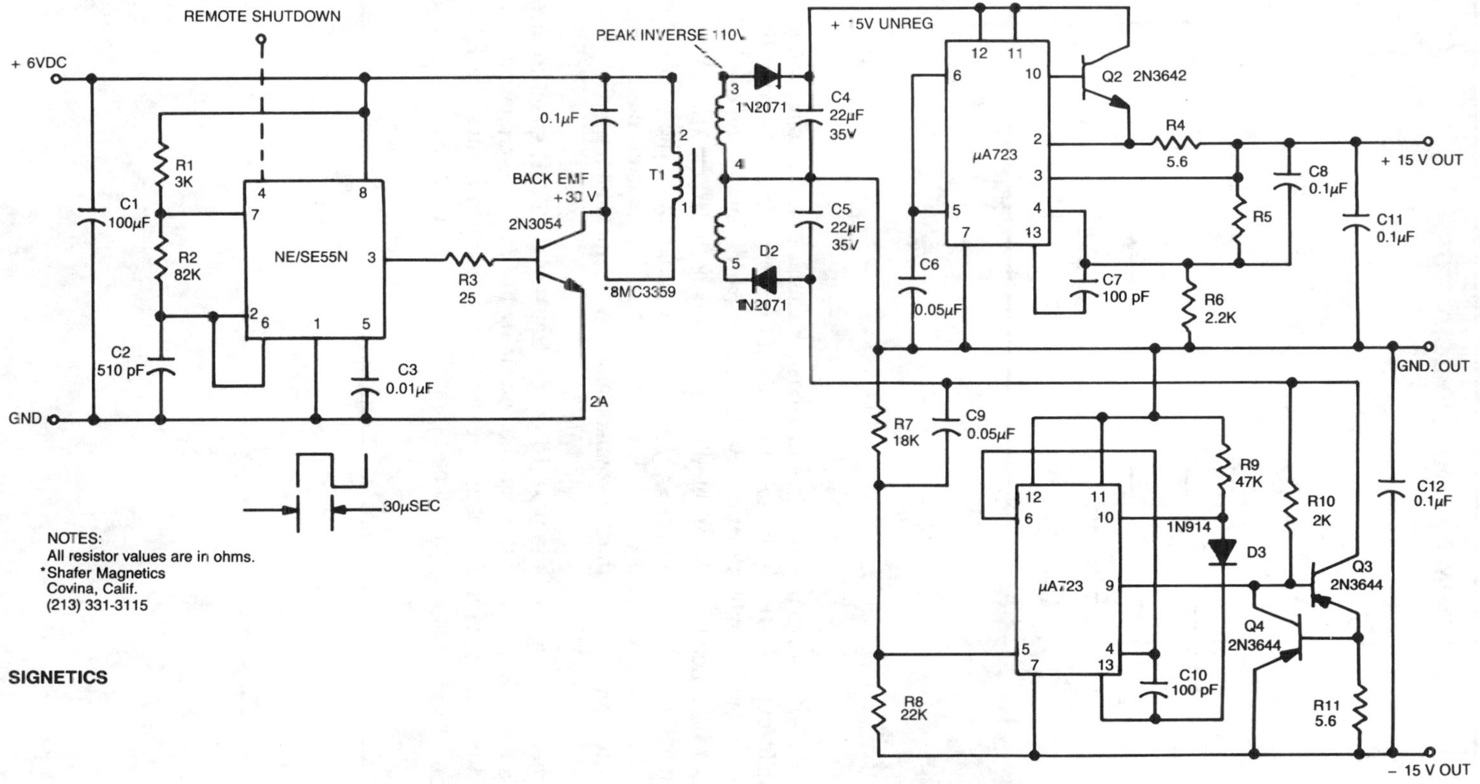

Fig. 19-16

The regulated dc-dc converter produces 15-Vdc outputs from a + 5 Vdc input. Line and load regulation is 0.1%.

NOTES:
All resistor values are in ohms.
*Shafer Magnetics
Covina, Calif.
(213) 331-3115

SIGNETICS

121

PULSE TRAIN-TO-SINUSOID CONVERTER

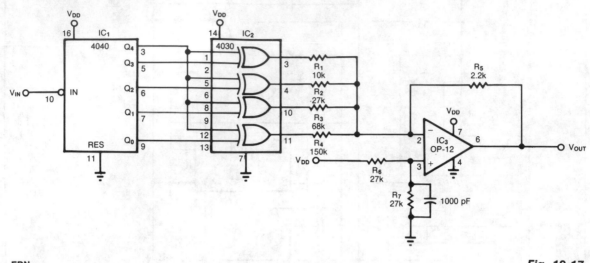

EDN

Fig. 19-17

The circuit lets you convert a serial pulse stream or sinusoidal input to a sinusoidal output at $1/32$ the frequency. By varying the frequency of V_{IN}, you can achieve an output range of $10^7:1$—from about 100 kHz to less than 0.01 Hz. The output resembles that of a 5-bit d/a converter operating on parallel digital data.

Counter IC1 generates binary codes that repeatedly scan the range from 00000 to 11111. The output amplifier adds the corresponding XOR gate outputs, V_{DD} or ground, weighted by the values of input resistors R1 through R4. The 16 counter codes 00000 to 01111, for instance, pass unchanged to the XOR gate outputs, and cause V_{OUT} to step through the half-sinusoidal cycle for maximum amplitude to minimum amplitude.

Counter output Q4 becomes high for the next 16 codes, causing the XOR gates to invert the Q0 through Q3 outputs. As a result, V_{OUT} steps through the remaining half cycle from minimum to maximum amplitude. The counter then rolls over and initiates the next cycle. You can change the R1 through R4 values to obtain other V_{OUT} waveforms. V_{DD} should be at least 12 V to assure maximum-frequency operation from IC1 to IC2.

20

Counters

The sources of the following circuits are contained in the Sources section beginning on page 782. The figure number contained in the box of each circuit correlates to the sources entry in the Sources section.

Low-Cost Frequency Counter

Up/Down Counter/Extreme Count Freezer

10-MHz Frequency Counter

Low-Power Wide-Range Programmable Counter

40-MHz Universal Counter

Frequency Counter Preamp

1.2-GHz Frequency Counter

LOW-COST FREQUENCY COUNTER

C1—30 pF NP0.
C2—3-30 pF trimmer.
D1-D4—1N4148 or equiv.
DS1—DL34M 4-digit, 7-segment, common-cathode MUX display.

Q1-Q5—2N2222, 2N3904 or equiv.
Q6—2N2907, 2N3906 or equiv.
U1—MM5369AA oscillator/divider.
U2, U6, U7—CD4017 CMOS decade counter/divider.

U3—CD4013 CMOS dual flip-flop.
U4—74HC132 HCMOS high-speed quad 2-input NAND.
U5—74HC4017 HCMOS high-speed decade counter/divider.

U8—74C926 CMOS 4-digit counter/MUX driver.
Y1—3.579545-MHz crystal.

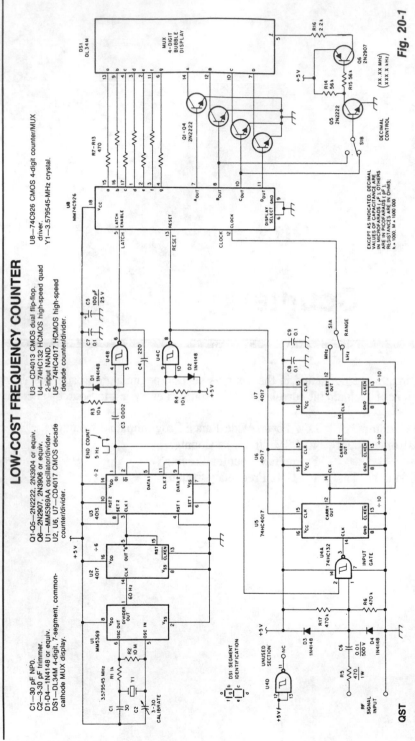

Fig. 20-1

QST

This counter uses a four-digit display, but with a flip of the range switch, it can display frequencies from 1 to 40 MHz, with a resolution of 100 Hz. The MM74C926 CMOS IC contains a four-digit decimal counter that can latch a given count and then use this information to drive a 7-segment, common-cathode multiplexed (MUX) display. The block diagram and schematic show the operation of the counter. Crystal-controlled timer U1 through U3 produces a 5-Hz square wave used for timing the frequency count. Y1 is a TV color-burst crystal operating in a reliable circuit that controls the oscillator frequency. U1 acts as the oscillator and also divides the fundamental operating frequency of 3.579545 MHz to produce a square-wave output of 60 Hz. U2 divides the 60-Hz output of U1 by six. In turn, the 10-Hz output of U2 is divided by two in U3, a dual flip-flop, to produce the 5-Hz pulse.

A quad, two-input, Schmitt-triggered NAND U4 is used for gating the rf-signal input and for generating the counter control pulses—Section U4D is unused. The 5-Hz output pulse of U3 is applied to the rf-signal input gate at U4A pin 2. When the timer output signal is low, the gate is closed.

124

UP/DOWN COUNTER/EXTREME COUNT FREEZER

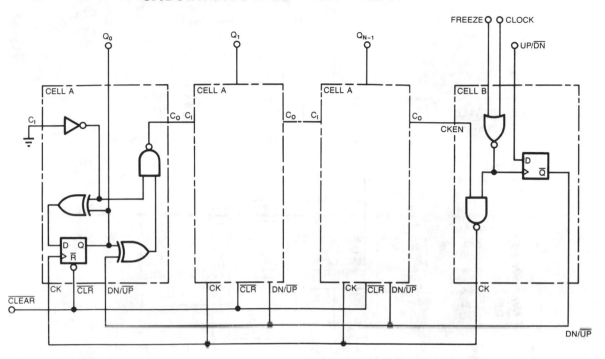

STATE TABLE

FREEZE	UP/$\overline{DN}$	CLOCK	CURRENT STATE $Q_{N-1} \cdots Q_0$	NEXT STATE $Q_{N-1} \cdots Q_0$
L	H	⌐⌐	$11 \ldots 10$	$11 \ldots 11$
L	H	⌐⌐	$11 \ldots 11$	$11 \ldots 11$
L	L	⌐⌐	$00 \ldots 01$	$00 \ldots 00$
L	L	⌐⌐	$00 \ldots 00$	$00 \ldots 00$
H	X	X	Q_X	Q_X

EDN

Fig. 20-2

The discrete-gate up/down-counter design has the unusual property of freezing, or saturating, when it reaches its lowest count in the down-count mode or its highest count in the count-up mode instead of rolling over and resetting as do most counters. This property proves especially useful in position-control systems. For example, you wouldn't want a robot's arm to slowly move to full extension as the counter counts up and then have it suddenly slam back to its rest position when the counter resets to zero.

You can cascade as many of the A cells as you need because the counter's outputs are synchronous. The B cell accepts the carry bit from the most significant bit's A cell and provides the clock control that stops the counter. Make sure that the freeze input to the B cell doesn't get asserted when the clock input is low; otherwise, the counter might make an extra count.

10-MHz FREQUENCY COUNTER

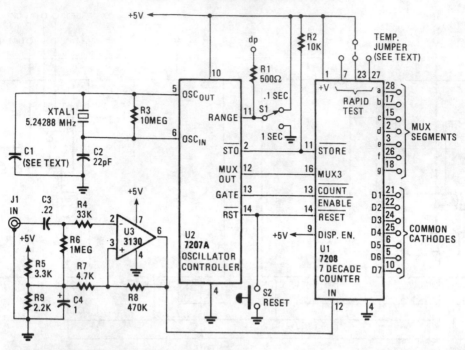

HANDS-ON ELECTRONICS/POPULAR ELECTRONICS

Fig. 20-3

The circuit consists of ICM7208 seven-decade counter U1, ICM7207A oscillator controller U2, and CA3130 biFET op amp U3. IC U1 counts input signals, decodes them to 7-segment format, and outputs signals that are used to drive a 7-digit display. IC U2 provides the timing for U1, while U3 conditions the input to U1. The 5.24288-MHz crystal frequency is divided by U2 to produce a 1280-MHz multiplexing signal at pin 12 of U2. That signal is input to U1 at pin 16 and used to scan the display digits in sequence.

LOW-POWER WIDE-RANGE PROGRAMMABLE COUNTER

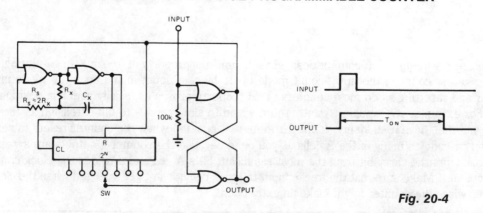

EDN

Fig. 20-4

LOW-POWER WIDE-RANGE PROGRAMMABLE COUNTER (*Cont.*)

This CMOS circuit can be used as a 1-shot time delay switch and general-purpose timer. The circuit consists of a gated oscillator and a latch made from one CD4001 quad 2-input NOR gate as shown and a CD4020 14-stage counter. T_{ON} is a function of the oscillator frequency from the $R_X C_X$ and the proper 2^N output from the counter. A pulse applied to the latch will "enable" the oscillator and counter. The latch output will remain high until the 2^N count resets the latch and disables the oscillator and counter. The circuit provides μs to hour interval timing. The extraordinarily long periods available from the CMOS oscillator, combined with the 14-stage counter, make this range possible. Further decoding is required for variations finer than a power of two.

40-MHz UNIVERSAL COUNTER

This circuit can be used to measure frequencies up to 40 MHz. To obtain the correct measured value, it is necessary to divide the oscillator frequency and the input frequency by four. In doing this, the time between measurements is also lengthened to 800 ms and the display multiplex rate is decreased to 125 Hz.

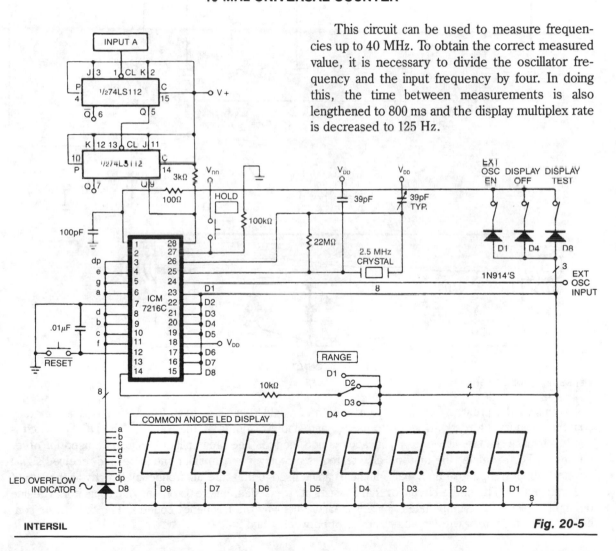

INTERSIL

Fig. 20-5

127

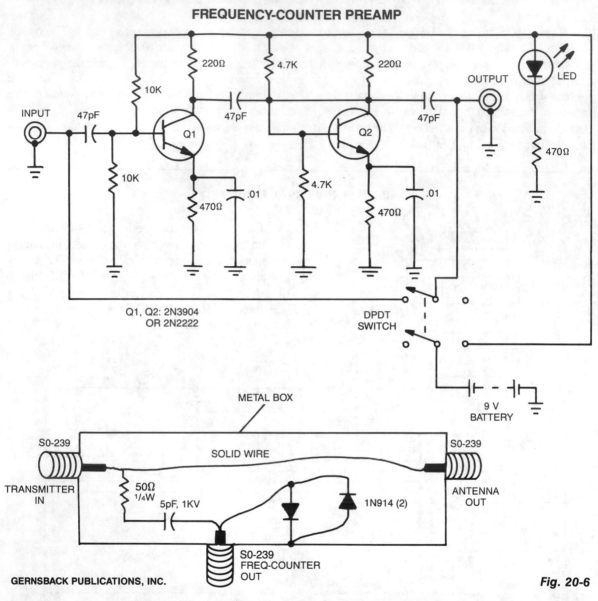

FREQUENCY-COUNTER PREAMP

INPUT

Q1, Q2: 2N3904 OR 2N2222

DPDT SWITCH

OUTPUT

LED

9 V BATTERY

METAL BOX

SOLID WIRE

SO-239

TRANSMITTER IN

50Ω $1/4W$

5pF, 1KV

SO-239 FREQ-COUNTER OUT

1N914 (2)

SO-239

ANTENNA OUT

GERNSBACK PUBLICATIONS, INC.

Fig. 20-6

By using the preamplifier with a short length of shielded cable and clip leads, signals that generally could not generate a readout, generate precise and stable readouts on the counter. The DPDT switch is used to bypass the circuit when amplification is not needed. The preamplifier can also be used for other purposes. For example, the unit was also tested as a receiver preamplifier and increased received signal strength about 6 S-units at 30 MHz. A line tap can be used to measure the frequency directly at the output of a transmitter. The entire circuit for that consists of two diodes, one resistor, and one capacitor. The line tap simply picks a low-amplitude signal for measurement by the frequency counter. The tap can be connected to transmitters with an output power of between 1 and 250 W.

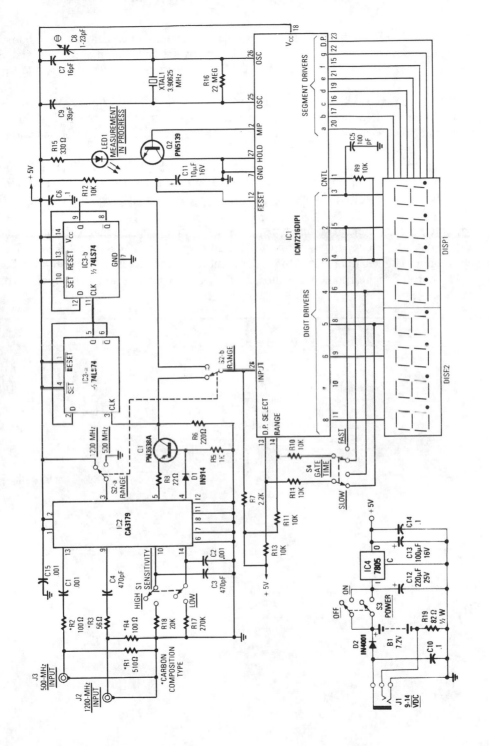

1.2-GHz FREQUENCY COUNTER

Fig. 20-7

Reprinted with permission from Radio-Electronics Magazine, 1987, R-E Experimenters Handbook.
Copyright Gernsback Publications, Inc., 1987.

129

The output of the CA3179 is fed through the D1/Q1 circuit. Those components serve to boost the 1-V output of the CA3179 to a standard TTL level. Then, depending on the position of range switch S2b, the signal is passed directly to the 7216, or through the divide-by-four circuit built from the two D flip-flops in IC3.

The other half of the range switch S2a controls the voltage at pin 3 of the CA3179. When pin 3 is high, the signal applied to pin 9 is fed through an extra internal divide-by-four stage before it is amplified and output on pins 4 and 5. When pin 3 is low, the signal on pin 13 is simply processed for output without being divided internally.

A 3.90625-MHz crystal provides the time base; the crystal yields a fast gate time of 0.256 second. The displayed frequency equals the input frequency divided by 1000 in the fast mode. In slow mode, gate time is 2.56 seconds. The displayed frequency equals the input frequency divided by 100 in the slow mode.

Switch S4, gate time, performs two functions. First it selects the appropriate gate time according to which digit output of IC1 the range input is connected to. Another of the 7216's inputs is also controlled by S4: the dp select input. The decimal point of the digit output to which that pin is connected will be the one that lights up. The correct decimal point illuminates, according to the position of S4, to provide a reading in MHz.

21

Crystal Oscillators

The sources of the following circuits are contained in the Sources section beginning on page 782. The figure number contained in the box of each circuit correlates to the sources entry in the Sources section.

Fundamental-Frequency Crystal Oscillator
Easy Start-Up Crystal Oscillator
Crystal Timebase
Low-Frequency Pierce Oscillator
1-MHz Pierce Oscillators
Simple CMOS Crystal Oscillators
Voltage-Controlled Crystal Oscillator

Two-Gate Quartz Oscillator
Crystal-Controlled Reflection Oscillator
Temperature-Compensated Crystal Oscillator
20-MHz VHF Crystal Oscillator
Marker Generator
100-MHz VHF Crystal Oscillator
50-MHz VHF Crystal Oscillator

FUNDAMENTAL-FREQUENCY CRYSTAL OSCILLATOR

FREQUENCY RANGE: 1.0 MHz to 20 MHz

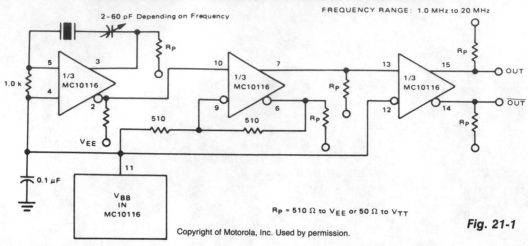

2 – 60 pF Depending on Frequency

R_P = 510 Ω to V_{EE} or 50 Ω to V_{TT}

Copyright of Motorola, Inc. Used by permission.

Fig. 21-1

For frequencies below 20 MHz, a fundamental-frequency crystal can be used and the resonant tank is no longer required. Also, at this lower frequency range the typical MECL 10,000 propagation delay of 2 ns becomes small compared to the period of oscillation, and it becomes necessary to use a noninverting output. Thus, the MC10116 oscillator section functions simply as an amplifier. The 1.0 KΩ resistor biases the line receiver near V_{BB} and the 0.1-μF capacitor is a filter capacitor for the V_{BB} supply. The capacitor, in series with the crystal, provides for minor frequency adjustments. The second section of the MC10116 is connected as a Schmitt-trigger circuit; this ensures good MECL edges from a rather slow, less than 20-MHz input signal. The third stage of the MC10116 is used as a buffer and to give complementary outputs from the crystal oscillator circuit. The circuit has a maximum operating frequency of approximately 20 MHz and a minimum of approximately 1 MHz; it is intended for use with a crystal which operates in the fundamental mode of oscillation.

EASY START-UP CRYSTAL OSCILLATOR

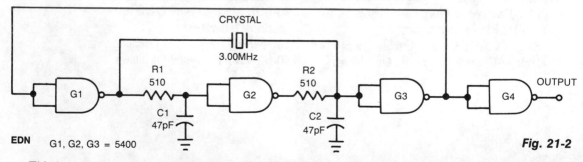

CRYSTAL

3.00MHz

EDN   G1, G2, G3 = 5400

Fig. 21-2

This low cost, crystal-controlled oscillator uses one TTL gate. Two factors ensure oscillator start-up: The connection of NAND gates G1, G2, and G3 into an unstable logic configuration and the high loop gain of the three inverters. Values of R1, R2, C1, and C2 aren't critical; select them so the oscillator operates at a frequency 70 to 90% higher than the crystal frequency when the crystal is disconnected. For 1 – 2 MHz operation, a low-power 54L00 IC is recommended; for a 2 – 6 MHz, a standard 5400 type; and for 6 – 20 MHz, a 54H00 or 54S00.

CRYSTAL TIMEBASE

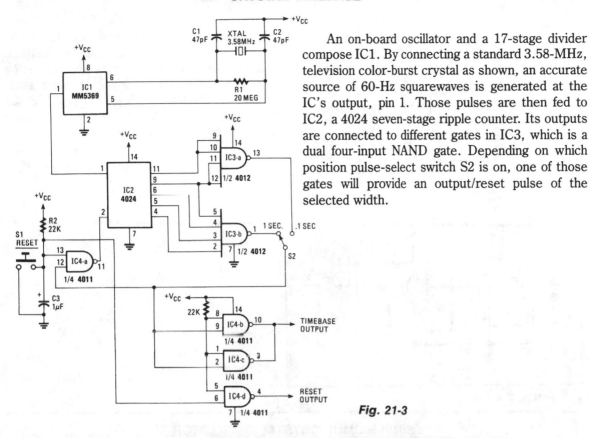

An on-board oscillator and a 17-stage divider compose IC1. By connecting a standard 3.58-MHz, television color-burst crystal as shown, an accurate source of 60-Hz squarewaves is generated at the IC's output, pin 1. Those pulses are then fed to IC2, a 4024 seven-stage ripple counter. Its outputs are connected to different gates in IC3, which is a dual four-input NAND gate. Depending on which position pulse-select switch S2 is on, one of those gates will provide an output/reset pulse of the selected width.

LOW-FREQUENCY PIERCE OSCILLATOR

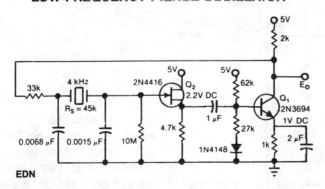

EDN

Fig. 21-4

The Pierce circuit oscillates at 4 kHz. At low frequencies, the crystal's internal series resistance R_S is quite high (45 K at 4 kHz). Therefore, an FET-based source follower is included to prevent Q1 from loading the crystal output.

1-MHz PIERCE OSCILLATORS

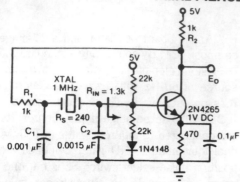

(a)

Simple network design is a key feature of the Pierce circuit, as these 1-MHz oscillators illustrate. Operating the crystal slightly above resonance (Fig. 21-5a) requires only one high-gain transistor stage. Operating it exactly at series resonance (Fig. 21-5b) requires an extra RC phase lag and two transistors which can have lower gain.

(b) EDN

Fig. 21-5

SIMPLE CMOS CRYSTAL OSCILLATOR

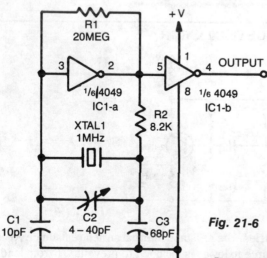

Fig. 21-6

The circuit is an inverter set up as a linear amplifier. Adding the crystal and capacitors to the feedback path, we turn the amplifier into an oscillator and force it to oscillate at, or least very near, the crystal's resonant frequency. Trimmer capacitor C2 adjusts the actual operating frequency of the circuit. The crystal should be a parallel-resonant type; maximum frequency will depend partly on supply voltage, but it should be possible to go to at least 1 MHz.

Reprinted with permission from Radio-Electronics Magazine February 1987.
Copyright Gernsback Publications, Inc., 1987.

VOLTAGE-CONTROLLED CRYSTAL OSCILLATOR

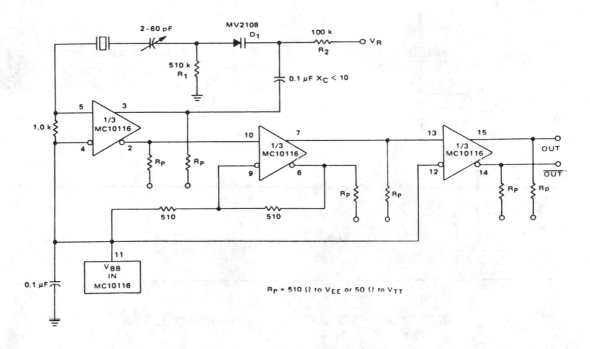

NOMINAL FREQUENCY	DEVIATION	
MHz	+PPM	–PPM
1.0000	57.0	48.0
1.8432	95.5	80.3
10.000	197.4	202.8
15.000	325.4	322.9

Fig. 21-7

A voltage-variable capacitance tuning diode is placed in series with the crystal feedback path. Changing the voltage on V_R varies the tuning diode capacitance and tunes the oscillator. The 510-KΩ resistor, R1, establishes a reference voltage for V_R—ground is used in this example. A 100-KΩ resistor, R2, isolates the tuning voltage from the feedback loop and 0.1-μF capacitor C2 provides ac coupling to the tuning diode. The circuit operates over a tuning range of 0 to 25 V. It is possible to change the tuning range from 0 to 25 V by reversing the tuning diode D1. Center frequency is set with the 2–60 pf trimmer capacitor. Deviation on either side of center is a function of the crystal frequency. The table in Fig. 21-7 shows measured deviation in parts per million for several tested crystals.

TWO-GATE QUARTZ OSCILLATOR

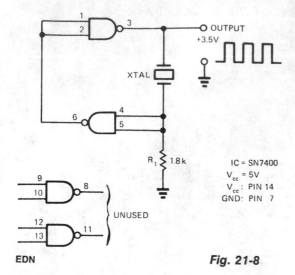

A SN7400 quartz crystal and a resistor provide a square-wave output of approximately 3.5 V. The circuit operates reliably at frequencies from 120 kHz to 4 MHz.

IC = SN7400
V_{cc} = 5V
V_{cc} : PIN 14
GND: PIN 7

EDN

Fig. 21-8

CRYSTAL-CONTROLLED REFLECTION OSCILLATOR

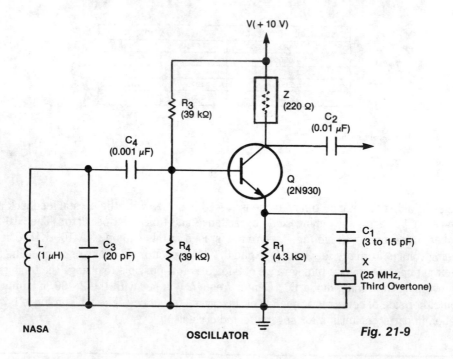

NASA

OSCILLATOR

Fig. 21-9

CRYSTAL-CONTROLLED REFLECTION OSCILLATOR (*Cont.*)

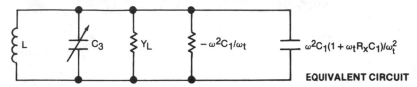

EQUIVALENT CIRCUIT

This unit is easily tunable and stable, consumes little power, and costs less than other types of oscillators that operate at the same frequencies. This unusual combination of features is made possible by a design concept that includes operation of the transistor well beyond the 3 dB frequency of its current-versus-frequency curve. The concept takes advantage of newly available crystals that resonate at frequencies up to about 1 GHz.

The emitter of transistor Q is connected with variable capacitor C1 and series-resonant crystal X. The emitter is also connected to ground through bias resistor R1. The base is connected to the parallel combination of inductor L and capacitor C3 through dc-blocking capacitor and C4 and is forward biased with respect to the emitter by resistors R3 and R4. Impedance Z could be the 220-Ω resistor shown or any small impedance that enables the extraction of the output signal through coupling capacitor C2. If Z is a tuned circuit, it is tuned to the frequency of the crystal.

TEMPERATURE-COMPENSATED CRYSTAL OSCILLATOR

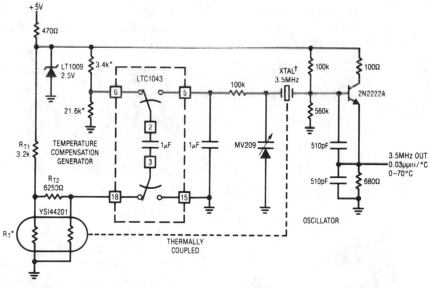

LINEAR TECHNOLOGY CORP.

Fig. 21-10

This circuit uses LTC1043 to differentiate between a temperature sensing network and a dc reference. The single-ended output biases a varactor-tuned crystal oscillator to compensate drift. The varactor crystal network has high dc impedance, eliminating the need for an LTC1043 output amplifier.

20-MHz VHF CRYSTAL OSCILLATOR

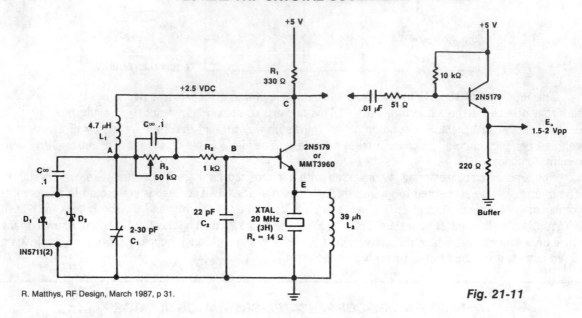

R. Matthys, RF Design, March 1987, p 31.

Fig. 21-11

A typical circuit at 20 MHz is shown. The crystal, which has an internal series resistance R_S of 14 Ω, oscillates at its third harmonic. The diode clamp D1 and D2 provides a constant amplitude control. The transistor operates continuously in a linear mode over a complete cycle of oscillation, and reflects a reasonably constant load across the crystal at all times.

MARKER GENERATOR

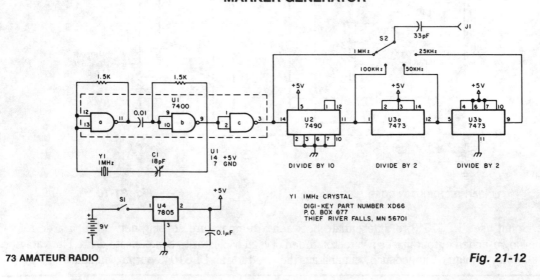

73 AMATEUR RADIO

Fig. 21-12

138

MARKER GENERATOR (*Cont.*)

The oscillator section uses three sections of a 7400 quad NAND gate integrated circuit. The 1-MHz signal from the oscillator is fed into a 7490 decade counter configured to divide by ten, providing the 100-kHz signal. To obtain the 50 and 25 kHz outputs, the 100-kHz signal is further divided by 7473 dual J-K flip-flop. The first half of the 7473 divides the 100-kHz signal by two, yielding the 50 kHz signal. The second half of the 7473 again divides by two, giving the 25 kHz signal. S2 selects the output, a square wave, rich in harmonics. The generator can be powered from any convenient 6 to 12 Vdc source. A 7805 fixed-voltage regulator supplies the regulated voltage for the oscillator and the divider chips. The generator described here is powered by a 9-V transistor radio battery.

100-MHz VHF CRYSTAL OSCILLATOR

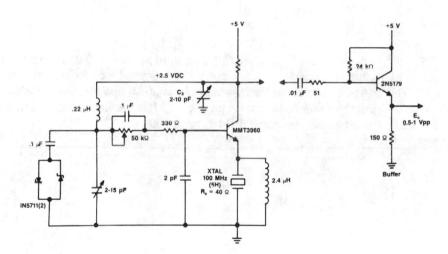

R. Matthys, RF Design, March 1987, p. 31.

Fig. 21-13

Figure 21-13 shows a 100-MHz oscillator operating on the fifth harmonic. Again to maintain the transistor's gain, note the increase in the collector's load resistance R1 because of the increase in the quartz crystal's internal series resistance R_S. C3 is needed at frequencies above 50 MHz to tune out the shunting effect of L1 on R1, to maintain a high load resistance for the transistor and get enough gain for oscillation. The equivalent series R_LC_L load across the crystal is 8.2 Ω (R_L) and 200 pF (C_L).

50-MHz VHF CRYSTAL OSCILLATOR

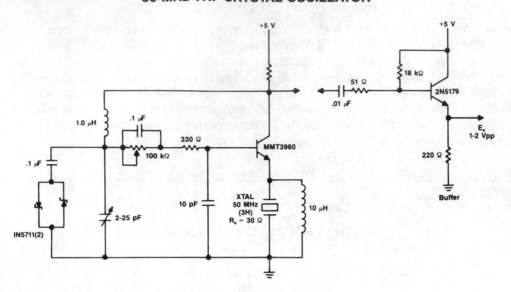

R. Matthys, RF Design, March 1987, p. 31.

Fig. 21-14

Figure 21-14 shows a 50-MHz oscillator operating on a third harmonic. The collector's load resistor R1 has been increased because the quartz crystal's internal series resistance R_S increases with frequency in the VHF range. The crystal's internal series resistance R_S is 30 Ω, and the transistor's minimum current gain H_{FE} is 100. Using the same technique as for the 20 MHz oscillator, the external series $R_L C_L$ equivalent load seen by the 50 MHz crystal is 5.6 Ω (R_L) and 1000 pF (C_L).

22

Decoders

The sources of the following circuits are contained in the Sources section beginning on page 782. The figure number contained in the box of each circuit correlates to the sources entry in the Sources section.

Second-Audio Program Adapter
Tone Decoder
Encoder/Decoder
Direction Detector/Decoder
Sound-Activated Decoder

SECOND-AUDIO PROGRAM ADAPTER

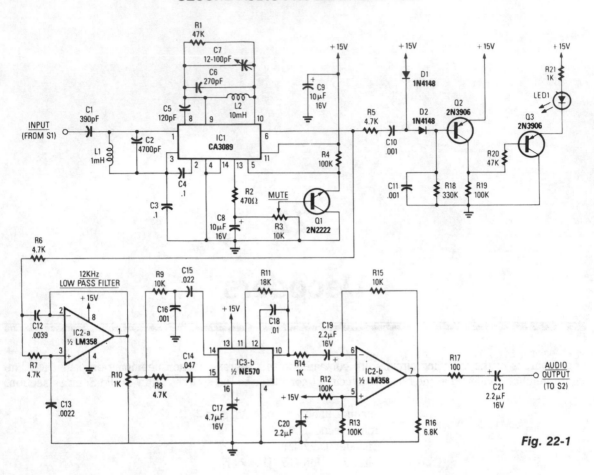

Reprinted with permission from Radio-Electronics Magazine 1989, R-E Experimenters Handbook. Copyright Gernsback Publications, Inc., 1989.

The baseband-audio input comes from the pole of switch S1 in the stereo decoder, and is coupled to IC1 (a CA3089) via a 78.6 kHz bandpass filter that consists of capacitors C1 and C2, and inductor L1. IC1 is a combination i-f amplifier and quadrature detector normally used for FM radio systems operating within an i-f of 10.7 MHz. The device works equally well at 78.6 kHz. Capacitors C6 and C7, and inductor L2 tune the detector section to 78.6 kHz, while C5 provides the necessary 90-degree phase shift for proper quadrature detector operation. The output voltage at pin 13 of IC1 is proportional to the level of the incoming signal. When the voltage at the wiper of potentiometer R3 reaches a predetermined threshold level, Q1 conducts, grounding pin 5 of IC1, enabling IC1's mute function.

Detected audio output from pin 6 of IC1 goes to IC2a, which is configured as a 12-kHz, −12 dB per octave, low-pass filter. The output of IC2a appears across potentiometer R10, which provides a means of adjusting the drive level into IC3b, the 2:1 compander.

SECOND-AUDIO PROGRAM ADAPTER *(Cont.)*

Audio from the wiper of R10 is split into two paths: a high-pass filter (C14 and R8) provides a path to the rectifier input of the compander, and a bandpass filter (R9, C16, and C15) that feeds the audio input of the compander. A fixed 390-μs de-emphasis network is formed by C18 and R11 in conjunction with IC3b. Corrected audio appears at pin 10 of IC3b and is coupled to IC2b, and output buffer amplifier.

Audio from pin 6 of IC1 is also coupled to an audio high-pass filter, R5 and C10, and fed to an audio rectifier, D1, D2, and C11. When a SAP signal is detected by IC1, it is rectified by D1 and D2; the resultant dc charges C11. An increasing positive voltage at the base of Q2 causes its current flow to decrease, so the voltage at Q2's collector also decreases. That in turn causes the base voltage of Q3 to drop, which causes Q3 to conduct, thereby lighting the LED.

TONE DECODER

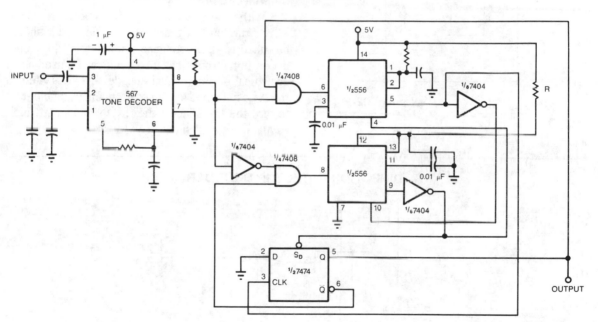

Fig. 22-2

Adding a pair of one shots to the output of a 567 tone decoder renders it less sensitive to out-of-band signals and noise. Without the one shots, the 567 is prone to spurious output chatter. Other protection schemes, such as feeding back outputs or using an input filter, do not work as well as the one shots. The output of the 567 is high in the absence of a tone and becomes low when it detects a tone. The tone decoder triggers a one shot via an AND gate. The one shot's period is set to slightly less than the duration of a tone burst.

When the output of the tone decoder decreases, it triggers the second one shot. The second one shot's period is set to slightly less than the interval between tone bursts. The flip-flop enables and disables the inputs to one shots so that spurious outputs from the tone decoder do not affect the output.

ENCODER/DECODER

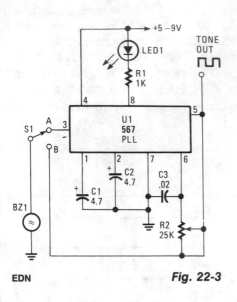

EDN

Fig. 22-3

The transducer circuit can be operated as either a tone encoder or decoder by changing the position of S1. The operating frequency of that dual-purpose circuit is determined by C3 and R2. Capacitors C1 and C2 are not critical and can be of almost any value between 1 and 5 mf. When the circuit is receiving an on-frequency signal, LED1 lights. Although a two wire piezo transducer with a resonance frequency of 2500 Hz was used in the circuit, any piezo unit should work—as long as the values of C3 and R2 are selected to tune to the transducer's operating frequency.

With power on and S1 in the B position, adjust R2 for the loudest tone output. The circuit should be tuned to the resonance frequency of the transducer. In that position, the circuit can be used as an acoustical or tone signal encoder. Next, switch to the A position and aim an on-frequency audible tone toward the transducer; the LED should light, indicating a decoded signal.

DIRECTION DETECTOR DECODER

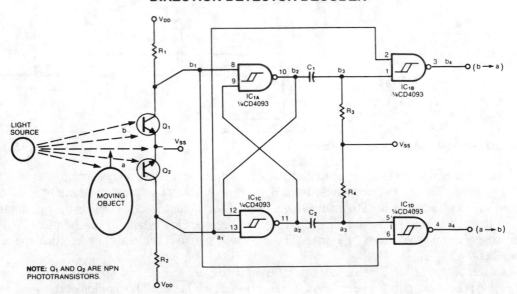

Fig. 22-4

DIRECTION DETECTOR DECODER (*Cont.*)

This circuit, which was developed to monitor the traffic of bumblebees in and out of the hive, differentiates a-to-b motion from b-to-b motion. When used with an optical decoder, the circuit distinguishes clockwise from counterclockwise rotation and provides a resolution of one output pulse per quadrature cycle.

Q1 and Q2 are mounted so that a moving object first blocks one phototransistor, then both, then the other. Depending on the direction in which the object is moving, either IC1B or IC1D emits a negative pulse when the moving object blocks the second sensor. An object can get as far as condition 3 and retreat without producing an output pulse; that is, the circuit ignores any probing or jittery motion. If an object gets as far as condition 4, however, a retreat will produce an opposite-direction pulse.

The time constants R3C1 and R4C2 set the output pulse width. A 100 KΩ/100pF combination, for example, produces 10-μs pulses. Select a value for pullup resistors R1 and R2 from the 10 K to 100 KΩ range, according to the sensitivity your application requires.

SOUND-ACTIVATED DECODER

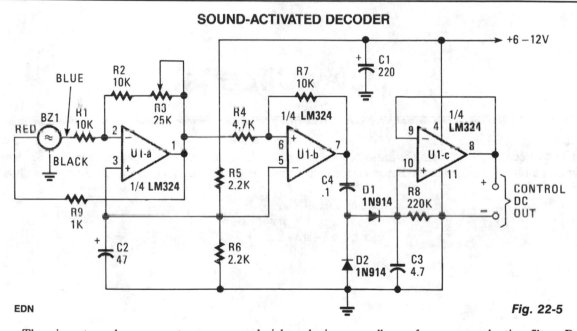

EDN

Fig. 22-5

The piezo transducer operates as a sound-pickup device as well as a frequency-selective filter. By controlling the gain of the op amps, the oscillator can be transformed into a sensitive and frequency-selective tone-decoder circuit. With the gain of U1a set just below the point of self oscillation, the transducer becomes sensitive to acoustically coupled audio tones that occur at or near its resonant frequency.

The circuit's output can be used to activate optocouplers, drive relay circuits, or to control almost any dc-operated circuit. The dc signal at the output of U1c varies with 0 to over 6 V, depending on the input-signal level. One unusual application for the sound-activated decoder would be in extremely high-noise environments, where normal broadband microphone pickup would be useless. Because piezo transducers respond only to frequencies within a very narrow bandwidth, little if any of the noise would get through the transducer.

23

Delay Circuits

The sources of the following circuits are contained in the Sources section beginning on page 782. The figure number contained in the box of each circuit correlates to the sources entry in the Sources section.

Leading-Edge Delay
Pulse Delay with Dual-Edge Trigger
Adjustable Delay

LEADING-EDGE DELAY

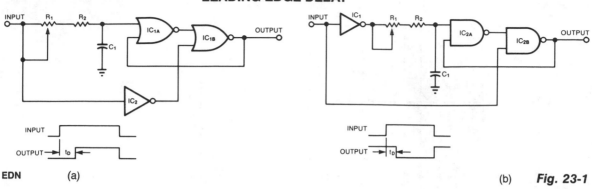

EDN   (a)

(b)   **Fig. 23-1**

Circuit (a) lets you delay the leading edge of a positive pulse while leaving the trailing edge almost unaffected. A positive input transition, inverted by IC2, has no effect on IC1B. However, when the positive transition reaches IC1A, (delayed by the adjustable network of R1, R2, and C1), it toggles both NOR gates, initiating the output pulse. When the input decreases IC1B follows suit, delayed only by the propagation through itself and IC2. Circuit (b) produces an inverted output pulse. Inverter IC1 serves as a buffer for the signal source—an advantage when driving a low-impedance (short-delay) network. Moreover, only the propagation delay of IC2B separates the output's trailing edge from that of the input. You can configure circuit (a) to handle negative pulses by using NAND instead of NOR gates. Similarly, circuit (b) will produce a delayed positive pulse in response to negative input pulse, if you substitute NOR gates for NAND gates.

PULSE DELAY WITH DUAL-EDGE TRIGGER

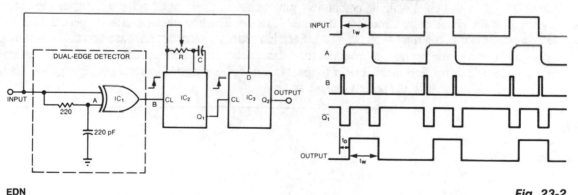

EDN

Fig. 23-2

A single monostable multivibrator delays a pulse train by a variable amount; nonetheless, this amount can be no less than the minimum allowed pulse width t_w. The exclusive-OR gate, IC1, generates a short pulse following every leading or falling edge of the input waveform. These pulses cause one-shot IC2 to produce a negative-going pulse with a duration equal to the desired time delay t_p, which you set by adjusting potentiometer R. Flip-flop IC3 then creates a delayed replica of the input pulse by latching the Q1 output of IC2 between positive-going transitions. You can independently control the output-pulse duration by cascading a second one shot with the first.

ADJUSTABLE DELAY

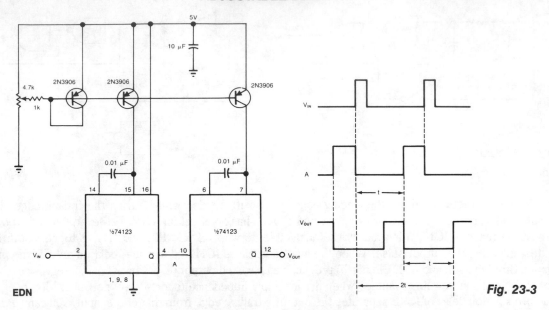

EDN

Fig. 23-3

You can obtain well over 360° of phase delay by cascading two monostable multivibrators. In a typical configuration, a single monostable multivibrator is used to introduce delay in a pulse train; the multivibrator triggers on each incoming pulse, provided it resets in time for the next pulse. Yet even when it resets in time, the single monostable multivibrator provides a maximum phase delay of less than 360°. However, with the cascaded-multivibrator approach, you can achieve 650° of phase delay by using an input-pulse spacing of 200 μs for example, with the component values shown. Every input pulse will trigger the circuit while you adjust the phase delay throughout its available range. The first multivibrator triggers the second one, whose reset marks the total delay time (2t). Each introduces a delay of t μs, based on 0.01-μF timing capacitors and equal charging currents from the three-transistor, dual-current source. The two multivibrator arrangement allows the first multivibrator to reset in time to be triggered by the next input pulse. Also, the variation of t is linear with the potentiometer setting.

24

Demodulator

The sources of the following circuits are contained in the Sources section beginning on page 782. The figure number contained in the box of each circuit correlates to the sources entry in the Sources section.

565 SCA Demodulator

565 SCA DEMODULATOR

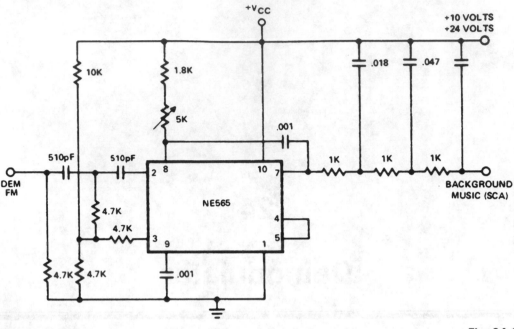

SIGNETICS

Fig. 24-1

This application involves demodulation of a frequency-modulated subcarrier of the main channel. This popular example uses the PLL to recover the SCA (Subsidiary Carrier Authorization or storecast music) signal from the combined signal of many commercial FM broadcast stations. The SCA signal is a 67 kHz frequency-modulated subcarrier which puts it above the frequency spectrum of the normal stereo or monaural FM program material. By connecting the circuit to a point between the FM discriminator and the deemphasis filter of an FM receiver and tuning the receiver to a station which broadcasts an SCA signal, you can obtain hours of commercial-free background music.

25

Detectors

The sources of the following circuits are contained in the Sources section beginning on page 782. The figure number contained in the box of each circuit correlates to the sources entry in the Sources section.

Wide-Range Peak Detector
Schmitt Trigger
Analog Peak Detector with Digital Hold
500-Hz Tone Detector
Audio Decibel Level Detector with Meter Driver
Precision Envelope Detector
Frequency-Boundary Detector
Low-Drift Peak Detector
Edge Detector
Null Detector

Precision Threshold Detector
Out-Of-Bounds Pulse-Width Detector
Digital Frequency Detector
Missing-Pulse Detector
Digital Peak Detector
High-Bandwidth Peak Detector
Wide-Bandwidth Peak Detector

WIDE-RANGE PEAK DETECTOR

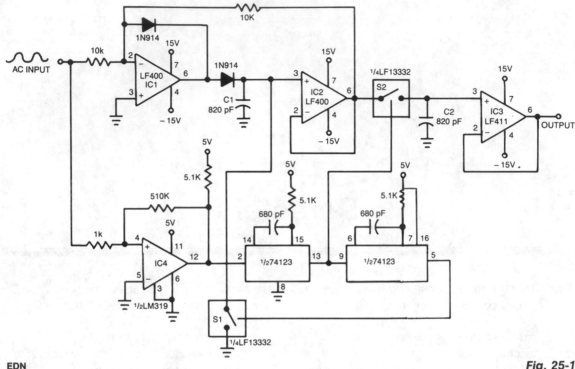

EDN

Fig. 25-1

IC1 and IC2 form an inverting half-wave precision-rectifier/peak-detector circuit. Negative input-signal, swings with peaks larger than the voltage on C1, cause this capacitor to charge to the new peak voltage. The capacitor holds this voltage until a larger signal peak arrives. When the input swings high, comparator IC4 detects the zero crossing and triggers the one-shot multivibrator. The one shot closes FET switch S2, thereby causing C2 to charge to the peak voltage held on C1, during the previous half cycle. The second one shot then produces a pulse that causes FET switch S1 to discharge C1. If the next negative signal-input peak is different from the previous one, the circuit captures it and it appears at IC3's output during the next half cycle. The peak detector thus resets itself once every input-waveform cycle. Note that the zero crossings are necessary to trigger the switches; therefore, the circuit is usable only with ac signals.

SCHMITT TRIGGER

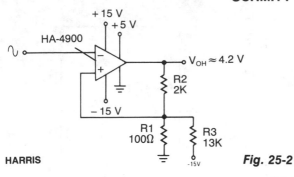

This circuit has a 100-mV hysteresis which can be used in applications where very fast transition times are required at the output, even though the signal input is very slow. The hysteresis loop also reduces false triggering because of noise in the input.

HARRIS

Fig. 25-2

ANALOG PEAK DETECTOR WITH DIGITAL HOLD

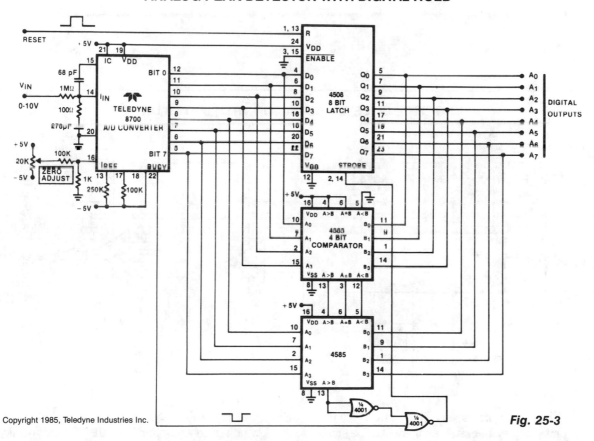

Copyright 1985, Teledyne Industries Inc.

Fig. 25-3

Analog peak detection is accomplished by repeatedly measuring the input signal with an a/d converter and comparing the current reading with the previous reading. If the current reading is larger than the previous, the current reading is stored in the latch and becomes the new peak value. Since the peak is stored in a CMOS latch, the peak can be stored indefinitely.

500-Hz TONE DETECTOR

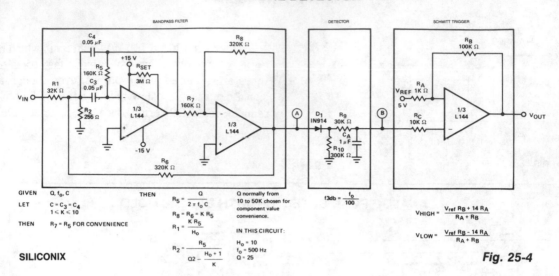

GIVEN Q, f_o, C

LET C = C_3 = C_4
$1 \le K \le 10$

THEN $R_7 = R_5$ FOR CONVENIENCE

THEN $R_5 = \dfrac{Q}{2\pi f_o C}$

$R_8 = R_6 = K\,R_5$

$R_1 = \dfrac{K\,R_5}{H_o}$

$R_2 = \dfrac{R_5}{Q2 - \dfrac{H_o + 1}{K}}$

Q normally from 10 to 50K chosen for component value convenience.

IN THIS CIRCUIT:

$H_o = 10$
$f_o = 500$ Hz
$Q = 25$

$f3db = \dfrac{f_o}{100}$

$V_{HIGH} = \dfrac{V_{ref}\,R_B + 14\,R_A}{R_A + R_B}$

$V_{LOW} = \dfrac{V_{ref}\,R_B - 14\,R_A}{R_A + R_B}$

SILICONIX

Fig. 25-4

AUDIO DECIBEL LEVEL DETECTOR WITH METER DRIVER

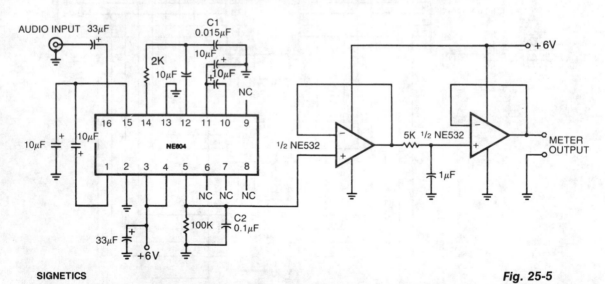

SIGNETICS

Fig. 25-5

This circuit draws very little power, less than 5 mA with a single 6-V power supply, making it ideal for portable battery-operated equipment. The small size and low power consumption belie the 90-dB dynamic range and 10.5-μV sensitivity. Dc output voltage proportional to the $\log_{10}$ of the input signal level. Thus, a standard 0–5 voltmeter can be linearly calibrated in decibels over a single 80-dB range. The circuit is within 1.5-dB tolerance over the 80-dB range for audio frequencies from 100 Hz to 10 kHz. Higher audio levels can be measured by placing an attenuator ahead of the input capacitor.

PRECISION ENVELOPE DETECTOR

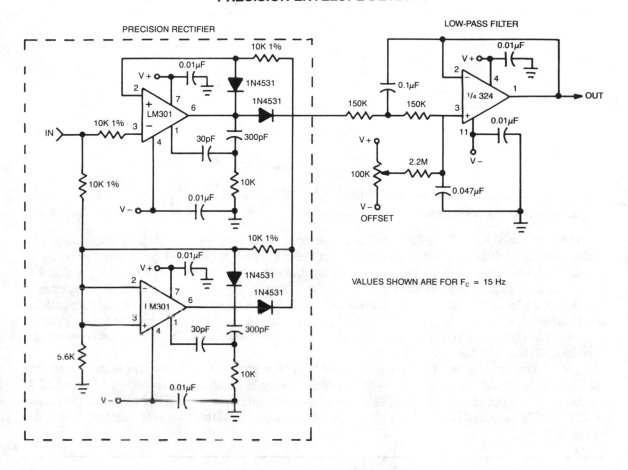

VALUES SHOWN ARE FOR F_c = 15 Hz

EDN

Fig. 25-6

This circuit is useful for signal-processing sonar data recorded on an instrumentation-quality analog tape recorder. The envelope detector utilizes ready available parts, and furnishes accuracy beyond 100 kHz. Two LM301 op amps connected as precision absolute-value circuits use 2-pole frequency compensation for increased slew rate. And one section of an LM324 quad op amp connected in a Butterworth LPF configuration subjects the rectifier's output to a low-pass filter.

$$f_c = 1/2\pi RC_1$$
$$C_2 = (1/2)\, C_1$$

FREQUENCY-BOUNDARY DETECTOR

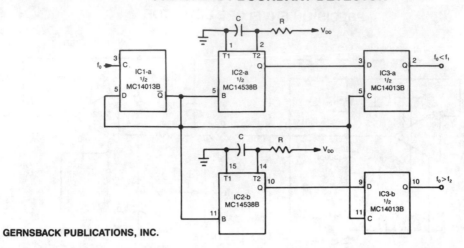

GERNSBACK PUBLICATIONS, INC.

Fig. 25-7

The circuit can be used to tell whether or not an input signal is within a certain frequency range. The device consists of three ICs, a dual monostable multivibrator, and two dual D-type flip-flops. The signal whose frequency is in question is fed to the clock input of one of the flip-flops. The Q output of that flip-flop (IC1a) is cross coupled to its data input so that it acts like a divide-by-two counter. The trailing edge of the Q output is used to trigger the one shots formed by IC2. The upper- and lower-frequency boundaries are determined by the two sections of IC2; the dual precision monostable multivibrator and their external rc networks. The upper-frequency boundary, f1, is set by the output of IC2a, and the lower-frequency boundary, f2, is set by the output of IC2b.

The frequency of the input to the circuit can be anywhere from dc to 100 kHz. The states of the outputs of IC2, which determine the upper- and lower-frequency boundaries, are latched by IC3a and IC3b respectively. The output of IC3a will be high only when the input frequency is less than that of the output of IC2a, f_1. The output of IC3b will be high only when the frequency of the input is greater than that of the output of IC2b, f_2.

LOW-DRIFT PEAK DETECTOR

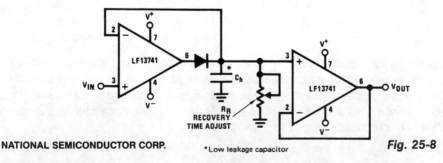

NATIONAL SEMICONDUCTOR CORP.

*Low leakage capacitor

Fig. 25-8

This circuit uses op amp U1 to compensate for the offset in peak detector diode D1. Across C_h is the exact peak voltage; U2 is used as a voltage follower to read this voltage.

EDGE DETECTOR

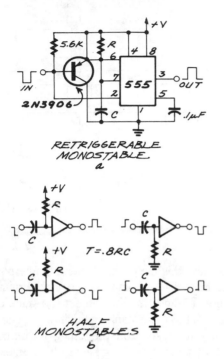

RETRIGGERABLE
MONOSTABLE.
a

HALF
MONOSTABLES
b

The 555 is a monostable that *wants* a negative-going trigger. If the pulse you're feeding it with is positive-going, you can run it through an inverter made up of either an inverting gate or, if you're tight on space, a single transistor. Both ways are shown. The circuits shown in Fig. 25-9b are edge detectors as well, and are usually referred to as half monostables, since they can't be used in every application. The width of the output pulse is determined by the RC value, but there are a few rules governing their use:

- The input pulse has to be wider than the output pulse
- The input pulse can't be glitchy
- The circuit can't be retriggered faster than the RC time

Fig. 25-9

PRECISION THRESHOLD DETECTOR

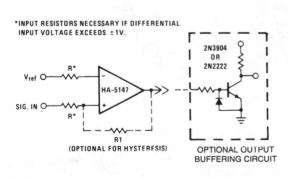

*INPUT RESISTORS NECESSARY IF DIFFERENTIAL INPUT VOLTAGE EXCEEDS ±1V.

R1
(OPTIONAL FOR HYSTERESIS)

OPTIONAL OUTPUT
BUFFERING CIRCUIT

HARRIS

Fig. 25-10

This circuit requires low noise, low and stable offset voltages, high open loop gain, and high speed. These requirements are met by the IIA-5147. The standard variations of this circuit can easily be implemented using the HA-5147. For example, hysteresis can be generated by adding R1 to provide small amounts of positive feedback. The circuit becomes a pulse width modulator if V_{ref} and the input signal are left to vary. Although the output drive capability of this device is excellent, the optional buffering circuit can be used to drive heavier loads, preventing loading effects on the amplifier.

OUT-OF-BOUNDS PULSE-WIDTH DETECTOR

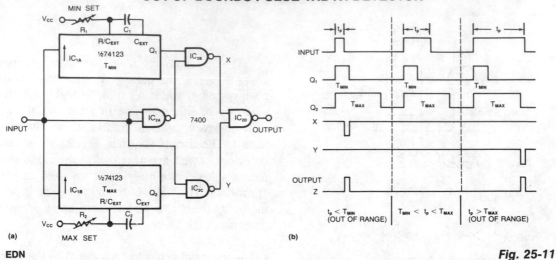

EDN

Fig. 25-11

Requiring only two ICs to monitor a train of positive pulses, this circuit produces a single positive output pulse for each input pulse whose duration is either too long or too short. You specify the minimum and maximum limits by adjusting the trimming potentiometers, R1 and R2. You can set the value of the acceptable pulse width from approximately 50 ns to 10 μs, for a 74123 monostable multivibrator. The leading edge of an input pulse triggers one shots IC1A and IC1B as you can see from the timing diagram. Each NAND-gate output is high unless either or both inputs are low, so outputs X and Y are high unless the circuit encounters an out-of-range pulse. IC2D then gates a negative pulse from IC2B or IC2C to produce the circuit's positive output pulse.

DIGITAL FREQUENCY DETECTOR

PARTS
IC1—SN7406 – HEX INVERTER
IC2—SN5400 – QUAD NAND

INPUT	POINT E
	LOGICAL HIGH
0 V	LOGICAL LOW
+5	LOGICAL LOW

EDN

Fig. 25-12

DIGITAL FREQUENCY DETECTOR (*Cont.*)

A simple inventer and NAND gate can be connected to yield a highly compact and reliable digital frequency detector. This circuit can detect frequencies up to 3 MHz with 50% duty cycles. When a frequency, f_i, appears at the input, points A and B detect a logical high dc level. Thereupon point E increases the latch sets and the LED lights. If the input frequency is absent and if the voltage is either at a constant high or low level, points A and B will be complementary and point E will decrease. This will reset the latch and extinguish the LED.

MISSING-PULSE DETECTOR

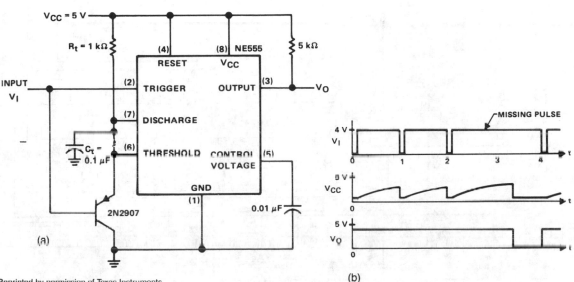

Reprinted by permission of Texas Instruments.

(b)

Fig. 25-13

This circuit will detect a missing pulse or abnormally long spacing between consecutive pulses in a train of pulses. The timer is connected in the monostable mode. The time delay should be set slightly longer than the timing of the input pulses. The timing interval of the monostable circuit is continuously retriggered by the input pulse train, V_I. The pulse spacing is less than the timing interval, which prevents V_C from rising high enough to end the timing cycle. A longer pulse spacing, a missing pulse, or a terminated pulse train will permit the timing interval to be completed. This will generate an output pulse, V_O as illustrated in Fig. 25-3b. The output remains high on pin 3 until a missing pulse is detected at which time the output decreases.

The NE555 monostable circuit should be running slightly slower, lower in frequency, than the frequency to be analyzed. Also, the input cannot be more than twice this free-running frequency or it would retrigger before the timeout and the output would remain in the low state continuously. The circuit operates in the monostable mode at about 8 kHz, so pulse trains of 8 to 16 kHz can be observed.

DIGITAL PEAK DETECTOR

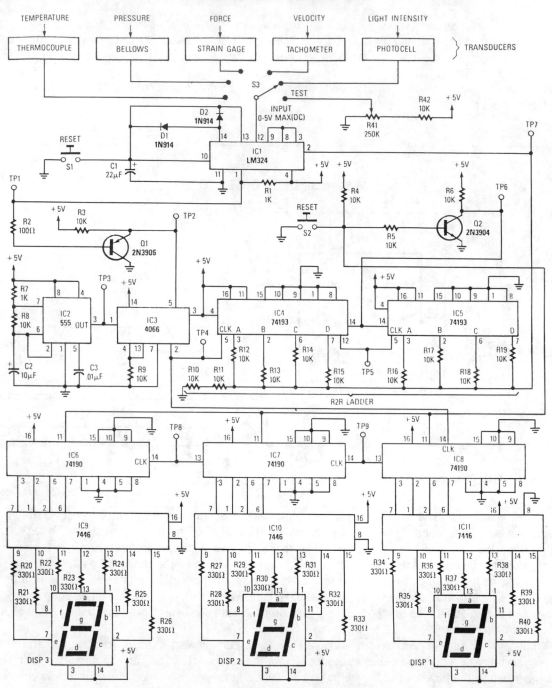

Reprinted with permission from Radio-Electronics Magazine, April 1989. Copyright Gernsback Publications, Inc., 1989.

Fig. 25-14

DIGITAL PEAK DETECTOR CONT.

The peak detector tracks and holds, using the charge-storing ability of a capacitor, the highest output voltage from a transducer. Initially, the voltage on the inverting input of the comparator is at ground level. As a small voltage (0 – 5 V) is captured by the peak detector and presented to the comparator's noninverting input, the output will swing high, which asserts the bilateral switch; clock pulses now pass through the switch to clock both the BCD and binary counters. The outputs of the binary counters are connected to an R2R ladder network, which functions as a digital-to-analog converter. As the binary count increases, the R2R ladder voltage also increases until it reaches a point slightly above the voltage of the peak detector; at that instant, the comparator output swings low, which disables the bilateral switch and stops the counters. The number displayed on the 7-segment LED's will represent a value equivalent to the transducer's output.

HIGH-BANDWIDTH PEAK DETECTOR

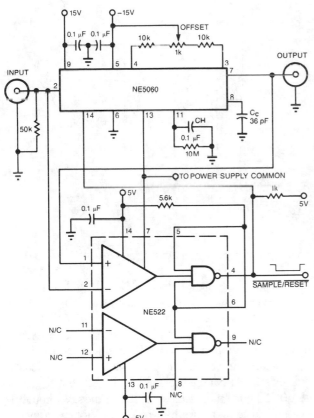

The high-speed peak detector uses a highly accurate, fast s/h amplifier controlled by a high-speed comparator. The s/h amplifier holds the peak voltage, until the comparator switches the amp to its sample mode, to capture a new, higher voltage level. The circuit handles all common-wave shapes and exhibits 5% accuracy from 50 Hz to 2 MHz.

The comparator's output decreases when the input signal exceeds the value of the currently held output. This transition puts the s/h amplifier into sample mode. Once the output reaches the value of the input, or the input signal falls below the output's level, the comparator's output increases; the high output brings the s/h amplifier back to the hold mode, thereby holding the peak value of the input signal. Reset the circuit by lowering the value of pin 4 of the NE522 comparator, which in turn allows the NE5060 s/h amplifier to acquire the input. The NE522 comparator has an open-collector output.

Fig. 25-15

NULL DETECTOR

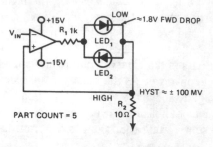

In this indicating comparator circuit, R2 sets the hysteresis. If the 741 saturates at ± 12 V, the current in R1 will be approximately ± 10 mA if 0.1 V hysteresis is desired. Then 0.1 V/10 mA = 10Ω = R2.

EDN

Fig. 25-16

WIDE-BANDWIDTH PEAK DETECTOR

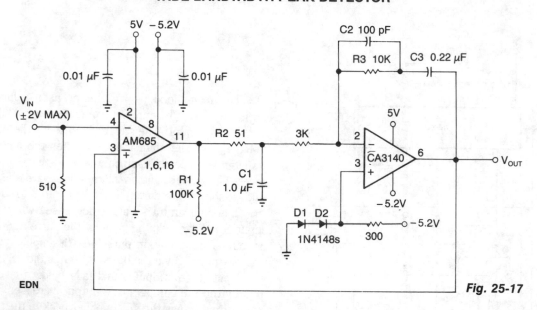

EDN

Fig. 25-17

This circuit can detect the positive peaks for signal frequencies higher than 5 MHz. It yields ±1% accuracy for 400 mV to 4 V pk-pk signal amplitudes on sine, square, and triangular waveforms. The Am685 comparator output increases whenever V_{IN} is a greater negative voltage than V_{OUT}; the high comparator output, in turn, charges C1 in a positive direction. The CA3140 op amp amplifies the C1 voltage with respect to the ECL-switching-threshold voltage (-1.3 V) developed by diodes D1 and D2. For repetitive waveforms, each cycle boosts V_{OUT} until it equals the peak input value. The peak-detection process is aided by the comparator's open-emitter output, which allows C1 to charge rapidly through R2, but to discharge slowly through R2 and R1. Reducing the value of C1 shortens system-response times. Although the circuit can't detect negative-going peaks, it can be modified to measure the pk-pk value of bipolar signals that are symmetric about ground. To do so, divide V_{OUT} by 2 using two 1-KΩ resistors and feed the comparator $V_{OUT}/2$ rather than V_{OUT}.

26

Digital-to-Analog Converters

The sources of the following circuits are contained in the Sources section beginning on page 782. The figure number contained in the box of each circuit correlates to the sources entry in the Sources section.

DIGITAL-TO-ANALOG CONVERTERS

10 Bit, 4 Quadrant Multiplying DAC
(Offset Binary Coding)

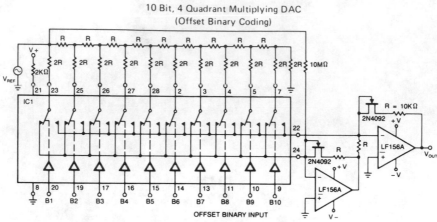

IC1: use five of either DG403, DG413, or DG423

Biopolar (Offset Binary)* Operation

DIGITAL INPUT	ANALOG OUTPUT
1 1 1 1 1 1 1 1 1 1	$- V_{REF} (1 - 2^{-9})$
1 0 0 0 0 0 0 0 0 1	$- V_{REF} (2^{-9})$
1 0 0 0 0 0 0 0 0 0	0
0 1 1 1 1 1 1 1 1 1	$V_{REF} (2^{-9})$
0 0 0 0 0 0 0 0 0 1	$V_{REF} (1 - 2^{-9})$
0 0 0 0 0 0 0 0 0 0	V_{REF}

NOTE: 1 LSB = $2^{-9} V_{REF}$

*Complementing B1 (MSB) will give 2's complement coding.

4 Bit Multiplying Current Switch D/A

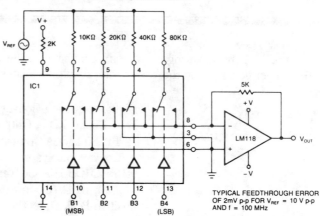

TYPICAL FEEDTHROUGH ERROR OF 2mV p-p FOR V_{REF} = 10 V p-p AND f = 100 MHz

IC1: use two of either DG403, DG413, or DG423

SILICONIX

Fig. 26-1

The following applications circuits are intended to illustrate the following points:

- A 2-KΩ resistor should be in series with V+ to limit supply current with negative ringing of the bit inputs
- Temperature compensation for $R_{DS(on)}$ can be provided in the feedback path of the op amp
- Bipolar reference voltages can be used in all configurations

FOUR-CHANNEL D/A OUTPUT AMPLIFIER

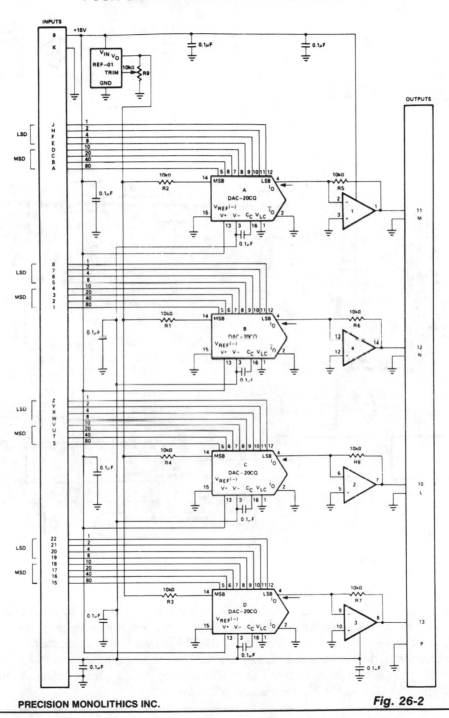

PRECISION MONOLITHICS INC.

Fig. 26-2

12-BIT BINARY 2s COMPLEMENT D/A CONVERSION SYSTEM

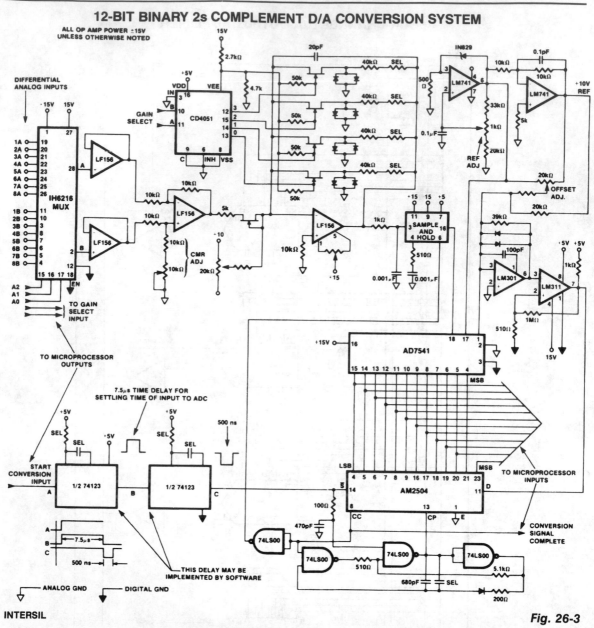

INTERSIL

Fig. 26-3

The front end of the DAC is configured differentially using dual eight-input IC multiplexer 1H6216 and three LM156 op amps. Following the differential amplifier is the programmable gain stage discussed earlier, with a low-pass filter on the output feeding the IH5110 sample and hold amplifier. The output of the IH5110 is connected to the comparator input, – input LM301, through the internal 10-KΩ feedback resistor of the 7541 multiplying d/a converter. The AD7541, along with a ±10-V reference and successive approximation logic, make up the 2's complement a/d converter.

9-BIT CMOS D/A CONVERTER

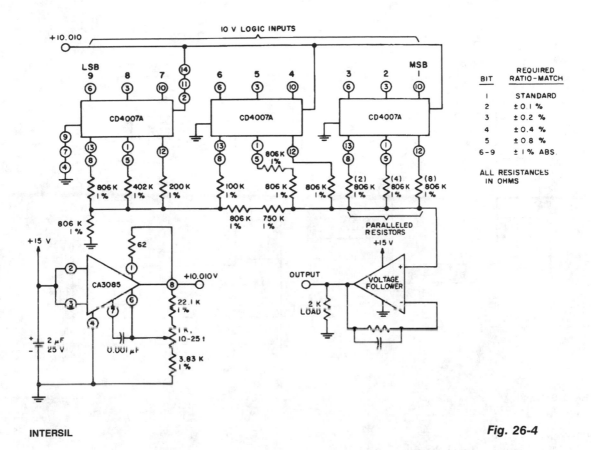

Fig. 26-4

Three CD4007A IC packages perform the switch function using a 10-V logic level. A single 15-V supply provides a positive bus for the follower amplifier and feeds the CA3085 voltage regulator. The scale adjust function is provided by the regulator output control, which is set to a nominal 10 V in this system. The line-voltage regulation (approximately 0.2%) permits 9-bit accuracy to be maintained with a variation of several volts in the supply. System power consumption ranges between 70 and 200 mW; a major portion is dissipated in the load resistor and op amp. The regulated supply provides a maximum current of 440 μA of which 370 μA flows through the scale adjusting. The resistor ladder is composed of 1% tolerance metal-oxide film resistors. The ratio match between resistance values is in the order of 2%. The follower amplifier has the offset adjustment nulled at approximately a 1 V output level.

MULTIPLYING D/A CONVERTER

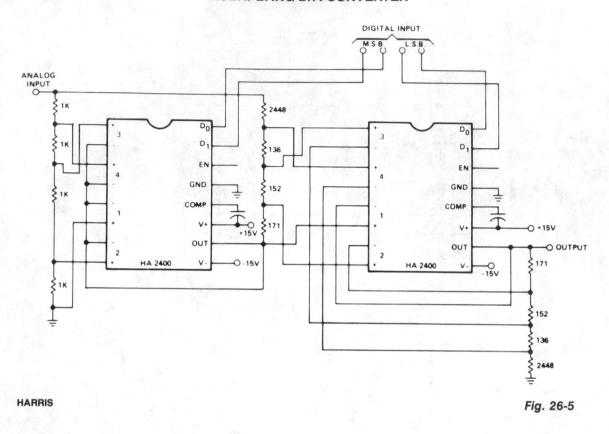

HARRIS

Fig. 26-5

The circuit performs the function:

$$V_{OUT} = V_{IN} \times \frac{N}{16}$$

where N is the binary number from 0 to 15 formed by the digital input. If the analog input is a fixed dc reference, the circuit is a conventional 4-bit D to ac signal, in which case the output is the product of the analog signal and the digital signal. The circuit on the left is a programmable attenuator with weights of 0, $1/4$, $1/2$, or $3/4$. The circuit on the right is a noninverting adder, which adds weights to the first output of 0, $1/16$, $1/8$, or $3/16$. If four quadrant multiplication is required, place a phase selector circuit in series with either the analog input or output. The D_O input of that stage becomes the + or − sign bit of the digital input.

POSITIVE PEAK DETECTOR

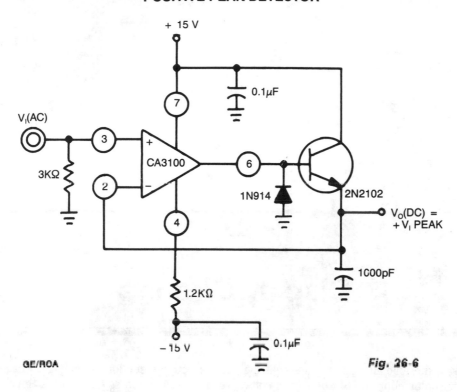

GE/RCA

Fig. 26-6

This peak detector uses a CA3100 BiMOS op amp as a wide-band noninverting amplifier to provide essentially constant gain for a wide range of input frequencies. The IN914 clips the negative half of V_{IN} $(R4)/(R3)$ $(R5)$. A 500-μA load current is constant for all load values and the output reflects only positive input peaks.

27

Display Circuits

The sources of the following circuits are contained in the Sources section beginning on page 782. The figure number contained in the box of each circuit correlates to the sources entry in the Sources section.

Two-Variable LED Matrix Display

TWO-VARIABLE LED MATRIX DISPLAY

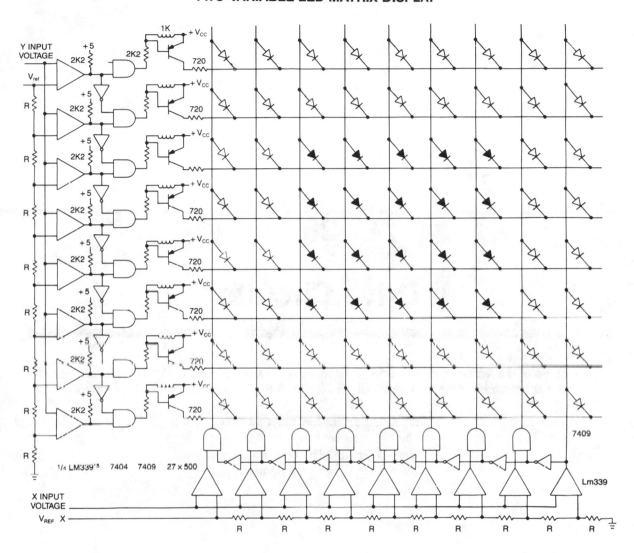

Fig. 27-1

This matrix can show the values of two variables, for example, frequency and voltage. The display is a graph made from a matrix of LEDs. The LEDs on each axis are color coded, red for *out of tolerance* and green for *in*, forming a red band around the inner green rectangle. The two input voltages proportional to the functions being measured are presented to the two columns of comparators. The other comparator input is a reference voltage derived from resistor ladder R1 to R_x. The output of each row of comparators is processed with an inverted and an AND gate to allow only one active output for any input value. The LED at the intersection of the active drives shows the relationship of the two inputs. The advantage of this display is the ease in reading, modification, and also its small size. All comparators are LM339 quads.

28

Drive Circuits

The sources of the following circuits are contained in the Sources section beginning on page 782. The figure number contained in the box of each circuit correlates to the sources entry in the Sources section.

Practical Current-Limiting Coil Driver
Line-Synchronized Driver
Low-Power RS-232C Driver
Totem-Pole Driver with Bootstrapping

PRACTICAL CURRENT-LIMITING COIL DRIVER

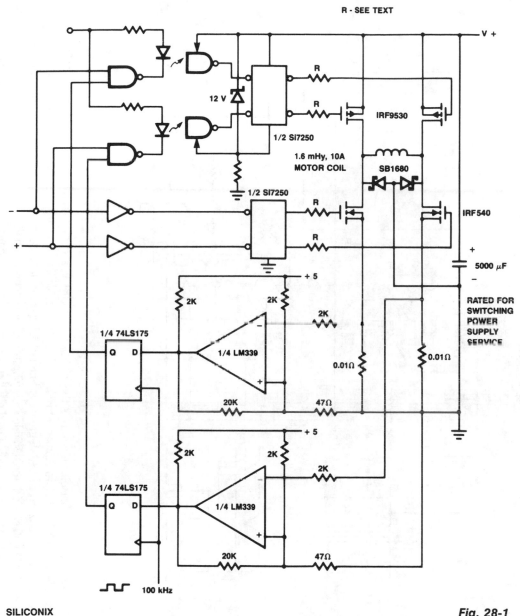

Fig. 28-1

The p-channel devices are switched off by current sensors when the coil current reaches 10 A. The operation is similar to that of a switching-type power supply. The Schottky diodes and resistors are for spike protection.

LINE-SYNCHRONIZED DRIVER

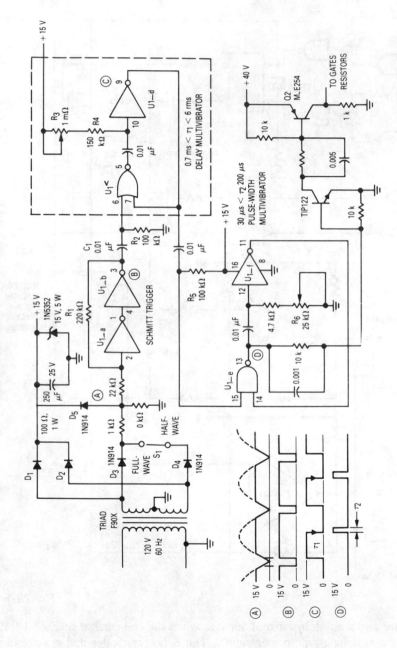

Fig. 28-2

The gate drive that phase controls the four parallel SCRs is accomplished with complementary MOS hex gate MC14572 and two bipolar transistors. This adjustable line-synchronized driver permits SCR conduction from near zero to 180 degrees. A Schmitt trigger clocks a delay monostable multivibrator that is followed by a pulse-width monostable multivibrator. Line synchronization is achieved through the half-wave section of the secondary winding of the full-wave, center-tapped transformer. This winding also supplies power to the circuit through rectifiers D1 and D2.

LOW-POWER RS-232C DRIVER

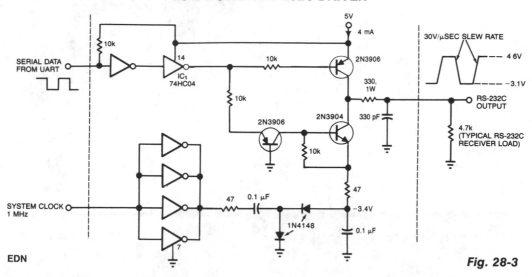

EDN

Fig. 28-3

This circuit draws only 4 mA from a 5-V supply while driving a standard RS-232C receiver. The system clock drives a dc-dc converter that produces −3.4 V. The frequency can range from 0.5 to 8 MHz, but a range of 0.5 to 1 MHz will minimize power dissipation. The circuit output withstands direct shorts to ground or to either of the supplies (±12 V max). In place of the 74HC04 high-speed CMOS driver shown, you may want to substitute miscellaneous spare gates; one noninverting buffer, for example, can replace the two inverting types that receive the UART signal.

TOTEM-POLE DRIVER WITH BOOTSTRAPPING

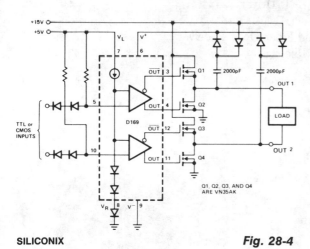

SILICONIX

Fig. 28-4

When driving MOSPOWER in a totem-pole output configuration, it is necessary to have the gate voltage 10 to 15 V positive with respect to the source in order to handle load currents near the MOSPOWER maximum ratings. The D169 lends itself to bootstrapping because of its high-voltage ratings. In the circuit shown, the voltage on the 2000-pF bootstrap capacitors is applied via diode OR gates to the V+ terminal. Therefore, regardless of which output is high, 30 V is present at V+. Maximum switching frequency is determined by the input capacitance of the MOSPOWER transistors used.

175

29

Fiber Optics Circuits

The sources of the following circuits are contained in the Sources section beginning on page 782. The figure number contained in the box of each circuit correlates to the sources entry in the Sources section.

Fiber Optic Transmitter
Digital Fiber Optic Receiver
50-Mb/s Fiber Optic LED Driver
Fiber Optic Link
Low-Cost 100-M Baud Fiber Optic Receiver
50-Mb/s Fiber Optic Receiver

FIBER OPTIC TRANSMITTER

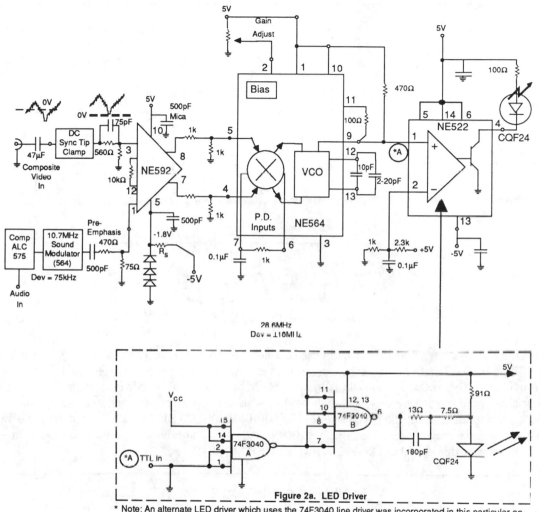

Figure 2a. LED Driver

* Note: An alternate LED driver which uses the 74F3040 line driver was incorporated in this particular application example. The 74F3040 has a higher current rating, but not the variable threshold capabilities of the NE522. The LED diode is operated in the saturated on-off mode for best signal to noise.

SIGNETICS

Fig. 29-1

This receiver circuit consists of wideband differential amplifier NE592, VCO NE564 and LED driver NE522—the high-speed comparator. The video signal is ac coupled into the modulator preamplifier and followed by a sync tip clamp to provide dc restoration on the composite video signal and to prevent variation of modulation deviation with varying picture content. A video signal level of 250 to 300 mV peak is required to maintain optimum picture modulation. Frequency compensation (preemphasis) is inserted in the form of a passive rc lead network at the input to the NE592 differential amplifier. The main FM modulator consists of an NE564 used only as a linear wideband VCO, but the other sections of the device are not used. Differential dc coupling to the VCO terminals is attained via the loop filter terminals, pins 4 and 5.

DIGITAL FIBER OPTIC RECEIVER

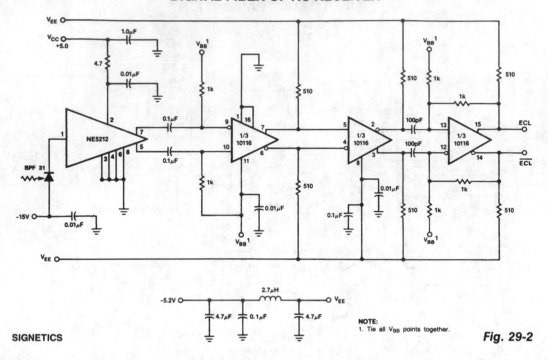

SIGNETICS

NOTE:
1. Tie all V_{BB} points together.

Fig. 29-2

This receiver uses the NE5212, the Signetics 10116 ECL line receiver, and the Phillips/Amperex BPF31 pin diode. The circuit is a capacitor-coupled receiver and utilizes positive feedback in the last stage to provide the hysteresis. The amount of hysteresis can be tailored to the individual application by changing the values of the feedback resistors to maintain the desired balance between noise immunity and sensitivity. At room temperature, the circuit operates at 50-M baud with a BER of 10E-10 and over the automotive temperature range at 40-M baud with a BER of 10E-9. Higher speed experimental diodes have been used to operate this circuit at 220-M baud with a BER of 10E-10.

50-Mb/s FIBER OPTIC LED DRIVER

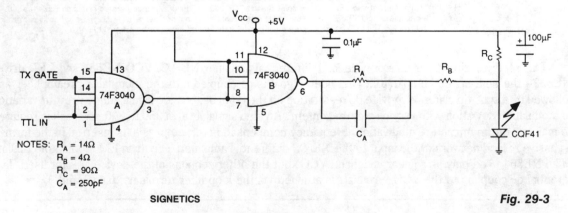

NOTES: $R_A = 14\Omega$
$R_B = 4\Omega$
$R_C = 90\Omega$
$C_A = 250pF$

SIGNETICS

Fig. 29-3

50-Mb/s FIBER OPTIC LED DRIVER (*Cont.*)

The pull-up transistor of the totem-pole output is used to turn on the LED and the pull-down transistor is used to turn off the LED. The lower impedance and higher current handling capability of the saturated pull-down transistor is used as an effective method of transferring the charge from the LED's anode to ground as its dynamic resistance increases during turn-off. The slightly higher output impedance of the pull-up stage ensures that the LED is not over peaked during the less difficult turn-on transition. This asymmetric current handling capability of the output stage with its variable impedance substantially reduces the pulse-width distortion and long-tailed response. As the signal propagates through two NAND gates, each transition passes through the high-to-low and low-to-high transition once, normalizing the total propagation delay through the circuit.

FIBER OPTIC LINK

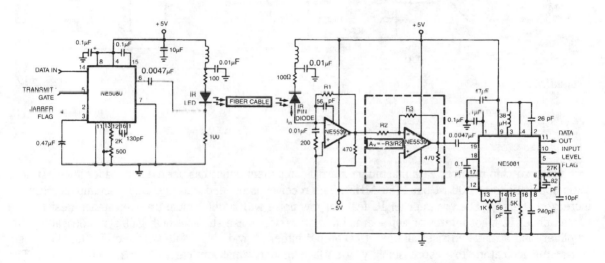

SIGNETICS

Fig. 29-4

The circuit shows a simplex fiber link between the NE5080 transmitter and the NE5081 receiver. The components shown are for a center frequency of 5 MHz, although this frequency can be increased to 20 MHz with proper selection of external component values. The NE5539 has a 530-MHz unity gain bandwidth which could limit maximum operating frequencies in some systems. Since the NE5081 can adequately accept signals below 10 mV at 5-MHz carrier, the gain stage within the dashed lines can be eliminated if the attenuation in the link is low. If the gain stage is used, be mindful of the bandwidth trade-off at higher gains. Refer to the NE5539 data sheet for details.

LOW-COST 100-M BAUD FIBER OPTIC RECEIVER

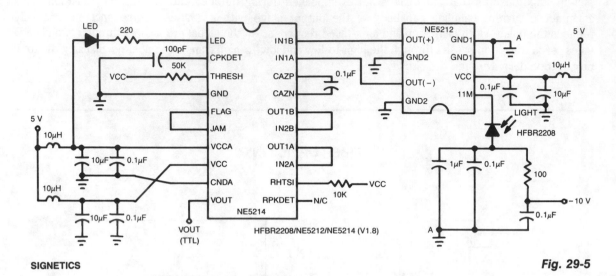

SIGNETICS

Fig. 29-5

This two-chip receiver with minimum external component count has been designed for low-cost fiber optic applications to 100-M baud (50 MHz). The receiver is divided into pre- and postamplifier ICs for increased stability. The preamplifier IC features low noise with a differential transresistance design. The postamplifier IC incorporates an auto-zeroed first stage with noise shaping, high-gain symmetrical-limiting amplifier, and a matched rise/fall time TTL output buffer. A wide-band full-wave rectifier functions as a link-status indicator. To ensure stability, a surface mount, small outline (SO), package is used. The received signal in the − 35 dBm optical (average) to − 9 dBm range is converted into a small unipolar current by the pin diode. The pin diode then feeds its signal current to a preamplifier, such as the NE5212. The preamplifier output is fed to a high-gain limiting amplifier, simply known as the post amp.

The NE5214/NE5217 postamplifiers are low-cost ICs that provide up to 60 dB of gain at 50 MHz to bring mV level signals up to TTL levels. The postamplifier IC incorporates an auto-zeroed first stage with noise shaping, a high-gain symmetrical-limiting amplifier, and a matched rise/fall time TTL output buffer. A secondary amplifier chain functions as a link-status indicator.

50-Mb/s FIBER OPTIC RECEIVER

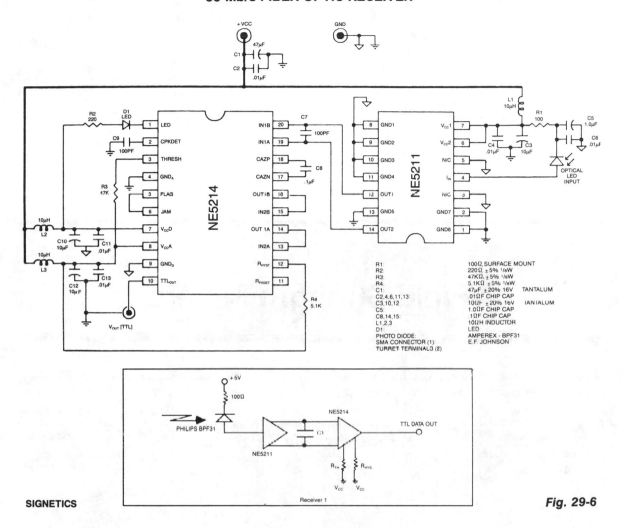

SIGNETICS

Receiver 1

Fig. 29-6

The optical signal is coupled to the pin diode. Current flowing in the diode also flows into the input of the NE5211 preamplifier. The preamplifier is a fixed-gain block that has a 28-KΩ differential transimpedance and does a single-ended to differential conversion. With the signal in differential form, greater noise immunity is assured. The second stage, or postamplifier NE5214, includes a gain block, auto-zero detection, and limiting. The auto-zero circuit allows dc coupling of the preamplifier and the postamplifier and cancels the signal dependent offset because of the optical-to-electrical conversion. The auto-zero capacitor must be 1000 pF or greater for proper operation. The peak detector has an external threshold adjustment, R_{TH}, allowing the system designer to tailor the threshold to the individual's need. Hysteresis included to minimize jitter introduced by the peak detector, and an external resistor, R_{HYS}, is used to set the amount of hysteresis desired. The output stage provides a single-ended TTL data signal with matched rise and fall times to minimize duty-cycle distortion.

30

Field-Strength Meters

The sources of the following circuits are contained in the Sources section beginning on page 782. The figure number contained in the box of each circuit correlates to the sources entry in the Sources section.

Sensitive Field-Strength Meter
Field-Strength Meter

SENSITIVE FIELD-STRENGTH METER

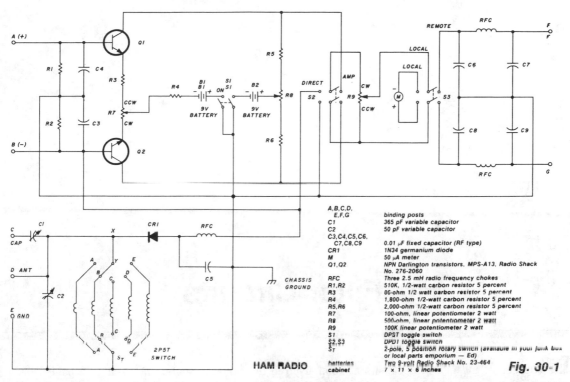

A,B,C,D, E,F,G	binding posts
C1	365 pF variable capacitor
C2	50 pF variable capacitor
C3,C4,C5,C6, C7,C8,C9	0.01 μF fixed capacitor (RF type)
CR1	1N34 germanium diode
M	50 μA meter
Q1,Q2	NPN Darlington transistors, MPS-A13, Radio Shack No. 276-2060
RFC	Three 2.5 mH radio frequency chokes
R1,R2	510K, 1/2-watt carbon resistor 5 percent
R3	86-ohm 1/2 watt carbon resistor 5 percent
R4	1,800-ohm 1/2-watt carbon resistor 5 percent
R5,R6	2,000-ohm 1/2-watt carbon resistor 5 percent
R7	100-ohm, linear potentiometer 2 watt
R8	500-ohm, linear potentiometer 2 watt
R9	100K linear potentiometer 2 watt
S1	DPST toggle switch
S2,S3	DPDT toggle switch
S$_T$	2-pole, 5 position rotary switch (available in your junk box or local parts emporium — Ed)
batteries	Two 9-volt Radio Shack No. 23-464
cabinet	7 x 11 x 6 inches

HAM RADIO   **Fig. 30-1**

The two-pole, five-position switch, coils and 365-pF variable capacitor cover a range from 1.5 to 30 MHz. The amplifier uses Darlington npn transistors whose high beta, 5000, provides high sensitivity with S1 used as the amplifier on/off switch. Switch S2 in the left position allows the output of the 1N34 diode to be fed directly into the 50-μA meter (M) for direct reading. When S2 is in the right position, the amplifier is switched into the circuit. Switch S3 is for local or remote monitoring. At full gain setting, the input signal is adjusted to give a full-scale reading of 50 mA on the meter. Then with the amplifier switched out of the circuit, the meter reading drops down to about 0.5 mA. A 2.5-mH rf choke and capacitors C3, C4, and C5 effectively keep rf out of the amplifier circuit.

FIELD-STRENGTH METER

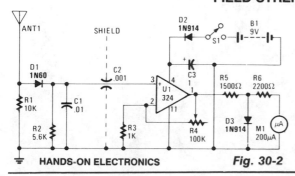

HANDS-ON ELECTRONICS   **Fig. 30-2**

The untuned, but amplified FSM can almost sense that mythical flea's whisper—from 3 through 148 MHz no less—and yet, is so immune to overload that the meter pointer won't pin. The key to the circuit is the amplifier, a 324 quad op amp, of which only one section is used. It's designed for a single-ended power supply, will provide at least 20-dB dc gain, and the output current is self-limiting. The pointer can't be pinned.

31

Filter Circuits

The sources of the following circuits are contained in the Sources section beginning on page 782. The figure number contained in the box of each circuit correlates to the sources entry in the Sources section.

Programmable Active Filters
Biquad Audio Filter
Low-Power Active Filter with Digitally
    Selectable Center Frequency
Glitch-Free Turbo Circuit
Voltage-Controlled Filter
Second-Order Biquad Bandpass Filter
Noisy Signal Filter
State-Variable Active Filter

Scratch Filter
Dynamic Noise Filter
State-Variable Filter with Multiple
    Filtering Outputs
Typical Active Bandpass Filter
Sixth-Order Elliptic High-Pass Filter
Fourth-Order Chebyshev High-Pass Filter
Fourth-Order Chebyshev Bandpass Filter
Rumble Filter

PROGRAMMABLE ACTIVE FILTERS

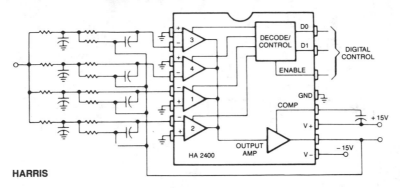

HARRIS

Fig. 31-1

This is a second-order, low-pass filter with programmable cutoff frequency. This circuit should be driven from a low-source impedance since there are paths from the output to the input through the unselected networks. Virtually any filter function which can be constructed with a conventional op amp can be made programmable with the HA-2400.

A useful variation would be to wire one channel as a unity gain amplifier, so that one could select the unfiltered signal, or the same signal filtered in various manners. These could be cascaded to provide a wide variety of programmable filter functions.

BIQUAD AUDIO FILTER

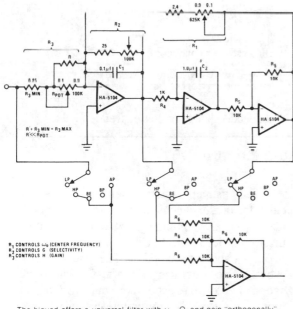

The biquad offers a universal filter with ω_o, Q, and gain "orthogonally" tuned.

HARRIS

Fig. 31-2

This universal filter offers low-pass, high-pass, bandpass, band elimination, and all-pass functions. The Biquad consists of two successive integration stages followed by an inverting stage. The entire group has a feedback loop from the front to the back consisting of R1 which is chiefly responsible for controlling the center frequency, ω_o. The first stage of integration is a *poor* integrator because R2 limits the range of integration. R2 and C form the time constant of the first stage integrator with R3 influencing gain H almost directly. The band-pass function is taken after the first stage with the low-pass function taken after the third stage. The remaining filter operations are generated by various combinations of three stages.

The Biquad is orthogonally tuned, meaning that ω_o, Q, and gain H can all be independently adjusted. The component values known will allow ω_o to range from 40 Hz to 20 kHz. The other component values give an adequate range of operation to allow for virtually universal filtering in the audio region. ω_o, Q, and gain H can all be independently adjusted by tuning R1 through R3 in succession.

LOW-POWER ACTIVE FILTER WITH DIGITALLY SELECTABLE CENTER FREQUENCY

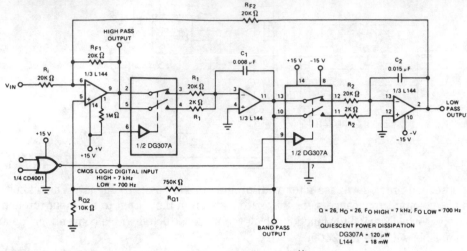

Table 1
Design Procedure for the State Variable Active Filter
Given: f_0 (Resonant Frequency),
H_0 (Gain at the Resonant Frequency) and Q_0

STANDARD DESIGN
(Assumes Infinte Op-Amp Gain)

1. CHOOSE $C_1 = C_2 = C$, A CONVENIENT VALUE

2. LET $R_1 = R_2 = R$

3. THEN $R = \dfrac{1}{2\pi \times f_o \times C}$

4. CHOOSE $R_{11} = R_{12} = KR$,
   WHERE R_{11}, R_{12} = A CONVENIENT VALUE

AND $K = \dfrac{H_o}{Q_o}$

IF H_o IS UNIMPORTANT (i.e., GAIN CAN BE ADDED BEFORE AND/OR AFTER THE FILTER), CHOOSE K = 1

5. LET R_{Q1} = A CONVENIENT VALUE

6. THEN $R_{Q2} = \dfrac{R_{Q1}}{(2 + K) \times Q_o - 1}$

$A (f_o)$ = THE NOMINAL OP AMP GAIN AT THE RESONANT FREQUENCY.

GBWP = THE NOMINAL GAIN-BANDWIDTH PRODUCT OF THE OPERATIONAL AMPLIFIER

SILICONIX

Fig. 31-3

The switchable center frequency active filter allows a decade change in center frequency.

GLITCH-FREE TURBO CIRCUIT

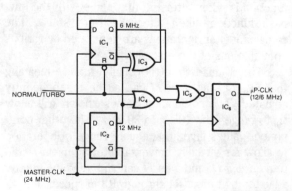

EDN

Fig. 31-4

This simple circuit generates a dual-speed clock for personal computers. The circuit synchronizes your asynchronous switch inputs with the master clock to provide glitch-free transitions from one clock speed to the other. The dual-speed clock allows some programs to run at the higher clock speed in order to execute more quickly. Other programs—for example, programs that use loops for timing—can still run at the lower speed as necessary. The circuit will work with any master-clock frequency that meets the flip-flops' minimum-pulse-width specs.

The two D two flip-flops, IC1 and IC2, and an XOR gate, IC3, form a binary divider that develops the 6- and 12-MHz clocks. When the NT signal is low, the reset pin forces the 6-MHz output low. On the other hand, when the NT signal is high, IC3 blocks the 12-MHz output. Therefore, only one of the two clock signals passes through IC3 and gets clocked into IC6. Because the master-clk signal clocks IC6, asynchronous switching of the NT signal can't generate an output pulse shorter than 41 μs ($1/24$ MHz). Also, the synchronization eliminates glitches.

VOLTAGE-CONTROLLED FILTER

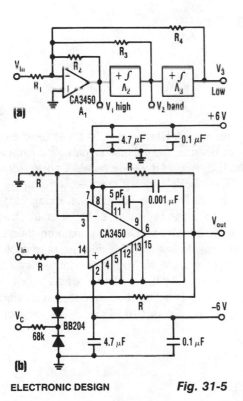

(a)

(b)

ELECTRONIC DESIGN   *Fig. 31-5*

The control voltage V_C easily sets the cutoff frequency ω_o of this state-variable filter to any desired value, from about 1.7 MHz up to 5 MHz, with a BB 204 varicap and $R - 100$ KΩ. V_C can range from 0 to 28 V. This range changes the capacitance of the varicap from about 4 to 12 pF.

The circuit consists of input summing circuit A1 and two noninverting integrators, A2 and A3. Both the integrators and the summing-amplifier circuits use CA3450 op amps. With them, cutoff frequencies up to 200 MHz are possible.

The circuit's cutoff frequency, it's Q-factor, and gain G are simply:

$$\omega_o = 2/CR, \quad Q = R_3/R_4,$$
$$\text{and } G = R_4/R_1$$

For a given value for R4, say 10 KΩ, Q depends only upon the resistance of R3. The Q can be any value, even 100, independently of both ω_o and G. Similarly, the gain then depends only on the resistance of R1 and can also be set as high as 100.

SECOND-ORDER BIQUAD BANDPASS FILTER

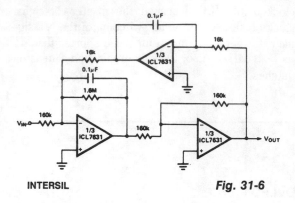

Note that I_Q on each amplifier might be different. $A_{VCL} = 10$, $Q = 100$, $f_o = 100$ Hz.

INTERSIL

Fig. 31-6

NOISY SIGNALS FILTER

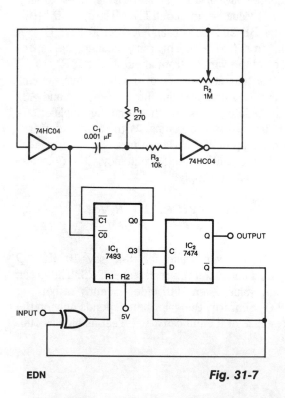

This circuit filters noise, such as glitches and contact bounce, from digital signals. You can easily adjust the circuit for a wide range of noise frequencies. The circuit's output changes state only if the input differs from the output long enough for the counter to count eight cycles. If the input changes before the counter reaches its maximum count, the counter resets without clocking the output of flip-flop, IC2. You use R2 to set the frequency of the two-inverter CMOS oscillator, which clocks the counter. Simply adjust the oscillator such that its period is one-eighth that of the noise you want to eliminate.

EDN

Fig. 31-7

STATE-VARIABLE ACTIVE FILTER

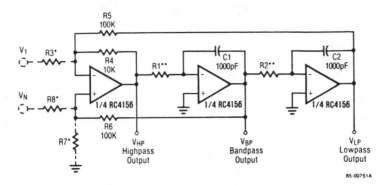

*Input connections are chosen for inverting or
non-inverting response. Values of R3, R7, R8
determine gain and Q.
**Values of R1 and R2 determine natural
frequency.

Reprinted with permission from Raytheon Co., Semiconductor Div.

Fig. 31-8

A generalized circuit diagram of the two-pole state-variable active filter is shown. The state-variable filter can be inverting or noninverting and can simultaneously provide three outputs: low-pass, bandpass, and high-pass. A notch filter can be realized by adding one summing op amp.

In the state-variable filter circuit, one amplifier performs a summing function and the other two act as integrators. The choice of passive component values is arbitrary, but must be consistent with the amplifier operating range and input signal characteristics. The values shown for C1, C2, R4, R5, and R6 are arbitrary. Preselecting their values will simplify the filter tuning procedures, but other values can be used if necessary.

SCRATCH FILTER

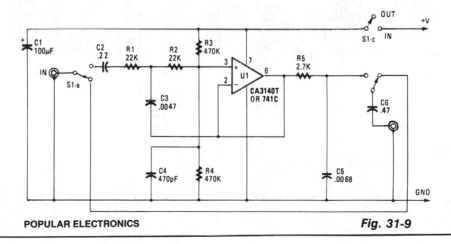

POPULAR ELECTRONICS

Fig. 31-9

DYNAMIC NOISE FILTER

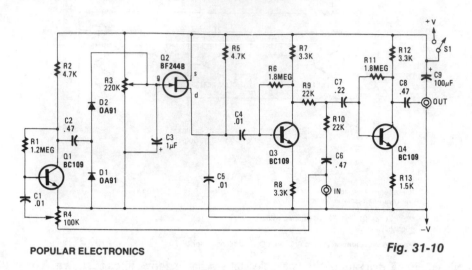

Fig. 31-10

STATE-VARIABLE FILTER WITH MULTIPLE FILTERING OUTPUTS

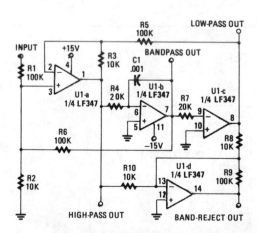

Fig. 31-11

TYPICAL ACTIVE BANDPASS FILTER

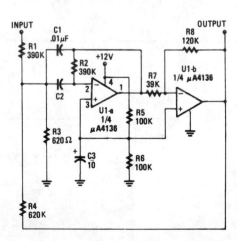

Fig. 31-12

SIXTH-ORDER ELLIPTIC HIGH-PASS FILTER

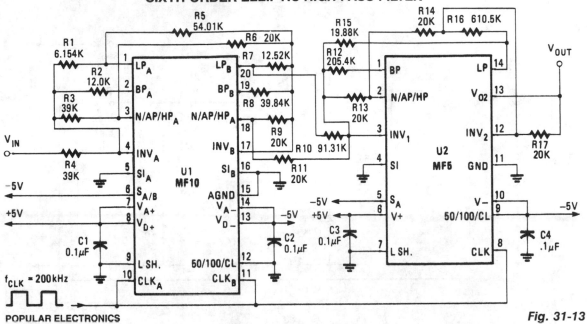

POPULAR ELECTRONICS

Fig. 31-13

FOURTH-ORDER CHEBYSHEV HIGH-PASS FILTER

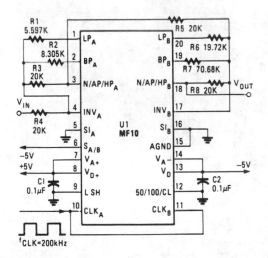

POPULAR ELECTRONICS

Fig. 31-14

FOURTH-ORDER CHEBYSHEV BANDPASS FILTER

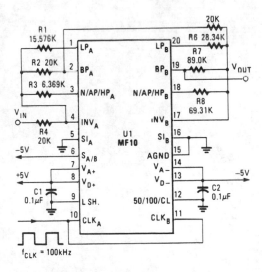

POPULAR ELECTRONICS

Fig. 31-15

RUMBLE FILTER

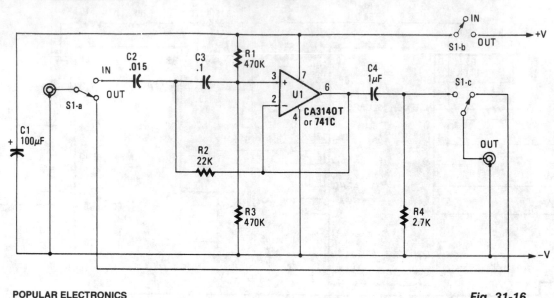

Fig. 31-16

32

Flashers and Blinkers

The sources of the following circuits are contained in the Sources section beginning on page 782. The figure number contained in the box of each circuit correlates to the sources entry in the Sources section.

Ring-Around LED Flasher
Three-Year LED Flasher
SCR Ring Counter
Astable Multivibrator
Ac Flasher
Single-Lamp Flasher
SCR Chaser
SCR Flasher
Incandescent Light Flasher

Five-Lamp Neon Flasher
Alternating LED Flasher
CMOS Flasher
60-W Flashing Light
Two-State Neon Oscillator
Alternating LED Flasher
Transistor Flasher
Minimum Component Flasher
Lamp Flasher

RING-AROUND LED FLASHER

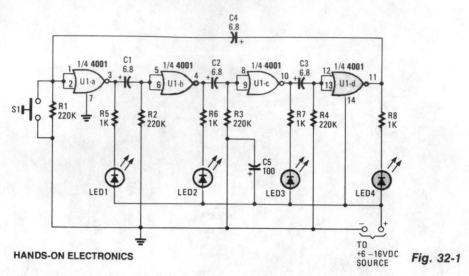

HANDS-ON ELECTRONICS

Fig. 32-1

When power is first turned on, two of the LEDs are on and the other two remain off until the timing cycle reverses. The LEDs flash in pairs, but by pressing and holding S1 closed until only one of the LEDs is on, and then releasing it, the four LEDs can be made to flash in sequential order. The number of LEDs flashing in a sequential ring can be easily increased to eight by adding another 4001 quad NOR gate. Just repeat the circuit and connect the additional circuit in series with the first—input to output—as an extension of the first circuit. When power is connected to the eight-LED flasher circuit, four LEDs will turn on at once and then flash off as the four remaining LEDs come on. As before, just press S1 and hold closed until all but one LED turns off; then the LEDs will begin their sequential march in a circle. You can connect as many circuits in series as you like.

THREE-YEAR LED FLASHER

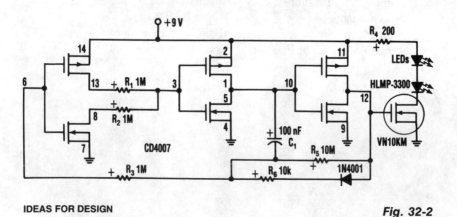

IDEAS FOR DESIGN

Fig. 32-2

THREE-YEAR LED FLASHER (*Cont.*)

Inserting two 1-MΩ resistors, R1 and R2, in the output stage of one of the circuit's inverters limits the current needed by the oscillator to no more than a few μA. This circuit includes a CD4007 package, which has three CMOS inverters. It forms a standard three-inverter oscillator. Resistors R1 and R2, in series with separate drains on inverter pins 8 and 13, limit the oscillator's supply current. Capacitor C1 and resistors R5 set the off time of the oscillator, C1; R6 sets the on time. A VN10KM small-power FET, current-limited by R4, drives two HLMP-3300 LEDs. The LEDs consume about 20 mA for 1 ms. Their average current determines battery life. Since the LEDs in the circuit flash at 1 Hz, the average current drain is about 1/1000 of 20 mA, or 20 μA. A 9-V battery should last about three years at the current drain—essentially the shelf life of an alkaline battery.

SCR RING COUNTER

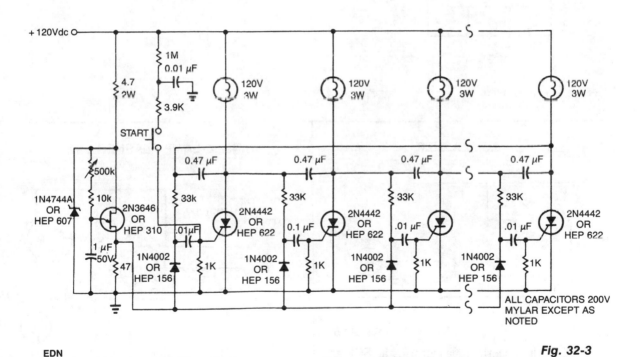

EDN

Fig. 32-3

One lamp at a time is lit in the string to give the appearance of a moving point of light.

ASTABLE MULTIVIBRATOR

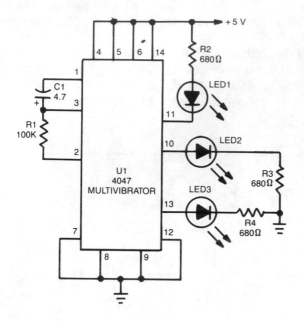

The 4047 is configured as a free-running, astable-multivibrator (oscillator) circuit. That configuration, offers three different outputs. The output pulses at the Q and $\overline{Q}$ output (pins 10 and 11, respectively) are the same as in the previous two circuits. The third output at pin 13 pulses twice as often as the outputs at 10 and 11. So, the circuit can be used to simultaneously provide both positive- and negative-trigger signals since the Q and $\overline{Q}$ output are never in the same state, and a clock frequency. Thus, the 4047 can replace both a simple oscillator (the 555, for instance) and a flip-flop in some applications.

POPULAR ELECTRONICS   Fig. 32-4

AC FLASHER

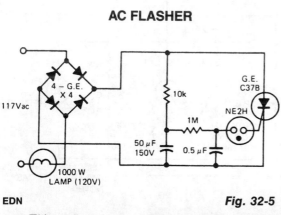

EDN   Fig. 32-5

This ac line-operated flasher uses an SCR and is capable of flashing a large lamp. Flashing rate is determined by the 10-KΩ resistor and the 50-mF capacitor. Increasing or decreasing the value of the capacitor has a corresponding effect on the flash rate.

SINGLE-LAMP FLASHER

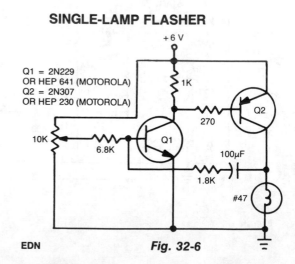

EDN   Fig. 32-6

The flash rate is controlled by a complementary multivibrator consisting of an npn and a pnp transistor.

SCR CHASER

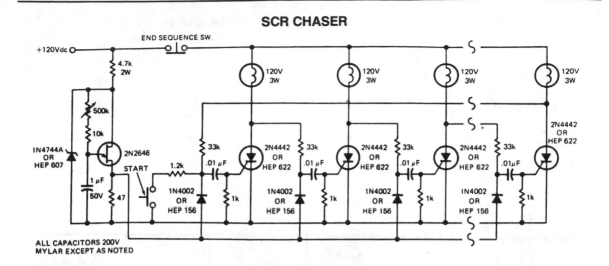

EDN

Fig. 32-7

Each lamp lights in succession to give the appearance of a growing column.

SCR FLASHER

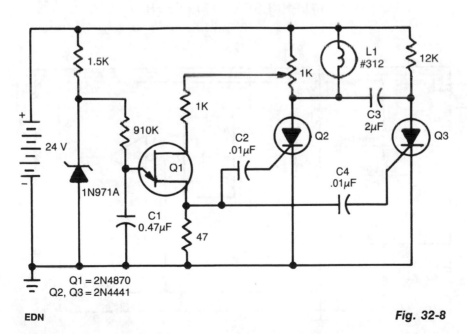

EDN

Fig. 32-8

This dc flasher uses two SCRs and a unijunction oscillator clock to set the flash rate, which can be varied by changing the value of C1.

197

INCANDESCENT LIGHT FLASHER

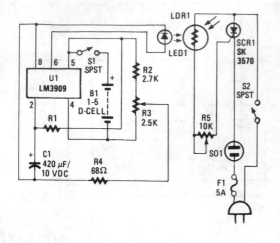

POPULAR ELECTRONICS

Fig. 32-9

FIVE-LAMP NEON FLASHER

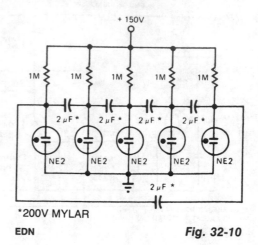

*200V MYLAR

EDN

Fig. 32-10

In this circuit, the number of lamps can be increased almost without limit.

ALTERNATING LED FLASHER

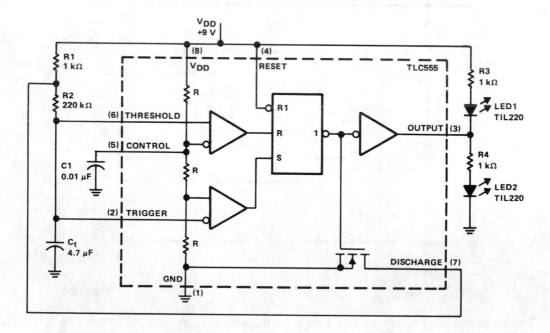

Fig. 32-11

ALTERNATING LED FLASHER (*Cont.*)

The timing components are R1, R2, and C_t. C1 is a bypass capacitor used to reduce the effects of noise. At start-up, the voltage across C_t is less than the trigger level voltage ($1/3\ V_{DD}$), causing the timer to be triggered via pin 2. The output of the timer at pin 3 increases, turning LED1 off, LED2 on, the discharge transistor at pin 7 off, and allowing C_t to charge through resistors R1 and R2. When capacitor C_t charges to the upper threshold voltage ($2/3\ V_{DD}$), the flip-flop is reset and the output at pin 3 decreases. LED1 is turned on, LED2 is turned off, and capacitor C_t discharges through resistor R2 and the discharge transistor. When the voltage at pin 2 reaches $1/3\ V_{DD}$, the lower threshold or trigger level, the timer triggers again and the cycle is repeated.

The totem-pole output at pin 3 is a square wave with a duty cycle of about 50%. The output alternately turns on each LED at slightly less than one blink per second. If the unit is battery operated, the TLC555 uses minimum current to produce this function. With a 9-V battery, the circuit draws 5 mA (no load) and 15 mA when turning on an LED. Most of the on current is for the LED.

CMOS FLASHER

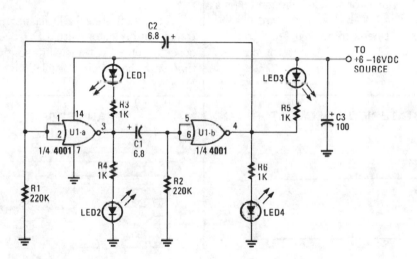

HANDS-ON ELECTRONICS

Fig. 32-12

Uses a low cost CMOS IC to turn four LEDs on and off at a rate that is set by the values of R1, R2, C1, and C2. The pulse rate for the component values given for R1 and R2 is about one cycle every four seconds. By lowering the values of R1 and R2 to 220 KΩ, the pulse rate increases to 1 Hz. The LEDs flash in pairs, with LED1 and LED4 turning together for one half of the time period, while LED2 and LED3 are on for the other half. The on/off duration of each pair of LEDs can be increased or decreased by changing the value of one of coupling capacitors C1 or C2. Increasing either capacitor's value by a factor of 10 will also increase the ON time of a pair of the LEDs for about the same factor.

60-W FLASHING LIGHT

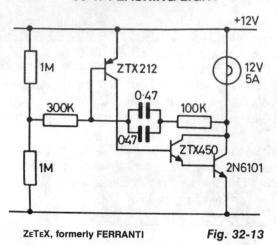

ZᴇTᴇX, formerly FERRANTI   *Fig. 32-13*

The 2N6101 transistor should be mounted on a small heatsink. The 300-KΩ resistor controls the off period and might need to be adjusted if transistor gains are high. The 100-KΩ resistor controls the on period.

ALTERNATING LED FLASHER

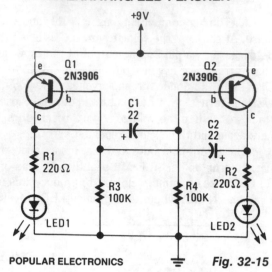

POPULAR ELECTRONICS   *Fig. 32-15*

The alternating LED flasher is simply a two-transistor oscillator with LEDs connected to the collector of each transistor, so that they light in time with the circuit's oscillations.

TWO-STATE NEON OSCILLATOR

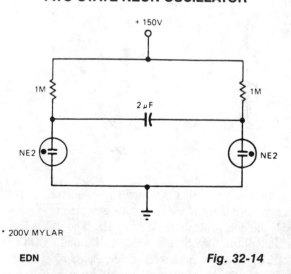

* 200V MYLAR

EDN   *Fig. 32-14*

The number of lamps is easily increased in this oscillator.

TRANSISTOR FLASHER

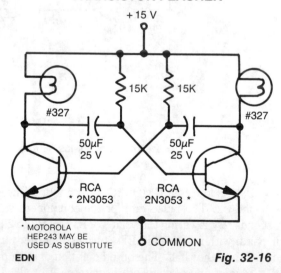

* MOTOROLA
HEP243 MAY BE
USED AS SUBSTITUTE

EDN   *Fig. 32-16*

This astable multivibrator uses incandescent lamps in place of collector load resistors. The lamps flash on and off alternately.

MINIMUM COMPONENT FLASHER

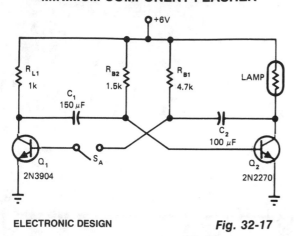

ELECTRONIC DESIGN

Fig. 32-17

LAMP FLASHER

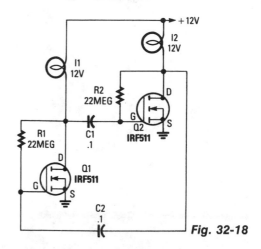

Fig. 32-18

POPULAR ELECTRONICS/HANDS-ON ELECTRONICS

Opening S_A, changes the indicator lamp from flashing to steady-lit condition. The 6-V incandescent lamp on the collector of Q2 requires about 0.3 A. A 1-KΩ load resistor limits Q1's collector current to about 6 mA. The circuit is, therefore, asymmetrical with respect to the on currents of the transistors, allowing use of a much smaller transistor for Q1 than for Q2.

The circuit is built around two power FETs, which are configured as a simple astable multivibrator to alternately switch the two lamps on and off. The rc values given sets the flash rate to about ¹/₃ Hz. By varying either the resistor or capacitor values, almost any flash rate can be obtained. Increase either C1 and C2, or R1 and R2, and the flash rate slows. Decrease them and the rate increases.

33

Flow Detector

The sources of the following circuits are contained in the Sources section beginning on page 782. The figure number contained in the box of each circuit correlates to the sources entry in the Sources section.

Low Flow-Rate Thermal Flowmeter

LOW FLOW-RATE THERMAL FLOWMETER

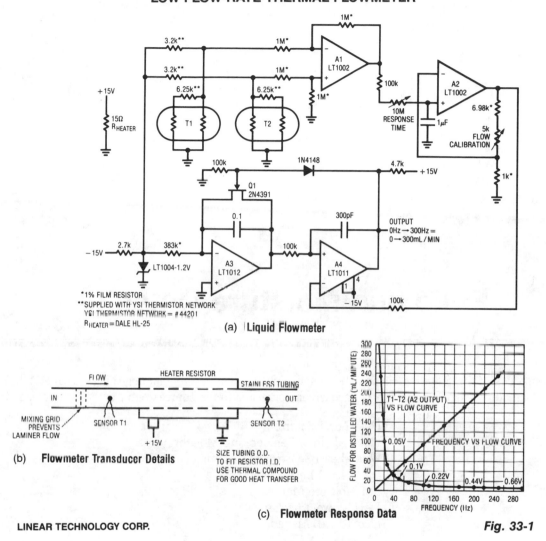

(a) Liquid Flowmeter

*1% FILM RESISTOR
**SUPPLIED WITH YSI THERMISTOR NETWORK
YSI THERMISTOR NETWORK = #44201
R_HEATER = DALE HL-25

(b) Flowmeter Transducer Details

(c) Flowmeter Response Data

LINEAR TECHNOLOGY CORP.

Fig. 33-1

This design measures the differential temperature between two sensors. Sensor T1, located before the heater resistor, assumes the fluid's temperature before it is heated by the resistor. Sensor T2 picks up the temperature rise induced into the fluid by the resistor's heating. The sensor's difference signal appears at Al's output. A2 amplifies this difference with a time constant set by the 10 MΩ adjustment. Fig. 33-1c shows A2's output versus flow rate. The function has an inverse relationship. A3 and A4 linearize this relationship, while simultaneously providing a frequency output. A3 functions as an integrator that is biased from the LT1004 and the 338-KΩ input resistor. Its output is compared to A2's output at A4. Large inputs from A2 force the integrator to run for a long time before A4 can increase, turning on Q1 and resetting A3. For small inputs from A2, A3 does not have to integrate long before resetting action occurs. Thus, the configuration oscillates at a frequency which is inversely proportional to A2's output voltage. Since this voltage is inversely related to flow rate, the oscillation frequency linearly corresponds to flow rate.

34

Fluid and Moisture Detectors

The sources of the following circuits are contained in the Sources section beginning on page 782. The figure number contained in the box of each circuit correlates to the sources entry in the Sources section.

Fluid-Level Control
Flood Alarm or Temperature Monitor
Water-Level Sensor and Control
Dual Liquid-Level Detector
Soil Moisture Meter
Liquid-Level Checker
Liquid-Level Monitor

FLUID-LEVEL CONTROL

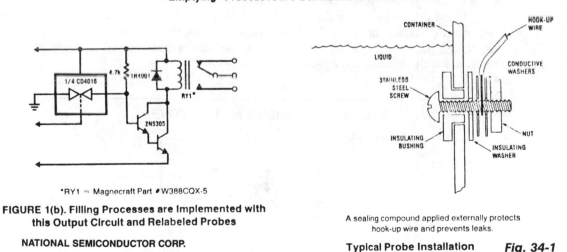

"Emptying" Processes are Controlled with this Circuit

*RY1 = Magnecraft Part #W388CQX-5

**FIGURE 1(b). Filling Processes are Implemented with
this Output Circuit and Relabeled Probes**

NATIONAL SEMICONDUCTOR CORP.

A sealing compound applied externally protects
hook-up wire and prevents leaks.

Typical Probe Installation   *Fig. 34-1*

This circuit is designed to detect the presence or absence of aqueous fluids. An ac signal generated on-chip is passed through two probes within the fluid. A detector determines the presence of the fluid by using the probes in a voltage divider circuit and measuring the signal level across the probes. An ac signal is used to prevent plating or dissolving of the probes as occurs when a dc signal is used. A pin is available for connecting an external resistance in cases where the fluid impedance is not compatible with the internal 13-KΩ divider resistance.

FLOOD ALARM OR TEMPERATURE MONITOR

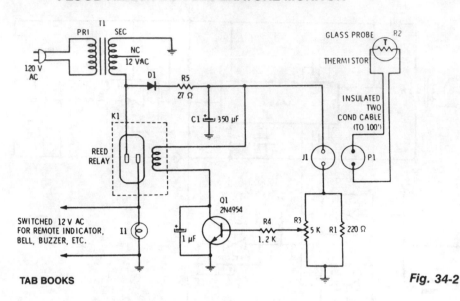

TAB BOOKS

Fig. 34-2

Filtered 15 Vdc is applied to a series circuit consisting of thermistor R2 and parallel combination of resistors R1 and R3. Transistor Q1 acts as a switch whose state is determined by the setting of potentiometer R3, which is first set so just enough current flows into the base to switch on when the thermistor is in contact with air. When the resistance of the thermistor decreases, the voltage at the base of Q1 rises. When the base current reaches the preset level, the transistor conducts and passes current through the reed relay coil, closing the reed relay contacts. Current at the base of transistor Q1 is determined by the environment into which the termistor is inserted.

WATER-LEVEL SENSOR AND CONTROL

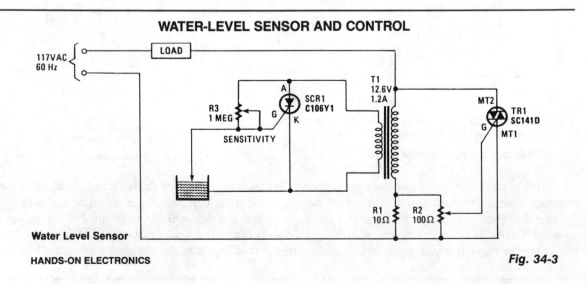

Water Level Sensor

HANDS-ON ELECTRONICS

Fig. 34-3

WATER-LEVEL SENSOR AND CONTROL (*CONT.*)

When the water level is low, the probe is out of the water and SCR1 is triggered on. It conducts and imposes a heavy load on transformer T1's secondary winding. That load is reflected back into the primary, gating triac TR1 on, which energizes the load. If the load is an electric valve in the water-supply line, it will open and remain open until the water rises and touches the probe; this shorts SCR1's gate and cathode, thereby turning off the SCR1, which effectively open-circuits the secondary. That open-circuit condition, when reflected back to the primary winding, removes the triac's trigger signal, thereby turning the water off.

DUAL LIQUID-LEVEL DETECTOR

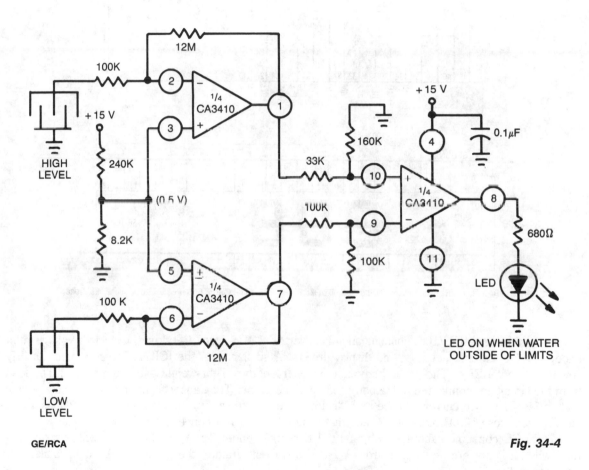

GE/RCA

Fig. 34-4

Uses CA3410 quad BiMOS op amp to sense small currents. Because the op amp's input current is low, a current of only 1 μA passing through the sensor will change the converter's output by as much as 10 to 12 V.

207

SOIL MOISTURE METER

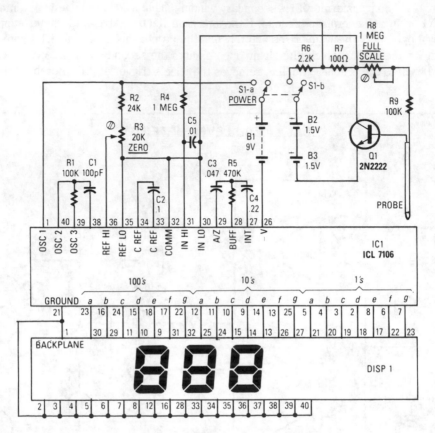

Fig. 34-5

IC1, an Intersil ICL7106, contains an a/d converter, a 3½-digit LCD driver, a clock, a voltage reference, seven segment decoders, and display drivers. A similar part, the ICL7107, can be used to drive seven segment LEDs. The probe body is a five-inch length of light-weight aluminum tubing. The leads from the circuit are connected to the body and tip of the probe. The sensor functions as a variable resistor that varies Q1's base current, hence its collector current. The varying collector current produces a varying voltage across 100 Ω resistor R7, and that voltage is what IC1 converts for display.

The LCD consumes about 25 µA, and IC1 consumes under 2 mA, so the circuit will run for a long time when it is powered by a standard 9-V battery. Current drain of the two 1.5-V AA cells is also very low: under 300 µA.

To calibrate, rotate R3 to the center of its range. Then place the end of the probe into a glass of water and adjust R8 for a reading of 100. When you remove the probe from the water, the LCD should indicate 000. You might have to adjust R3 slightly for the display to indicate 000. If so, readjust R8 with the probe immersed. Check for a reading of 000 again with probe out of water.

LIQUID-LEVEL CHECKER

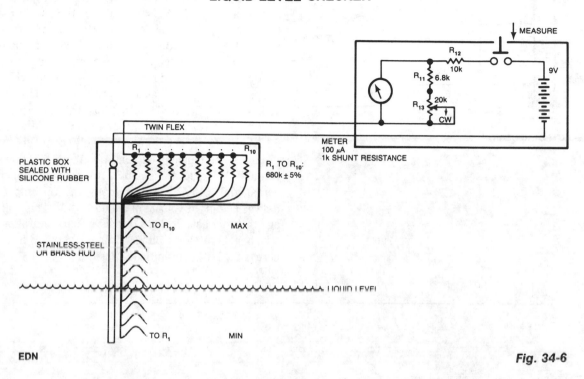

Fig. 34-6

Although many circuits use the varying-capacitance method for checking liquid levels, this simple resistive circuit is much easier to construct. Even a tank of a liquid, such as water, has sufficient conductive salts in solution for this method to work. The probe uses a metal rod that supports 10 insulated wires, which have stripped ends pointing down. As the level of liquid rises, resistors R1 through R10 are successively brought into circuit, each drawing an extra 10 μA through the meter. Shunt resistors R11 and R13 calibrate the meter for a full-scale reading when the tank is full. Resistor R12 limits the current through the meter. If tank isn't rectangular—ie, if the volume of the liquid it contains isn't directly proportional to the liquid's depth—space the resistors accordingly or use a nonlinear progression of resistor values and retain constant resistor spacing.

LIQUID-LEVEL MONITOR

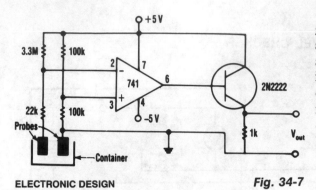

ELECTRONIC DESIGN

Fig. 34-7

This monitor uses a common 741 amp configured as a comparator and a low cost nontransistor as an output driver. With no liquid detected, a voltage of about 2.92 V is present in the op amp's inverting input at pin 2. The 100-KΩ resistors establish a reference voltage of +2.5 V at the non-inverting input at pin 3 of the op amp. Under those conditions, the op amp's output is −3.56 V, which keeps the 2N2222 transistor turned off and the voltage across its 1-KΩ output load resistor at 0 V. When liquid reaches the probes, the 3.3-MΩ and 22-KΩ resistor circuit conductively connects to ground. When enough current, about 1.4 μA, flows through the liquid, the small 30 mV drop developed across the 22-KΩ resistor drives the op amp to deliver an output voltage of about 4.42 V. This voltage then drives a 2N2222 transistor into saturation, which generates a voltage drop of about 3.86 V across its 1-KΩ output load resistor.

35

Followers

The sources of the following circuits are contained in the Sources section beginning on page 782. The figure number contained in the box of each circuit correlates to the sources entry in the Sources section.

High-Frequency Inverting Follower
High-Frequency Noninverting Follower
Simple Follower
Voltage Follower

HIGH-FREQUENCY INVERTING FOLLOWER

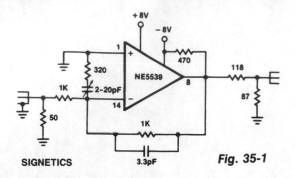

SIGNETICS

Fig. 35-1

HIGH-FREQUENCY NONINVERTING FOLLOWER

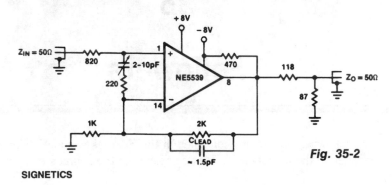

SIGNETICS

Fig. 35-2

SIMPLE FOLLOWER

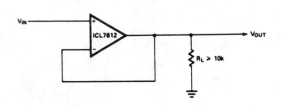

INTERSIL

Fig. 35-3

VOLTAGE FOLLOWER

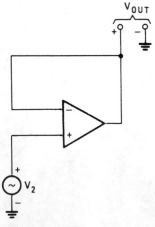

HANDS-ON ELECTRONICS

Fig. 35-4

36

Frequency Multipliers and Dividers

The sources of the following circuits are contained in the Sources section beginning on page 782. The figure number contained in the box of each circuit correlates to the sources entry in the Sources section

Pulse-Width Multiplier
Frequency Doubler
Digital Frequency Doubler
Divide-by-1$^1/_2$ Counter Divider
Odd-Number Counter Divider
Single-Chip Frequency Doubler

PULSE-WIDTH MULTIPLIER

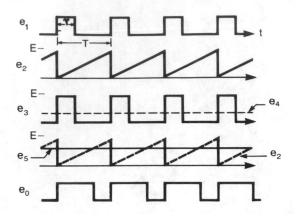

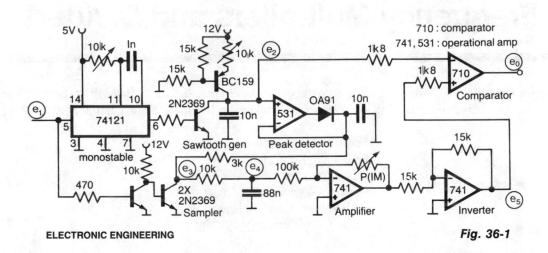

ELECTRONIC ENGINEERING

Fig. 36-1

This circuit for multiplying the width of incoming pulses by a factor greater or less than unity is simple to build and has the feature that the multiplying factor can be selected by adjusting one potentiometer only. The multiplying factor is determined by setting the potentiometer in the feedback of the 741 amplifier. The input pulses e_1, width T and repetition period T is used to trigger a sawtooth generator at its rising edges to produce the waveform e_2 having a peak value of E volts. This peak value is then sampled by the input pulses to generate the pulse train e_3 having an average value of e_4 ($=E_T/T$) which is proportional to T and independent on T. The dc voltage e_4 is amplified by a factor k and compared with sawtooth waveform e_2 giving output pulses of duration k_T. The circuit is capable of operating over the frequency range 10 kHz – 100 kHz. Note that k should be chosen less than T/T to ensure accurate pulse-width multiplication.

214

FREQUENCY DOUBLER

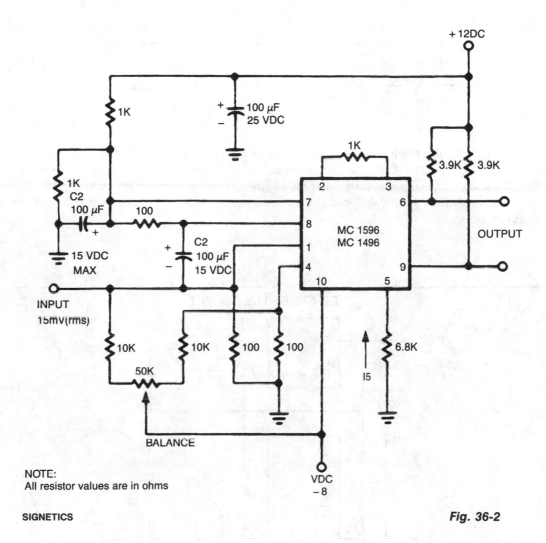

NOTE:
All resistor values are in ohms

SIGNETICS

Fig. 36-2

The output contains the sum component, which is twice the frequency of the input, since both input signals are the same frequency.

DIGITAL FREQUENCY DOUBLER

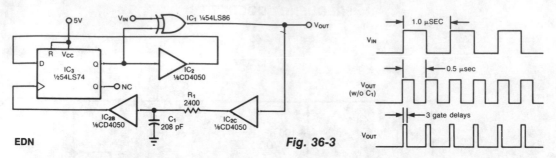

EDN

Fig. 36-3

The circuit doubles the frequency of a digital signal by operating on both signal edges. Each transition causes exclusive-OR gate IC1 to produce a pulse, which clocks flip-flop IC3 after propagating through buffers IC2C and IC2B. If you remove capacitor C1, the circuit produces narrow output pulses. By including C1, you can obtain a desired duty cycle for a given input frequency f_{IN}. The C1 value for an approximate 50% duty cycle is:

$$C1 = \frac{1}{2R_1\,f_{IN}}$$

DIVIDE-BY-1½ CIRCUIT

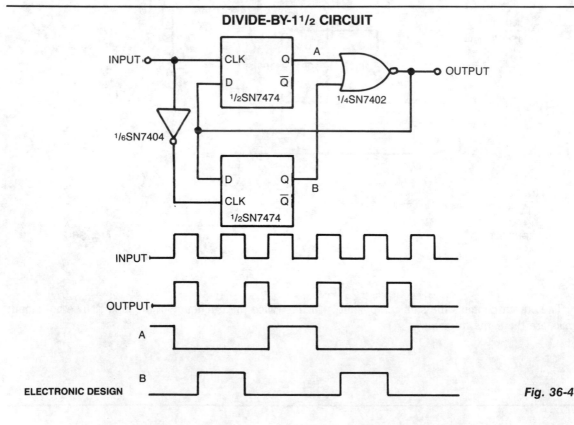

ELECTRONIC DESIGN

Fig. 36-4

An input signal drives both SN7474 D-type flip-flops, which are positive edge-triggered devices. A low-to-high input signal transition triggers the A flip-flop, while a high-to-low input signal transition triggers the B flip-flop via the SN7404 inverter. Either flip-flop in the high state will cause the output to decrease via the SN7402 NOR gate. This in turn disables the opposite flip-flop from going to the high state. The flip-flop in the high state remains there for one clock period, then it is clocked low. With both flip-flops low, the output increases, enabling the opposite flip-flop to be clocked high one-half clock cycle later. This alternate enabling and disabling action of the flip-flops results in a divide-by-1½ function. That is, three clock pulses in, produce two evenly spaced clock pulses out. The circuit has no lock-up states and no inherent glitches. Replacing the NOR gate with an SN7400 NAND gate inverts the A, B, and output signals. By adding simple binary or BCD counters, counting chains, such as divide-by-3, -6, -12, -24, -15, -30, etc., can be generated using the divide-by-1½ circuit as a basis.

ODD-NUMBER COUNTER DIVIDER

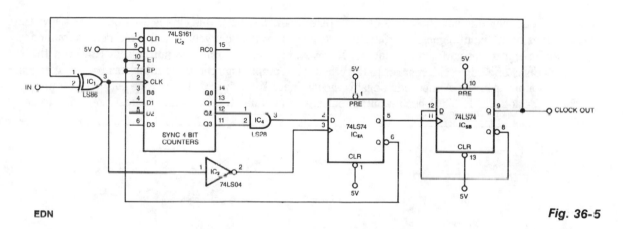

EDN

Fig. 36-5

This circuit, shown symmetrically, divides an input by virtually any odd number. The circuit counts $n + \frac{1}{2}$ clocks twice to achieve the desired divisor. By selecting the proper n, which is the decoded output of the LS161 counter, you can obtain divisors from 3 to 31. The circuit, as shown, divides by 25; you can obtain higher divisors by cascading additional LS161 counters. The counter and IC5A form the $n + \frac{1}{2}$ counter. Once the counter reaches the decoded count, n, IC5A ticks off an additional $\frac{1}{2}$ clock, which clears the counter and puts it in hold. Additionally, IC5A clocks IC5B, which changes the clock phasing through the XOR gate, IC1. The next edge of the input clocks IC5A, which reenables the counter to start counting for an additional $n + \frac{1}{2}$ cycles. Although the circuit has been tested at 16 MHz, a worst-case timing analysis reveals that the maximum input frequency is between 7 and 8 MHz.

SINGLE-CHIP FREQUENCY DOUBLER

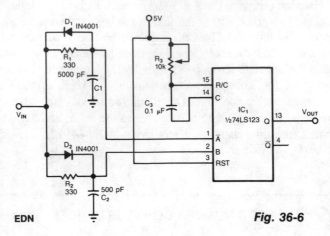

EDN

Fig. 36-6

The frequency doubler uses only one IC. Like other doublers, this circuit uses both the rising and falling edges of the input signals to produce digital pulses, thus effectively doubling the input's frequency.

Without the rc networks at IC1 inputs, IC1 would not produce any output pulses. However, the rc networks delay one edge with respect to the other. The A input lags the B input for positive-going edges, and the B input lags the A input for negative-going ones. You can vary the output duty cycle from 0 to 100% by varying R3. IC1's minimum output pulse width defines the maximum frequency of this circuit.

37

Frequency-to-Voltage Converters

The sources of the following circuits are contained in the Sources section beginning on page 782. The figure number contained in the box of each circuit correlates to the sources entry in the Sources section.

FREQUENCY-TO-VOLTAGE CONVERTER

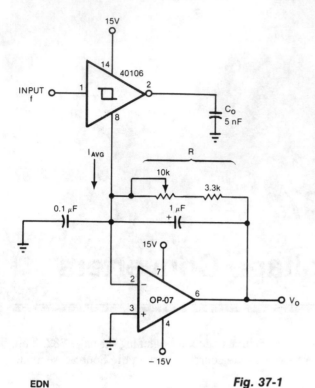

Six components can configure a circuit whose output voltage is proportional to its input frequency. The average current (I_{AVG} from the 40106 Schmitt trigger inverter's ground pin 8 is linearly dependent on the frequency at which $C0$ is discharged into the op amp's summing junction. The op amp forces this current to flow through the 13.33-KΩ feedback resistor, producing a corresponding voltage drop. This frequency-to-voltage converter yields 0 to -10 V output with 0 to 10 kHz input frequencies.

EDN

Fig. 37-1

38

Function Generators

The sources of the following circuits are contained in the Sources section beginning on page 782. The figure number contained in the box of each circuit correlates to the sources entry in the Sources section.

Precision One Shot
Linear Triangle-Wave Timer
Four-Output Waveform Generator
Function Generator
Classic Op Amp Astable Multivibrator
Programmable Triangle-/Square-Wave
   Generator
Noninteger Programmable Pulse Divider
XOR Gate Complementary Signal Generator
Low-Cost FSK Generator
Harmonics Generator
Low-Frequency FM Generators
Positive-Triggered Monostable
Precision Audio Waveform Generator
Monostable Multivibrator
Versatile 2ϕ Pulse Generator
Fixed-Frequency Generator
Single-Supply Multivibrator
Easily Tuned Sine-/Square-Wave Oscillator
Astable

Two-Function Signal Generator
Triangle Generator
Monostable Operator
Programmable-Frequency Free-Running
   Multivibrator
Function Generator
Programmable-Frequency Astable
Linear-Ramp Monostable
Low-Frequency Multivibrator
Retriggerable One Shot
Astable Multivibrator
Single-Control Function Generator
Triangle-/Square-Wave Generator
Variable Duty Cycle Timer
Basic Function Generator
Wide-Range Tunable Function Generator
Sawtooth and Pulse Generator
Precise Triangle/Square-Wave Generator
Wide-Range Triangle/Square-Wave Generator

PRECISION ONE SHOT

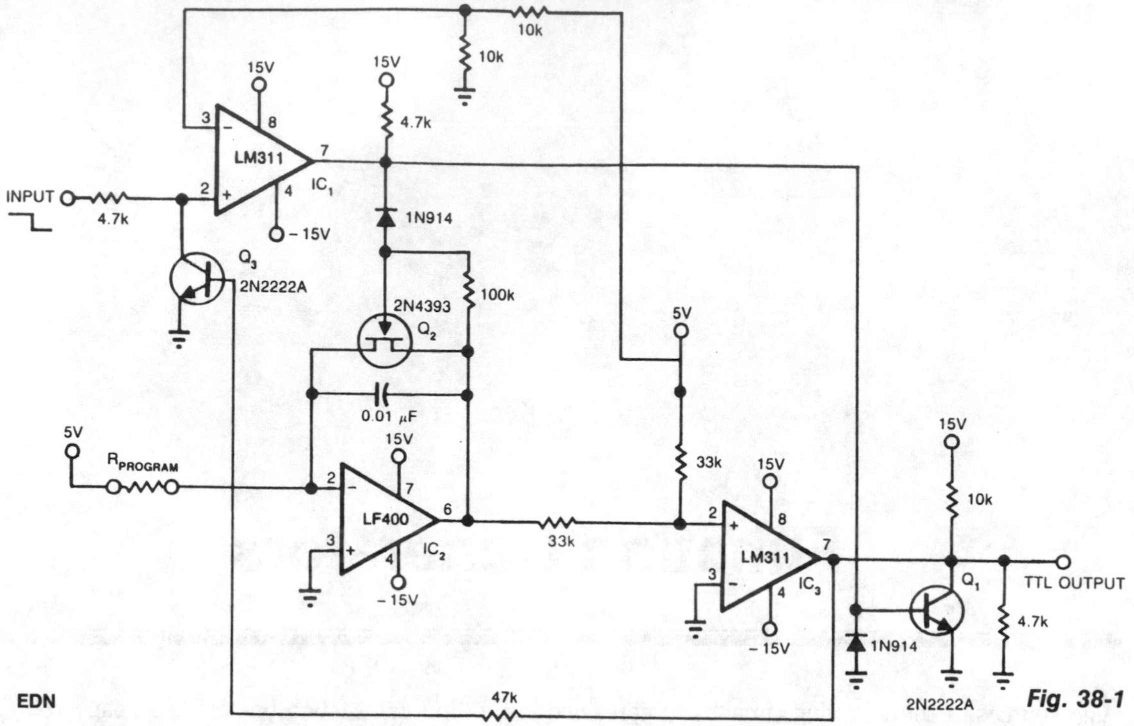

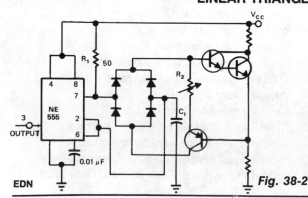

EDN

Fig. 38-1

If you need a wide-range, resistor-programmable monostable multivibrator, you can program the circuit for pulse widths from 1 μs to 10s—10^7:1 range. A high-to-low transition at the input causes IC1's output to switch low, thereby turning off Q1 and Q2. With the latter transistor turned off, IC3's output increases and the output of IC2 begins to ramp toward the negative supply level at a rate determined by the 0.01-μF capacitor and the programming resistor. When IC2's output voltage reaches −5 V, IC3's output switches low. If you anticipate input pulses shorter than the desired output pulses, Q3 is necessary. This transistor keeps IC1s input low while an output pulse is present, thereby preventing inadvertent resetting of the one shot.

LINEAR TRIANGLE-WAVE TIMER

Using one current source for the charge and discharge path in this circuit ensures identical rise and fall times at the capacitor terminal. A Darlington pair ensures identical biasing of the IC during the charge and discharge cycles. The period of the triangle wave is: T $\approx$ 0.46V$C$1/$R$2. V_{CC} must be at least 8 V to maintain linearity. At the output at pin 3 of the IC timer, a 50% duty-cycle square wave, frequency tunable by R2 alone, appears.

EDN

Fig. 38-2

FOUR-OUTPUT WAVEFORM GENERATOR

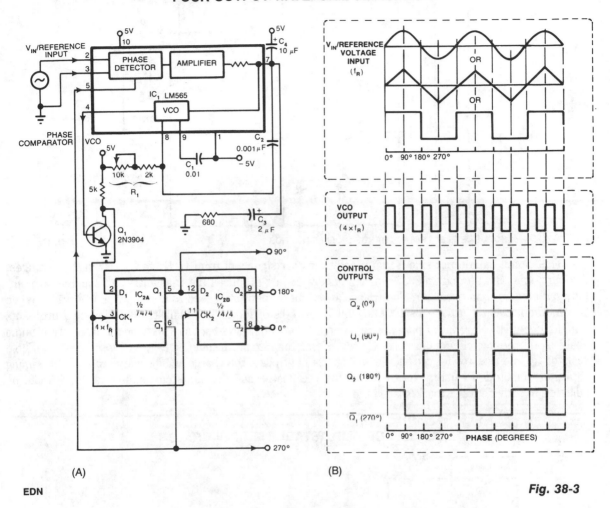

EDN

Fig. 38-3

Many applications require control signals that have phase shifts with reference to an input signal. Circuit accepts a sine, square, or triangular wave as an input reference signal and produces square-wave outputs with 0°, 90°, 180°, and 270° phase shifts with respect to the input. Figure 38-3B shows the input and output waveforms. The circuit contains two ICs: an LM565 phase-locked loop and a 7474 dual-D positive edge-triggered flip-flop. R1 and C1 set the free-running frequency of the LM565's VCO. You should adjust R1 so that the frequency is approximately four times that of the input reference signal. The LM565 responds to input signals greater than 10 mV pk-pk; 3 V pk-pk is the chip's maximum allowable input level. Q1 matches the LM565's output to the flip-flops' inputs. The flip-flops' outputs provide the TTL-compatible square-wave signals with 0°, 90°, 180°, and 270° phase shift with reference to the input signal.

FUNCTION GENERATOR

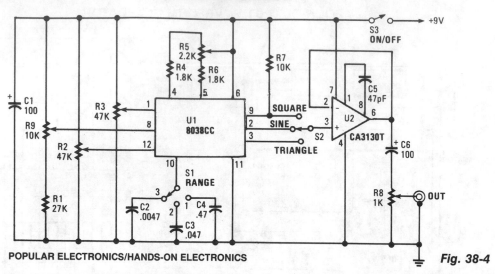

Fig. 38-4

This circuit can output sine, square, and triangular signals of from 15 Hz to 25 kHz in three ranges. The circuit is built around an 8038 function generator that produces the triangular- and square-wave outputs directly from an oscillator. The triangular output is then processed to develop the sine wave. While that method doesn't provide a sufficiently low level of distortion to let you make distortion measurements on audio gear, the degree of purity is high enough for frequency-response tests and a lot of other audio analysis. Three switched capacitors, C2 to C4, set the circuits frequency range via switch S1. Variable resistor R9 and resistor R1 provide the voltage for controlling the charge and discharge rates of the timing capacitor selected. Resistors R4 to R6 control the charge and discharge currents. Resistor R5 can be adjusted to provide a 1.1 mark/space ratio.

CLASSIC OP AMP ASTABLE MULTIVIBRATOR

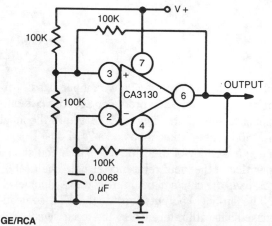

Uses CA3130 BiMOS op amp that operates at a frequency of 1 kHz. With rail-to-rail output swing, frequency is independent of supply voltage, device, and temperature. Only the temperature coefficient of R_T and C_T enters into circuit stability.

ALL RESISTANCE VALUES ARE IN OHMS   Fig. 38-5

PROGRAMMABLE TRIANGLE-/SQUARE-WAVE GENERATOR

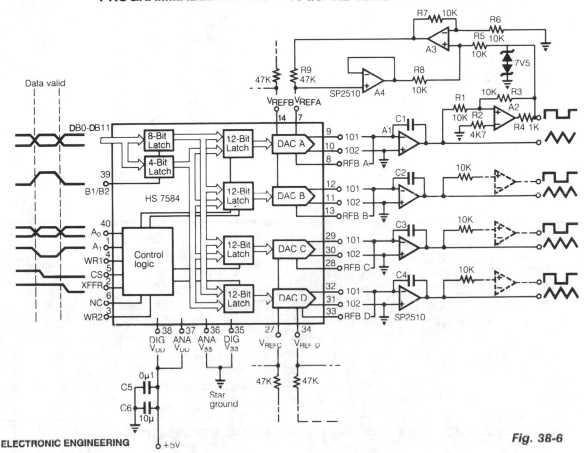

Fig. 38-6

The programmable multiple output generator provides the control signals for data converter ATE. Major performance criteria are simple, interfaces to a number of microprocessor systems, low power consumption, stable output timing relationships combined with a minimum of board space. For schematic simplicity only, one output circuit is shown in full.

The monolithic HS7584 provides four current output DAC's with four quadrant multiplication, individual reference input and a feedback resistor. The digitally controlled integrator's frequency is determined by:

$$f = \frac{\text{Digital Input}}{4\,R\,C}$$

C is the value of C1 to C4 and R is the resistance of the DAC. With the four DACS on a single chip, the resistance matching is good, which results in stable timing relationships of the generator outputs. The output of the comparator A2 determines whether the constant current source provided by A3 and A4 is positive or negative.

NONINTEGER PROGRAMMABLE PULSE DIVIDER

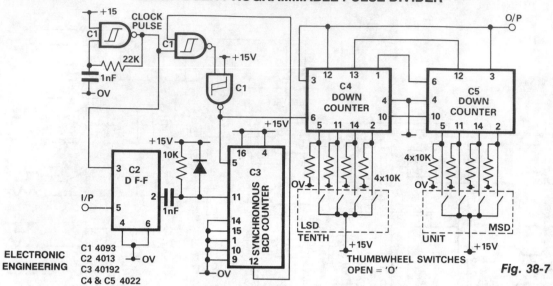

Fig. 38-7

ELECTRONIC ENGINEERING

C1 4093
C2 4013
C3 40192
C4 & C5 4022

The purpose of D-type flip-flop IC2 is to synchronize the input signal with the clock pulse. When the clock pulse changes from low to high and the input is high, IC2 output is high. Subsequently, IC3 resets to zero and starts counting up. Until the counter counts to ten, the counter is inhibited. Thus, the number of pulses of the output of IC3 is ten times input pulse. The designed frequency of the clock pulse must be ten times higher than the maximum frequency of the input. IC4 and IC5 are cascaded to form a two decade programmable down counter. Since the number of pulses appearing at the input of the down counter is ten times the input to the divider, the effective range of the divisor for this divider is 0.1 to 9.9.

XOR GATE COMPLEMENTARY SIGNALS GENERATOR

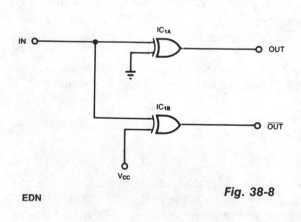

Fig. 38-8

EDN

Some applications, such as driving three-state buffers for data multiplexers or for biphase clocks in high-speed systems, require complementary signals having a small-time skew and nearly simultaneous transitions. Here, XOR gates function as both inverting and noninverting gates. For CMOS systems, practically any type of XOR gate will work. However, the advanced-CMOS logic (ACL) families have the greatest drive capability, the shortest gate delays, and the tightest manufacturing tolerances. For TTL systems, compatible CMOS types such as the ACT or S/AS86 families are preferable. Do not use low-power TTL versions (LS or ALS), because they have large propagation delay differences when one XOR gate is inverting and the other is noninverting.

LOW-COST FSK GENERATOR

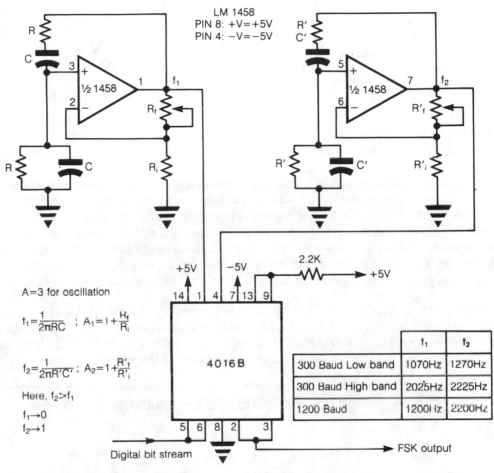

LM 1458
PIN 8: +V=+5V
PIN 4: −V=−5V

A=3 for oscillation

$$f_1 = \frac{1}{2\pi RC} \quad ; \quad A_1 = 1 + \frac{R_f}{R_i}$$

$$f_2 = \frac{1}{2\pi R'C'} \quad ; \quad A_2 = 1 + \frac{R'_f}{R'_i}$$

Here, $f_2 > f_1$

$f_1 \rightarrow 0$

$f_2 \rightarrow 1$

4016B

	f_1	f_2
300 Baud Low band	1070Hz	1270Hz
300 Baud High band	2025Hz	2225Hz
1200 Baud	1200Hz	2200Hz

Digital bit stream

FSK output

ELECTRONIC ENGINEERING

Fig. 38-9

In FSK, two discrete frequencies are used to represent the binary digits 0 and 1. The heart of the circuit consists of two Wien-bridge oscillators built using a dual op amp LM 1458, for the two frequencies. The two frequencies are enabled corresponding to digital data using two switches in SCL 4016. The control lines of these switches are logically inverted with respect to each other using one of the switches in SCL 4016 as an inverter, so as to enable only one oscillator output at a time. The digital bit stream is used to control the analog switches as shown. Since the switching frequency limit of SCL 4016 is 40 MHz, high-data rates can be easily accommodated. This method comes in handy when expensive FSK generator chips are not readily available; also, the components used in this circuit are easily available off the shelf and are quite cheap.

HARMONICS GENERATOR

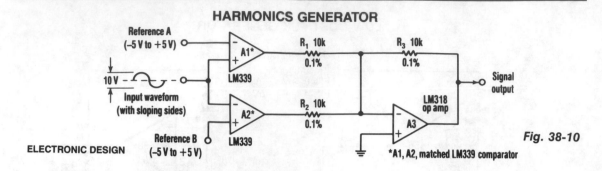

ELECTRONIC DESIGN

Fig. 38-10

*A1, A2, matched LM339 comparator

Two comparators and a summing amplifier that generate differential harmonic spectra comprise a simple frequency multiplier. The resulting circuit can extract harmonics from a sine, triangle, sawtooth, or any other sloping-sided waveform.

With a sloped-input waveform, a comparator produces an output pulse width that's proportional to the input amplitude plus a reference voltage. Changing the reference can vary the pulse width from 0 to 100%. As the pulse width changes, the harmonic spectrum changes, but two comparators combined in the adder eliminate harmonics, depending on the duty cycle. For example, a 50% pulse will lack all the even-numbered harmonics. Similarly, a 25% duty-cycle pulse will be missing multiples of the fourth harmonic and deliver the second, sixth, and tenth harmonics. Accordingly, the circuit generates multiples of the input frequency that might not have existed in the input waveform. Adjusting the references can create virtually any harmonic.

Because comparators A1 and A2 supply differential inputs to the added A3, the adder cancels out equal harmonics. Therefore, both A1 and A2 should have identical ac characteristics, and A3 should have good common-mode rejection and a high slew rate. In particular, R1, R2, and R3 should match within 0.1%. Of course, the accuracy of the circuit depends heavily on the amplitude stability of the input.

LOW-FREQUENCY FM GENERATORS

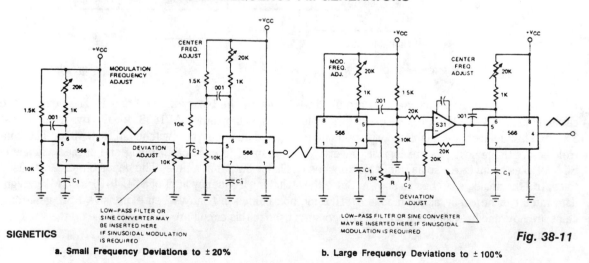

SIGNETICS

a. Small Frequency Deviations to ±20%

b. Large Frequency Deviations to ±100%

Fig. 38-11

228

LOW-FREQUENCY FM GENERATORS (*Cont.*)

Here are two FM generators for low frequency, less than 0.5 MHz center frequency, applications. Each uses a 566 function generator as a modulation generator and a second 566 as the carrier generator. Capacitor C1 selects the modulation frequency adjustment range and C1 selects the center frequency. Capacitor C2 is a coupling capacitor which only needs to be large enough to avoid distorting the modulating waveform. If a frequency sweep in only one direction is required, the 566 ramp generators given in this section can be used to drive the center generator.

POSITIVE-TRIGGERED MONOSTABLE

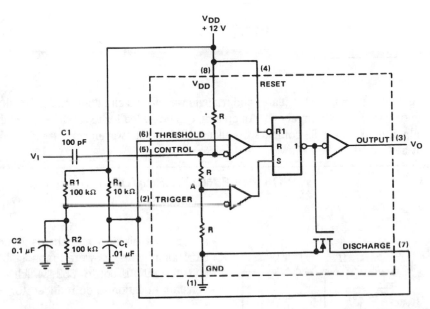

Reprinted by permission of Texas Instruments.

Fig. 38-12

A positive-going trigger pulse can be used to start the timing cycle with the circuit shown. In this design, trigger input pin 2 is biased to 6 V ($1/2$ V_{DD}) by divider R1 and R2. Control input pin 5 is biased to 8 V ($2/3$ V_{DD}) by the internal divider circuit. With no trigger voltage applied, point A is at 4 V ($1/3$ V_{DD}). To turn the timer on, the voltage at point A has to be greater than the 6 V present on pin 2. Positive 5-V trigger pulse V_I applied to the control input pin 5 is ac coupled through capacitor C1, adding the trigger voltage to the 8 V already on pin 5; this results in 13 V with respect to ground. The output pulse width is determined by the values of R_t and C_t.

When voltage at point A is increased to 6.5 V, which is greater than the 6 V on pin 2, the timer cycle is initialized. The output of timer pin 3 increases, turning off discharge transistor pin 7 and allowing C_t to charge through resistor R_t. When capacitor C_t charges to the upper threshold voltage of 8 V ($2/3$ V_{DD}), the flip-flop is reset and output pin 3 decreases. Capacitor C_t then discharges through the discharge transistor. The timer is not triggered again until another trigger pulse is applied to control input pin 5.

PRECISION AUDIO WAVEFORM GENERATOR

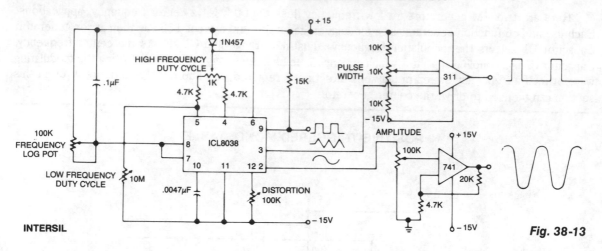

INTERSIL

Fig. 38-13

This circuit generates sinusoidal, square, and triangle waveforms simultaneously. Set the frequency to a particular value or vary it, as shown above. An op amp can be added for extra drive capability and simplified amplitude adjustment. A simple comparator, slicing the triangle waveform, provides continuous duty cycle adjustment at a constant frequency.

MONOSTABLE MULTIVIBRATOR

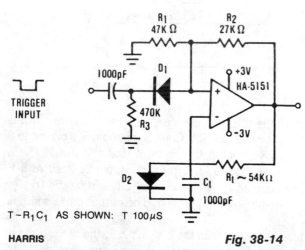

T – R_1C_1 AS SHOWN: T 100 μS

HARRIS

Fig. 38-14

The circuit illustrates the usefulness of the HA-5151 as a battery-powered monostable. In this circuit, the ratio is set to .632, which allows the time constant equation to be reduced to:

$$T = R_t C_t$$

D2 is used to force the output to a defined state by clamping the negative input at +0.6 V. Triggering is set by C1, R3, and D2. An applied trigger pulls the positive input below the clamp voltage, +0.6 V, which causes the output to change state. This state is held because the negative input cannot follow the change because of $R_t \cdot C_t$. This particular circuit has an output pulse width set at approximately 100 μs. Use of potentiometers for R_t and variable capacitors for C_t will allow for a wide variation in T.

VERSATILE 2φ PULSE GENERATOR

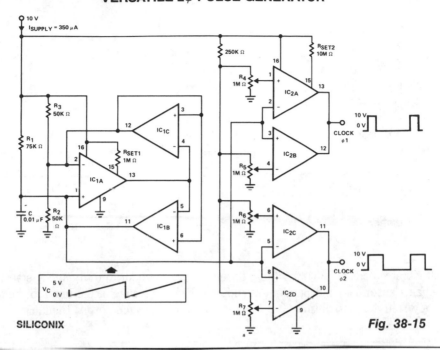

SILICONIX

Fig. 38-15

FIXED-FREQUENCY GENERATOR

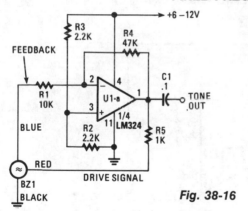

Fig. 38-16

HANDS-ON ELECTRONICS/POPULAR ELECTRONICS

A single op amp, one fourth of an LM324 quad op amp, is configured as a standard inverting amplifier. At power up, a positive voltage is applied to the noninverting input of U1, via R3, forcing its output high. That high output travels along three paths. The first path is the tone output. Along the second path, by way of R5, that high is used as the drive signal for BZ1. In the third path, the high output of U1 is fed back, via R4, to the inverting input of U1. That forces U1's output to go low. And that low, when fed back to the inverting input of U1, causes the op amp output again to a high, and the cycle repeats itself. As configured, U1 provides a voltage gain of 4.7 (gain = R4/R1).

The outer ring of the piezo element is usually connected to the circuit ground. The large inner circle serves as the driven area, and the small elongated section supplies the feedback signal. Resistor R5 sets BZ1's output-volume level. That level can be increased by decreasing R5 for example, to 470 Ω. To decrease the volume, increase R5 to about 2.2 KΩ, or so.

Resistors R2 and R3 set the bias for op amp U1's positive input pin 3 to half of the supply-voltage level. That allows for a maximum voltage swing at U1's output. Although a quad op amp is specified, almost any similar low cost single or dual op amp will work for U1a.

231

SINGLE-SUPPLY MULTIVIBRATOR

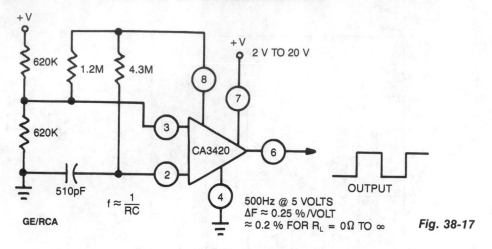

$$f \approx \frac{1}{RC}$$

500Hz @ 5 VOLTS
ΔF ≈ 0.25 %/VOLT
≈ 0.2 % FOR R_L = 0Ω TO ∞

GE/RCA

Fig. 38-17

This multivibrator uses a CA3420 BiMOS op amp to provide improved frequency stability. The output frequency remains essentially independent of supply voltage. Because of the inherent buffering action of pin 6, frequency shift is approximately 0.2% when R_L varies between zero Ω to infinity.

EASILY TUNED SINE-WAVE/SQUARE-WAVE OSCILLATOR

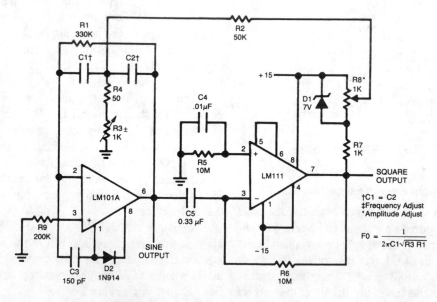

†C1 = C2
‡Frequency Adjust
*Amplitude Adjust

$$F0 = \frac{1}{2\pi C1 \sqrt{R3\ R1}}$$

NATIONAL SEMICONDUCTOR CORP.

Fig. 38-18

EASILY TUNED SINE-WAVE / SQUARE-WAVE OSCILLATOR (*Cont.*)

The circuit will provide both a sine- and square-wave output for frequencies from below 20 Hz to above 20 kHz. The frequency of oscillation is easily tuned by varying a single resistor. This is a considerable advantage over Wien-Bridge circuits where two elements must be tuned simultaneously to change frequency. Also, the output amplitude is relatively stable when the frequency is changed. An amp is used as a tuned circuit, driven by square wave from a voltage comparator. The frequency is controlled by R1, R2, C1, C2, and R3, with R3 used for tuning. Tuning the filter does not affect its gain or bandwidth, so the output amplitude does not change with frequency.

A comparator is fed with the sine-wave output to obtain a square wave. The square wave is then fed back to the input of the tuned circuit to cause oscillation. Zener diode, D1, stabilizes the amplitude of the square wave fed back to the filter input. Starting is insured by R6 and C5 which provide dc negative feedback around the comparator. This keeps the comparator in the active region. Distortion ranges between 0.75% and 2% depending on the setting of R3. Although greater tuning range can be accomplished by increasing the size of R3 beyond 1 KΩ, distortion becomes excessive. Decreasing R3 lower than 50 Ω can make the filter oscillate by itself.

ASTABLE

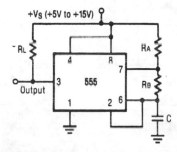

This astable will trigger itself and run free as a multivibrator. The external capacitor charges through R_A and R_B and discharges through R_B only. Thus, the duty cycle is set by the ratio of these two resistors, and the capacitor charges and discharges between $1/3$ V_S and $2/3$ V_S. The charge and discharge times, and therefore frequency, are independent of supply voltage. The free-running frequency versus R_A, R_B and C is shown in the graph.

Free Running Frequency vs. R$_A$, R$_B$ and C

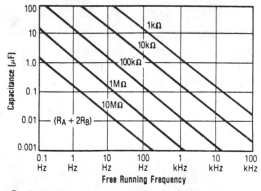

Reprinted with permission from Raytheon Co., Semiconductor Division.

Fig. 38-19

TWO-FUNCTION SIGNAL GENERATOR

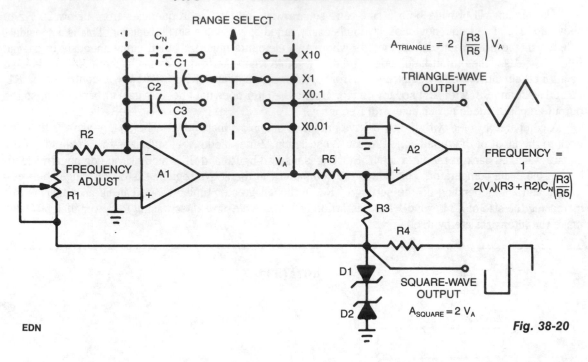

$$A_{\text{TRIANGLE}} = 2 \left(\frac{R3}{R5} \right) V_A$$

$$\text{FREQUENCY} = \frac{1}{2(V_A)(R3 + R2)C_N \left(\frac{R3}{R5} \right)}$$

$$A_{\text{SQUARE}} = 2 V_A$$

EDN

Fig. 38-20

You can continuously vary the frequencies of the triangle and square waves produced by this circuit over a full decade. If R5 = R3, the amplitude of the two waveforms will be equal ($A_{\text{SQUARE}} = A_{\text{TRIANGLE}}$).

TRIANGLE GENERATOR

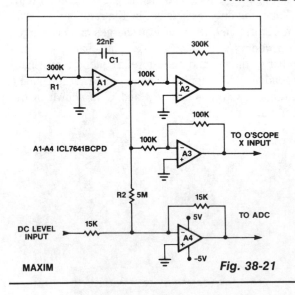

A1-A4 ICL7641BCPD

MAXIM

Fig. 38-21

This circuit generates a symmetrical, 10-mV pk-pk triangle waveform which is summed with a dc level and connected to the a/d analog input for noise/DNL testing. The dc level input offsets the triangle waveform over the input range of the ADC. The 10-mV amplitude amounts to an 8 LSB span for a 12-bit, 5-V, full-scale ADC.

MONOSTABLE OPERATION

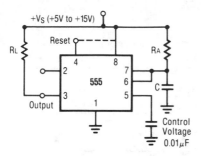

Time Delay vs. R$_A$, R$_B$ and C

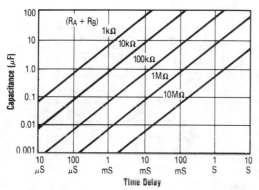

Reprinted with permission from Raytheon Co., Semiconductor Division.

Fig. 38-22

In this mode, the timer functions as a one shot. The external capacitor is initially held discharged by a transistor internal to the timer. Applying a negative trigger pulse to pin 2 sets the flip-flop, driving the output high, and releasing the short circuit across the external capacitor. The voltage across the capacitor increases with the time constant $r = R_A C$ to $2/3$ V_S, where the comparator resets the flip-flop and discharges the external capacitor. The output is now in the low state.

Circuit triggering takes place when the negative-going trigger pulse reaches $1/3$ V_S; the circuit stays in the output high state until the set time elapses. The time the output remains in the high state is 1.1 $R_A C$ and can be determined by the graph. A negative pulse applied to pin 4 (reset) during the timing cycle will discharge the external capacitor and start the cycle over again beginning on the positive-going edge of the reset pulse. If reset function is not used, pin 4 should be connected to V_S to avoid false resetting.

PROGRAMMABLE-FREQUENCY, FREE-RUNNING MULTIVIBRATOR

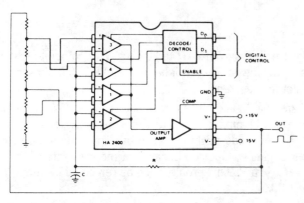

HARRIS

This is the simplest of any programmable oscillator circuit, since only one stable timing capacitor is required. The output square wave is about 25 V pk-pk, and has rise and fall times of about 0.5 μs. If a programmable attenuator circuit is placed between the output and the divider network, 16 frequencies can be produced with two HA-200's and still only one timing capacitor.

Fig. 38-23

FUNCTION GENERATOR

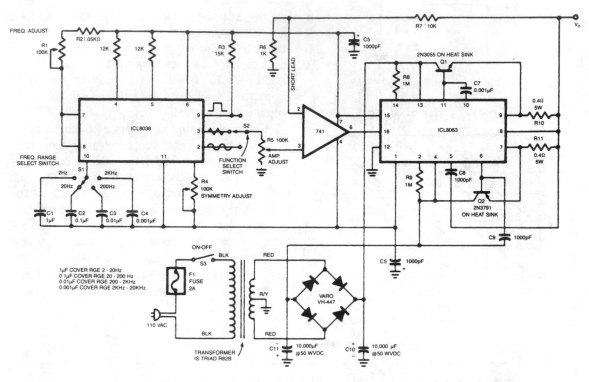

INTERSIL

Fig. 38-24

This generator will supply sine, triangular, and square waves from 2 Hz to 20 kHz. This complete test instrument can be plugged into a standard 110 Vac line for power. V_{OUT} will be up to ± 25 V (50 V pk-pk across loads as small as 10 Ω (about 2.5 A maximum output current).

Capacitor working voltages should be greater than 50 Vdc and all resistors should be $1/2$ W, unless otherwise indicated. The interconnecting leads from the 741 pins 2 and 3 to their respective resistors should be kept short, less than 2 inches if possible; longer leads might result in oscillation. Full output swing is possible to about 5 kHz; after that the output begins to taper off because of the slew rate of the 741, until at 20 kHz the output swing will be about 20 $V_{pp} \pm 10$ V. This problem can be remedied by simply using an op amp with a higher slew rate, such as the LF356.

PROGRAMMABLE-FREQUENCY ASTABLE

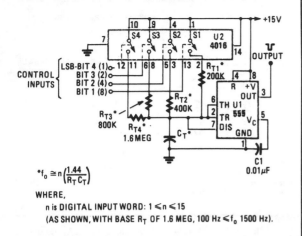

$$^*f_o \cong n\left(\frac{1.44}{R_T C_T}\right)$$

WHERE,
  n is DIGITAL INPUT WORD: $1 \leqslant n \leqslant 15$
  (AS SHOWN, WITH BASE R_T OF 1.6 MEG, 100 Hz $\leqslant f_o$ 1500 Hz).

POPULAR ELECTRONICS   *Fig. 38-25*

LINEAR-RAMP MONOSTABLE

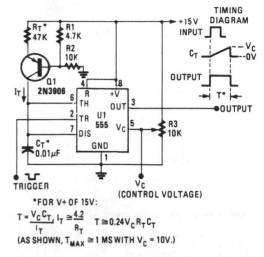

*FOR V+ OF 15V:

$$T = \frac{V_C C_T}{I_T}, \quad I_T \cong \frac{4.2}{R_T} \quad T \cong 0.24 V_C R_T C_T$$

(AS SHOWN, $T_{MAX} \cong 1$ MS WITH V_C = 10V.)

HANDS-ON ELECTRONICS   *Fig. 38-26*

LOW-FREQUENCY MULTIVIBRATOR

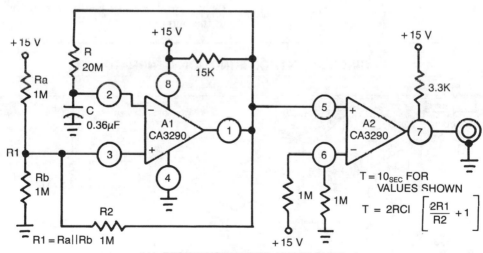

$$T = 2RCl\left[\frac{2R1}{R2} + 1\right]$$

T = 10 SEC FOR VALUES SHOWN

R1 = Ra||Rb 1M

ALL RESISTANCE VALUES ARE IN OHMS

GE/RCA   *Fig. 38-27*

This circuit uses half the CA3290 BiMOS dual voltage comparator as conventional multivibrator. The second half maintains frequency against effects of output loading. Large values of timing resistor, R1, assure long time delays with low-leakage capacitors.

RETRIGGERABLE ONE SHOT

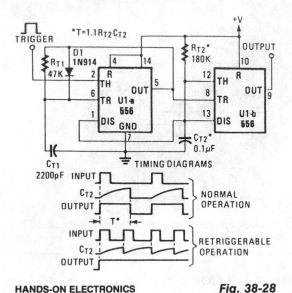

HANDS-ON ELECTRONICS   *Fig. 38-28*

ASTABLE MULTIVIBRATOR

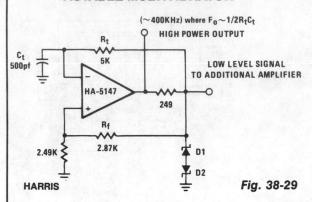

HARRIS   *Fig. 38-29*

The power bandwidth of the HA-5147 extends the circuit's frequency range to approximately 500 kHz. R_t can be made adjustable to vary the frequency if desired. Any timing errors because of V_{OS} or I_{bias} have been minimized by the precision characteristics of the HA-5147. D1 and D2, if used, should be matched to prevent additional timing errors. These clamping diodes can be omitted by tying R_t and positive feedback resistor R_f directly to the output.

SINGLE-CONTROL FUNCTION GENERATOR

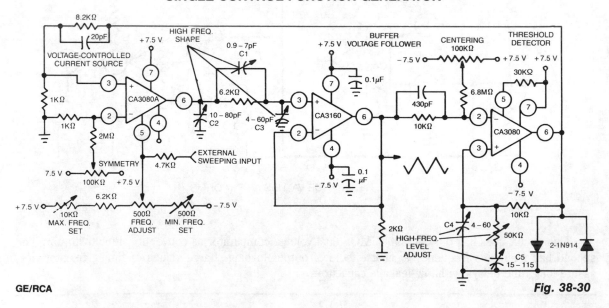

GE/RCA   *Fig. 38-30*

SINGLE-CONTROL FUNCTION GENERATOR (*Cont.*)

This function generator, with an adjustment range in excess of 1,000,000 to 1, uses a CA3160 BiMOS op amp as a voltage follower, a CA3080 OTA as a high-speed comparator, and a CA3080 as a programmable-current source. Three variable capacitors, C1, C2, and C3 shape the triangular signal between 500 kH and 1 MHz. Capacitors C4 and C5, and the trimmer potentiometer in series with C5, maintain essentially constant (±10%) amplitude to 1 MHz.

TRIANGLE-/SQUARE-WAVE GENERATOR

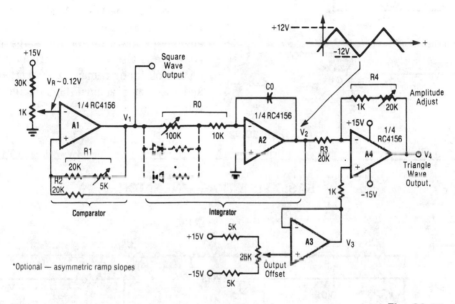

Reprinted with permission from Raytheon Co., Semiconductor Division.

Fig. 38-31

This circuit uses a positive-feedback loop closed around a combined comparator and integrator. When power is applied, the output of the comparator will switch to one of two states, to the maximum positive or maximum negative voltage. This applies a peak input signal to the integrator, and the integrator output will ramp either down or up, opposite of the input signal. When the integrator output, which is connected to the comparator input, reaches a threshold set by R1 and R2, the comparator will switch to the opposite polarity. This cycle will repeat endlessly, the integrator charging positive then negative, and the comparator switching in a square-wave fashion.

VARIABLE DUTY CYCLE TIMER

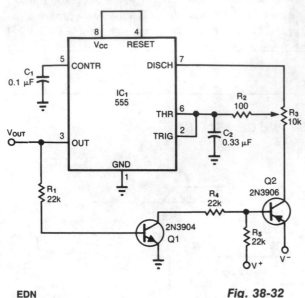

When configured as a free-running multivibrator, a 555 timer provides no more than a 50% duty cycle. By adding two transistors, however, you can obtain a variable 5 to 95% duty cycle without changing the sum of the on and off times. When V_{OUT} decreases, Q1 is on and Q2 is off, disconnecting V+ while timing capacitor C2 discharges into pin 7 of the timer. When V_{OUT} increases, Q2 reconnects V+ for recharging C2.

Adjusting linear trimming potentiometer R3 to increase the charging resistance increases the on time, but decreases the off time by the same amount by lowering the discharge resistance (the converse is also true). As a result, the sum of the on and off times remains constant. R2 protects Q2 and the timer against high-charge/discharge currents.

EDN   Fig. 38-32

BASIC FUNCTION GENERATOR

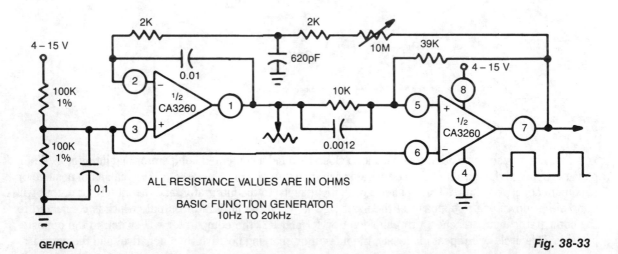

ALL RESISTANCE VALUES ARE IN OHMS

BASIC FUNCTION GENERATOR
10Hz TO 20kHz

GE/RCA   Fig. 38-33

This function generator uses a CA3260 BiMOS op amp to perform both the integrator and switching functions. A 620-pF capacitor and 2-KΩ resistor shape feedback square wave to reduce spikes. Full audio spectrum, 10 Hz to 20 kHz, is covered with a single 10Ω potentiometer. Requires 9-V battery.

WIDE-RANGE TUNABLE FUNCTION GENERATOR

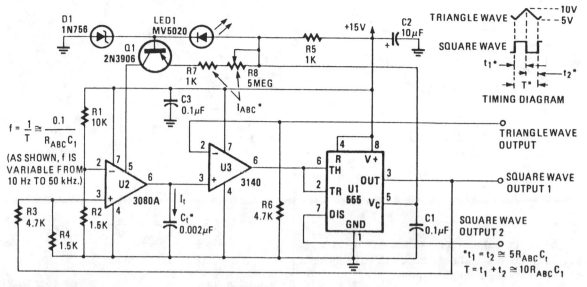

$$f = \frac{1}{T} \cong \frac{0.1}{R_{ABC}C_1}$$

(AS SHOWN, f IS VARIABLE FROM 10 Hz TO 50 kHz.)

$^*t_1 = t_2 \cong 5R_{ABC}C_t$

$T = t_1 + t_2 \cong 10R_{ABC}C_1$

Fig. 38-34

SAWTOOTH AND PULSE GENERATOR

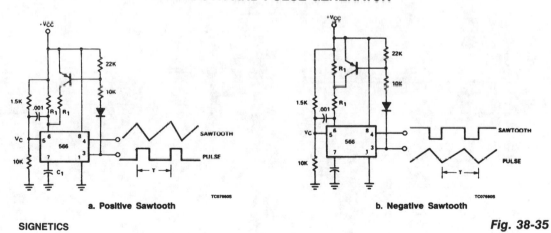

a. Positive Sawtooth

b. Negative Sawtooth

Fig. 38-35

The pin 3 output of the 566 can be used to provide different charge and discharge currents for C1 so that a sawtooth output is available at pin 4 and a pulse at pin 3. The pnp transistor should be well saturated to preserve good temperature stability. The charge and discharge times can be estimated by using the formula:

$$T = \frac{R_T C_1 V_{CC}}{5(V_{CC} - V_C)}$$

where R_T is the combined resistance between pin 6 and V_{CC} for the interval considered.

PRECISE TRIANGLE-/
SQUARE-WAVE GENERATOR

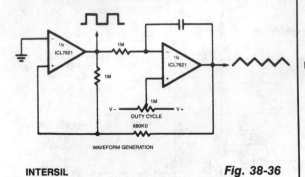

WAVEFORM GENERATION

INTERSIL

Fig. 38-36

Since the output range swings exactly from rail to rail, frequency and duty cycle are virtually independent of power supply variations.

WIDE-RANGE TRIANGLE-/
SQUARE-WAVE GENERATOR

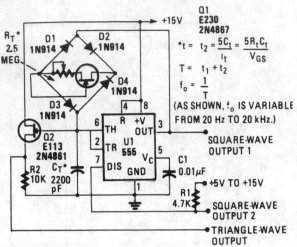

Q1
E230
2N4867

$$*t = t_2 = \frac{5C_t}{I_t} = \frac{5R_t C_t}{V_{GS}}$$

$$T = t_1 + t_2$$

$$f_o = \frac{1}{T}$$

(AS SHOWN, f_o IS VARIABLE FROM 20 Hz TO 20 kHz.)

HANDS-ON ELECTRONICS

Fig. 38-37

39

Games

The sources of the following circuits are contained in the Sources section beginning on page 782. The figure number contained in the box of each circuit correlates to the sources entry in the Sources section.

Coin Flipper
Who's First
Electronic Dice

COIN FLIPPER

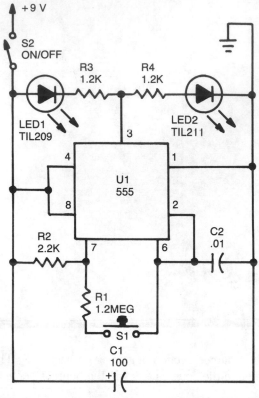

The circuit is basically a 555 astable circuit that divides two LEDs, LED1 and LED2. LED2 is switched on when the output of U1 is high, and LED1 is activated when its output is low. When U1 oscillates, LED1 and LED2 switch on alternately as the output of U1 switches from state to state. Resistor R1's value is high in comparison to R2, so the waveform at the output is a square wave with a mark/space ratio of nearly one-to-one. When you release S1, you break the circuit and U1 latches whatever the output state happens to be at the time.

Fig. 39-1

POPULAR ELECTRONICS/HANDS-ON ELECTRONICS

WHO'S FIRST

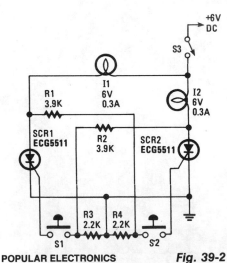

Using two SCRs, this control circuit is designed to lock out the other SCR when one has been triggered, so only one lamp will light. Indicator lamp I1 is controlled by SCR1. The operator simply presses switch S1. Lamp I2 is similarly controlled by S2 and SCR2. With both switches open, neither lamp is lit. The result is insufficient gate current to trigger SCR2 into conduction, so lamp I2 does not light. If S2 is pressed first, the reverse situation occurs. Once one of the SCRs is activated, it is necessary to open S3 to turn the light off.

POPULAR ELECTRONICS   *Fig. 39-2*

ELECTRONIC DICE

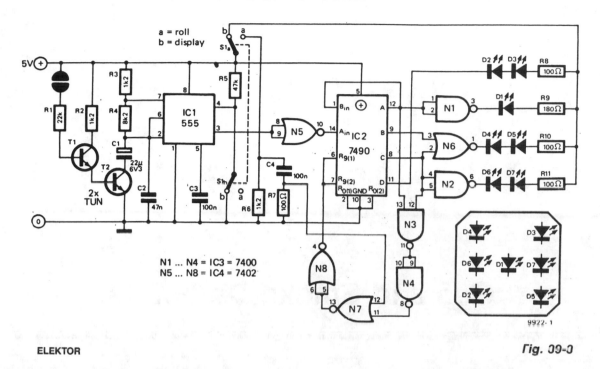

ELEKTOR

Fig. 09-3

The basic die circuit is given. A 555 timer, IC1, is connected as an astable multivibrator. This feeds clock pulses to divide-by-six counter IC2 the outputs of which are decoded by gates N1 to N6 to drive an array of LEDs in the familiar die pattern. When switch S1 is in position b, the reset input of IC1 decreases and the oscillator is inhibited. Power is fed to the LEDs via S1b so that the display is activated. When the die is *rolled* by switching S1 to position a, the display is blanked. C4 is connected to positive supply via S1a, producing a short pulse which resets IC2 via N7 and N8. The reset input of IC1 is pulled high via R5, so the multivibrator begins to oscillate and feeds clock pulses to IC2 via N5. When S1 is switched back to position a, the multivibrator is again inhibited. Then, the counter stops and power is applied to the LEDs which display the value of the *throw*.

40

Gas and Smoke Detectors

The sources of the following circuits are contained in the Sources section beginning on page 782. The figure number contained in the box of each circuit correlates to the sources entry in the Sources section.

Smoke Detector

Furnace Exhaust Gas Temperature
    Monitor with Low-Supply Detection

Methane Concentration Detector with
    Linearized Output

Smoke/Gas/Vapor Detector

Gas/Smoke Detector

Smoke Detector

Smoke Detector

Gas/Smoke Detector

SCR Smoke Alarm

SMOKE DETECTOR

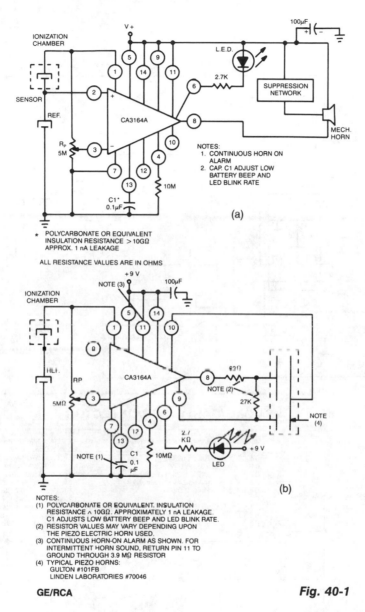

GE/RCA

Fig. 40-1

Use CA3164A BiMOS detector/alarm system. For operation as smoke detector with electromechanical horn (Fig. 40-1a), the output of driver at terminal 8 is used. Large npn transistor Q3, with an active pull-up and transistor Q2 provide over 300 mA of drive current. For operation as a smoke detector with a piezoelectric horn (Fig. 40-1b), the circuit requires output from a second inverting amplifier at terminal 10, as well as the output from terminal 8.

METHANE CONCENTRATION DETECTOR WITH LINEARIZED OUTPUT

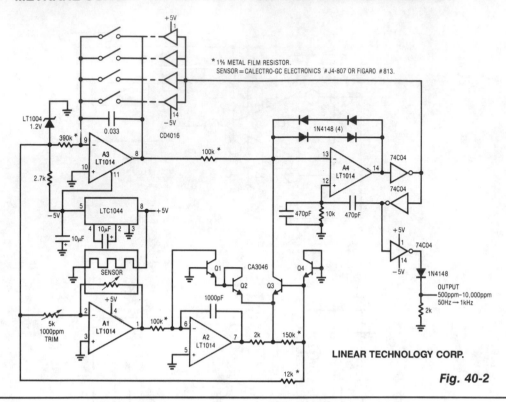

* 1% METAL FILM RESISTOR.
SENSOR = CALECTRO-GC ELECTRONICS #J4-807 OR FIGARO #813.

LINEAR TECHNOLOGY CORP.

Fig. 40-2

SMOKE/GAS/VAPOR DETECTOR

PARTS LIST

V_B	9 Volt Transistor Battery	R3	4.7, ½W Resistor
C1	47µF, 35 V	R4	5 Ohms, 12 Watt
C2, C3	100 pf, 600V	R5	68k ½W Resistor
C4	0.01µF, 50V	R6	1.3k ½W Resistor
D1	1N662 Silicon Diode	R7	Fenwal GB32J2 or Veco 35D6
D2	1N750 Zener Diode, 4.7V	S1	Toggle Switch SPST
D3, D4, D5	1N4002 Diode	F1	Fuse, 6/10 A Slow-Blow, Littlefuse A/3AG/MDL
Q1	2N1132 PNP Transistor	T1	Transformer 117 VAC 60 Hz Primary
Q2	2N4096 SCR		12.6V CT @ 1.2 A Secondary
R1	25k Potentiometer Ohmite Style 53C1		Allied Electronics 6K94HF
R2	5.1k, ½W Resistor	TGS	Sensor TGS-202

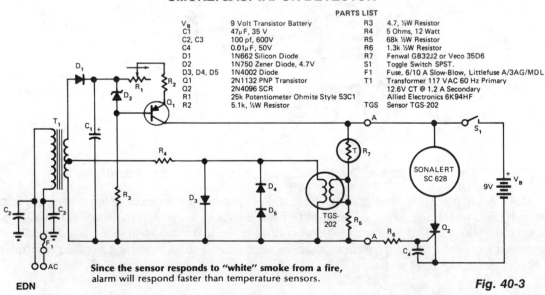

Since the sensor responds to "white" smoke from a fire, alarm will respond faster than temperature sensors.

EDN

Fig. 40-3

SMOKE/GAS/VAPOR DETECTOR (*Cont.*)

Transformer T1 supplies power to the heater of the sensor. Since the sensor is fairly sensitive to heater voltage, diodes D3, D4, and D5 regulate the heater voltage. T1, together with D1 and C2, forms a dc power supply, whose current is regulated by Q1 and adjusted by R1. The constant current from Q1 feeds a variable resistance, consisting of thermistor R7 and the parallel combination of R5 and the sensor resistance. When a hazard causes the voltage at A-A to drop, the net voltage at the SCR gate turns positive, triggering the SCR on and operating the alarm. The alarm draws a small amount of current, so the battery will last a long time. Switch S1 turns off the alarm and resets the SCR.

GAS/SMOKE DETECTOR

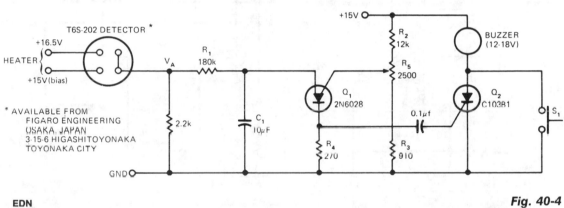

EDN

Fig. 40-4

The sensor is based on the selective absorption of hydrocarbons by an n-type metal-oxide surface. The heater in the device serves to burn off the hydrocarbons once smoke or gas is no longer present in the immediate area; hence, the device is reuseable. When initially turned on, a 15 minute warm-up period is required to reach equilibrium ($V_A \cong 0.6$ V) in a hydrocarbon-free environment. When gas or smoke is introduced near the sensor, V_A will quickly rise (rate and final equilibrium depend on the type of gas and concentration) and trigger Q1, a programmable unijunction transistor. The voltage pulse generated across R4 triggers Q2, sounding the buzzer until S1 resets the unit. R1 and C1 give a time delay to prevent small transient waves of smoke, such as from a cigarette, from triggering the alarm. Triggering threshold is set by R5, R2, and R3; with the components shown, between 50 and 200 ppm of hydrocarbons can be easily detected. Since it is somewhat sensitive to heater voltage, a regulated supply should be used. Power requirements are 1.5 V at 500 mA for the heater and 15 V at 30 mA, depending on type of buzzer, for the bias supply.

FURNACE EXHAUST GAS TEMPERATURE MONITOR WITH LOW SUPPLY DETECTION

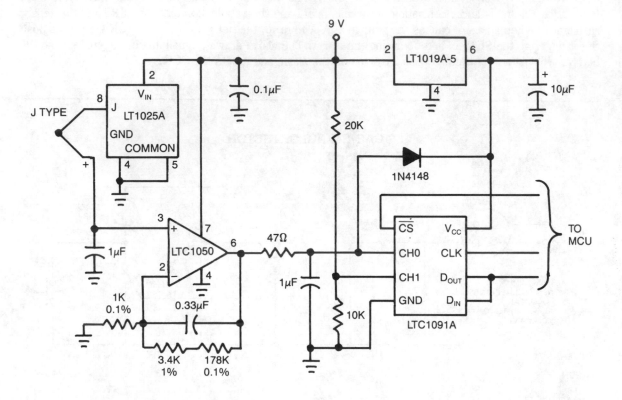

Fig. 40-5

This circuit can be used to measure exhaust gas temperature in a furnace. The 10-bit LTC1091A gives 0.5°C resolution over a 0°C to 500°C range. The LTC1050 amplifies and filters the thermocouple signal, the LT1025A provides cold junction compensation and the LT1019A provides an accurate reference. The J-type thermocouple characteristic is linearized digitally inside the MCU. Linear interpolation between known temperature points spaced 30°C apart introduces less than 0.1°C error. The 20-K/10 KΩ divider on CH1 of the LTC1091 provides low supply voltage detection. Remote location is easy, with data transferred from the MCU to the LTC1091 via the three-wire serial part.

SMOKE DETECTOR

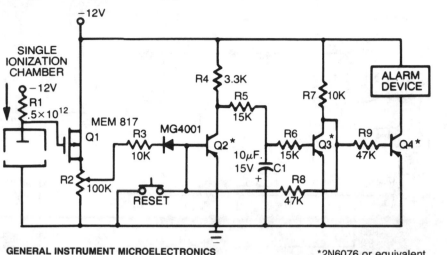

GENERAL INSTRUMENT MICROELECTRONICS

*2N6076 or equivalent

Fig. 40-6

This smoke detector uses a MEM 817 p-channel enhancement mode MOSFET as its buffer amplifier. Operation of the sensor is based on a decrease in the current when smoke enters the chamber, thereby causing a negative voltage excursion at the gate of the buffer MOSFET. Quiescent voltage values at the output of the chamber vary from about −4 V to −6 V, and detection of smoke will result in an excursion of about −4 V. The MOSFET is connected as a source follower.

SCR SMOKE ALARM

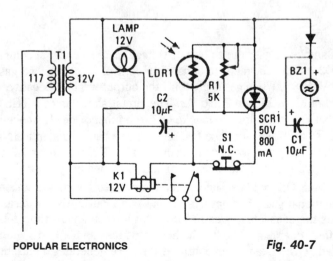

POPULAR ELECTRONICS

Fig. 40-7

GAS/SMOKE DETECTOR

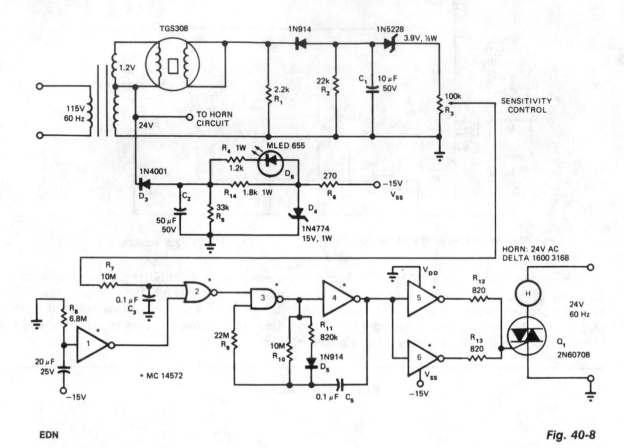

EDN

Fig. 40-8

In the presence of smoke or gas, the ac output voltage increases and becomes rectified, filtered and zener-diode coupled (D2 for thresholding) to sensitivity control R3. Under no gas condition, the output equals approximately 0 V (high). When gas is present, the output will be a negative value (low) sufficient to overcome the threshold of McMOS gate 2 and D2. The circuit shown uses a TGS 308 sensor, a general-purpose gas detector that is not sensitive to smoke or carbon monoxide. If smoke is the primary element to be detected, use the TGS 202 sensor. The two sensors are basically identical; the main differences lie in the heater voltage and the required warm up time delay. The TGS requires a 1.2 V heater and a 2 minute delay, whereas the TGS 202 requires 1.5 V and 5 minutes, respectively.

The system uses a McMOS gated oscillator directly interfacing with a triac-controlled ac horn. Using the MC14572 HEX functional gate, four inverters, one two-input NAND gate and one two-input NOR gate, the circuit provides the complete gas/smoke detector logic functions time delay, gated astable multivibrator control and buffers operation. The 24-Vac horn produces an 85/90-dB sound level output at a distance of 10 ft. Controlled by the astable multivibrator, the horn generates a pulsating alarm—a signal that may be advantageous over a continuous one in some noise environments.

SMOKE DETECTOR

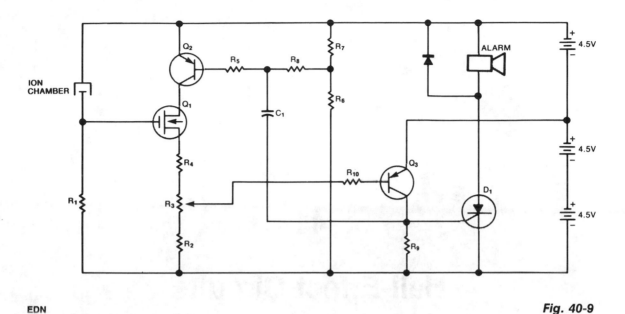

EDN

Fig. 40-9

This circuit comes from U.S. Patent 3,778,800, granted to BRK Electronics in Aurora, IL. The circuit provides a smoke detector with an alarm for both smoke and low batteries. The R6/R7 voltage divider monitors the battery and will turn Q2 and Q1 off when the battery voltage falls too low. The smoke-detector chamber will also cut Q1 off when it senses smoke. Q1 via Q3, triggers SCR D1 and sounds the alarm. Capacitor C1 provides feedback that causes the alarm to sound intermittently. The smoke detector and low-battery circuits sound the alarm at two different rates.

41

Hall-Effect Circuits

The sources of the following circuits are contained in the Sources section beginning on page 782. The figure number contained in the box of each circuit correlates to the sources entry in the Sources section.

Current Monitor
Security Door Ajar Alarm
Hall-Effect Switches
Hall-Effect Compass

CURRENT MONITOR

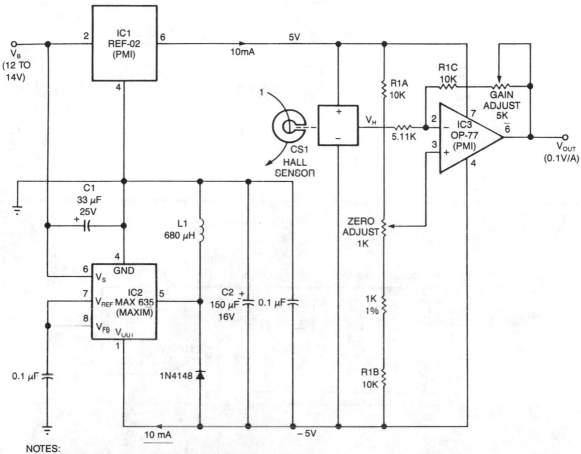

NOTES:
1. C1 AND C1 ARE 199D TANTALEX CAPACITORS FROM SPRAGUE
2. L1 IS A 6860-23 INDUCTOR FROM CADDELL-BURNS
3. R1A, R1B, AND R1C ARE PART OF A THIN-FILM RESISTOR NETWORK SUCH AS THE CADDOCK T914-10K.
4. CS I IS A HALL-EFFECT CURRENT SENSOR (CSLA1CD) FROM MICROSWITCH.

EDN

Fig. 41-1

This circuit uses a Hall-effect sensor, consisting of an IC that resides in a small gap in a flux-collector torrid, to measure dc current in the range of 0 to 40 A. You wrap the current-carrying wire through the toroid; the Hall voltage V_H is then linearly proportional to current I. The current drain from V_B is less than 30 mA.

To monitor an automobile alternator's output current, for example, connect the car battery between the circuit's V_B terminal and ground, and wrap one turn of wire through the toroid. Or, you could wrap 10 turns—if they fit—to measure 1 A full scale. When $I = 0$ V current sensor CS_I's V_H output equals one-half of its 10 V bias voltage. Because regulators IC1 and IC2 provide a bipolar bias voltage, V_H and V_{OUT} are zero when I is zero; you can then adjust the output gain and offset to scale V_{OUT} at 1 V per 10 A.

SECURITY DOOR AJAR ALARM

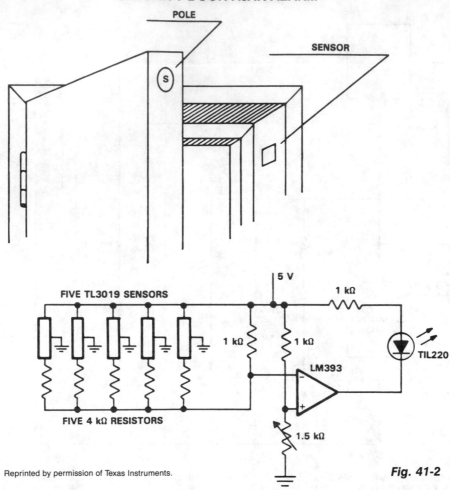

Reprinted by permission of Texas Instruments.

Fig. 41-2

In operation, the TL3019 device will activate, or become low, when a south pole of a magnet comes near the chip face of the device. The example shows five doors. Each door has a magnet embedded in its edge with the south pole facing the outer surface. At the point where the magnet is positioned with the door closed, a TL3019 sensor is placed in the door jamb. With the door closed, the Hall devices will be in a logic low state. This design has five doors and uses five TL3019 devices. Each TL3019 has a 4-KΩ resistor in series and all door sensor and resistor sets are in parallel and connected to the inverting input of an LM393 comparator. With all doors closed, the effective resistance will be about 800 Ω and produce 2.2 V at the inverting input. The noninverting input goes to a voltage divider network which sets the reference voltage. The 1.5-KΩ potentiometer is adjusted so the indicator goes out with all doors closed. This will cause 2.35 V to appear at the noninverting input of the comparator. When a door opens, the voltage at the inverting input will go to 2.5 V which is greater than V_{REF}, and the LED will light. A large number of doors and windows can be monitored with this type of circuit. Also, it could be expanded to add an audible alarm in addition to the visual LED.

HALL-EFFECT SWITCHES

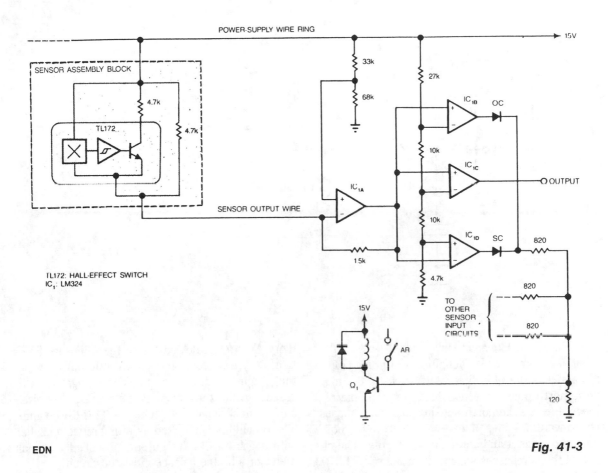

TL172: HALL-EFFECT SWITCH
IC_1: LM324

EDN

Fig. 41-3

Hall-effect switches have several advantages over mechanical and optically coupled switches. They're insensitive to environmental light and dirt, they don't bind, and they don't sustain mechanical wear. Their major drawback is that they require three wires per device. The circuit shown, however, reduces this wire count to $N + 1$ wires for N devices.

Amplifier IC1A is configured as a current-to-voltage converter. It senses the sensor assembly's output current. When the Hall-effect switch is actuated, the sensor's output current increases to twice its quiescent value. Amplifier IC1B, configured as a comparator, detects this increase. The comparator's output decreases when the Hall-effect switch turns on.

The circuit also contains a fault-detection function. If any sensor output wire is open, its corresponding LED will turn on. If the power-supply line opens, several LEDs will turn on. A short circuit will also turn an LED on. Every time an LED turns on, Q1 turns on and the alarm relay is actuated.

HALL-EFFECT COMPASS

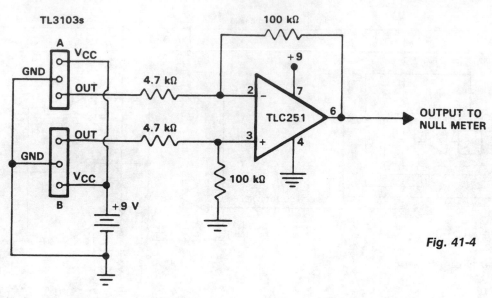

Reprinted by permission of Texas Instruments.

The TL3103 linear Hall-effect device can be used as a compass. By definition, the north pole of a magnet is the pole that is attracted by the magnetic north pole of the earth. The north pole of a magnet repels the north-seeking pole of a compass. By convention, lines of flux emanate from the north pole of a magnet and enter the south pole. The circuit of the compass is shown. By using two TL3103 devices instead of one, we achieve twice the sensitivity. With each device facing the opposite direction, device A would have a position output while the output of device B would be negative with respect to the zero magnetic field level. This gives a differential signal to apply to the TLC251 op amp. The op amp is connected as a difference amplifier with a gain of 20. Its output is applied to a null meter or a bridge balance indicator circuit.

42

High-Frequency Amplifiers

The sources of the following circuits are contained in the Sources section beginning on page 782. The figure number contained in the box of each circuit correlates to the sources entry in the Sources section.

2 – 30 MHz 140-W (PEP) AMATEUR RADIO LINEAR AMPLIFIER

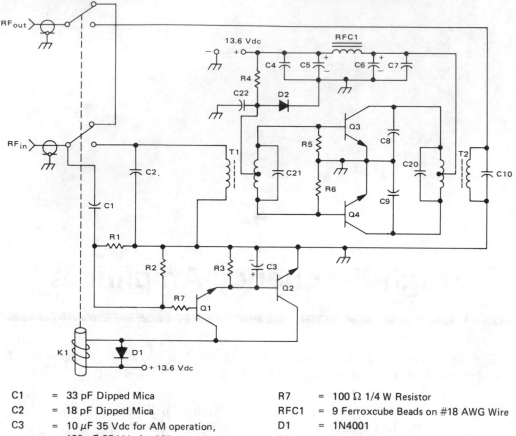

C1	=	33 pF Dipped Mica		R7	=	100 Ω 1/4 W Resistor
C2	=	18 pF Dipped Mica		RFC1	=	9 Ferroxcube Beads on #18 AWG Wire
C3	=	10 μF 35 Vdc for AM operation,		D1	=	1N4001
		100 μF 35 Vdc for SSB operation.		D2	=	1N4997
C4	=	.1 μF Erie		Q1, Q2	=	2N4401
C5	=	10 μF 35 Vdc Electrolytic		Q3, 4	=	MRF454
C6	=	1 μF Tantalum		T1, T2	=	16:1 Transformers
C7	=	.001 μF Erie Disc		C20	=	910 pF Dipped Mica
C8, 9	=	330 pF Dipped Mica		C21	=	1100 pF Dipped Mica
R1	=	100 kΩ 1/4 W Resistor		C10	=	24 pF Dipped Mica
R2, 3	=	10 kΩ 1/4 W Resistor		C22	=	500 μF 3 Vdc Electrolytic
R4	=	33 Ω 5 W Wire Wound Resistor		K1	=	Potter & Brumfield
R5, 6	=	10 Ω 1/2 W Resistor				KT11A 12 Vdc Relay or Equivalent

Fig. 42-1

The amplifier operates across the 2 – 30 MHz band with relatively flat gain response and reaches gain saturation at approximately 210 W of output power. Both input and output transformers are 4:1 turns ratio (16:1 impedance ratio) to achieve low input SWR across the specified band and a high saturation capability. When using this design, it is important to interconnect the ground plane on the bottom of the board to the top, especially at the emitters of the MRF454s.

80-W (PEP) 3 – 30 MHz 12.5 – 13.6 V AMPLIFIER

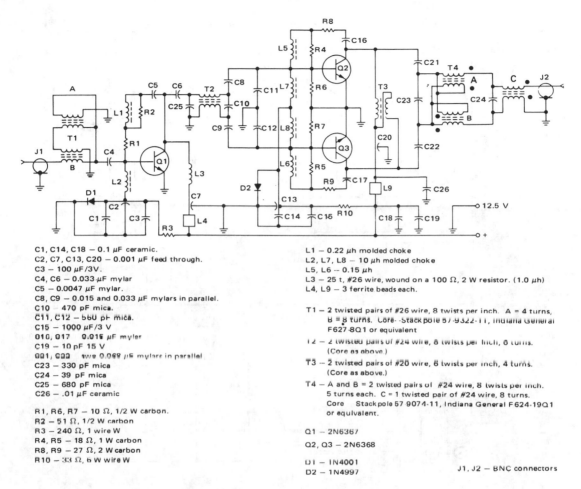

C1, C14, C18 — 0.1 µF ceramic.
C2, C7, C13, C20 — 0.001 µF feed through.
C3 — 100 µF/3V.
C4, C6 — 0.033 µF mylar
C5 — 0.0047 µF mylar.
C8, C9 — 0.015 and 0.033 µF mylars in parallel.
C10 — 470 pF mica.
C11, C12 — 560 pF mica.
C15 — 1000 µF/3 V
C16, C17 — 0.015 µF mylar
C19 — 10 pF 15 V
C21, C22 — two 0.068 µF mylars in parallel
C23 — 330 pF mica
C24 — 39 pF mica
C25 — 680 pF mica
C26 — .01 µF ceramic

R1, R6, R7 — 10 Ω, 1/2 W carbon.
R2 — 51 Ω, 1/2 W carbon
R3 — 240 Ω, 1 wire W
R4, R5 — 18 Ω, 1 W carbon
R8, R9 — 27 Ω, 2 W carbon
R10 — 33 Ω, 6 W wire W

L1 — 0.22 µh molded choke
L2, L7, L8 — 10 µh molded choke
L5, L6 — 0.15 µh
L3 — 25 t, #26 wire, wound on a 100 Ω, 2 W resistor. (1.0 µh)
L4, L9 — 3 ferrite beads each.

T1 — 2 twisted pairs of #26 wire, 8 twists per inch. A = 4 turns,
      B = 8 turns. Core—Stackpole 57-9322-11, Indiana General
      F627-8Q1 or equivalent
T2 — 2 twisted pairs of #24 wire, 8 twists per inch, 6 turns.
      (Core as above.)
T3 — 2 twisted pairs of #20 wire, 8 twists per inch, 4 turns.
      (Core as above.)
T4 — A and B = 2 twisted pairs of #24 wire, 8 twists per inch.
      5 turns each. C = 1 twisted pair of #24 wire, 8 turns.
      Core — Stackpole 57 9074-11, Indiana General F624-19Q1
      or equivalent.

Q1 — 2N6367

Q2, Q3 — 2N6368

D1 — 1N4001
D2 — 1N4997   J1, J2 — BNC connectors

Fig. 42-2

This amplifier utilizes a 2N6367 and a pair of 2N6368 transistors. The 2N6367 transistor is employed as a driver and is specified for up to 9 W (PEP) output. In the amplifier design the driver must supply on 5 W (PEP) at 30 MHz with a resulting IMD performance of about −37 to −38 dB. At lower operating frequencies, drive requirements drop to the 2 – 3 W (PEP) range and IMD performance improves to better than 40 dB. Two 2N6368 transistors are employed in the final stage of the transmitter design in a push-pull configuration. These devices are rated at 40 W (PEP) and −30 dB maximum IMD, although −35 dB performance is more typical for narrowband operation. Without frequency compensation, the completed amplifier can deliver 90 W (PEP) in the 25 – 30 MHz band with IMD performance down −30 dB. If only the power amplifier stage is frequency compensated, 95 W (PEP) can be obtained at 6 – 10 MHz.

29-MHz AMPLIFIER

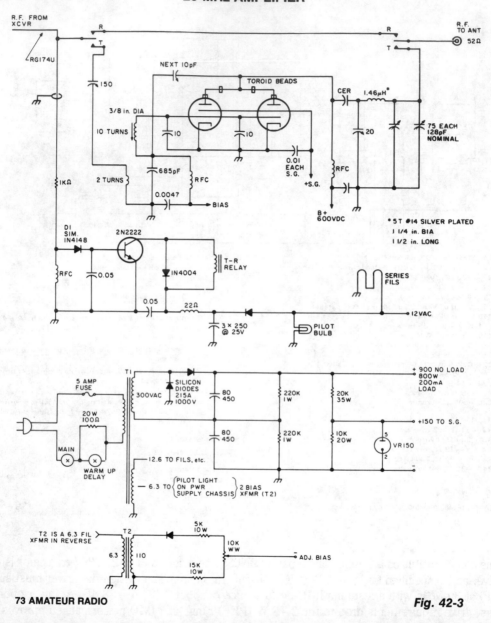

73 AMATEUR RADIO

Fig. 42-3

The only adjustments that require close attention are input, output, and neutralization. The 150-pF capacitor in the input line compensates for impedance mismatch. You tune for maximum signal transfer from exiter to final with an in-line meter or external field strength meter. The final is a conventional pi network. When neutralized, the plate current dip should be at about the same setting of the 20-pF plate capacitor as maximum output. Adjust bias to let tubes idle at about 30 mA.

28-dB NONINVERTING AMPLIFIER

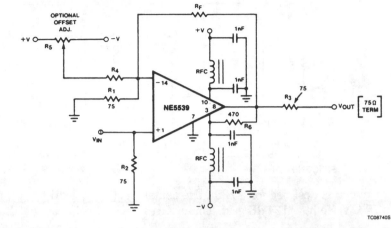

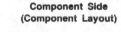

NOTES:
$R_1 = 75\Omega$ 5% CARBON
$R_2 = 75\Omega$ 5% CARBON
$R_3 = 75\Omega$ 5% CARBON
$R_4 = 36k$ 5% CARBON

$R_5 = 20k$ TRIMPOT (CERMET)
$R_F = 1.5k$ (28dB GAIN)
$R_6 = 470\Omega$ 5% CARBON

RFC 3T # 26 BUSS WIRE ON
FERROXCUBE VK 200 09/3B CORE
BYPASS CAPACITORS
1nF CERAMIC
(MEPCO OR EQUIV.)

Top Plane Copper[1] **(Component Side)**	**Component Side** **(Component Layout)**	**Bottom Plane** **Copper[1]**
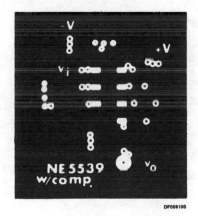	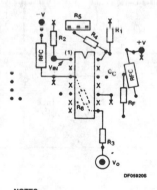	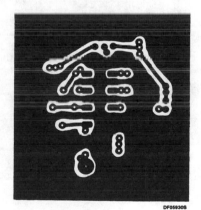

NOTES:
(X) indicates ground connection to top plane.
*R_6 is on bottom side.

NOTE:
1. Bond edges of top and bottom ground plane copper.

SIGNETICS

Fig. 42-4

The physical circuit layout is extremely critical. Breadboarding is not recommended. A double-sided copper-clad printed circuit board will result in more favorable system operation.

WIDEBAND UHF AMPLIFIER WITH HIGH-PERFORMANCE FETS

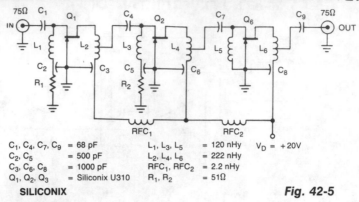

C_1, C_4, C_7, C_9	= 68 pF	L_1, L_3, L_5	= 120 nHy	V_D = +20V
C_2, C_5	= 500 pF	L_2, L_4, L_6	= 222 nHy
C_3, C_6, C_8	= 1000 pF	RFC_1, RFC_2	= 2.2 nHy
Q_1, Q_2, Q_3	= Siliconix U310	R_1, R_2	= 51Ω

SILICONIX

Fig. 42-5

The amplifier circuit is designed for a 225 MHz center frequency, 1 dB bandwidth of 50 MHz, low-input VSWR in a 75-Ω system, and 24 dB gain. Three stages of U310 FETs are used, in a straight-forward design.

BROADCAST BAND RF AMPLIFIER

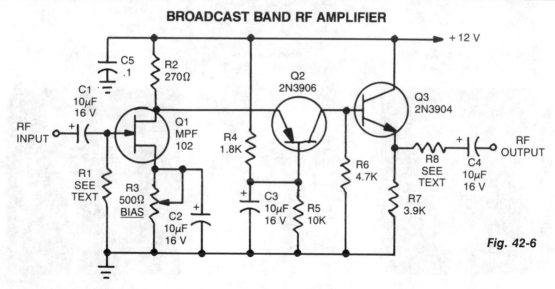

Fig. 42-6

Reprinted with permission of Radio-Electronics Magazine, 1989 R-E Experimenters Handbook. Copyright Gernsback Publications, Inc., 1989.

The circuit has a frequency response ranging from 100 Hz to 3 MHz; gain is about 30 dB. Field-effect transistor Q1 is configured in the common-source self-biased mode. Optional resistor R1 allows you to set the input impedance to any desired value; commonly, it will be 50 Ω.

The signal is then direct coupled to Q2, a common-base circuit that isolates the input and output stages and provides the amplifier's exceptional stability. Last, Q3 functions as an emitter follower, to provide low output impedance at about 50 Ω. If you need higher output impedance, include resistor R8. It will affect impedance according to this formula: $R8 \approx R_{OUT} - 50$. Otherwise, connect output capacitor C4 directly to the emitter of Q3.

MINIATURE WIDEBAND AMPLIFIER

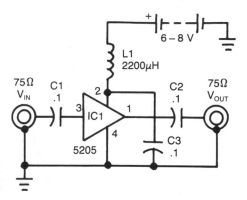

Fig. 42-7

This wideband amplifier uses only five components. External signals enter pin 3 of IC1 via ac coupling capacitor C1. Following amplification, the boosted signals from IC1 pin 1 are coupled to the output by capacitor C2. Capacitor C3 decouples the dc power supply, while rf current is isolated from the power supply by rf choke L1.

The NE5205's low current consumption of 25 mA at 6 Vdc makes battery-powered operation a reality. Although the device is rated for a 6 to 8 V power supply, 6 V is recommended for normal operation. From 6 V an internal bias of 3.3 V results, which permits a 1.4 V pk-pk output swing for video applications.

WIDEBAND 500 kHz – 1 GHz HYBRID AMPLIFIER

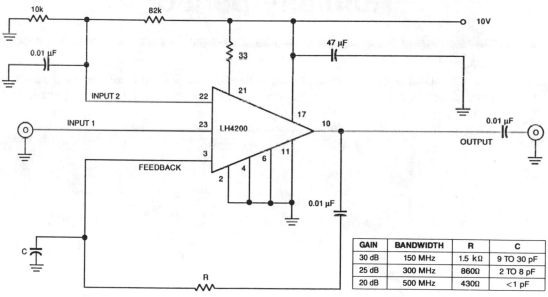

GAIN	BANDWIDTH	R	C
30 dB	150 MHz	1.5 kΩ	9 TO 30 pF
25 dB	300 MHz	860Ω	2 TO 8 pF
20 dB	500 MHz	430Ω	<1 pF

EDN

Fig. 42-8

The amplifier's input stage is a dual-gate GaAs FET, which provides low input capacitance and high transconductance. The dual-gate structure accepts the signal on input 1. Input 2 controls the gain of the amplifier. The amplifier has a third input for use in series feedback. The output feeds back to pin 3 via a single resistor, which controls the overall power gain of the amplifier. At 10 MHz, the output is capable of delivering 12 dBm into a 50-Ω load with 1 dB of signal compression. The ac-coupled amplifier has a gain of 37 dB at 100 MHz and 3 dB at 1 GHz.

43

Humidity Sensor

The sources of the following circuits are contained in the Sources section beginning on page 782. The figure number contained in the box of each circuit correlates to the sources entry in the Sources section.

Low-Cost Humidity Sensor

LOW-COST HUMIDITY SENSOR

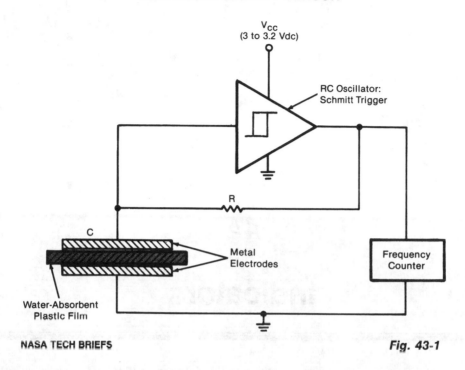

NASA TECH BRIEFS

Fig. 43-1

The sensor is an RC oscillator in which a water-absorbent plastic film is the insulator in the capacitive element. The capacitance of the film increases with the amount of water it absorbs from the air, and thus reduces the oscillation frequency. A frequency counter produces a digital output that represents the change in frequency and hence the change in relative humidity. The sensor can be used to measure humidity in the atmosphere, in the soil, and in industrial gases, for example. A Schmitt-trigger-type IC is connected to the capacitor, which consists of a film of a commercially produced sulfonated fluorocarbon polymer, 2 in. (5.08 cm) square, sandwiched between perforated metal plates. The oscillation frequency decreases almost linearly from about 100 to 16 kHz as the relative humidity increases from about 20 to 76%.

44

Indicators

The sources of the following circuits are contained in the Sources section beginning on page 782. The figure number contained in the box of each circuit correlates to the sources entry in the Sources section.

Stereo Indicator
On-the-Air Indicator
Receiver Signal Alarm
Rf-Actuated Relay
Visual-Level Indicator

STEREO INDICATOR

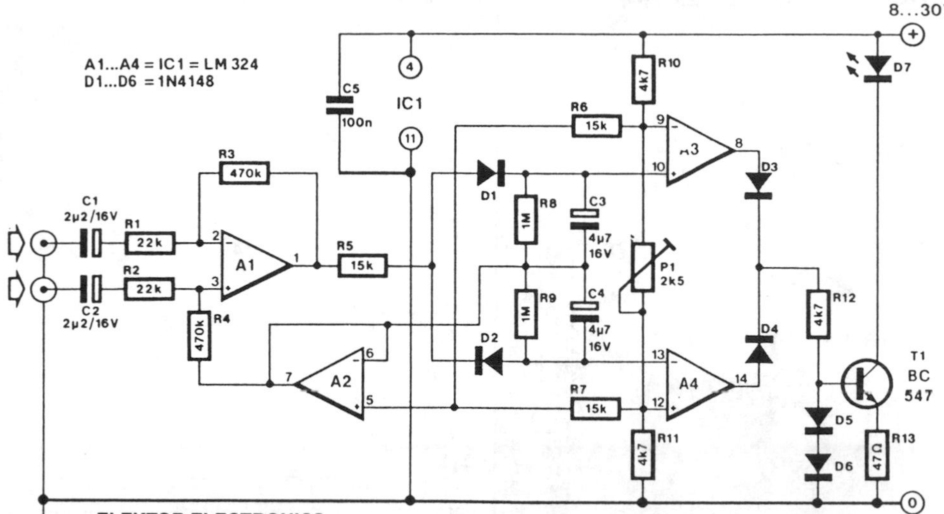

A1...A4 = IC1 = LM 324
D1...D6 = 1N4148

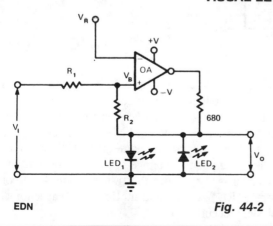

ELEKTOR ELECTRONICS

Fig. 44-1

On most FM tuners, the stereo indicator lights upon detection of the 19-kHz pilot tone. However, this doesn't mean that the program is actually stereophonic, since the pilot tone is often transmitted with mono programs also. A similar situation exists on stereo amplifiers, where the stereo LED is simply controlled from the mono/stereo switch.

The LED-based stereo indicator described here lights only when a true stereo signal is fed to the inputs. Differential amplifier A1 raises the difference between the L and R input signals. When these are equal, the output of A1 remains at the same potential as the output of A2, which forms a virtual ground rail at half the supply voltage. When A1 detects a difference between the L and R input signals, it supplies a positive or negative voltage with respect to the virtual ground rail, and so causes C3 to be charged via D1 or C4 via D2. Comparator A3/A4 switches on the LED driver via OR circuit D3/D4. The input signal level should not be less than 100 mV to compensate for the drop across D1 or D2. The sensitivity of the stereo indicator is adjustable with P1.

VISUAL LEVEL INDICATOR

Fig. 44-2

EDN

This indicator is basically a switch with hysteresis characteristics. If the input voltage momentarily (or permanently) exceeds the most positive reference level, LED1 is switched on. If, on the other hand, the voltage falls below the negative, or least positive, reference level, LED1 will be switched off and LED2 switched on. The output voltage, V_O is clamped either to the diode voltage V_{D1}, or V_{D2} depending on which LED is conducting. For V_O to be positive, V_B has to be positive with respect to the reference voltage V_R; for V_O to be negative, V_B has to be negative with respect to V_R.

RECEIVER SIGNAL ALARM

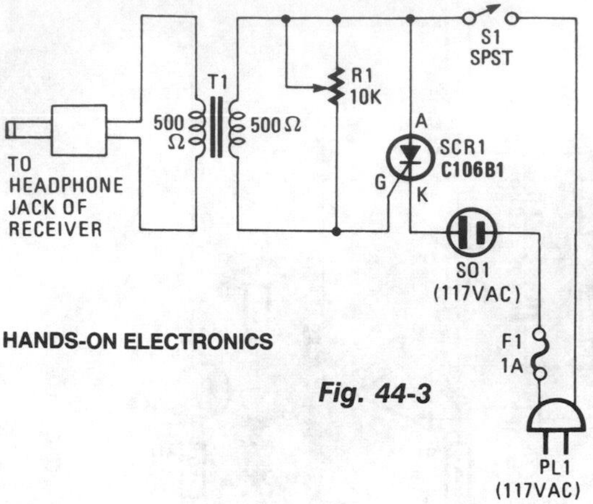

RF-ACTUATED RELAY

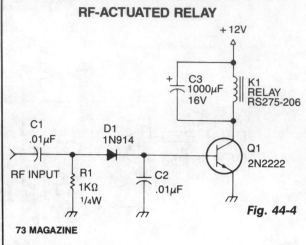

TO HEADPHONE JACK OF RECEIVER

HANDS-ON ELECTRONICS

Fig. 44-3

73 MAGAZINE

Fig. 44-4

Automatic antenna switching or rf power indication can be achieved with this circuit. Relay will key with less than 150 mW drive on 2 m.

ON-THE-AIR INDICATOR

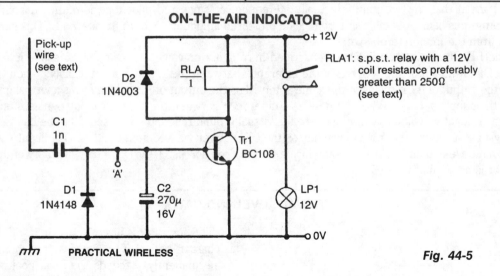

RLA1: s.p.s.t. relay with a 12V coil resistance preferably greater than 250Ω (see text)

PRACTICAL WIRELESS

Fig. 44-5

The circuit is a simple rf-actuated switch which will respond to any strong field in the region of the pickup wire. The length of the wire will depend on how much coupling is needed, but a 250-mm length wrapped around the outside of the coaxial cable feeding the antenna should suffice for most power levels. If only one band is used, the wire can be made a resonant length—495 mm for 144 MHz band operation for example. When rf energy is picked up by the device, diode D1 will conduct on the negative half-cycles, but will be cut off on the positive half-cycles. The result will be a net positive voltage at the base of transistor Tr1, forward biasing it into conduction. On ssb and cw transmissions, where the transmission is not continuous, that bias would be constantly varying and the relay RLA would chatter. However, capacitor C2 holds the bias voltage steady until a long gap in transmissions occurs.

45

Infrared Circuits

The sources of the following circuits are contained in the Sources section beginning on page 782. The figure number contained in the box of each circuit correlates to the sources entry in the Sources section.

Infrared Wireless Speaker System Digital IR Transmitter
Long-Range Object Detector   Simple IR Detector
IR Receiver   Infrared Transmitter
IR Transmitter   Infrared Transmitter

INFRARED WIRELESS SPEAKER SYSTEM

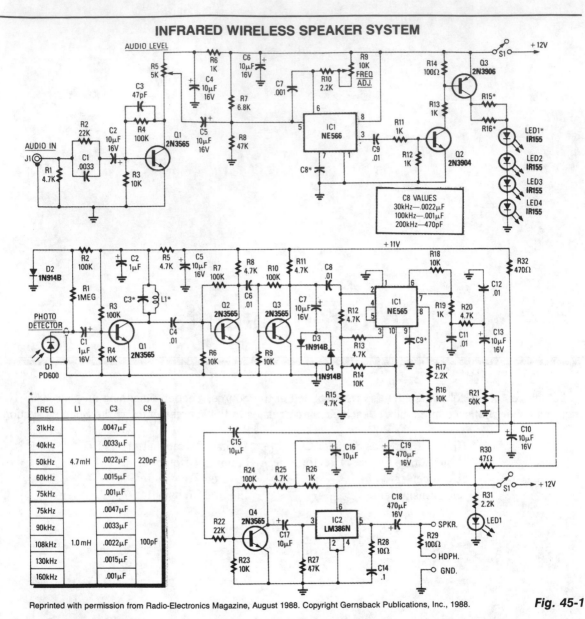

Reprinted with permission from Radio-Electronics Magazine, August 1988. Copyright Gernsback Publications, Inc., 1988.

Fig. 45-1

Although the IR region is free from radio interference, it is subject to interference from incandescent lamps, fluorescent lamps, stray reflections, and other sources.

A simple way to overcome that problem is to create a *carrier* by chopping the IR radiation at a rate of 100 kHz. The audio then modulates the carrier by modulating the chopping rate. A receiver then detects the IR beam as a 100-kHz FM signal. The only disadvantage is that instead of a simple audio amplifier, a high-gain FM receiver is necessary. However, with the ICs that are now available, an FM receiver is easy to build, and contains little more circuitry than a high-gain audio amplifier. The kit is available from North Country Radio, P.O. Box 53, Wykagyl Station, NY 10804.

LONG-RANGE OBJECT DETECTOR

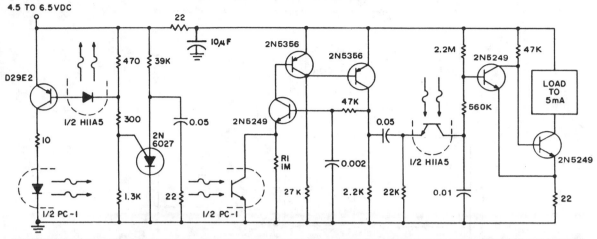

PC-1 SELECTION	TRANSMISSION RANGE	REFLECTIVE RANGE
H23A1	5″	1″
LED56 and L14Q1	12″	3″
LED56 and L14G1	18″	4½″
LED55C and L14G1	32″	8″
1N6266 and L14G3	48″	12″
F5D1 and L14G3	80″	20″
F5D1 and L14P2	200″	50″

GE

When long ranges must be worked with IR light sources, and when high system reliability is required, pulsed-mode operation of the IR is required. Additional reliability of operation is attained by synchronously detecting the photodetector current, as this circuit does. PC-1 is an IR and phototransistor pair which detect the presence of an object blocking the transmission of light from the IR to the phototransistor. Relatively long-distance transmission is obtained by pulsing the IR, with about 10-μs pulses, at a 2-ms period, to 350 mA via the 2N6027 oscillator. The phototransistor current is amplified by the 2N5249 and 2N5356 amplifier to further increase distance and allows use of the H11A5, also pulsed by the 2N6027, as a synchronous detector, providing a fail-safe, noise immune signal to the 2N5249 pair forming a Schmitt-trigger output.

This design was built for battery operation, with long battery life a primary consideration. Note that another stage of amplification driving the IR can boost the range limited by the IR V_F, by 5 to 10 times. A higher supply voltage for the IR can double this range.

Today, optoelectronics are mostly used to transmit electronic information over light beams. These applications range from the use of optocouplers transmitting information between IC logic circuits and power circuits, between power lines and signal circuits, between telephone lines and control circuitry, to the pulse-modulated systems which transmit information through air or fiber optics over relatively great distances.

IR RECEIVER

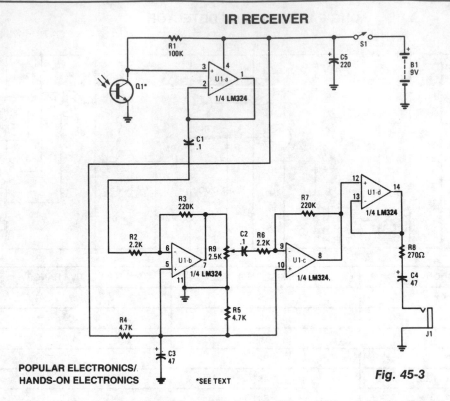

**POPULAR ELECTRONICS/
HANDS-ON ELECTRONICS**

*SEE TEXT

Fig. 45-3

Infrared emissions detected by Q1 are fed through U1a to U1b, which amplifies the signal by a factor of 100. The amplified output of U1b is fed to U1c through R9, C2, and R6. Potentiometer R9 serves as a volume control. With R9 set to pass the maximum signal, U1c provides a gain of 100, for a total system gain of 10,000 dB. The output of U1c is connected to voltage follower circuit U1d to better match and drive headphones that can be plugged into J1.

IR TRANSMITTER

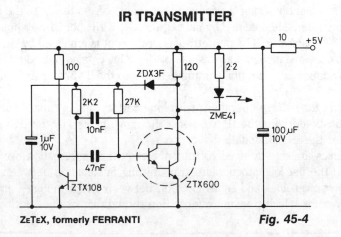

ZETEX, formerly FERRANTI   *Fig. 45-4*

IR TRANSMITTER (Cont.)

The transmitter consists of an oscillator which drives a high output IR emitting diode. The oscillator is a sure start multivibrator circuit that provides an output of 15 to 1000 mark to space ratio at a frequency of 1 kHz. This large mark to space ratio allows the IR diode to be operated at a high peak current, provided by the ZTX600 Darlington transistor, to maximize the transmitter range. A decoupling network is included in the power supply lead to isolate it from any logic circuitry using the same 5-V power supply source. The transmitter supply current is approximately 65 mA.

DIGITAL IR TRANSMITTER

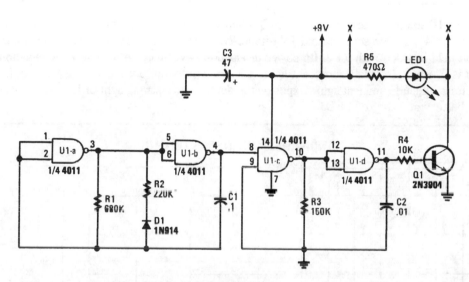

POPULAR ELECTRONICS/HANDS-ON ELECTRONICS

Fig. 45-5

Gates U1a and U1b are configured as a low-frequency oscillator. The output waveform at pin 11 is nonsymmetrical with the positive portion of the signal, making up only 20% of the time period.

Diode D1, a 1N914 general-purpose unit, together with C1, R1, and R2, determine the on time for the positive portion of the output waveform. The off, or negative portion of the output waveform, depends mainly on the values of R1 and C1. The operating frequency of that oscillator is about 11 Hz. The second oscillator consists of U1c and U1d, which outputs on almost symmetrical waveform at a frequency of about 400 Hz. The output of first oscillator U1a/U1b is fed to pin 8 of U1c to key second oscillator U1c/U1d on and off at about 11 Hz, with the on time limited to about 20% of the time period (about 15 ms).

The output waveform of the second oscillator is fed to the base of Q1, which is used to drive IR diode LED1 in short bursts. Pulsing LED1 helps to save battery power, and also allows each circuit to be given its own special sound footprint.

By changing any of the values of R1, R2, R3, C1, or C2, the sound footprint can be varied. As the component values are increased, the oscillator's frequency goes down, and as the values are decreased, the frequency goes up.

SIMPLE IR DETECTOR

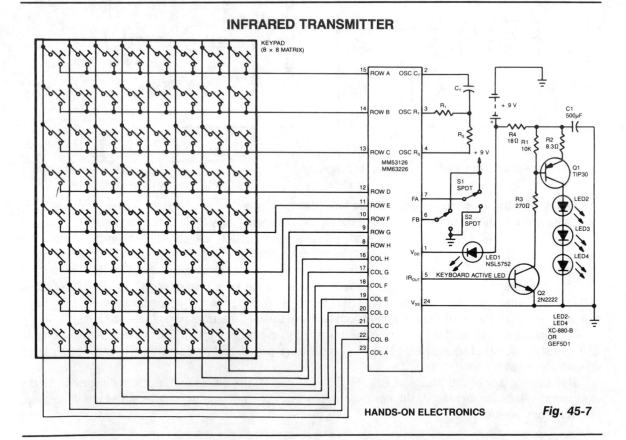

NOTES:
D_1 IS AN ID5531-UR
Q_1 IS A 2N3638
Q_2 IS A TRW 802

EDN

Fig. 45-6

This simple IR detector turns on a real LED when Q2 is exposed to invisible IR radiation found in fiber-optics systems, position sensors, and TV remote-control units. The device can be built on top of a 9-V battery and held in place with RTV. Its power dissipation is virtually zero, unless IR radiation or high ambient light is present. Normal fluorescent lighting is not a problem, but if necessary add an IR filter to the Q2 detector to exclude ambient light. Exposing the detector to a strong light or IR source gives a quick check of the battery and the red LED.

INFRARED TRANSMITTER

HANDS-ON ELECTRONICS

Fig. 45-7

INFRARED TRANSMITTER

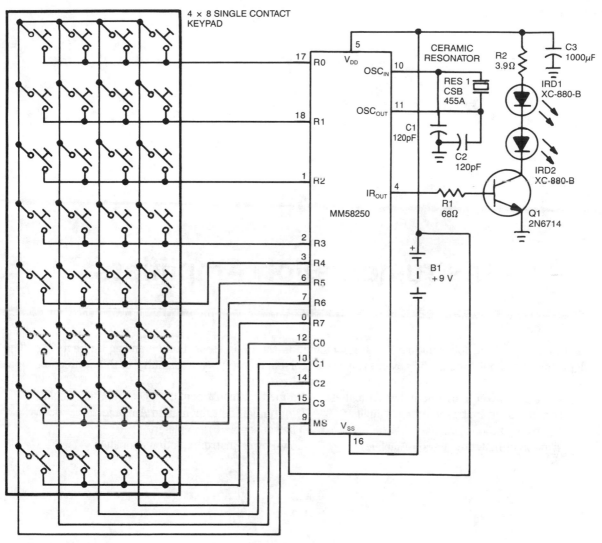

Fig. 45-8

46

Instrumentation Amplifiers

The sources of the following circuits are contained in the Sources section beginning on page 782. The figure number contained in the box of each circuit correlates to the sources entry in the Sources section.

Ultra-Precision Instrumentation Amplifier
Strain Guage Instrumentation Amplifier
Instrumentation Amplifier
Wideband Instrumentation Amplifier

Biomedical Instrumentation Differential Amplifier
Differential Instrumentation Amplifier
Thermocouple Preamplifier
Low-Power Instrumentation Amplifier

ULTRA-PRECISION INSTRUMENTATION AMPLIFIER

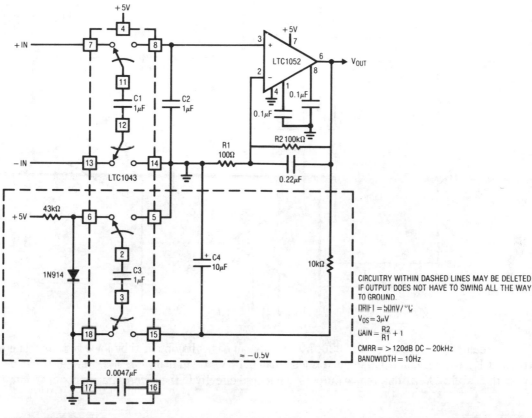

LINEAR TECHNOLOGY

Fig. 46-1

This circuit will run from a single 5 V power supply. The LTC1043 switched-capacitor instrumentation building block provides a differential-to-single-ended transition using a flying-capacitor technique. C1 alternately samples the differential input signal and charges ground referred C2 with this information. The LTC1052 measures the voltage across C2 and provides the circuit's output. Gain is set by the ratio of the amplifier's feedback resistors. Normally, the LTC1052's output stage can swing within 15 mV of ground. If operation all the way to zero is required, the circuit shown in dashed lines can be employed. This configuration uses the remaining LTC1043 section to generate a small negative voltage by inverting the diode drop. This potential drives the 10-KΩ, pull-down resistor, forcing the LTC1052's output into class A operation for voltages near zero. Note that the circuit's switched-capacitor front-end forms a sampled-data filter allowing the common-mode rejection ratio to remain high, even with increasing frequency. The 0.0047µF unit sets front-end switching frequency at a few hundred Hz.

STRAIN GAUGE INSTRUMENTATION AMPLIFIER

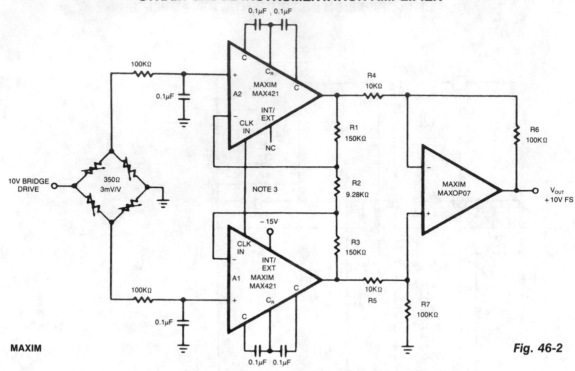

MAXIM

Fig. 46-2

This circuit has an overall gain of 320. More gain can easily be obtained by lowering the value of R2. Untrimmed V_{OS} is 10 μV, and V_{OS} tempco is less than 0.1 μV/°C. In many circuits, the OP07 can be omitted, with the two MAX421 differential outputs connected directly to the differential inputs of an integrating a/d.

INSTRUMENTATION AMPLIFIER

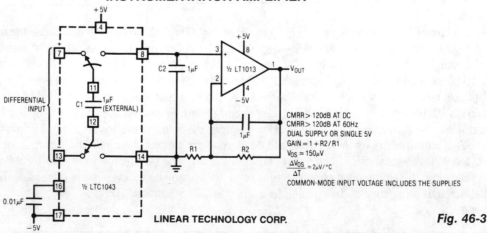

CMRR > 120dB AT DC
CMRR > 120dB AT 60Hz
DUAL SUPPLY OR SINGLE 5V
GAIN = 1 + R2/R1
$V_{OS} \approx 150\mu V$
$\frac{\Delta V_{OS}}{\Delta T} \approx 2\mu V/°C$
COMMON-MODE INPUT VOLTAGE INCLUDES THE SUPPLIES

LINEAR TECHNOLOGY CORP.

Fig. 46-3

INSTRUMENTATION AMPLIFIER (*Cont.*)

LTC1043 and LT1013 dual op amps are used to create a dual instrumentation amplifier using just two packages. A single DPDT section converts the differential input to a ground-referred single-ended signal at the LT1013's input. With the input switches closed, C1 acquires the input signal. When the input switches open, C2's switches close and C2 receives charge. Continuous clocking forces C2's voltage to equal the difference between the circuit's inputs. The 0.01-μF capacitor at pin 16 sets the switching frequency at 500 Hz. Common-mode voltages are rejected by over 120 dB and drift is low.

WIDEBAND INSTRUMENTATION AMPLIFIER

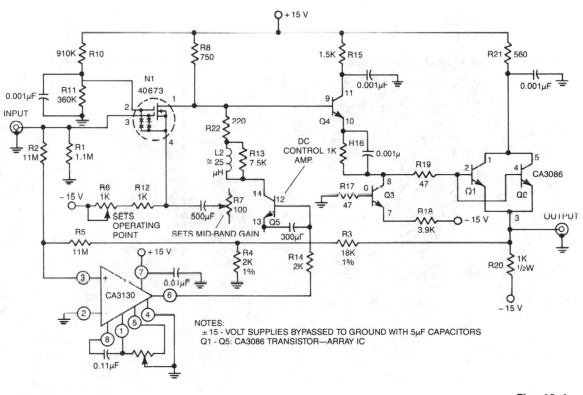

NOTES:
±15 - VOLT SUPPLIES BYPASSED TO GROUND WITH 5μF CAPACITORS
Q1 - Q5: CA3086 TRANSISTOR—ARRAY IC

GE/RCA

Fig. 46-4

Has an input resistance of 1-MΩ, a bandwidth from dc to about 35 MHz, and a gain of 10 times. Low-frequency gain is provided by a CA3130 BiMOS op amp operated as a single-supply amplifier. High-frequency gain is provided by a 40673 dual-gate MOSFET. The entire amplifier is nulled by shorting the input to ground and adjusting R9 for zero dc output voltage.

BIOMEDICAL INSTRUMENTATION DIFFERENTIAL AMPLIFIER

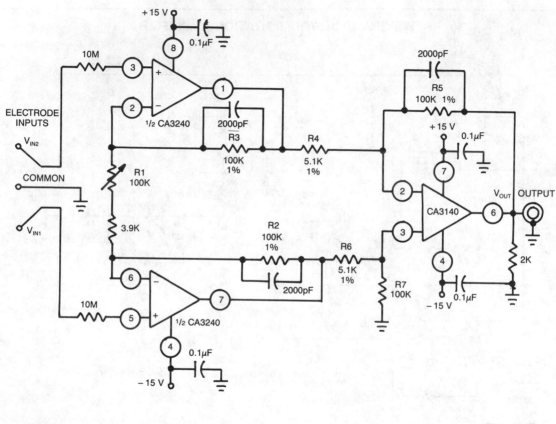

GE/RCA

Fig. 46-5

This differential amplifier uses the isolated high-impedance inputs of the CA3420 BiMOS op amp. Because the CA3240's input current is only 50 pA maximum, 10-MΩ resistors can be used in series with the input probes to limit the current to 2 μA under a fault condition.

DIFFERENTIAL INSTRUMENTATION AMPLIFIER

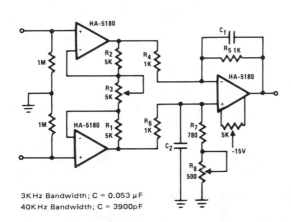

3KHz Bandwidth; C = 0.053 μF
40KHz Bandwidth; C = 3900pF

HARRIS   **Fig. 46-6**

This circuit relies on extremely high input impedance for effective operation. The HA-5180 with its JFET input stage, performs well as a preamplifier. The standard three amplifier configuration is used with very close matching of the resistor ratios $R5/R4$ and $(R7 + R8)/R6$, to insure high common-mode rejection (CMR). The gain is controlled through $R3$ and is equal to $2R1/R3$. Additional gain can be had by increasing the ratios $R5/R4$ and $(R7 + R8)/R6$. The capacitors C1 and C2 improve the ac response by limiting the effects of transients and noise. Two suggested values are given for maximum transient suppression at frequencies of interest. Some of the faster DVM's are operating at peak sampling frequency of 3-kHz, hence the 4-kHz, low-pass time constant. The 40-kHz, low-pass time constant for ac voltage ranges is an arbitrary choice, but should be chosen to match the bandwidth of the other components in the system. C1 and C2 might however, reduce CMR for ac signals if not closely matched. Input impedances have also been added to provide adequate dc bias currents for the HA-5180 when open-circuited.

THERMOCOUPLE PREAMPLIFIER

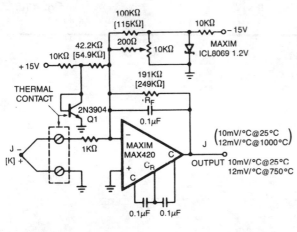

MAXIM   **Fig. 46-7**

The MAX420 is operated at a gain of 191 to convert the 52 μV/°C output of the type J thermocouple to a 10 mV/°C signal. The −2.2 mV/°C tempco of the 2N3904 is added into the summing junction with a gain of 42.2 to provide cold-junction compensation. The ICL8069 is used to remove the offset caused by the 600-mV initial voltage of the 2N3904. Adjust the 10-KΩ trimpot for the proper reading with the 2N3904 and isothermal connection block at a temperature near the center of the circuit's operating range. Use the component values shown in parentheses when using a type K thermocouple.

LOW-POWER INSTRUMENTATION AMPLIFIER

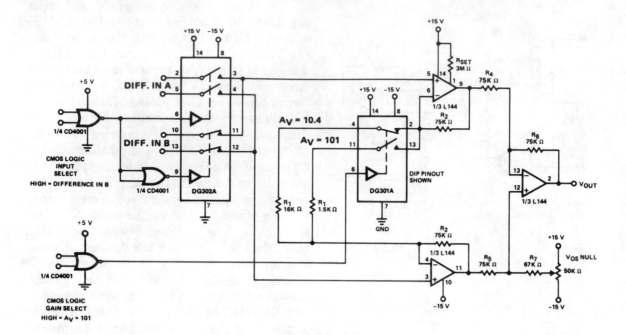

R_{SET} programs L144 power dissipation, gain-bandwidth product. Refer to AN73-6 and the L144 data sheet.

Voltage gain of the instrumentation amplifier is:

$$A_V = 1 + \frac{2R_2}{R_1} \quad \text{(In the circuit shown, } A_{V1} = 10.4, A_{V2} = 101\text{)}$$

SILICONIX

Fig. 46-8

284

47

Integrator Circuits

The sources of the following circuits are contained in the Sources section beginning on page 782. The figure number contained in the box of each circuit correlates to the sources entry in the Sources section.

Resettable Integrator
Integrator with Programmable Reset Level

RESETTABLE INTEGRATOR

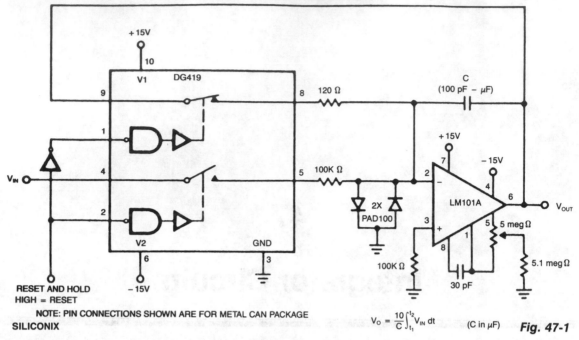

$$V_O = \frac{10}{C}\int_{t_1}^{t_2} V_{IN}\, dt \quad (C\ in\ \mu F)$$

Fig. 47-1

The low $r_{DS(on)}$ and high peak current capability of the DG419 makes it ideal for discharging an integrator capacitor. A high logic input pulse disconnects the integrator from the analog input and discharges the capacitor. When the logic input lowers, the integrator is triggered. D1 and D2 prevent the capacitor from charging to over 15 V.

INTEGRATOR WITH PROGRAMMABLE RESET LEVEL

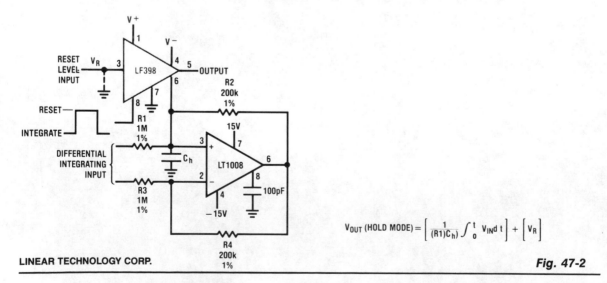

$$V_{OUT\ (HOLD\ MODE)} = \left[\frac{1}{(R1)C_h}\int_0^t V_{IN}d\,t\right] + \left[V_R\right]$$

LINEAR TECHNOLOGY CORP.

Fig. 47-2

48

Intercom Circuits

The sources of the following circuits are contained in the Sources section beginning on page 782. The figure number contained in the box of each circuit correlates to the sources entry in the Sources section.

Personal Pocket Pager
Bidirectional Intercom System
Intercom
Hands-Off Intercom
Two-Way Intercom

PERSONAL POCKET PAGER

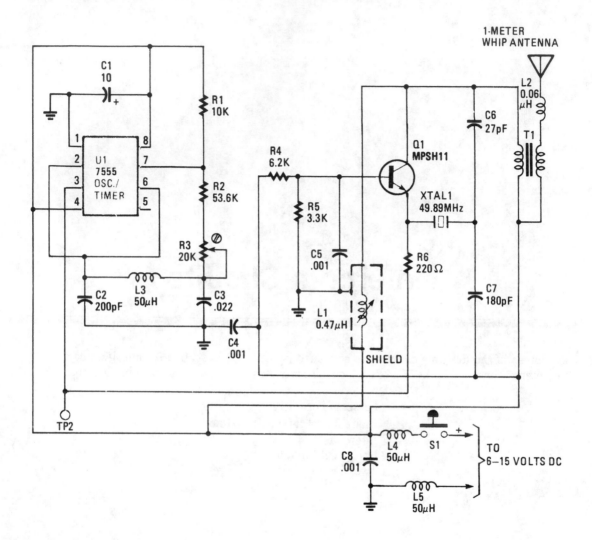

Fig. 48-1

When activated, the transmitter sends out a 49.890-MHz, AM rf carrier. The receiver detects, amplifies, and decodes the rf signal, which, in turn, activates a piezo buzzer. The receiver is small enough to carry in a pocket or sit on your workbench. The transmitter is also small and fits easily into a pocket for quick access.

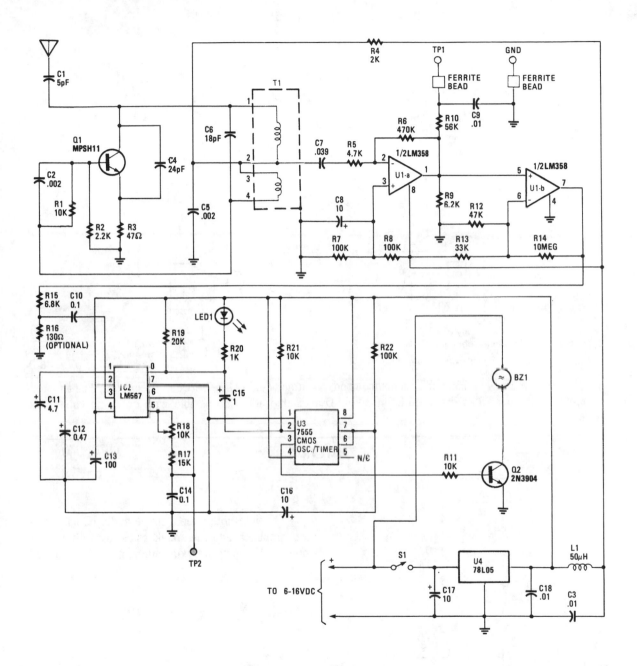

BIDIRECTIONAL INTERCOM SYSTEM

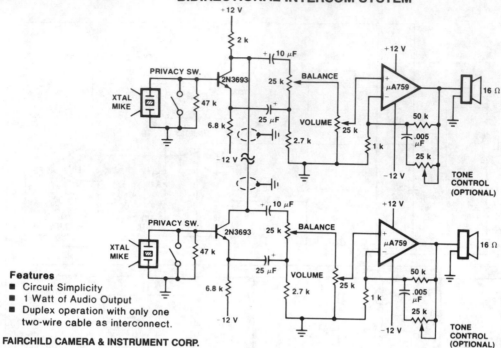

FAIRCHILD CAMERA & INSTRUMENT CORP.

Features
- Circuit Simplicity
- 1 Watt of Audio Output
- Duplex operation with only one two-wire cable as interconnect.

Fig. 48-2

This system uses µA759 audio IC devices and a common connection between the preamps as an interconnect. Either mike can drive either speaker. Duplex operation is possible with only one cable (two wires).

INTERCOM

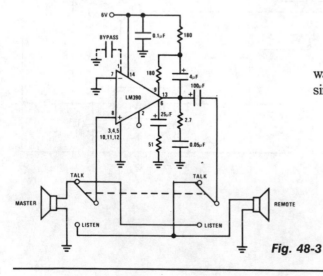

This intercom uses a single audio IC as a two-way amplifier, and the speakers as microphones. A single 6-V supply provides adequate audio volume.

NATIONAL SEMICONDUCTOR CORP.

Fig. 48-3

HANDS-OFF INTERCOM

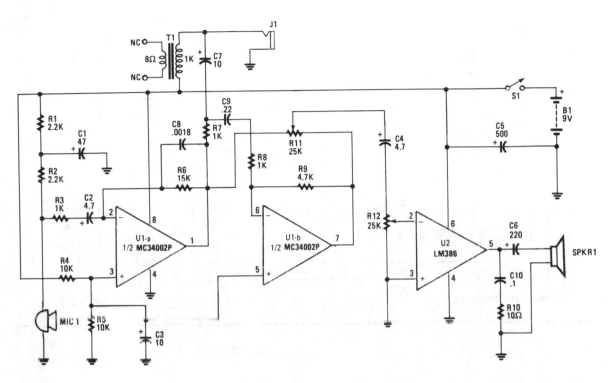

HANDS-ON ELECTRONICS/POPULAR ELECTRONICS

Fig. 48-4

Amplifier A increases the microphone's output to a usable level. The output signal is fed to op amp B, which inverts the signal 180°. A balance-control potentiometer connects across the outputs of amplifiers A and B. If an audio tone is fed into the microphone and the balance potentiometer's wiper is all the way over to the A output position, the tone will be heard at a high level. As the wiper is rotated toward the B output, the audio level will decrease until it just about disappears near the center of the potentiometer's range. As you continue to rotate the wiper, the signal will begin to increase once again.

With the balance control set for a minimal output, the intercom's tendency to self-oscillate from acoustical feedback between the microphone and speaker is kept to a minimum. The microphone's amplified signal at A's output is fed to the other intercom through the audio in/out cable. Since both intercom units are alike, the audio information coming from one unit feeds the other at the input of op amp B. The incoming audio is amplified slightly by op amp B and the output signal is sufficiently increased by the power amp to drive the speaker.

TWO-WAY INTERCOM

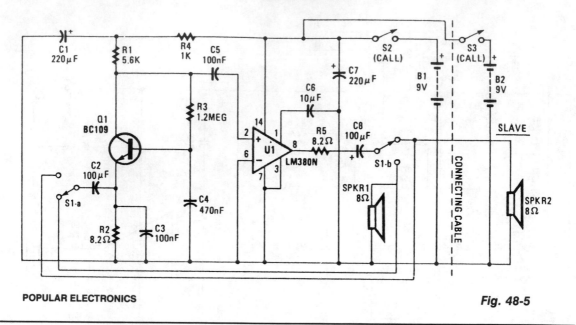

Fig. 48-5

49

Inverters

The sources of the following circuits are contained in the Sources section beginning on page 782. The figure number contained in the box of each circuit correlates to the sources entry in the Sources section.

12 Vdc-to-117 Vac at 60 Hz Power Inverter
Power MOSFET Inverter
Medium Power Inverter
Complementary Output Variable Frequency
    Inverter
Precision Voltage Inverter
Power Inverter

12 VDC-TO-117 VAC AT 60 Hz POWER INVERTER

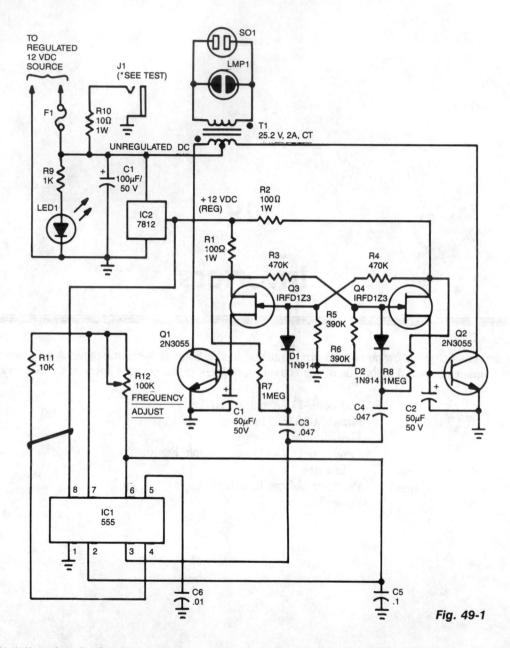

Fig. 49-1

Capacitor C5 and potentiometer R12 determine the frequency of the output signal at pin 3 of IC1, the 555 oscillator. The output signal is differentiated by C3 and C4 before it's input to the base of power transistors Q1 and Q2 via diodes D1 and D2, respectively. The signal from IC1 is adjusted to 120 Hz, because the flip-flop formed by transistors Q3 and Q4 divides the frequency by 2.

When Q3 is on, the base of Q1 is connected via R1 to the regulated 12-V supply. Then, when the flip-flop changes states, Q4 is turned on and the base of Q2 connected to the 12-V supply through R2. The 100 mA base current allows Q1 and Q2 to alternately conduct through their respective halves to the transformer's secondary winding.

To eliminate switching transients caused by the rapid switching of Q3 and Q4, capacitors C1 and C2 filter the inputs to the base of Q1 and Q2 respectively. Power for the unit comes from an automobile's 12-V system or from a storage battery. The power is regulated by IC2, a 7812 regulator. LED1, connected across the 12-V input, can be used to indicate whether power is being fed to the circuit. The neon pilot lamp, LMP1, shows a presence or absence of output power.

POWER MOSFET INVERTER

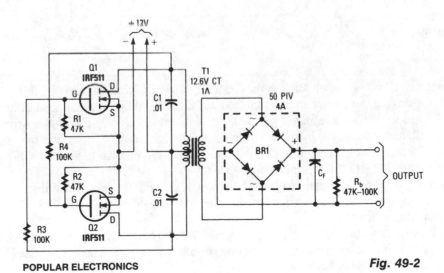

POPULAR ELECTRONICS

Fig. 49-2

This inverter can deliver high-voltage ac or dc, with a rectifier and filter, up to several hundred volts. The secondary and primary of T1—a 12.6 to 440 V power transformer, respectively—are reversed; e.g., the primary becomes the secondary and the secondary becomes the primary. Transistors Q1 and Q2 can be any power FET. Be sure to heat sink Q1 and Q2. Capacitors C1 and C2 are used as spike suppressors.

MEDIUM POWER INVERTER

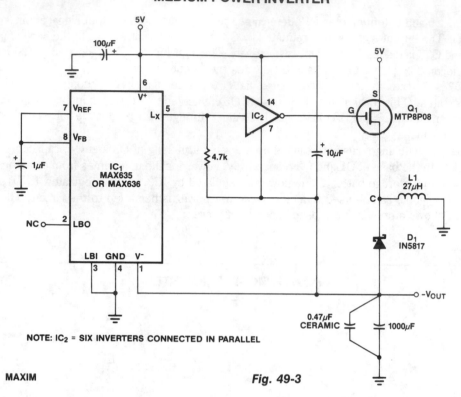

NOTE: IC$_2$ = SIX INVERTERS CONNECTED IN PARALLEL

MAXIM

Fig. 49-3

In this circuit, a CMOS inverter, such as the CD4069, is used to convert the open drain L_x output to a signal suitable for driving the gate of an external P MOSFET. The MTP8P03 has a gate threshold voltage of 2.0 V to 4.5 V, so it will have a relatively high resistance if driven with only 5 V of gate drive. To increase the gate drive voltage, and thereby increase efficiency and power handling capability, the negative supply pin of the CMOS inverter is connected to the negative output, rather than to ground. Once the circuit is started, the P MOSFET gate drive swings from +5 V to $-V_{OUT}$. At start up, the $-V_{OUT}$ is one Schottky diode drop above ground and the gate drive to the power MOSFET is slightly less than 5 V. The output should be only lightly loaded to ensure start up, since the output power capability of the circuit is very low until $-V_{OUT}$ is a couple of volts.

This circuit generates complementary output signals from 50 to 240 Hz. Digital timing control ensures a separation of 10 to 15° between the fall time of one output and the rise time of the complementary output.

The digital portion of inverter U1 to U4 controls the drive to Q1 and Q2, both MTE60N20 TMOS devices. These devices are turned on alternately with 11.25° separation between complementary outputs. A +12-V supply for CMOS gates U1 to U4 is developed by T1, D3, D4, C7, and U6. The power supply for the TMOS frequency generator is derived from the diode bridge, U5, and capacitor C7; it is applied to the center tap of T2.

COMPLEMENTARY OUTPUT VARIABLE FREQUENCY INVERTER

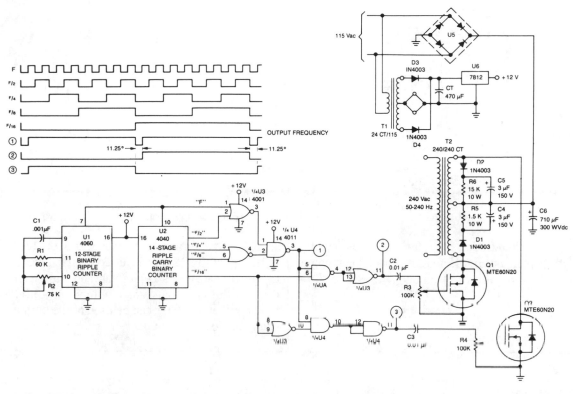

Fig. 49-4

U1 is a 4060 12-stage binary ripple counter that is used as a free-running oscillator; its frequency of oscillation is: $1/2.2 \, C1R2$. The output of U1 is applied to U2, a 14-stage binary ripple counter that provides square-wave outputs of $1/2$, $1/4$, $1/8$, and $1/16$ of the clock frequency. These signals are combined in U3 and U4 to provide a complementary drive for Q1 and Q2.

Outputs from U3 and U4 are ac-coupled to Q1 and Q2 via C2 and C4, respectively. R3 and R4 adjust the gate drive to Q1 and Q2. Q1 and Q2 alternately draw current through opposing sides of the primary to synthesize an ac input voltage at a given frequency. Only one side of the primary of T2 is driven at one time, so maximum power output is half of the transformer rating.

PRECISION VOLTAGE INVERTER

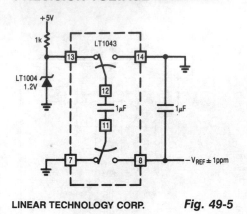

LINEAR TECHNOLOGY CORP. **Fig. 49-5**

This circuit allows a reference to be inverted with 1 ppm accuracy, features high input impedance, and requires no trimming.

POWER INVERTER

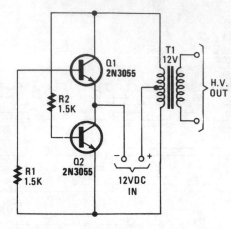

POPULAR ELECTRONICS **Fig. 49-6**

The transformer can be any 6.3 or 12.6 V type. Apply the 12-Vdc input so the positive goes to the transformer's center tap and the negative goes to the two transistor emitters. Any bridge-type rectifier and filter can be used at the output, if you need dc.

50

Lamp-Control Circuits

The sources of the following circuits are contained in the Sources section beginning on page 782. The figure number contained in the box of each circuit correlates to the sources entry in the Sources section.

Halogen Lamp Dimmer
Pseudorandom Sequencer
Light Modulator
Lamp Life Extender
Phase Control
Triac Light Dimmer
800-W Soft-Start Light Dimmer

Solid-State Light Dissolver
Line-Voltage Operated Automatic
    Night Light
8-W Fluorescent Lamp Inverter
Pulse-Width Modulation Lamp
    Brightness Controller
Constant Brightness Control

HALOGEN LAMP DIMMER

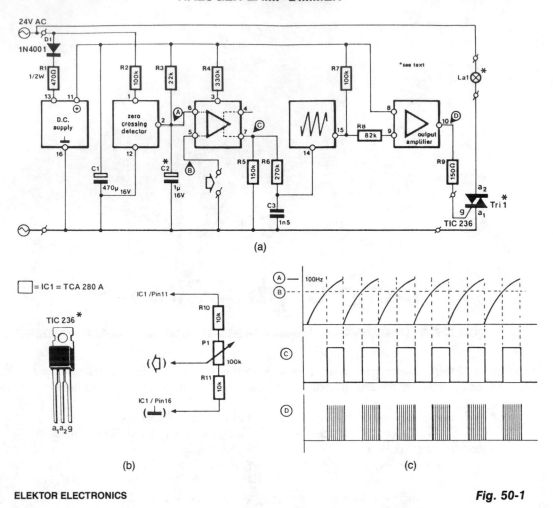

Fig. 50-1

This circuit is suitable for fitting into slide projectors without a dimmer facility as with 24-Vac fed halogen lamps. With a few small alterations, it can also be used for dimming 12-V halogen lamps, but not those in a car, because these are fed from a dc source. The circuit shown in Fig. 50-1a is intended for operation from a 24-Vac supply, and can handle a lamp load of up to 150 W. For loads up to 250 W, the TIC236 should be replaced by a TIC246.

Figure 50-1b shows detail of the connection of a potentiometer to the intensity control input of the TCA280A. Voltage divider R10-P1-R11 is fitted externally and can be fed from the stabilized voltage available at pin 11 of IC1. The minimum and maximum intensities of the lamp are determined by R10 and R11, respectively.

PSEUDORANDOM SEQUENCER

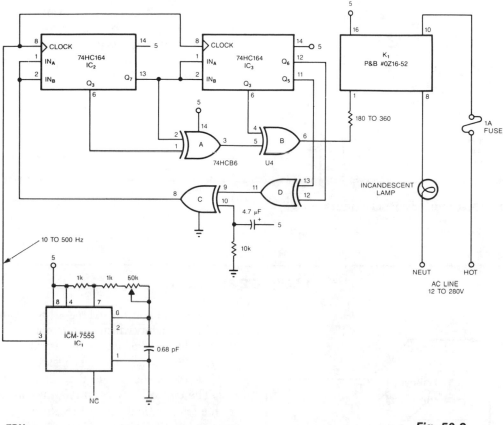

Fig. 50-2

This pseudorandom sequencer drives a solid-state relay. If you power a low-wattage lamp from the relay, the lamp will appear to flicker like a candle's flame in the wind; using higher-wattage lamps allows you to simulate the blaze of a fireplace or campfire. You can enhance the effect by using three or more such circuits to power an array of lamps.

The circuit is comprised of an oscillator IC1 and a 15-stage, pseudorandom sequencer, IC2 through IC4. The sequencer produces a serial bit stream that repeats only every 32, 767 bits. Feedback from the sequencer's stages 14 and 15 go through IC4D and back to the serial input of IC2. Note the rc network feeding IC4C; the network feeds a positive pulse into the sequencer to ensure that it won't get stuck with all zeros at power-up. The leftover XOR gates IC4A and IC4B further scramble the pattern. The serial stream from IC4B drives a solid-state relay that features zero-voltage switching and can handle loads as high as 1 A at 12 to 280 Vac.

LIGHT MODULATOR

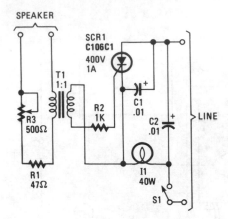

SPEAKER

SCR1
C106C1
400V
1A

T1
1:1

R2
1K

C1
.01

C2
.01

LINE

R3
500Ω

R1
47Ω

I1
40W

S1

HANDS-ON ELECTRONICS/POPULAR ELECTRONICS

Fig. 50-3

The lights seem to dance in time with the music. Line-voltage lamps of about 40 to 100 W do nicely. The current for the lamp is from an SCR. When low-level audio is present across T1, SCR1 is not triggered into conduction. A louder signal, however, triggers the SCR so that the lamp lights and follows the sounds. Since SCR1 is operated by an alternating current, the rectifier moves out of the avalanche condition when the gate current is low. Potentiometer R3 lets you adjust the power reaching transformer T1, so that with normal operating volume, SCR1 triggers again and again, except during quiet passages.

LAMP LIFE EXTENDER

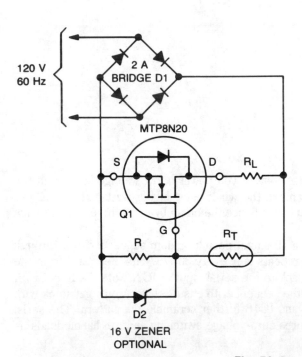

120 V
60 Hz

2 A
BRIDGE D1

MTP8N20

S   D R_L

Q1

G R_T

R

D2

16 V ZENER
OPTIONAL

Fig. 50-4

Lamp life can be extended by improving the conditions under which its filament is operated. This includes eliminating the inrush overcurrent surge and reducing the mechanical stress (vibration) on the filament caused by an ac source.

The circuit shown controls the inrush current to the lamp without the 10 to 15 times-rated current stage that normally occurs when power is applied to a cold lamp. It does so by adjusting the inrush current over time to the inverse of the value normally experienced.

R_L is a standard tungsten lamp in the range of 15 to 250 W, R is 10-Ω and R_T is a negative temperature coefficient resistance that is initially 1.65 MΩ and decreases, by self-heating, to 150 KΩ in approximately 0.5 s. Use of the TMOS device allows high Ω values for R and R_T, keeping drive power at a negligible level.

This circuit has a number of advantages: very low power dissipation, long life, low-cost components, no significant effect on lamp ratings, negligible effect on efficiency, negligible RFI, and it can be used in hazardous environments.

PHASE CONTROL

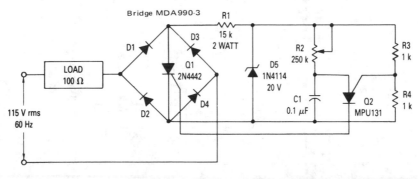

Bridge MDA990-3

Fig. 50-5

This circuit uses a PUT for phases control of an SCR. The relaxation oscillator formed by Q2 provides conduction control of Q1 from 1 to 7.8 ms or 21.6° to 168.5°. This constitutes control of over 97% of the power available to the load. Only one SCR is needed to provide phase control for both the positive and negative portion of the sine wave by putting the SCR across the bridge—composed of diodes D1 through D4.

TRIAC LIGHT DIMMER

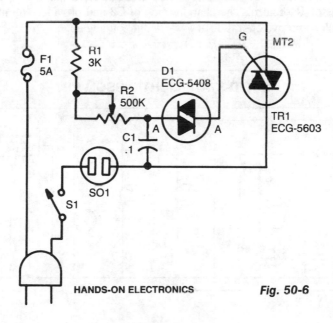

HANDS-ON ELECTRONICS   *Fig. 50-6*

800-W SOFT-START LIGHT DIMMER

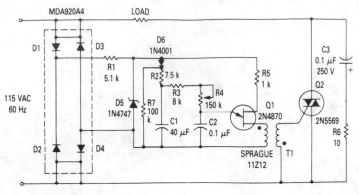

Fig. 50-7

The zener provides a constant voltage of 20 V to unijunction transistor Q1, except at the end of each half-cycle of the input when the line voltage drops to zero. Initially, the voltage across capacitor C1 is zero and capacitor C2 cannot charge to trigger Q1. C1 will begin to charge, but because the voltage is low, C2 will be charged to a voltage adequate to trigger C1 only near the end of the half cycle. Although the lamp resistance is low at this time, the voltage applied to the lamp is low and the inrush current is small. Then the voltage on C1 rises, allowing C2 to trigger Q1 earlier in the cycle. At the same time, the lamp is being heated by the slowly increasing applied voltage. By the time the peak voltage applied to the lamp has reached its maximum value, the bulb has been heated sufficiently to keep the peak inrush current at a reasonable value. Resistor R4 controls the charging rate of C2 and provides the means to dim the lamp. Diode D6 and resistor R7 improve operation at low-conduction angles.

SOLID-STATE LIGHT DISSOLVER

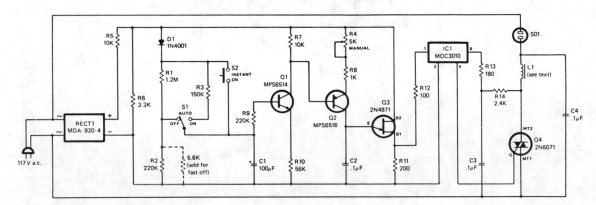

MODERN ELECTRONICS

Fig. 50-8

SOLID-STATE LIGHT DISSOLVER (*Cont.*)

The dimming action is controlled by varying the amount of current passed through triac Q4 and, thus, the lamp plugged into ac receptacle SO1. Unijunction transistor Q3 operates as a relaxation oscillator whose output pulse frequency depends on how fast capacitor C2 recharges after firing. Transistors Q1 and Q2 furnish the charging current, with the R3/C1 and R1/R2/C1 time-constant networks controlling the turn-on and turn-off times. Inside IC1 is a LED, a detector, and a small triac. In circuit, the low-level pulses coming from Q3 make the LED in IC1 emit short bursts of light that are picked up and converted into electrical current pulses by the internal detector. This small current triggers the internal triac, which then outputs the pulses to the gate of power triac Q4, triggering it on so that it delivers current to the lamp. Potentiometer R4 serves as a master control of the pulse rate and provides both manual control and a limit in the brightness of the lamp plugged into SO1. Momentarily pressing S2 causes the lamp to instantly turn on. Choke L1 suppresses any spikes produced by the power triac and limits interference with AM radio reception. No safeguards against interference need to be made for FM and TV reception, since these media are immune to this type of noise.

PHASE CONTROL

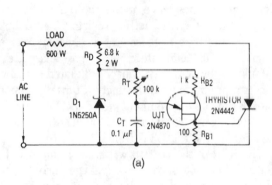

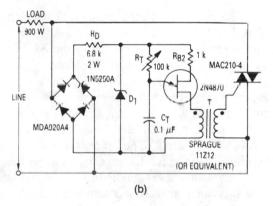

(a)

(b)

Fig. 50-9

The most elementary application is a half-wave control circuit. The thyristor is acting both as a power control device and as a rectifier, providing variable power to the load during the positive half cycle and no power to the load during the negative half cycle. The circuit is designed to be a two terminal control which can be inserted in place of a switch. If full-wave power is desired as the upper extreme of this control, a switch can be added which will short circuit the SCR when R_T is turned to its maximum power position. Full-wave control might be realized by the addition of a bridge rectifier, a pulse transformer, and by changing the thyristor from an SCR to a TRIAC, shown in Fig 50-9b. In this circuit, R_{B1} is not necessary since the pulse transformer isolates the thyristor gate from the steady-state UJT current. Occasionally, a circuit is required to provide constant output voltage regardless of line voltage changes. Adding potentiometer P1 to the circuits will provide an approximate solution to this problem. The potentiometer is adjusted to provide reasonably constant output over the desired range of line voltage.

LINE-VOLTAGE OPERATED AUTOMATIC NIGHT LIGHT

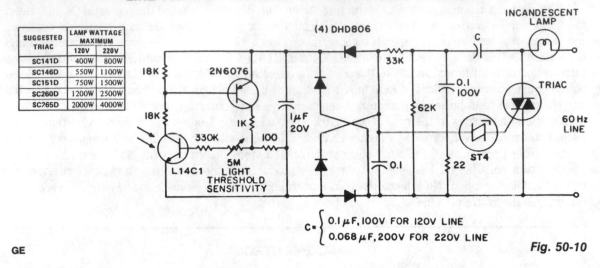

SUGGESTED TRIAC	LAMP WATTAGE MAXIMUM	
	120V	220V
SC141D	400W	800W
SC146D	550W	1100W
SC151D	750W	1500W
SC260D	1200W	2500W
SC265D	2000W	4000W

$$C = \begin{cases} 0.1\,\mu F, 100V \text{ FOR } 120V \text{ LINE} \\ 0.068\,\mu F, 200V \text{ FOR } 220V \text{ LINE} \end{cases}$$

GE

Fig. 50-10

This circuit has stable threshold characteristics from its dependence on the photo diode current in the L14C1 to generate a base emitter voltage drop across the sensitivity setting resistor. The double phase shift network supplying voltage to the ST-4 trigger insures triac triggering at line voltage phase angles small enough to minimize RFI problems with a lamp load. This eliminates the need for a large, expensive inductor, contains the dV/dt snubber network, and utilizes lower voltage capacitors than the snubber of rfi suppression network normally used. The addition of a programmable unijunction timer can modify this circuit to turn the lamp on for a fixed time interval each time its environment gets dark. Only the additions to the previous circuit are shown in the interest of simplicity. When power is applied to the lamp, the 2N6028 timer starts. Upon completion of the time interval, the H11C3 is triggered and turns off the lamp by preventing the ST-4 from triggering the triac. The SCR of the H11C3 will stay on until the L14C1 is illuminated and allows the 2N6076 to commutate it off. Because of capacitor leakage currents, temperature variations and component tolerances, the time delay may vary considerably from nominal values.

8-W FLUORESCENT LAMP INVERTER

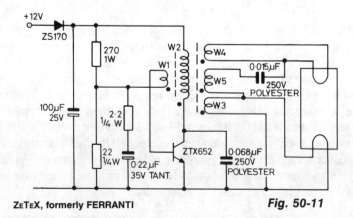

ZᴇTᴇX, formerly FERRANTI

Fig. 50-11

8-W FLUORESCENT LAMP INVERTER (*Cont.*)

This circuit has been designed to drive an 8-W fluorescent lamp from a 12-V source using an inexpensive inverter based on the ZTX652 transistor. The inverter will operate from supplies in the range of 10 to 16.5 V, thus making it suitable for use in on-charge systems such as caravanettes as well as periodically charged systems, such as camping lights, outhouse lights, etc. Other features of the inverter are an inaudible 20-kHz oscillator and reverse polarity protection.

PULSE-WIDTH MODULATION LAMP BRIGHTNESS CONTROLLER

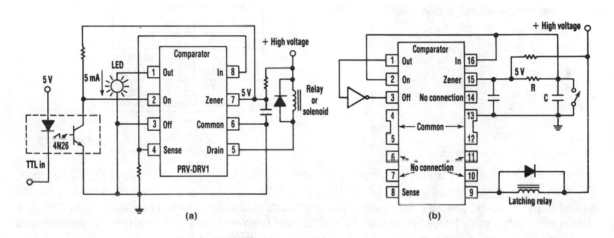

ELECTRONIC DESIGN

Fig. 50-12

At half brightness, the lamp current is pulsed on and off (Fig. 50-12b) by the voltage developed across the resistor and capacitor at the current-sense output. Lamp current is sensed by the current-sense output. A simple pulse-width moduliation lamp-brightness control circuit can also be built with the device. When the device powers up, the sense output is low, pulling the comparator output and the on input low, and turning the FET switch on. When the switch is on, current from the sense output charges the capacitor in the rc timing network to the 200-mV comparator threshold voltage. The comparator trips, turning the switch off. The charge then leaks off the capacitor, its voltage drops below 100 mV, and the FET is again turned on. The average current through the load is basically a function of the resistor value. The pulse-width modulation frequency on the other hand, is a function of the capacitor value.

CONSTANT BRIGHTNESS CONTROL

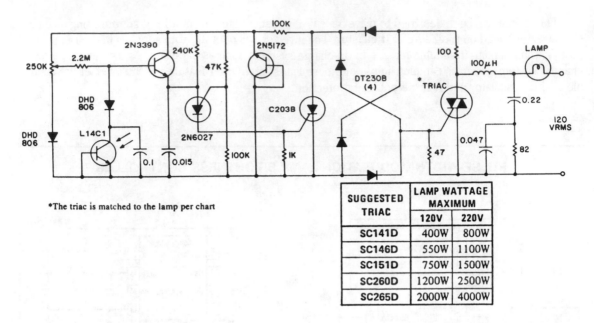

*The triac is matched to the lamp per chart

SUGGESTED TRIAC	LAMP WATTAGE MAXIMUM	
	120V	220V
SC141D	400W	800W
SC146D	550W	1100W
SC151D	750W	1500W
SC260D	1200W	2500W
SC265D	2000W	4000W

GE

Fig. 50-13

An automatic control maintains a lamp at a constant brightness over a wide range of supply voltages. This circuit utilizes the consistency of photodiode response to control the phase angle of power line voltage applied to the lamp and can vary the power between that available and $\approx 30\%$ of available. This provides a candlepower range from 100% to less than 10% of nominal lamp output. The 100-μH choke, resistor, and capacitors form an rlc filter network that is used to eliminate conducted RFI.

51

Laser Circuits

The sources of the following circuits are contained in the Sources section beginning on page 782. The figure number contained in the box of each circuit correlates to the sources entry in the Sources section.

Visible Red Continuous Laser Gun
Laser Diode Pulsers

VISIBLE RED CONTINUOUS LASER GUN

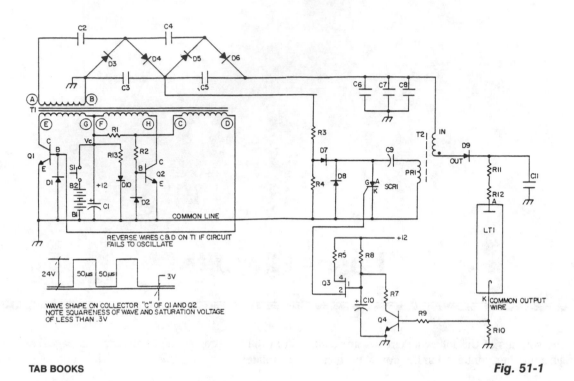

Fig. 51-1

Q1 and Q2 switch the primary windings of transformer T1 via a square wave at a frequency determined by its magnetic properties. Diodes D1 and D2 provide base return paths for the feedback current of Q1 and Q2. The output winding of T1 is connected to a multiple section voltage multiplier. That multiplier consists of capacitors C1 through C5 and diodes D3 through D6. Resistors R3 and R4 divide the 800 V taken off at the junction of C3, and C5 for charging the dump capacitor C9 in the ignitor circuit. The ignitor, consisting of the T2 pulse transformer and capacitor discharge circuitry, provides the high-voltage dc pulse to ignite the laser the SCR1 dumping the energy of capacitor C9 into the primary of T2. The high-voltage pulses in capacitor C11 through rectifier diode D9. When C11 is charged to a LT1, ignition takes place and a current now flows that is sufficient to sustain itself at the lower voltage output of the voltage multiplier section. The path for this sustaining current is through the secondary of T2 and ballast resistors R11 and R12. The ignitor circuit is now deactivated by the clamping of Q3 emitter via Q4 being turned on by the voltage drop occurring across R10. This voltage drop will only occur when the laser tube is ignited and causes the SCR1 to cease firing; otherwise, the ignitor circuit would continue to operate, unnecessarily drawing on the limited power available.

LASER DIODE PULSERS

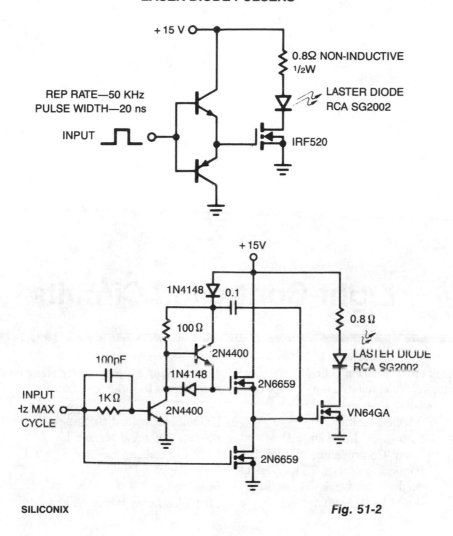

SILICONIX

Fig. 51-2

The laser diode pulser is a simple drive circuit capable of driving the laser diode with 10-A, 20-ns pulses. For a 0.1% duty cycle, the repetition rate will be 50 kHz. A complementary emitter follower is used as a driver. Switching speed is determined by the f_T of the bipolar transistors used and the impedance of the drive source. A faster drive circuit is shown. It can supply higher peak gate current to switch the IRF520 very quickly. This circuit uses a MOSPOWER totem-pole stage to drive the high power switch. The upper MOSFET is driven by a bootstrap circuit. Typical switching times for this circuit are about 10 ns for both turn-on and turn-off.

52

Light-Controlled Circuits

The sources of the following circuits are contained in the Sources section beginning on page 782. The figure number contained in the box of each circuit correlates to the sources entry in the Sources section.

Flame Monitor
Low-Light Level Drop Detector
Light-Controlled Lamp Switch
Optical Sensor-to-TTL Interface
Light-Sensitive Audio Oscillator
Light Level Detector

Lighted Display and Brightness Control
Warning Light and Marker Light
Light-Controlled One-Shot Timer
Solar-Triggered Switch
Sun Tracker
Photoelectric Ac Power Switch

FLAME MONITOR

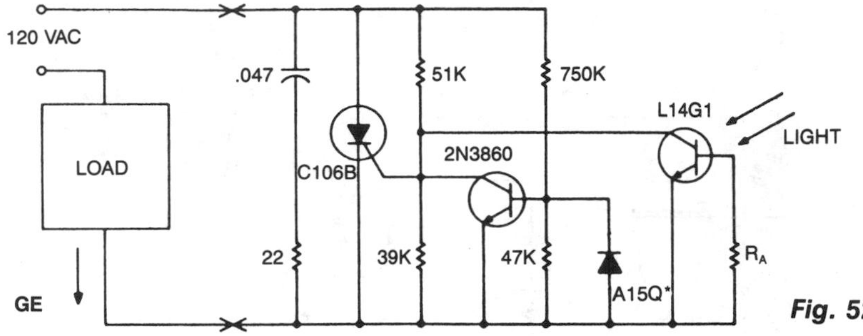

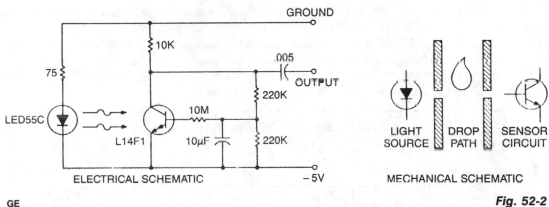

Fig. 52-1

Monitoring a flame and directly switching a 120-V load is easily accomplished using the L14G1 for point sources of light. For light sources which subtend over 10° of arc, the L14C1 should be used and the illumination levels raised by a factor of 5. This circuit provides zero voltage switching to eliminate phase controlling.

LOW-LIGHT LEVEL DROP DETECTOR

Fig. 52-2

This self-biasing configuration is useful any time small changes in light level must be detected, for example, when monitoring very low flow rates by counting drops of fluid. In this bias method, the photodarlington is dc bias stabilized by feedback from the collector, compensating for different photodarlington gains and LED outputs. The 10-μF capacitor integrates the collector voltage feedback, and the 10-MΩ resistor provides a high base-source impedance to minimize effects on optical performance. The detector drop causes a momentary decrease in light reaching the chip, which causes collector voltage to momentarily rise, generating an output signal. The initial light bias is small because of output power constraints on the LED and mechanical spacing system constraints. The change in light level is a fraction of this initial bias because of stray light paths and drop translucence. The high sensitivity of the photodarlington allows acceptable output signal levels when biased in this manner. This compares with unacceptable signal levels and bias point stability when biased conventionally, i.e., base open and signal output across the collector bias resistor.

LIGHT-CONTROLLED LAMP SWITCH

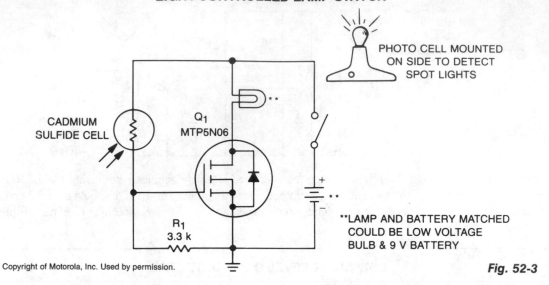

CADMIUM
SULFIDE CELL

Q₁
MTP5N06

R₁
3.3 k

PHOTO CELL MOUNTED
ON SIDE TO DETECT
SPOT LIGHTS

**LAMP AND BATTERY MATCHED
COULD BE LOW VOLTAGE
BULB & 9 V BATTERY

Fig. 52-3

A school drama needed lamps that automatically turned on and off when spot lights did the same. Lamp switching had to be wireless, durable, dependable, simple and inexpensive.

With stage and spot lights off, very little light falls on the CdS photocell, so its internal resistance is several megohms and R1 keeps the gate of Q1 at nearly zero volts, which keeps it off. When a spot or stage light hits the photocell, its resistance drops to several hundred ohms, raising Q1's gate voltage, which turns it on and applies power to the lamp.

OPTICAL SENSOR-TO-TTL INTERFACE

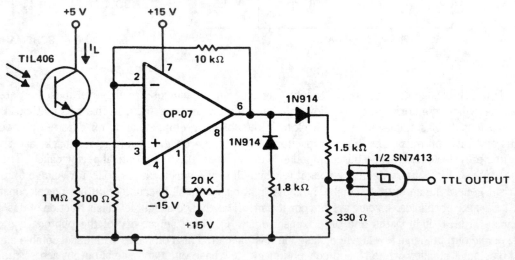

+5 V

+15 V

TIL406

I_L

OP-07

10 kΩ

1N914

1N914

1.5 kΩ

1/2 SN7413

1.8 kΩ

TTL OUTPUT

1 MΩ

100 Ω

20 K

−15 V

+15 V

330 Ω

Fig. 52-4

OPTICAL SENSOR-TO-TTL INTERFACE (*Cont.*)

This circuit is designed to detect a low light level at the sensor, amplify the signal, and provide a TTL-level output. When the optical sensor detects low-level light, on condition, its output is small and must be amplified. An amp with very low input bias current and high input resistance must be used to detect the on condition. When sensor TIL406 is in the on condition, its output is assumed to be 250 nA (allowing a safety margin). This results in a 250-mV signal being applied to the noninverting input of amplifier OP-07. Because of the circuit configuration, the OP-07 provides a gain of 100 and its output is in positive saturation. The OP-07 output level is applied to a loading network that provides the basic TTL level.

LIGHT-SENSITIVE AUDIO OSCILLATOR

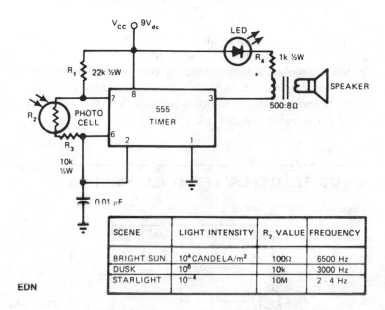

SCENE	LIGHT INTENSITY	R_2 VALUE	FREQUENCY
BRIGHT SUN	10^4 CANDELA/m²	100Ω	6500 Hz
DUSK	10^0	10k	3000 Hz
STARLIGHT	10^{-4}	10M	2 - 4 Hz

EDN

Fig. 52-5

This circuit's frequency of oscillation increases directly with light intensity. The greater the light intensity, the higher the frequency of the oscillator. The 555 timer operates in the astable oscillator mode where frequency and duty cycle are controlled by two resistors and one capacitor. The capacitor charges through R1 and R2, and discharges through R2, a standard photo cell. Resistor R3 limits the upper frequency of oscillation to the audio range. The lower range of approximately 1 pps is set by the value of R2, approximately 10-MΩ, with the photo cell almost totally dark.

A loudspeaker provides audio output, and an LED is used as a pilot light that flashes when the frequency falls below about 10 to 12 Hz. Extremely sensitive, especially on the dark end of the photocell resistance range, the unit can detect lightning many miles away, providing a rapid frequency increase with each flash of lightning. When used with a flashlight at night, the device becomes a simple optical radar for the blind, showing angular direction to a light-reflecting object, as well as height and distance to the object when hand scanned back and forth.

This light-sensitive audio oscillator can also serve as an audible horizontal level device by noting the position of a liquid bubble illuminated by a light source. Thus, you can sense fluid levels as well as the vibration state of a fluid surface level.

LIGHT LEVEL DETECTOR

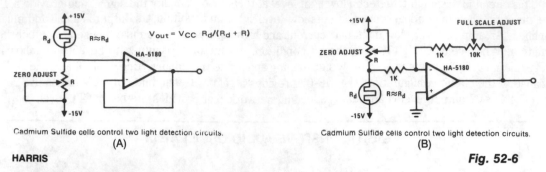

$$V_{out} = V_{CC} R_d/(R_d + R)$$

Cadmium Sulfide cells control two light detection circuits.

(A)

Cadmium Sulfide cells control two light detection circuits.

(B)

HARRIS

Fig. 52-6

If R, the sensor matching resistor, is equal to the "dark" resistance of the cadmium sulfide cell, the amplifier output will range from 0 to ≈ 2 as the light level ranges from "dark" to "bright." The circuit in Fig. B operates similarly, but use the standard noninverting configuration instead of the voltage-follower configuration; this allows for variable gain. Although the "dark" resistance of the cadmium sulfide cell is only ≈ 7 KΩ, the principles of operation apply to other types of detectors which require the high-input impedance of the HA-5180 for reasonable linearity and useability.

LIGHTED DISPLAY BRIGHTNESS CONTROL

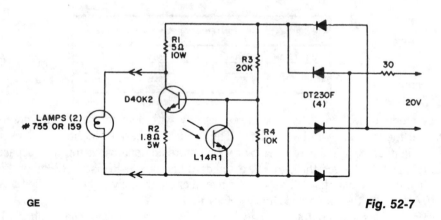

GE

Fig. 52-7

This circuit provides a very low cost method of controlling light levels. Circuit power is obtained from a relatively high source impedance transformer or motor windings, normally used to drive the low-voltage lamps used in these functions. It should be noted that the bias resistors are optimized for the 20-V, 30-Ω source, and they must be recalculated for other sources. The L14R1 is placed to receive the same ambient illumination as the display and should be shielded from the light of the display lamps. The illumination level of lighted displays should be lowered as the room ambient light dims, to avoid undesirable or unpleasant visual effects.

WARNING LIGHT AND MARKER LIGHT

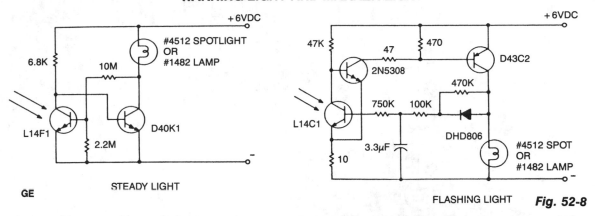

GE

STEADY LIGHT

FLASHING LIGHT   *Fig. 52-8*

A flashing light of high brightness and short duty cycle is often desired to provide maximum visibility and battery life. This necessitates using an output transistor, which can supply the cold filament surge current of the lamp while maintaining a low saturation voltage. The oscillation period and flash duration are determined in the feedback loop, while the use of a phototransistor sensor minimizes sensitivity variations.

LIGHT-CONTROLLED ONE-SHOT TIMER

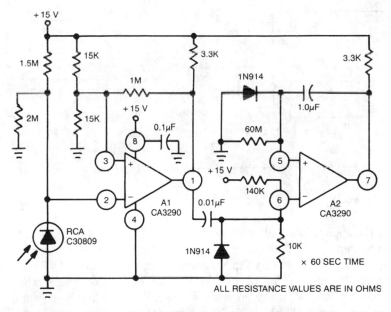

ALL RESISTANCE VALUES ARE IN OHMS

GE/RCA   *Fig. 52-9*

This circuit uses A1 of the CA3290 BiMOS dual voltage comparator to sense a change in light diode current. A2, a one-shot timer, is triggered by the A1 output. If the light source to the photodiode is interrupted, the circuit output switches to a low state for approximately 60 s.

SOLAR-TRIGGERED SWITCH

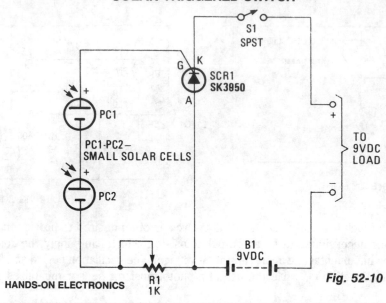

HANDS-ON ELECTRONICS

Fig. 52-10

SUN TRACKER

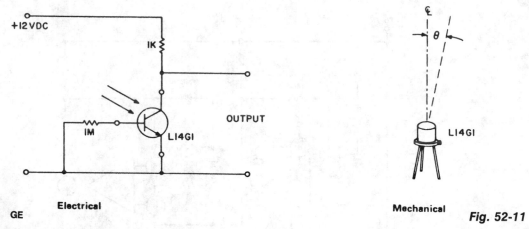

Electrical

GE

Mechanical

Fig. 52-11

In solar cell array applications and solar instrumentation, it is desirable to monitor the approximate position of the sun to allow efficient automatic alignment. The L14G1 lens can provide about 15° of accuracy in a simple level sensing circuit, and a full hemisphere can be monitored with about 150 phototransistors. The sun provides $\approx$ 80 mW/cm² to the L14G1 when on the centerline. This will keep the output down to $\leq$ 0.5 V for $\theta \leq 7.5°$. The sky provides $\approx$ 0.5 mW/cm² to the L14G1 and will keep the output greater than 10 V when viewed. White clouds viewed from above can lower this voltage to $\approx$ 5 V on some devices. This circuit can directly drive TTL logic by using the 5-V supply and changing the load resistor to 430 Ω. Different bright objects can also be located with the same type of circuitry simply by adjusting the resistor values to provide the desired sensitivity.

PHOTOELECTRIC AC POWER SWITCH

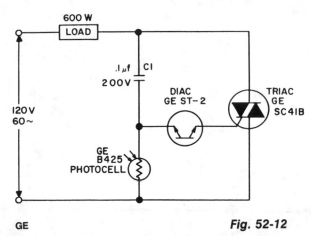

For a dark photocell, high resistance, the voltage across the diac rises rapidly with the line voltage due to the current through C1, triggering the diac early in the cycle. When the photocell resistance is less than about 2000 Ω, the drop across it is limited to less than the diac triggering voltage, and the load power is shut off.

GE   *Fig. 52-12*

53

Limiters

The sources of the following circuits are contained in the Sources section beginning on page 782. The figure number contained in the box of each circuit correlates to the sources entry in the Sources section.

Noise Limiter
Dynamic Noise Reduction Circuit
Output Limiter

NOISE LIMITER

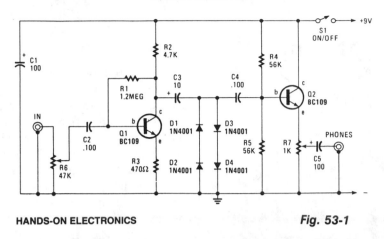

Fig. 53-1

This circuit is fed from the earphone jack of your receiver and goes to limiter control R6 and is then amplified by Q1: a common-emitter stage that has a voltage gain of only about 10, because of the negative feedback introduced by R3. The output of Q1 is fed to a simple clipping circuit, consisting of diodes D1 through D4. The diodes, connected in pairs, act like Zeners with an avalanche rating of about 1 V. The two pairs are connected opposite in polarity to each other, so that the audio signal is clipped at about 1 V. The signal is then coupled to the output socket through an emitter-follower buffer stage built around Q2 and an output attenuator control R7.

DYNAMIC NOISE REDUCTION CIRCUIT

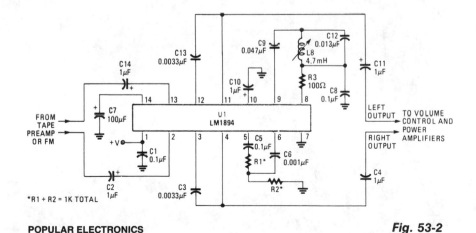

Fig. 53-2

OUTPUT LIMITER

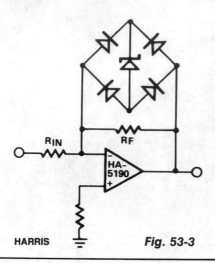

HA-5190 is rated for ± 5 V output swing, and saturates at ± 7 V. As with most op amps, recovery from output saturation is slow compared to the amplifier's normal response time. Some form of limiting, either of the input signal or in the feedback path, is desirable if saturation might occur. The circuit illustrates a feedback limiter, where gain is reduced if the output exceeds $\pm (V_Z + 2V_f)$. A 5-V zener with a sharp knee characteristic is recommended.

HARRIS   *Fig. 53-3*

54

LVDT Circuit

The sources of the following circuits are contained in the Sources section beginning on page 782. The figure number contained in the box of each circuit correlates to the sources entry in the Sources section.

LVDT Driver Demodulator

LVDT DRIVER DEMODULATOR

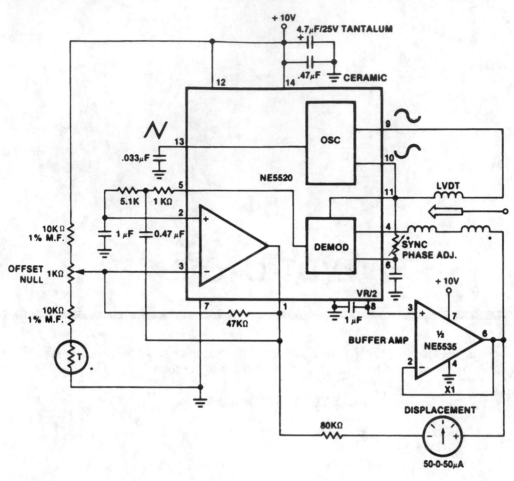

Fig. 54-1

A very simple motion transducer can be constructed using the circuit shown. The output is biased to one-half the supply voltage. This requires special interface circuitry for the signal readout. One simple method is to use a zero center meter in a bridge configuration. Displacement now can be measured as a positive or negative meter reading. Readout sensitivity is a function of the particular LVDT and of the gain of the error amplifier. Dc offsets can be nulled by using a simple offset adjustment circuit as indicated.

55

Mathematical Circuits

The sources of the following circuits are contained in the Sources section beginning on page 782. The figure number contained in the box of each circuit correlates to the sources entry in the Sources section.

Divide/Multiply with Only One Trim
Adder
Subtractor

DIVIDE/MULTIPLY WITH ONLY ONE TRIM

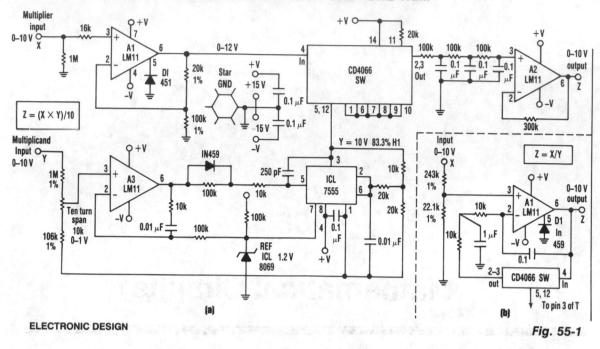

$$Z = (X \times Y)/10$$

$$Z = X/Y$$

ELECTRONIC DESIGN

Fig. 55-1

This relatively simple, inexpensive circuit requiring one trimming operation can multiply or divide with a consistent accuracy of greater than 1 part in 1,000. An inexpensive CMOS version of standard 555 timer chip T, in conjunction with low-drift LM11 error amplifier A3, an inexpensive analog chopper switch SW, form a unique voltage-to-duty-cycle converter to produce the difficult transfer function necessary for accurate conversion.

An unknown multiplicand voltage applied to the A3 error op amp circuit's Y input controls the duty cycle of the timer through its pin 5 modulation input. The network between the sink-and-source output of the timer, pin 3, and the state trigger inputs, pins 2 and 6, cause the timer to oscillate. An error feedback signal from the timer's discharge output, pin 7, represents the duty cycle. Integrating this duty-cycle signal with voltage reference REF representing full scale, and applying the result to the inverting input of A3, closes the feedback loop and insures high accuracy.

Multiplier X feeds into another LM11 op amp, A1, which acts as a input buffer and scaler. A third LM11, A2, filters and buffers the Z output. Between A1 and A2, the timer's duty-cycle output modulates the analog switches of a CD4066 to achieve the desired multiplier output. To perform division instead of multiplication, reconfigure the op amp A1 circuit with the use of jumpers. Amplifier A2 isn't required in the division configuration.

To calibrate the circuit, connect the X and Y inputs together and apply 10 V. Then adjust the 10-turn span potentiometer to achieve a 10-V output at Z for multiplication, or 1 V for the division configuration. Also check for zero output at a zero multiplier input. The circuit is scaled for 0–10 V inputs and outputs with a small overrage capability, but other scalings are possible. Star grounding or a heavy ground bus should be used to reduce offset problems that are unavoidable in this design.

ADDER

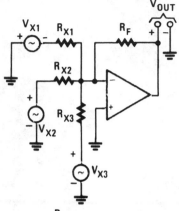

$$V_{OUT} = - \frac{R_F}{R_X} (V_{X1} + V_{X2} + V_{X3})$$

$$(R_{X1} = R_{X2} = R_{X3})$$

HANDS-ON ELECTRONICS

Fig. 55-2

SUBTRACTOR

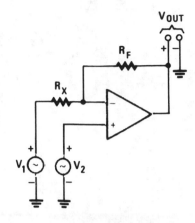

$$V_{OUT} = V_2 + \frac{R_F}{R_X} (V_2 - V_1)$$

HANDS-ON ELECTRONICS

Fig. 55-3

56

Measuring and Test Circuits

The sources of the following circuits are contained in the Sources section beginning on page 782. The figure number contained in the box of each circuit correlates to the sources entry in the Sources section.

DUTY CYCLE MONITOR

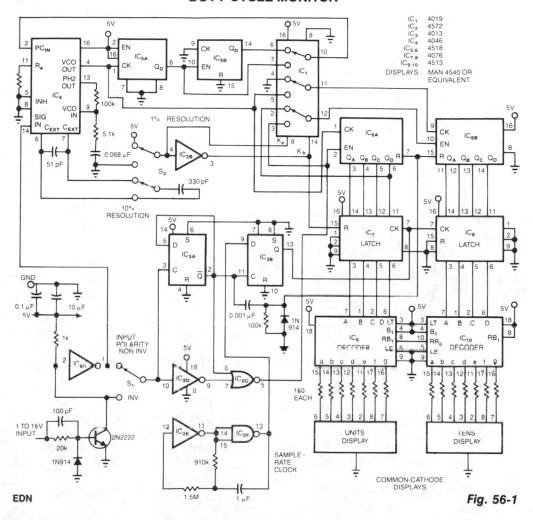

EDN

Fig. 56-1

The circuit monitors and displays a digital signal's duty cycle and provides accuracy as high as $\pm 1\%$. Using switch S2, you can choose a frequency range of either 250 Hz to 2.5 kHz at $\pm 1\%$ accuracy or 2 kHz to 50 kHz at $\pm 10\%$ accuracy. The common-cathode display gives the signal's duty-cycle percentage. Phase-locked loop IC4 and counters IC5A and IC5B multiply the input frequency by a factor of either 10 or 100, depending on switch S2's setting. IC6A and IC6B count this multiplied frequency during the incoming signal's mark interval. IC7 and IC8 then latch this count and display it at the clock's sample rate. For example, if you select a 1% resolution, when the signal's mark period is 40% of the total period, the circuit will enable the counter comprising IC6A and IC6B for 40 counts. To obtain space-interval sampling, you can reverse the input polarity using switch S1. IC2A samples the input signal's period and enables gate IC2C and resets the counter. IC2E and IC2F form the sample-rate clock; IC3B synchronizes the clock's output with the input, so that the circuit can update latches IC7 and IC8.

3-IN-1 TEST SET

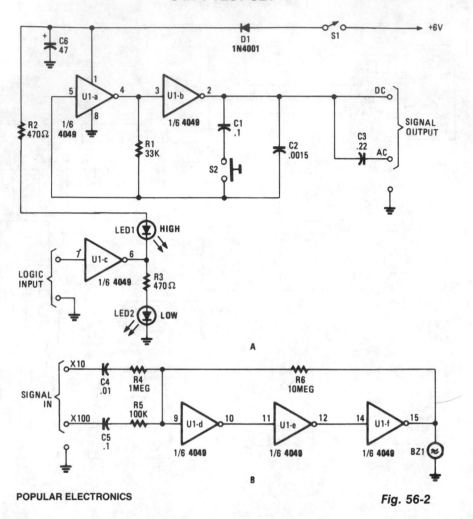

Fig. 56-2

This circuit is designed around a 4049 hex inverter/buffer. Two inverters are used in a dual-frequency signal-injector circuit, another inverter is used as a logic probe, and the remaining three inverters are used as a sensitive dual-input, audio-signal tracer.

The signal-injector portion gates are configured as a two-frequency, pulse-generator circuit. Under normal conditions, the generator's output frequency is around 10 kHz, but when S2 is closed, the output frequency drops to about 100 Hz. The logic-probe portion is made up of U1c, the output of the inverter decreases. The low output of U1c reverse biases LED2, so it remains off. That low output also forward biases LED1, causing it to light. But when a logic low presented U1c's input, the situation is reversed, so LED2 lights and LED1 darkens.

The audio-signal tracer portion is made up of the three remaining inverters which are configured as a linear audio amplifier to increase the input signal level by a factor of 10 or 100. The amplified output signal feeds a miniature piezo element of audible detection.

STEREO POWER METER

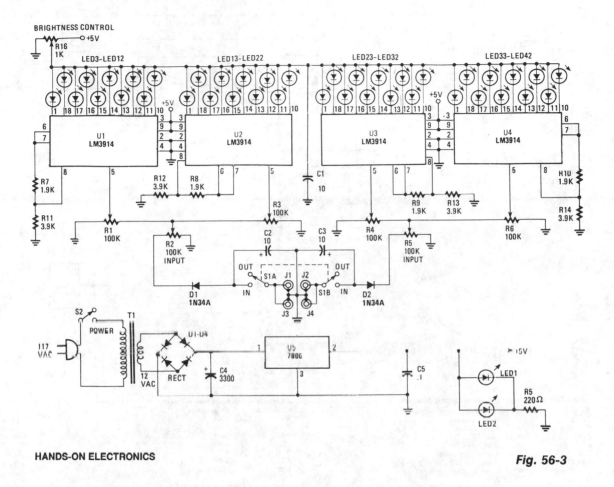

Fig. 56-3

The Stereo Power Meter is made up of two identical circuits and a power supply. Each circuit contains two LM3914 display chips which contain 10 voltage comparators, a 10-step voltage divider, a reference-voltage source, and a mode-select circuit that selects a bar or dot display via a logic input at pin 9. The brightness of the LEDs is controlled by the 1900-Ω resistors and the reference voltage is controlled by the 3900-Ω resistors. The 10-step voltage divider within the chips is connected between the reference voltage and ground. Since each step of the voltage divider is separated by a 1-KΩ resistor, each comparator senses a voltage 10% greater than the preceding comparator. The signal is applied to pin 5, which is buffered through a resistor-diode network and then amplified as it passes to each of the 10 comparators. Each LED is grounded through the comparators as the input signal voltage matches the reference voltage. That results in one to 10 LEDs illuminating as the signal voltage increases.

WIDE-RANGE RF POWER METER

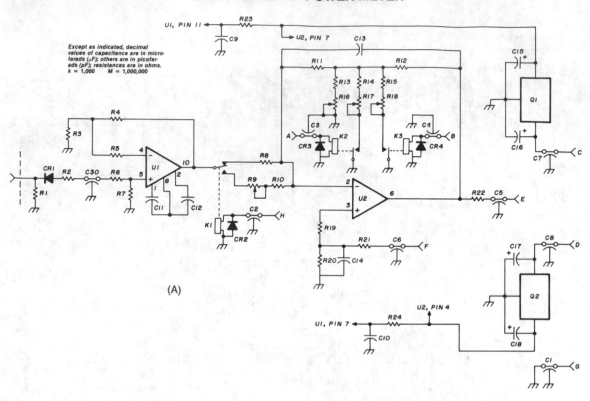

Except as indicated, decimal
values of capacitance are in microfarads (µF); others are in picofarads (pF); resistances are in ohms.
k = 1,000   M = 1,000,000

(A)

table 1. RF power meter and power supply parts list

C1 thru C8	1000 pF feedthru (Erie, Cambion)
C9,10,15,18	1 µF 10wvdc tantalum
C11-12	0.1 µF metalized film
C13	500 pF disc
C14	0.01 µF disc ceramic
C16,17	2.2 µF 25 wvdc tantalum
C19,21	100 µF 15 wvdc electrolytic
C20	500 µF 15 wvdc electrolytic
C22,23	0.01 µF disc
C30	100pF chip capacitor
CR1	HSCH-3486 Hewlett-Packard
CR2,3,4,9,10	1N914 or equivalent
CR5,6,7,8	1N4003 or equivalent
K1	SPDT reed Magnecraft W172-DIP5 (internal diode — CR2 not used)
K2,3,4,5	SPST reed EAC EAC Z610-ND
M1	1 mA DC meter with dB scale
Q1,4	78LO5 regulator
Q2	79LO5 regulator
Q3	78L12 regulator
R1,2	50 ohm 1/8 watt carbon film

All resistors 1% metal film 1/4 watt

R3,6,14,22	1k
R5,7	100k
R10	120k
R4	150k
R8,19	4.99k
R11,12	20k
R13	2.74k
R15	165 ohm

All resistors 5% carbon film 1/4 watt

R20	100 ohm
R21	1 megohm
R23, 24	10 ohm
R27	1.5k
R9	50k Panasonic CEG54 trimpot
R16	500 ohm Panasonic CEG52 trimpot
R17	200 ohm Panasonic CEG22 trimpot
R18	100 ohm Panasonic CEG12 trimpot
R25,26	10k potentiometer
S1	DP6T rotary switch
T1,T2	6.3 VAC transformers
U1	ICL76508CPD Intersil
U2	LM11CLH National
Box	CU-124 BUD
Chassis	9 1/2 x 5 x 2 chassis BUD Ac-403

WIDE-RANGE RF POWER METER (Cont.)

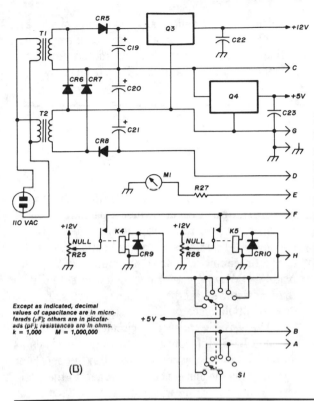

(D)

Except as indicated, decimal values of capacitance are in microfarads (µF); others are in picofarads (pF); resistances are in ohms.
k = 1,000   M = 1,000,000

The Hewlett-Packard HSCH-3486 zero-bias Schottky diode is used as the detector. To avoid using a modulation method of detection, a chopper-stabilized op amp is used. The chopper op amp basically converts the input dc voltage to ac, amplifies it, and converts it back to dc. Amplifying the dc output from the detector 150 times with a chopper op amp puts the signal at a level that simpler op amps, such as the LM11, can handle. Offset voltages in the amplifier are nulled with two pots, one for the high range and one for the lower three ranges.

LED PEAKMETER

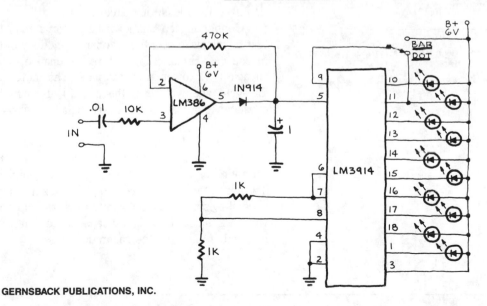

Fig. 56-5

GERNSBACK PUBLICATIONS, INC.

The circuit includes a peak detector that immediately drives the readout to any new higher signal level and slowly lowers it after the signal drops to zero. The readout is a moving dot or expanding bar display. The circuit can be expanded for a longer bar readout. Tapping five or more LED peakmeters into a frequency equalizer or series of audio filters should give a unique result. The bottom LED remains on with no signal at the input, thus providing a pilot light for the unit.

LC CHECKER

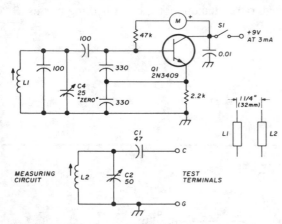

LI- 30T. NO.28E CLOSE-WOUND ON 3/8" (IOmm) SLUG-TUNED FORM. APPROX. 7μH.
L2-50T. SAME AS LI. APPROX. 30μH.
QI- 2N3904 OR SIMILAR.
M- 0 TO IOO OR 0 TO 200μA.
SI- SPST TOGGLE OR SLIDE SWITCH.

HAM RADIO   *Fig. 56-6*

The circuit is based on the *grid-dip* or *absorption effect*, which occurs when a parallel resonant circuit is coupled to an oscillator of the same frequency. Q1 operates in a conventional Colpitts oscillator circuit at a fixed frequency of approximately 4 MHz. A meter connected in series with the transistor's base-bias resistor serves as the dip or absorption indicator.

The variable measuring circuit consists of C1, C2, and L2 and is connected to panel terminals as shown. L2 is loosely coupled to L1 in the oscillator circuit. This measuring circuit is tuned to the oscillator frequency with variable capacitor C2 set at full capacitance. When power is applied to the oscillator, the meter shows a dip caused by power absorption from the measuring circuit.

Connecting an unknown capacitor across the test terminals lowers the resonant frequency of the measuring circuit. To restore resonance, tune capacitor C2 lower in capacitance. The meter will dip again when you reach this point. Determine the capacitance across the text terminals by calibrating the dial settings of C2.

Capacitor C4, a small variable trimmer in the oscillator circuit, compensates for drift or other variations and is normally set at half capacitance. The capacitor is a panel control, labeled zero, and it is used to set the oscillator exactly at the dip point when C2 is set at maximum capacitance. This corresponds to zero on the calibration scale.

TACHOMETER AND DIRECTION-OF-ROTATION CIRCUIT

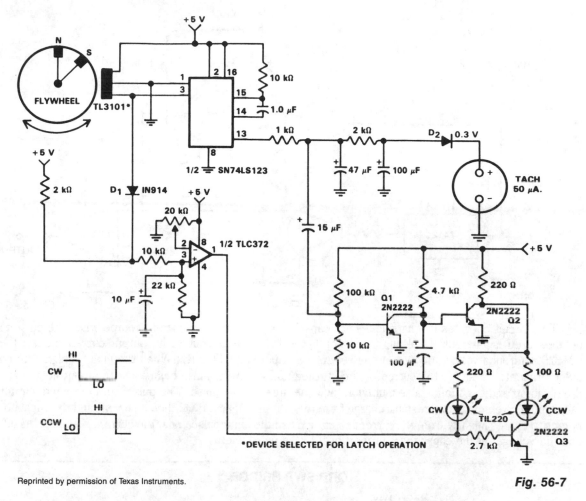

Reprinted by permission of Texas Instruments.

Fig. 56-7

In machine and equipment design, some applications require measurement of both the shaft speed and the direction of rotation. Fig. 56-4 shows the circuit of a tachometer, which also indicates the direction of rotation. The flywheel has two magnets embedded in the outer rim about 45° apart. One magnet has the north pole toward the outside and the other magnet has the south pole toward the outside rim of the flywheel. Because of the magnet spacing, a short on pulse is produced by the TL3101 in one direction and a long on pulse in the other direction. A 0 – 50 μA meter is used to monitor the flywheel speed while the LEDs indicate the direction of rotation. The direction-of-rotation circuit can be divided into three parts:

- TLC372 device for input conditioning and reference adjustment.
- Two 2N2222 transistors which apply the V_{CC} to the two LEDs when needed.
- The two TIL220 LEDs which indicate clockwise (CW) or counterclockwise (CCW) direction of rotation.

VERY SHORT PULSE-WIDTH MEASURER

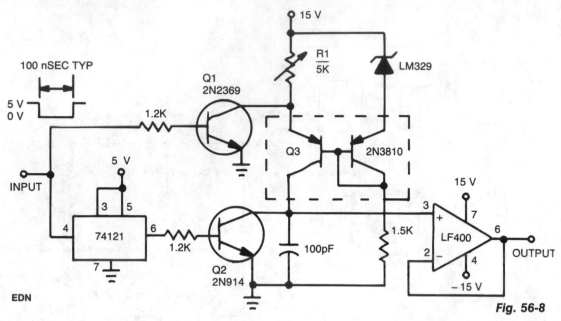

Fig. 56-8

This circuit operates by charging a small capacitor from a constant-current source when the pulse to be measured is present. Dual pnp transistor Q3 is the current source; its output current equals the LM329 reference voltage divided by the resistance of potentiometer R1. When the input is high with no pulse present, Q1 keeps the current source turned off. When the pulse begins and the input decreases, Q1 turns off and the monostable multivibrator generates a short pulse. The pulse from the multivibrator turns on Q2, removing any residual charge from the 100-pF capacitor. Q2 then turns off, and the capacitor begins to charge linearly from the current source. When the input pulse ends, the current source turns off, and the voltage on the capacitor is proportional to the pulse width.

QRP SWR BRIDGE

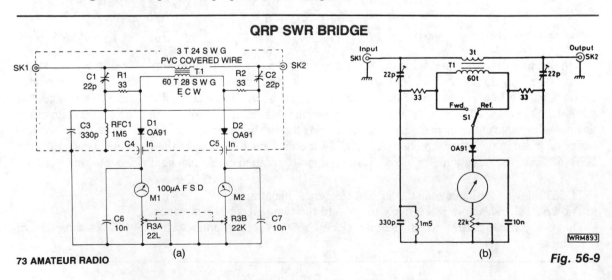

73 AMATEUR RADIO

(a)   (b)   **Fig. 56-9**

QRP SWR BRIDGE *(Cont.)*

The design shown is a simple unit for QRP operation on all authorized frequencies up to 30 MHz, based on a toroidal transformer T1. The secondary winding of T1 samples a small amount of rf power, both forward and reflected, which is divided by the bridge circuit and rectified by diodes D1 and D2. Forward and reflected readings are obtained simultaneously on the two meters M1 and M2, and the bridge is matched and balanced at the required load impedance by C1 and C2. See Fig. 56-9b for an alternative, less expensive, single meter version. The bridge also measure forward power.

ELECTROSTATIC DETECTOR

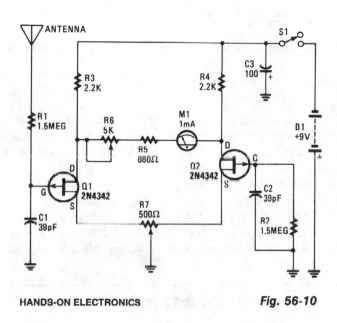

HANDS-ON ELECTRONICS   *Fig. 56-10*

The heart of the electroscope is the two junction FETs Q1 and Q2 connected in a balanced-bridge circuit. The gate input of Q1 is connected to the wire pick-up antenna, while Q2's gate is tied to the circuit's common ground through R2. That type of bridge circuit offers excellent temperature stability; therefore, Q1 is allowed to operate in an open-gate configuration. Potentiometer R7 is used to balance the bridge circuit, and R6 sets the maximum meter swing. Capacitors C1 and C2 help to reduce the 60-Hz pickup and add to the short-term stability of the circuit.

CURRENT MONITOR AND ALARM

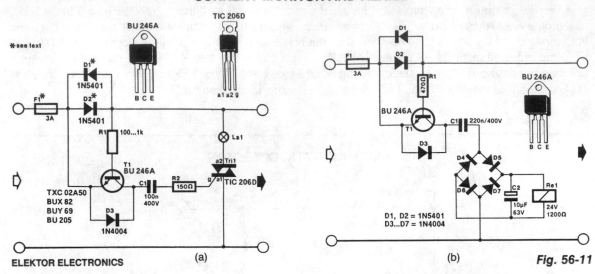

(a)

(b)

Fig. 56-11

The circuit in Fig. 56-11a lights the signal lamp upon detecting a line current consumption of more than 5 mA, and handles currents of several amperes with appropriate diodes fitted in the D1 and D2 positions. Transistor T1 is switched on when the drop across D1/D2 exceeds a certain level. Diodes from the well-known 1N400x series can be used for currents of up to 1 A, while 1N540x types are rated for up to 3 A. Fuse F1 should suit the particular application.

The circuit in Fig. 56-11b is a current-triggered alarm. Rectifier bridge D4 through D7 can only provide the coil voltage for Rel when the current through D1/D2 exceeds a certain level, because then series capacitor C1 passes the alternating main current. Capacitor C1 needs to suit the sensitivity of the relay coil. This is readily effected by connecting capacitors in parallel until the coil voltage is high enough for the relay to operate reliably.

PICOAMMETER CIRCUIT

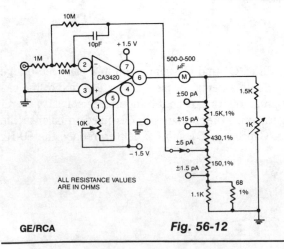

Fig. 56-12

This circuit uses the exceptionally low input current 0.1 pA of the CA3420 BiMOS op amp. With only a single 10-MΩ resistor, the circuit covers the range from ± 50 pA to a maximum full-scale sensitivity of ± 1.5 pA.

PAPER SHEET DISCRIMINATOR FOR PRINTING AND COPYING MACHINES

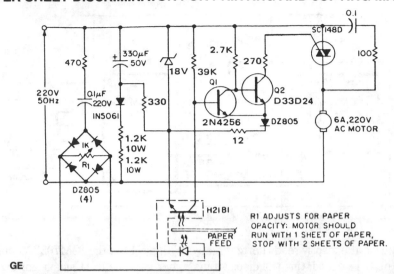

R1 ADJUSTS FOR PAPER
OPACITY; MOTOR SHOULD
RUN WITH 1 SHEET OF PAPER,
STOP WITH 2 SHEETS OF PAPER.

GE

Fig. 56-13

The circuit outputs power to the drive motor when one or no sheets are being fed, but interrupts motor power when two or more superimposed sheets pass through the optodetector slot. The optodetector can be either an H2aB darlington interruptor module or an H23B matched emitter-detector pair. The output from the optodarlington is coupled to a Schmitt trigger, comprising transistors Q1 and Q2 for noise immunity and minor paper opacity variation immunity. When the Schmitt is on, gate current is applied to the SC148D output device. The dc power supply for the detector and Schmitt is a simple rc diode half-wave configuration chosen for its low cost (fewer diodes and no transformers) and minimum bulk. While such a supply is directly coupled to the power triac, this is precluded by current drain considerations (50 mA dc for the gate drive alone). Note that direct coupling of the Schmitt to the output triacs is preferred, since RFI is virtually eliminated with the quasi-dc gate drive.

STUD FINDER

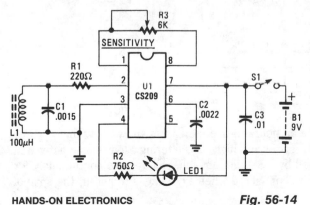

HANDS-ON ELECTRONICS

Fig. 56-14

The CS209 is designed to detect the presence or proximity of magnetic metals. It has an internal oscillator that, along with its external lc resonant circuit, provides oscillations whose amplitude is dependent upon the Q of the lc network. Close proximity to magnetic material reduces the Q of the tuned circuit, thus the oscillations tend to decrease in amplitude. The decrease in amplitude is detected and used in turn on LED1, indicating the presence of a magnetic material (i.e., nail or screw).

PRECISION FREQUENCY COUNTER/TACHOMETER

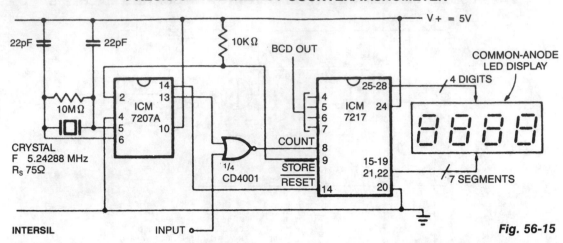

Fig. 56-15

INTERSIL

In this configuration, the display reads hertz directly. With pin 11 of the ICM7027A connected to V_{DD}, the gating time will be 0.1 second; this will display tens of hertz as the least significant digit. For shorter gating times, an ICM7207 can be used with a 6.5536-MHz crystal, giving a 0.01 second gating with pin 11 connected to V_{DD}, and a 0.1 second gating with pin 11 open.

MOTOR HOUR METER

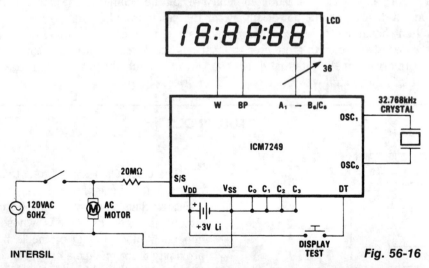

Fig. 56-16

INTERSIL

In this application, the ICM7249 is configured as an hours-in-use meter and shows how many whole hours of line voltage have been applied. The 20-MΩ resistor and high-pass filtering allow ac line activation of the S/S input. This configuration, which is powered by a 3-V lithium cell, will operate continuously for $2^{1}/_{2}$ years. Without the display, which only needs to be connected when a reading is required, the span of operation is extended to 10 years.

LOW-POWER MAGNETIC CURRENT SENSOR

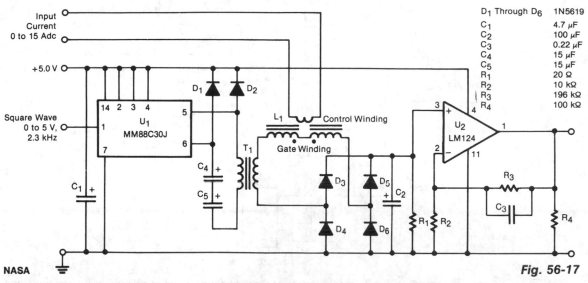

D₁ Through D₆	1N5619
C₁	4.7 µF
C₂	100 µF
C₃	0.22 µF
C₄	15 µF
C₅	15 µF
R₁	20 Ω
R₂	10 kΩ
R₃	196 kΩ
R₄	100 kΩ

NASA

Fig. 56-17

A transducer senses a direct current magnetically, providing isolation between the input and the output. The detecting-and-isolating element is a saturable reactor, in which the input current, to be measured, passes through a one-turn control coil. The transducer provides an output of 0 to 3 Vdc, an input current of 0 to 15 Adc, and consumes 22 mW at 10 Adc input.

Line driver U1 excites the saturable reactor L1 by feeding a 2.3-kHz square wave through transformer T1. The output of L1 is rectified by the bridge rectifier composed of diodes D3 through D6, then amplified by op amp U2, which has a gain of 20.

Diodes D1 and D2 commutate the reactive current fed back to the primary of T1 from L1. Without these diodes, large reactive voltage spikes on the primary would waste power and could destroy U1. Filter capacitor C1 stores the energy fed back through D1 and D2.

To minimize core losses, the core of T1 is made of an alloy of 80% nickel and 20% iron. To minimize capacitance, the primary and secondary windings are interleaved and progressively wound 350°. The primary and secondary windings consist of 408 and 660 turns, respectively, of #34 wire.

LINE-CURRENT MONITOR

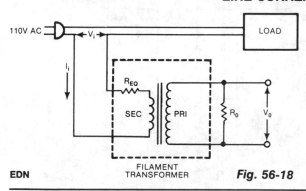

EDN

Fig. 56-18

A low-cost filament transformer provides a linear indication of the load current in an ac line. This method causes a slight series voltage drop over a wide range of load currents.

S METER

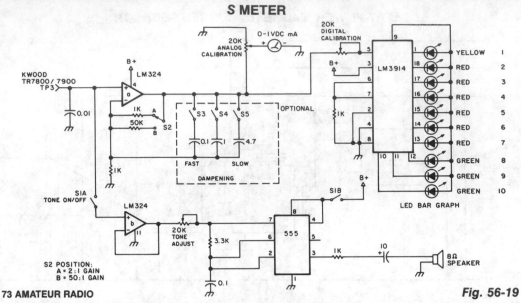

73 AMATEUR RADIO

Fig. 56-19

This design is for an external signal strength meter that is analog, digital, and audible for mobile transmitter hunters. The S meter also incorporates a gain circuit. The digital LED bar graph display has a very fast response time. The 3.3-KΩ resistor near LM3914 can be replaced with a 5-K pot to control LED brightness. The S2A position gives a 2:1 gain and the S2B position gives about a 50:1 gain. The calibration pots control the amount of meter action relative to the gain. The optional dampening circuit is used for the averaging of a transmitted signal that has modulated power or when a dip on the voice peaks occur. The capacitors can be switched one by one or switched into a very slow response using 5.8 μF total capacitance.

HOT-WIRE ANEMOMETER

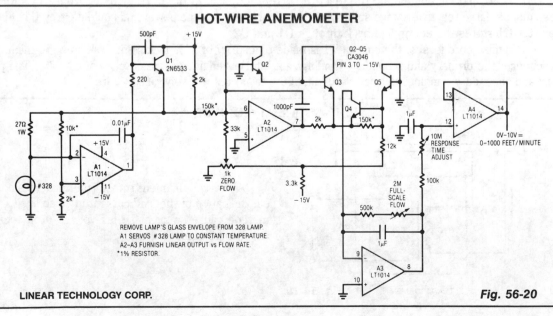

LINEAR TECHNOLOGY CORP.

Fig. 56-20

AUDIBLE LOGIC TESTER

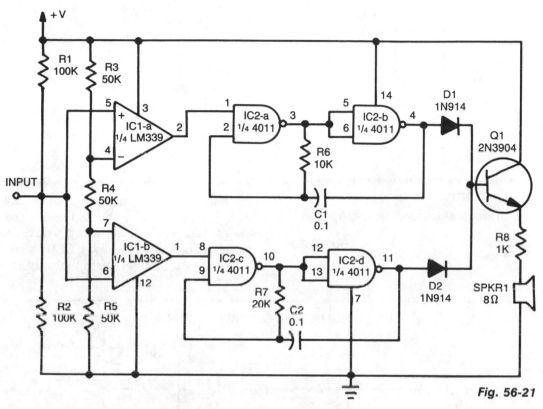

Fig. 56-21

This tester provides an audible indication of the logic level of the signal presented to its input. A logic high is indicated by a high tone, a logic low is indicated by a low tone, and oscillation is indicated by an alternating tone. The input is high impedance, so it will not load down the circuit under test. The tester can be used to troubleshoot TTL or CMOS logic. The input consists of two sections of an LM339 quad comparator. IC1a increases when the input voltage exceeds 67% of the supply voltage. The other comparator increases when the input drops below 33% of the supply.

The tone generators consist of two gated astable multivibrators. The generator built around IC2a and IC2b produces the high tone. The one built around IC2c and IC2d produces the low tone. Two diodes, D1 and D2, isolate the tone-generator outputs. Transistor Q1 is used to drive a low-impedance speaker.

SCR TESTER

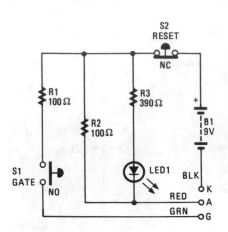

HANDS-ON ELECTRONICS/POPULAR ELECTRONICS

Fig. 56-22

The DUT's(Device Under Test) cathode, anode, and gate are connected to the unit's K, A, and G terminals, respectively. Pressing switch S1 feeds a gate current to the DUT, which triggers it on. Resistor R1 limits the gate current to the appropriate level. Resistor R3 limits the current through the LED to about 20 mA, which, with the current through R2, results in a latching current of about 110 mA. The LED is used to monitor the latching current. If the DUT is good, once the gate is triggered with S1, the LED will remain lit, indicating that the device is conducting. To end the test, turn off the device by interrupting the latching current flow using switch S2. The LED should turn off and remain off. The preceding procedure will work with SCRs and triacs. To check LEDs and other diodes, connect the anode and cathode leads to the anode and cathode of the diode; LED1 should light. When the leads are reversed, the LED should remain off.

DIGITAL FREQUENCY METER

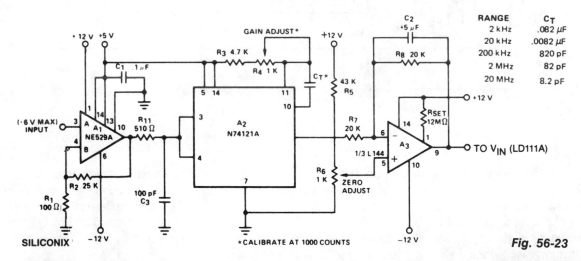

Fig. 56-23

The circuit converts frequency to voltage by taking the average dc value of the pulses from the 74121 monostable multivibrator. The one shot is triggered by the positive-going ac signal at the input of the 529 comparator. The amplifier acts as a dc filter, and also provides zeroing. This circuit will maintain an accuracy of 2% over 5 decades of range. The input signal to the comparator should be greater than 0.1 V pk-pk, and less than 12 V pk-pk for proper operation.

LOW-CURRENT MEASUREMENT SYSTEM

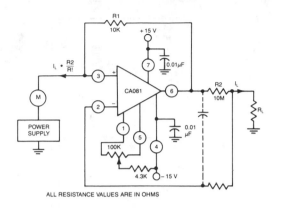

ALL RESISTANCE VALUES ARE IN OHMS

GE/RCA

Fig. 56-24

This circuit uses a CA018 BiMOS op amp. Low current, supplied at input potential as power supply to load resistor R_L, is increased by R2/R1, when load current I_L is monitored by power supply meter M. Thus, if I_L is 100 nA, with values shown, I_L presented to supply will be 100 μA.

SIMPLE CONTINUITY TESTER

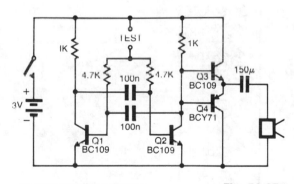

ELECTRONIC ENGINEERING

Fig. 56-25

The pitch of the tone is dependent upon the resistance under test. The tester will respond to resistances of hundreds of kilohms, yet it is possible to distinguish differences of just tens of ohms in low-resistance circuits. Q1 and Q2 form a multivibrator, the frequency of which is influenced by the resistance between the test points. The output stage, Q3 and Q4, will drive a small loudspeaker or a telephone earpiece. The unit is powered by a 3-V battery, and draws very little current when not in use.

SOUND-LEVEL METER

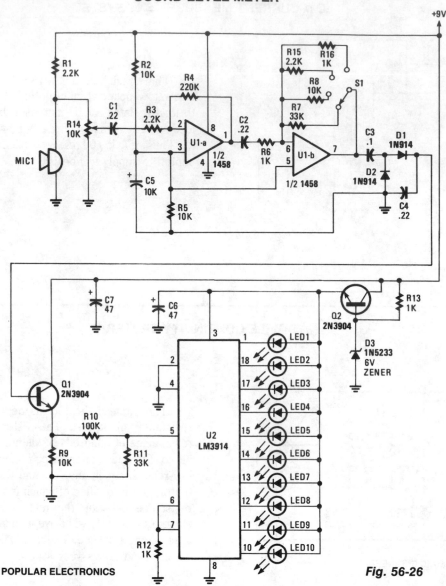

POPULAR ELECTRONICS

Fig. 56-26

Sounds are picked up by MIC1 and fed to the input of the first op amp. The signal is then fed to the input of second op amp U1b, where it is boosted again by a factor of between 1 and 33, depending upon the setting of range switch S1.

With the range switch set in the A position, R6 is 1 KΩ and R7 is 33 KΩ, so that stage has a gain of 33. In the B position, the gain is 10 Ω; in the C position, the gain is 22 Ω; and in the D position the gain is 1 Ω.

As the signal voltage fed to the input of U2 at pin 5 varies, one of ten LEDs will light to correspond with the input-voltage level. At the input's lowest operating level, U2 produces an output at pin 1, causing LED1 to light. The highest input level presented to the input of U2, about 1.2 V, causes LED10 to turn on.

LED PANEL METER

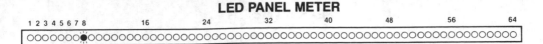

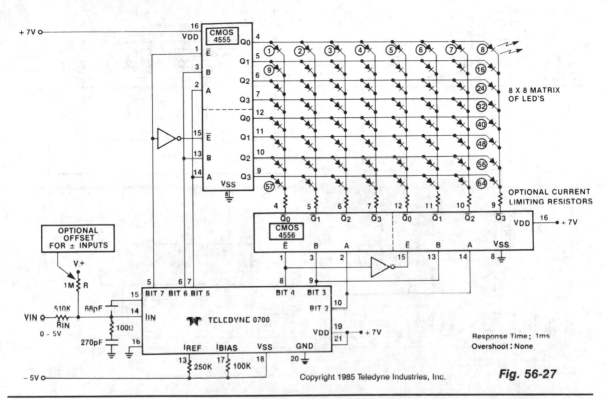

Response Time : 1ms
Overshoot : None

Copyright 1985 Teledyne Industries, Inc.

Fig. 56-27

OPTICAL PICK-UP TACHOMETER

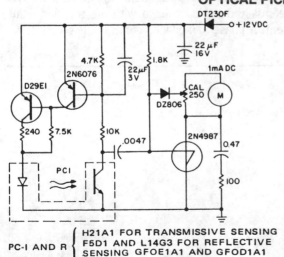

PC-I AND R { H21A1 FOR TRANSMISSIVE SENSING
F5D1 AND L14G3 FOR REFLECTIVE
SENSING GFOE1A1 AND GFOD1A1
GE { FOR FIBER OPTIC PROBE **Fig. 56-28**

Remote, noncontact, measurement of the speed of rotating objects is the purpose of this simple circuit. Linearity and accuracy are extremely high and normally limited by the milliammeter used and the initial calibration. This circuit is configured to count the leading edge of light pulses and to ignore normal ambient light levels. It is designed for portable operation since the tachometer is not sensitive to supply voltage within the supply voltage tolerance. Full scale at the maximum sensitivity of the calibration resistance is read at about 300 light pulses per second. A digital voltmeter can be used, on the 100-mV full-scale range, in place of the milliammeter. Shunt its input with a 100-Ω resistor in parallel with a 100-μF capacitor. This rc network replaces the filtering supplied by the analog meter.

PEAK-dB METER

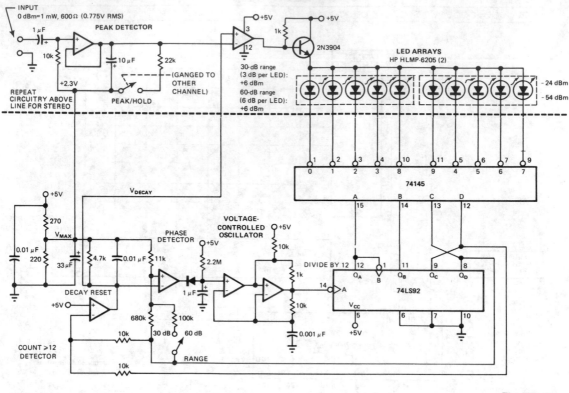

EDN

Fig. 56-29

This circuit compares a rectified input, V_{IN}, with a voltage that decays exponentially across a 4.7-KΩ resistor and a 0.01-μF capacitor. Comparing the exponentially decaying voltage with the rectified input provides a peak-level indication that requires no adjustment. A phase-locked loop controls the scan rate so that each LED represents 6 dB in the 30 dB range.

57

Medical Electronics Circuits

The sources of the following circuits are contained in the Sources section beginning on page 782. The figure number contained in the box of each circuit correlates to the sources entry in the Sources section.

Three-Chip EKG Simulator
Breath Monitor
Stimulus Isolator
Constant-Current Stimulator

THREE-CHIP EKG SIMULATOR

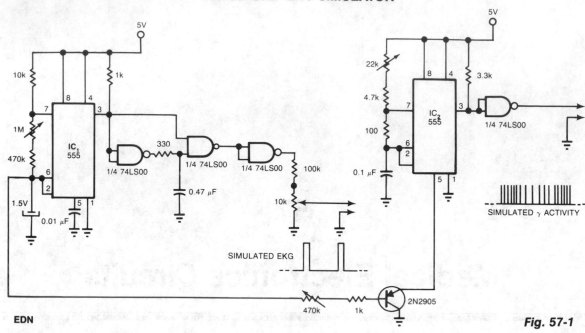

EDN

Fig. 57-1

Two 555s and a quad NAND-gate IC can simulate an electrocardiograph signal and a γ-wave radioisotope signal for applications in nuclear medicine. This circuit synchronizes the radioisotope signal to the EKG signal. You can use the circuit's outputs to test, for example, microprocessor-based software that calculates the left ventricular ejection fraction before you use the software in clinical applications. IC1, a 555 timer, provides a positive-going pulse train that simulates an EKG signal. A 10-KΩ potentiometer provides frequency adjustment. The other 555 timer, configured as a pulse-position modulator, provides the simulated γ-wave activity.

BREATH MONITOR

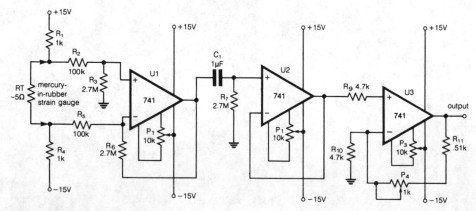

ELECTRONIC ENGINEERING

Fig. 57-2

BREATH MONITOR (*Cont.*)

The mercury-in-rubber strain gauge is the detector of breathing. In the model device, the strain gauge produced by Medimatic, Demark was used. The change in the length of the strain gauge wrapped around the chest during breathing causes a varying electrical resistance of about 0.2 Ω/cm. The constant current passing R1, R_T and R4 gives the constant component on the output of U1 differential amplifier. The change in the resistance of the strain gauge, which results from breathing, produces proportional varying in output voltage of U1 amplifier. The voltage follower, based on U2, separates the output voltage of the filter from the input stage amplifier. U3 works as a noninverter amplifier with regulated gain. This device should be supplied with a stabilized constant voltage of ± 15 V.

STIMULUS ISOLATOR

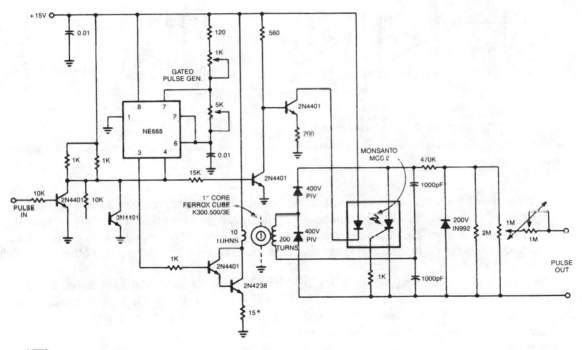

NOTES:
All resistor values in ohms
*Power rating depends on duty cycle from 1/2w for 20-25% duty cycle to 15-20W for 75-90% duty cycle

SIGNETICS

Fig. 57-3

This stimulus isolator uses a photo-SCR and a toroid for shaping pulses of up to 200 V at 200 μA.

CONSTANT-CURRENT STIMULATOR

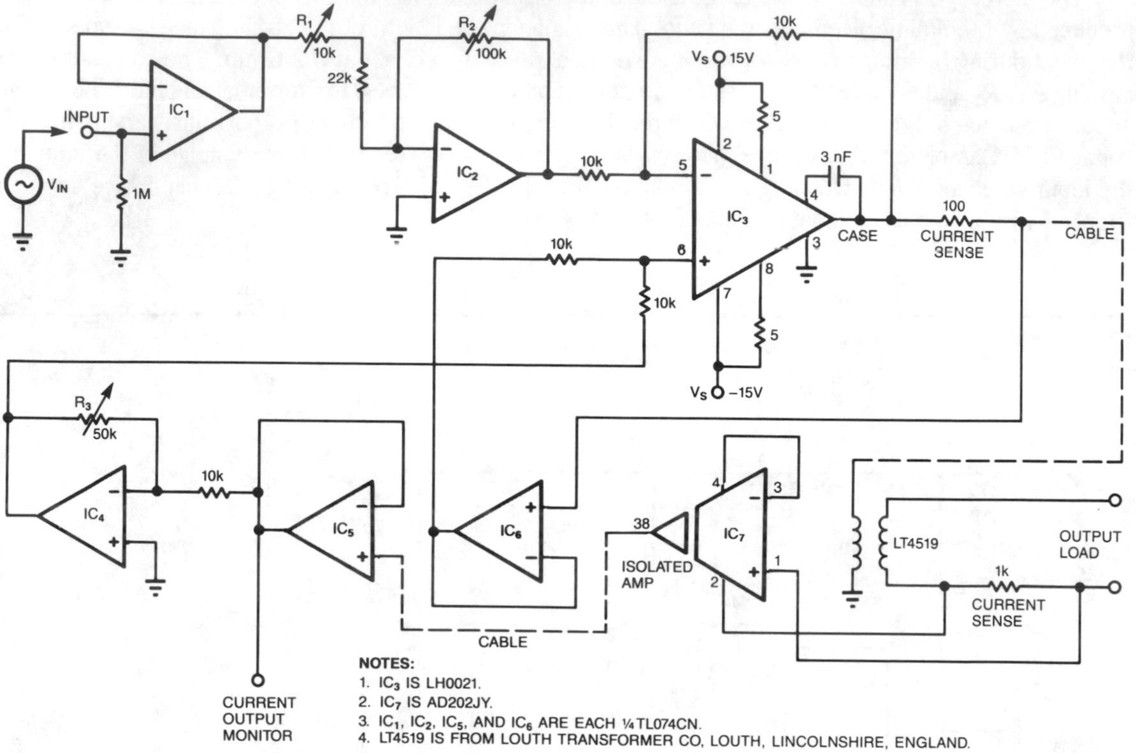

NOTES:
1. IC$_3$ IS LH0021.
2. IC$_7$ IS AD202JY.
3. IC$_1$, IC$_2$, IC$_5$, AND IC$_6$ ARE EACH ¼ TL074CN.
4. LT4519 IS FROM LOUTH TRANSFORMER CO, LOUTH, LINCOLNSHIRE, ENGLAND.

EDN

Fig. 57-4

Most circuits that provide an electrical stimulus for research subjects are constant-voltage designs; this circuit is a constant-current design. Stimulator circuits must be isolated for two reasons: to ensure safety and to minimize interference. Isolated stimulators are essentially two-terminal devices; output currents can flow only between the two output terminals and can at no time flow through any other path, such as the power ground.

The circuit's bandwidth ranges from 50 Hz to 5 kHz when a ± 1 V sinusoidal input drives the circuit. Output loads can range from a short circuit to 100 KΩ and have as much as 0.033 μF of parallel capacitance. The transformer and associated circuitry conveniently connect to the main circuit via a cable. Note: This circuit is not approved for use on human beings.

Op amps, IC1 and IC2, buffer and set the gain of the circuit, respectively. You adjust trimmer R1 so that R2, a 10-turn pot, yields output currents ranging from 0 to 1 mA/V_{IN}. IC3 is a power op amp. Its output drives the primary of a transformer that has a current gain of 0.1, or a voltage gain of 10. Operating from a $+15$ V supply, the transformer therefore has a voltage compliance of ± 150 V.

The circuit senses not only the current supplied to the transformer but also the current in the transformer's secondary. IC7, a fully isolated, medical-grade amplifier, provides the isolated feedback signal because the op amp has its own built-in isolation transformer. Trimmer R3 sets the feedback gain precisely at 27 KΩ nominal.

58

Metronome

The sources of the following circuits are contained in the Sources section beginning on page 782. The figure number contained in the box of each circuit correlates to the sources entry in the Sources section.

Downbeat-Emphasized Electronic Metronome

DOWNBEAT-EMPHASIZED ELECTRONIC METRONOME

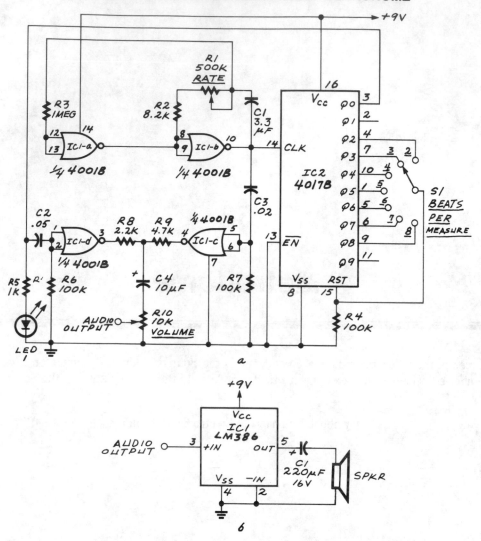

Fig. 58-1

IC1a and IC1b form an astable multivibrator. The astable's signal is fed to IC1c, also to the clock input of IC2, a 4017B decade counter. That IC's Q0 through Q9 outputs become high one at a time for each successive clock pulse received at pin 14. Switch S1 feeds one of those outputs to the 4017B's reset input. Whenever the selected output becomes high, the 4017B restarts its counting cycle; that determines the number of beats per measure. The network composed of C2 and R6 sharpens the downbeat pulse, and the network composed of C3 and R7 sharpens the free-running pulses. By making C2 larger than C3, the downbeat receives greater emphasis.

59

Miscellaneous Treasures

The sources of the following circuits are contained in the Sources section beginning on page 782. The figure number contained in the box of each circuit correlates to the sources entry in the Sources section.

Closed-Loop Tracer
Pulse Amplitude Discriminator
Tracer Receiver
Central Image Canceller
Bug Tracer
Breath Alert Alcohol Tester
Automatic Electronic Music
Positive Input/Negative Output Charge Pump
Acid Rain Monitor

Simple Low-Cost Rf Switch
Burst Power Control
Flame Ignitor
Bar Code Scanner
50-MHz Trigger
Air-Motion Detector
Bug Detector
Door Opener

CLOSED-LOOP TRACER

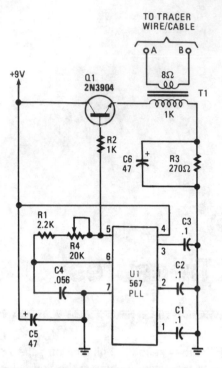

The tracer, consisting of both a transmitter and receiver, is designed to follow a closed-loop wire or cable system. It follows an induced voltage path created by feeding a low-current, audio-frequency signal through the cable. When the pick-up coils come within close proximity of the current-carrying cable, a small voltage is generated in each coil, and that induced voltage is then processed by the receiver's circuitry.

The circuit is built around a 567 phase-locked loop (PLL) configured as a variable-frequency, audio-generator circuit, which is designed to produce a square-wave output at pin 5. Potentiometer R4 allows the oscillator to be easily tuned to the receiver's frequency. Transistor Q1 isolates the oscillator from the load and matches the impedance of the primary of T1. Resistor R3 limits current flow through Q1. The low-impedance secondary of T1 supplies the cable drive signal.

POPULAR ELECTRONICS/HANDS-ON ELECTRONICS

Fig. 59-1

PULSE AMPLITUDE DISCRIMINATOR

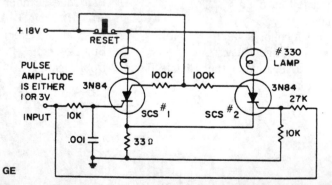

GE   *Fig. 59-2*

A 1-V amplitude pulse triggers SCS1, but has insufficient amplitude to trigger SCS2. A 3-V input pulse is delayed in reaching SCS1 by the 10-KΩ and .001-μF integrating network. Instead, it triggers SCS2, then raises the common emitter voltage to prevent SCS1 from triggering. The 100-KΩ resistors suppress the rate effect.

TRACER RECEIVER

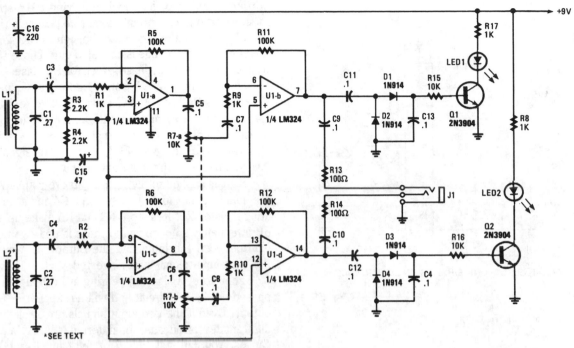

POPULAR ELECTRONICS/HANDS-ON ELECTRONICS

Fig. 59-3

The tracer receiver is a stereo audio amplifier/detector circuit operating near 1 kHz. Inductors L1 and L2—hand-wound coils, consisting of 200 turns of #26 wire on 2-inch ferrite cores— are tuned to the operating frequency of the amplifier/detector. The received signal strength of each individual receiver is indicated by an LED. The audio output of the receiver is fed to a stereo headphone. That dual-receiver scheme helps in locating and tracking the hidden wire or cable by giving a directional output that indicates the cable's path.

The 1-kHz signal is picked up by L1 and coupled to the input of op amp U1a, which provides a gain of about 100 dB. The output of op amp U1a is fed through volume-control potentiometer R7 to the input of U1b, which magnifies the already amplified signal 100 times more. That puts the maximum gain of the receiver at about 10,000 dB. The output of U1b follows two paths: in the first path, the signal is couple through C9 and R13 to J1, and is used to drive one half of a stereo headphone.

In the other path, the signal is fed through a voltage doubling/detector circuit—consisting of D1, D2, C11, and C13—that converts the amplified 1-kHz signal to the dc voltage that's used to drive Q1. When Q1 is turned on, LED1 lights, indicating a received signal.

CENTRAL IMAGE CANCELLER

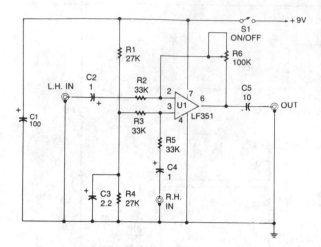

Fig. 59-4

The circuit allows you to eliminate the vocal portion of an audio signal, while leaving the instrumental portion. The circuit mixes two channels that must be 180° out of phase, so the signals that form the center-stereo image is canceled out. Those signals usually appear in phase. Resistor R3 biases the noninverting input of U1 from a center tap formed by resistors R1 and R4, and capacitor C3. Resistor R4, capacitor C3, and potentiometer R6 form a negative-feedback circuit that establishes the closed-loop voltage gain of U1 at unity. The signal is inverted between the input and output.

Signals applied to the right input are coupled to the noninverting input of U1 through C4 and attenuating resistor R5. Resistors R3 and R5 make up a 6 dB attenuator, so once again, there is unity voltage gain between the input and the output. However, the right input signal is not inverted.

Therefore, a signal appearing at both inputs is phased out by the circuit and will not appear at the output. Even if the two input signals are at slightly different levels because of different source impedances, you can still adjust for full cancellation by carefully tweaking R6.

BUG TRACER

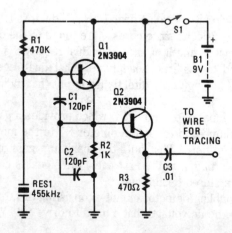

Fig. 59-5

The bug tracer is made up of a simple rf-injector circuit consisting of Q1 and Q2, and a pocket-size, AM broadcast receiver. The two-transistor rf-injector circuit supplies a constant rf signal to one end of a cable. Then the AM receiver is used as a detector, allowing you to trace the wire to its source.

Transistor Q1, along with piezoelectric ceramic resonator RES1, make up a simple rf oscillator that operates either at or near the AM-radio, 455-kHz, i-f frequency. That means that the second or third harmonic signal can easily be picked up by the receiver. Transistor Q2 is connected to an emitter-follower circuit to protect the oscillator from output loading; that helps to stabilize the output frequency and signal level.

BREATH ALERT ALCOHOL TESTER

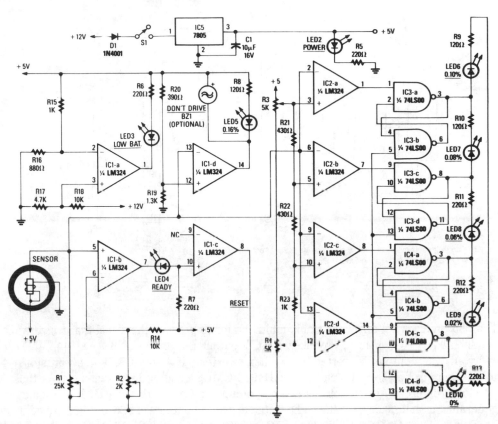

Reprinted with permission from Radio-Electronics Magazine, October 1988. Copyright Gernsback Publications, Inc., 1988.

Fig. 59-6

When power is applied to the circuit, the heater coil in the sensor is energized by the 5-V output of IC5, a 7805 voltage regulator. Breathing into the sensor with alcohol on your breath will lower the sensor's resistance; consequently, the input voltage to the detector circuit, will change. The detector circuit consists of quad op amp, IC2 and its associated circuitry. All sections of the detector circuit are calibrated via R3 and R4, and the inputs to each section are controlled by the voltage-divider network R21 through R23. As each section is triggered, the outputs decrease, and sample-and-hold circuits, IC3 and IC4, will latch onto the highest input value and drive the appropriate LED. The different colored LEDs represent alcohol levels from 0 to 0.16%.

If the level of alcohol is above the legal limit, or 0.16%, part of another quad op amp, IC1d, will turn on both the optional buzzer and LED5. That is an indication of a high level of alcohol present in your blood, and you definitely should not drive.

After a test is taken, the sensor takes a few seconds to ready itself for another test. When the sensor is ready, its input to IC1b, adjusted via R2 to a threshold of 0.5 V, causes LED4 (ready) to light. That, in turn, causes IC1c to reset the rest of the circuitry. The last section of IC1 is biased via R15 and R16, and used to indicate a low-battery condition—when the battery voltage drops below 6.8 V—which could result in an inaccurate breath test.

ELECTRONIC MUSIC

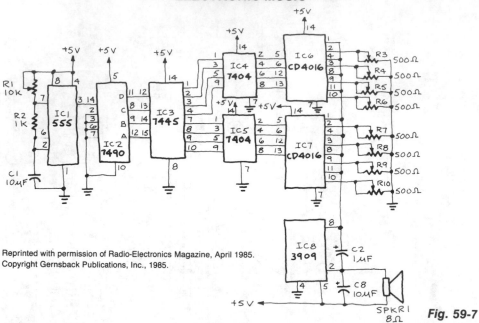

Fig. 59-7

IC1, a 555 timer, is set up as an astable multivibrator to produce the signal that triggers IC2, a 7490 decade counter. That IC, in turn, produces a BCD output that is fed to IC3, a 7445 BCD-to-decimal decoder/driver. IC3's output is inverted by two hex inverters, IC4 and IC5. The outputs of IC4 and IC5 are inputted to control pins on IC6 and IC7, CD4016 CMOS quad bilateral switches. As those switches open and close, different resistances (as set by potentiometers R3 through R10) are inserted into the sound-generating circuit made from IC8. The frequency at the outputs of IC6 and IC7 are adjusted to various rates, using potentiometers R3 through R10, to produce the desired tones. Capacitors can be placed in series with the potentiometers to produce a sloping sound instead of a straight tone. The negative-going output signals of IC6 and IC7 are fed through a common bus to pin 8 of IC8.

POSITIVE INPUT/NEGATIVE OUTPUT CHARGE PUMP

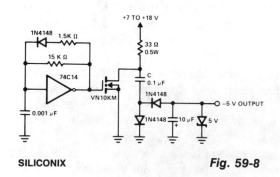

SILICONIX

Fig. 59-8

A charge pump is a simple means of generating a low-power voltage supply of opposite polarity from the main supply. The 74C14 IC is a self-oscillating driver for the MOSFET power switch. It produces a pulse width of 6.5 μs at a repetition frequency of 100 kHz. When the MOSFET device is off, capacitor C is charged to the positive supply. When the power through the MOSFET switches on, C delivers a negative voltage through the series diode to the output. The zener serves as a dissipative regulator. Because the MOSFET switches fast, operation at high frequencies allows the capacitors in the system to be small.

ACID RAIN MONITOR

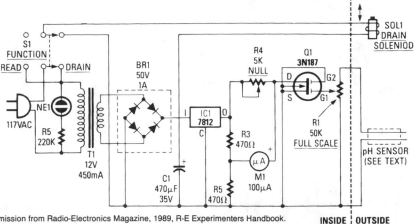

Fig. 59-9

A simple bridge rectifier and a 12-V regulator powers the MOSFET sensing circuit. The unregulated output of the bridge rectifier operates the drain solenoid via switch S1. The sensor itself is built from two electrodes: one made of copper, the other of lead. In combination with the liquid trapped by the sensor, the electrodes form a miniature lead/acid cell whose output is amplified by MOSFET Q1. The maximum output produced by our prototype cell was about 50 μA.

MOSFET Q1 serves as the fourth leg of a Wheatstone bridge. When sensed acidity causes the sensor to generate a voltage, Q1 turns on slightly, so its drain-to-source resistance decreases. That resistance variation causes an imbalance in the bridge, and that imbalance is indicated by meter M1.

SIMPLE LOW-COST RF SWITCH

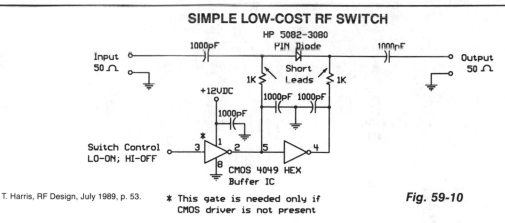

T. Harris, RF Design, July 1989, p. 53.

* This gate is needed only if
CMOS driver is not present

Fig. 59-10

When the digital logic level at the control input is low, the PIN diode is forward-biased by the CMOS gates. The two 1-KΩ bias resistors limit this current to the PIN diode's safe forward current limit. In this state, the switch is on. When the control input is high, the diode is reverse-biased and the switch is off. This switch is well-suited for electronically steered antenna arrays, multiple path switching, and other applications requiring small, low-cost rf switches. This particular design was used in a four-pole rotary switch for a Doppler-shift radio direction-finder operating at 144 MHz.

BURST POWER CONTROL

GE

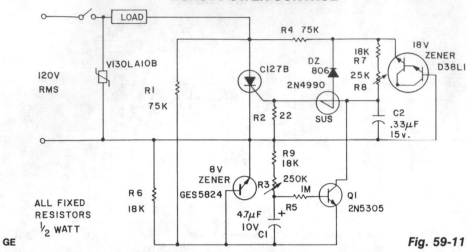

Fig. 59-11

Industrial applications sometimes require that power be only briefly applied to a load following the closure of a switch, such as a microswitch or foot switch. The load could be a heating element for use in sealing plastic bags, a dc motor that is indexed or stepped with each application of power, or a dc solenoid which is to be energized for a brief time—for example, a staple gun.

The phase angle at which the SCR triggers on is determined by the charging rate of C2 through R7 and potentiometer R8. When the breakover voltage of the gate triggering device, a silicon unilateral switch, is reached, the SCR turns on. The average dc voltage to the load is determined by the setting of R8.

FLAME IGNITOR

T1 Primary – 8T No. 16 Litz
Secondary – 800T No. 28
Core – 1/2" dia. x 2" Ferrite Rod

For 220V Operation:
GE-MOV® V275LA15
SCR C106D
D1 A14N
C1 .33 µF 400V NP
R3 4.7 M
R5 10. M

GE

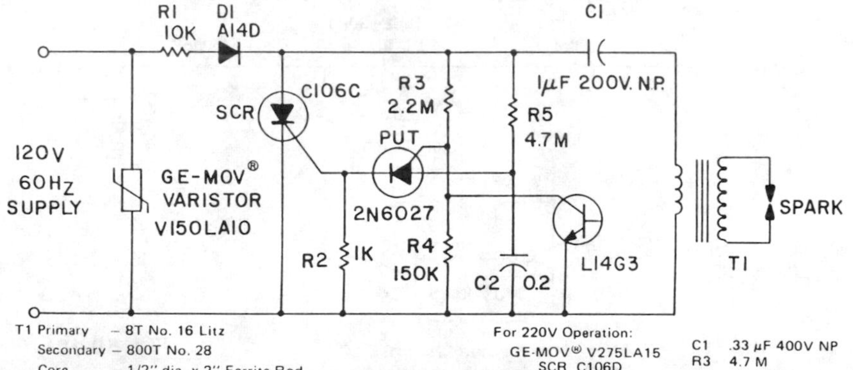

Fig. 59-12

The spark developed by the circuit is suitable for a gas ignitor. Capacitor C1 is charged through R1 and D1 toward peak line voltage. C2 is simultaneously being charged at a slower rate through R5. When the charge on C2 is sufficient to trigger the PUT, the SCR is triggered on, providing a rapid discharge path for C1 through the transformer primary. The SCR is triggered about 20 times per second with the component values shown. The L14G3 serves as a flame sensor. When ignition is achieved and sensed by the photodetector, the low V_{CE} prevents further SCR triggering.

BAR-CODE SCANNER

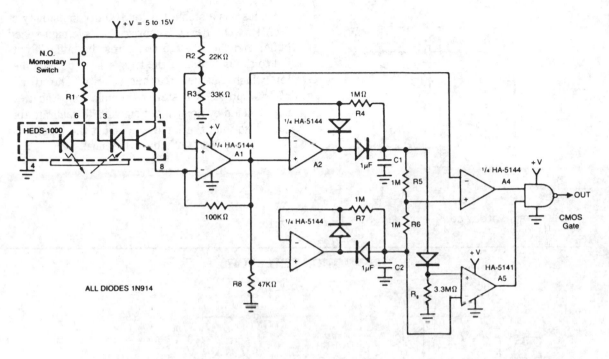

HARRIS

Fig. 59-13

The circuit illustrates a method of interfacing a HEDS-1000 emitter-detector pair with a HA-5144 for use as a bar-code scanner circuit. The HA-5144 is used as an amplifier system which converts the bar and space widths of the printed bar code into a pulse-width modulated digital signal. Amplifier A1 is used to amplify the current output of the detector. The output of A1 is passed to two precision peak-detector circuits which detect the positive and negative peaks of the received signal. Amplifier A4 is used as a comparator whose reference is maintained at the midpoint of the peak-to-peak signal by resistors R5 and R6. This provides a more accurate edge detection and less ambiguity in bar width. Amplifier A5 is used as an optional noise gate which only allows data to pass through the gate when the peak-to-peak modulation signal is larger than one diode drop. This circuit is operated by a single supply voltage with low-power consumption which makes it ideal for battery-operated data entry systems.

50-MHz TRIGGER

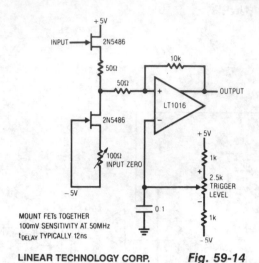

MOUNT FETs TOGETHER
100mV SENSITIVITY AT 50MHz
t_{DELAY} TYPICALLY 12ns

LINEAR TECHNOLOGY CORP.   *Fig. 59-14*

This has a stable trigger 100 mV sensitivity at 50 MHz. The FETs comprise a simple high-speed buffer and the LT1016 compares the buffer's output to the potential at the trigger level potentiometer, which can be of either polarity. The 10-KΩ resistor provides hysteresis, eliminating "chattering" caused by noisy input signals. To calibrate this circuit, ground the input and adjust the input zero control for 0 V at Q2's drain terminal.

AIR-MOTION DETECTOR

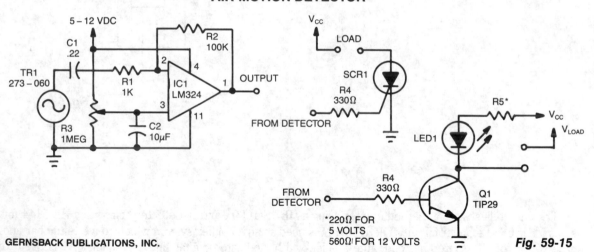

GERNSBACK PUBLICATIONS, INC.   *Fig. 59-15*

*220Ω FOR 5 VOLTS
560Ω FOR 12 VOLTS

When a current of air hits the piezo element, a small signal is generated and is fed through C1 and R1 to inverting input pin 2 of one section of the LM324. That causes output pin 1 to increase. Resistor R3 is used to adjust the sensitivity of the detector. The circuit can be set so high as to detect the wave of a hand or so low that blowing on the element as hard as you can will produce no output. Resistor R2 is used to adjust the level of the output voltage at pin 1.

The detector circuit can be used in various control applications. For example, an SCR can be used to control 117-Vac loads as shown. Also, an npn transistor, such as a TIP29, can be used to control loads as shown.

BUG DETECTOR

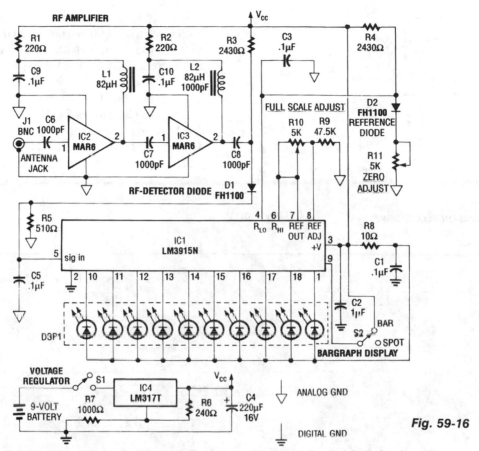

Fig. 59-16

Reprinted with permission from Radio-Electronics Magazine, June 1989. Copyright Gernsback Publications, Inc., 1989.

This rf detector can locate low-power transmitters (bugs) that are hidden from sight. It can sense the presence of a 1-mW transmitter at 20 feet, which is sensitive enough to detect the tiniest bug. As you bring the rf detector closer to the bug, more and more segments of its LED bar-graph display light, which aids in direction finding.

The front end has a two-stage wideband rf amplifier, and a forward-biased hot-carrier diode for a detector. After detection, the signal is filtered and fed to IC1, an LM3915N bar-graph driver having a logarithmic output. Each successive LED segment represents a 3-dB step.

DOOR OPENER

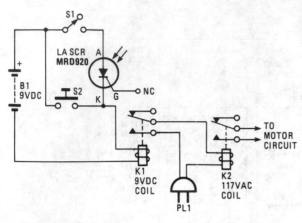

The door opener derives its power from a 9-V battery. A momentary-contact switch, S2, is provided in the event that manual opening and closing is required. Relay K1 is a 9-V type and relay K2 is a 117-Vac latching-type, which automatically latches with the first burst of current and opens on the second burst. The gate lead of the LASCR is not used; a light source triggers the LASCR unit into conduction, causing current to flow in the coil of the relay. That, in turn, causes K1's contacts to close, thereby energizing K2 (closing its contacts), and operating the garage door motor.

HANDS-ON ELECTRONICS   *Fig. 59-17*

60

Mixers

The sources of the following circuits are contained in the Sources section beginning on page 782. The figure number contained in the box of each circuit correlates to the sources entry in the Sources section.

Signal Combiner
Simple Mixer
Input-Buffered Mixer
Four-Channel Mixer
Universal Mixer Stage

SIGNAL COMBINER

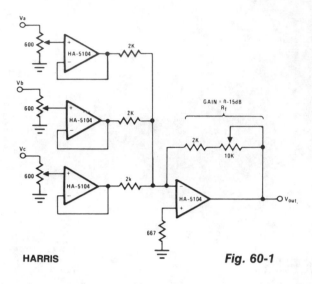

This circuit uses buffer stages to prevent channel crosstalk back through the mixer resistor network. The potentiometers used for each stage allow for convenient signal strength adjustment, while maintaining input impedance matching at the 600-Ω audio standard. The feedback resistor R_f will permit the output signal gain to be as high as 15 dB.

HARRIS

Fig. 60-1

SIMPLE MIXER

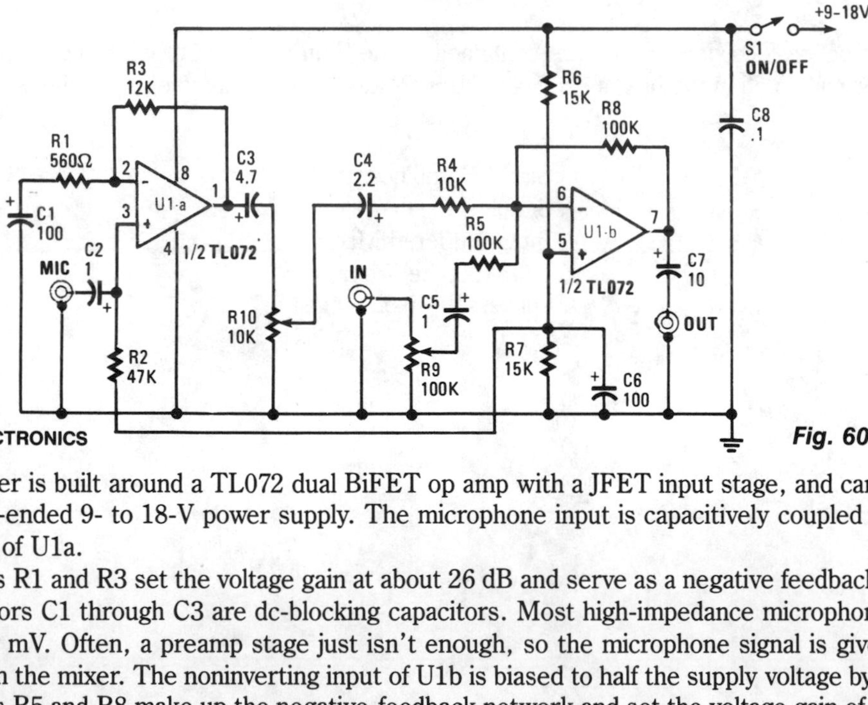

POPULAR ELECTRONICS

Fig. 60-2

This mixer is built around a TL072 dual BiFET op amp with a JFET input stage, and can be powered from a single-ended 9- to 18-V power supply. The microphone input is capacitively coupled to the noninverting input of U1a.

Resistors R1 and R3 set the voltage gain at about 26 dB and serve as a negative feedback network for U1a. Capacitors C1 through C3 are dc-blocking capacitors. Most high-impedance microphones have outputs of a few mV. Often, a preamp stage just isn't enough, so the microphone signal is given a boost of about 20 db in the mixer. The noninverting input of U1b is biased to half the supply voltage by R6, R7, and C6. Resistors R5 and R8 make up the negative-feedback network and set the voltage gain of U1b at unity. Capacitor C5 is for dc blocking at this input.

INPUT-BUFFERED MIXER

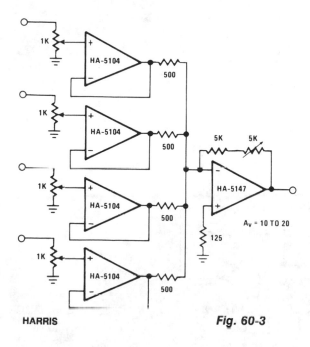

A high signal-to-noise ratio is important in signal construction and combination. The HA-5147 aids in lowering overall system noise and thereby raises system sensitivity. The signal combination circuit incorporates input buffering with several other features to form a relatively efficient mixer stage with a minimum of channel crosstalk. The potentiometer used for each channel allows for both variable input levels and a constant impedance for the driving source. The buffers serve mainly to prevent reverse crosstalk back through the resistor network. This buffering allows for the combination of varying strength signals without reverse contamination. The gain of the final stage is set at a minimum of 10 and can be adjusted to as much as 20. This allows a great amount of flexibility when combining a vast array of input signals

HARRIS   *Fig. 60-3*

FOUR-CHANNEL MIXER

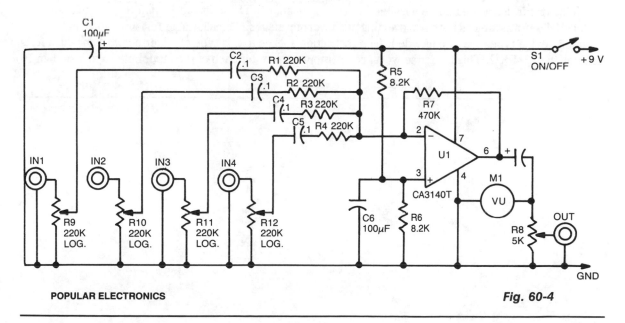

POPULAR ELECTRONICS   *Fig. 60-4*

369

UNIVERSAL MIXER STAGE

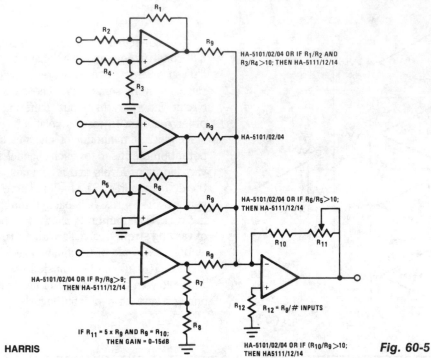

HA-5101/02/04 OR IF R_1/R_2 AND $R_3/R_4>$10; THEN HA-5111/12/14

HA-5101/02/04

HA-5101/02/04 OR IF $R_6/R_5>$10; THEN HA-5111/12/14

HA-5101/02/04 OR IF $R_7/R_8>$9; THEN HA-5111/12/14

$R_{12} = R_9/\#$ INPUTS

IF $R_{11} = 5 \times R_9$ AND $R_9 = R_{10}$; THEN GAIN = 0-15dB

HARRIS

HA-5101/02/04 OR IF ($R_{10}/R_9>$10; THEN HA5111/12/14

Fig. 60-5

This circuit illustrates some possible buffer combinations. These include a differential input stage, a voltage follower as well as both noninverting and inverting stages. The allowable resistor ratios and recommended device types are also included. One restriction applies to this type of mixer network in which R_g is greater than 2.4 KΩ. This limits the worst case output current for each of the input buffers to less than 10 mA.

61

Modulators

The sources of the following circuits are contained in the Sources section beginning on page 782. The figure number contained in the box of each circuit correlates to the sources entry in the Sources section.

RF MODULATOR

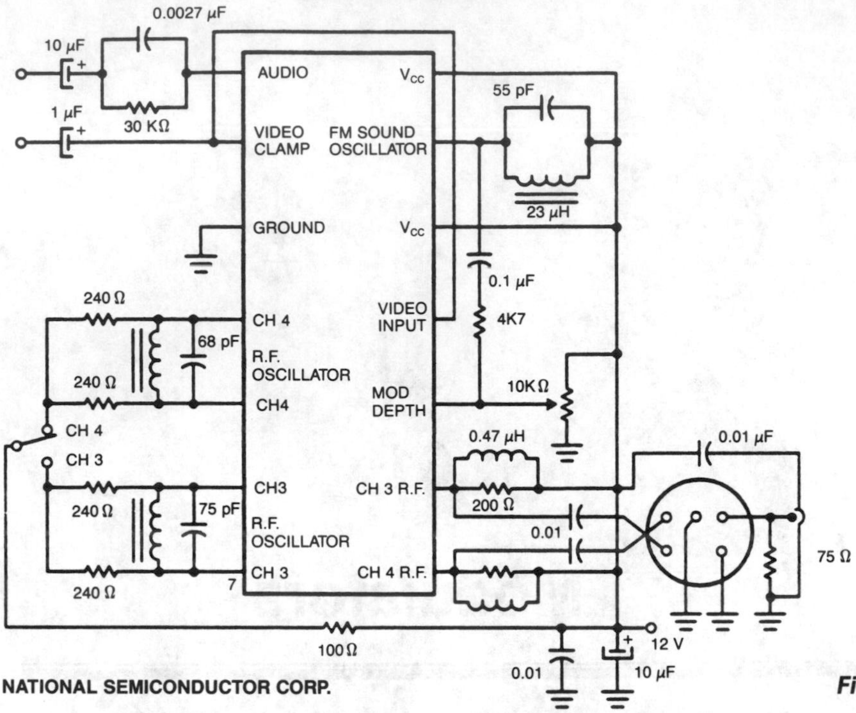

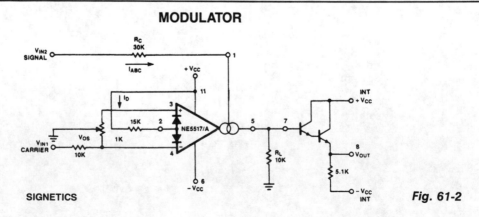

NATIONAL SEMICONDUCTOR CORP.

Fig. 61-1

Two IC rf modulators convert a suitable baseband video and audio signal up to a low VHF modulated carrier—Channel 2 through 6 in the U.S., and 1 through 3 in Japan—the LM1889 and the LM2889. Both ICs are identical regarding the rf modulation function, including pin outs, and can provide either of two rf carriers with dc switch selection of the desired carrier frequency. The LM1889 includes a crystal-controlled chroma subcarrier oscillator and balanced modulator for encoding R through Y or U and V color difference signals. A sound intercarrier frequency lc oscillator is modulated using an external varactor diode. The LM2889 replaces the chroma subcarrier function of the LM1889 with a video dc restoration clamp and an internal frequency-modulated sound intercarrier oscillator.

MODULATOR

SIGNETICS

Fig. 61-2

Because the transconductance of an operational transconductance amplifier is directly proportional to I_{ABC}, the amplification of a signal can be controlled easily. The output current is the product from transconductance X input voltage. The circuit is effective up to approximately 200 kHz. Modulation of 99% is easy to achieve.

SAW OSCILLATOR MODULATOR

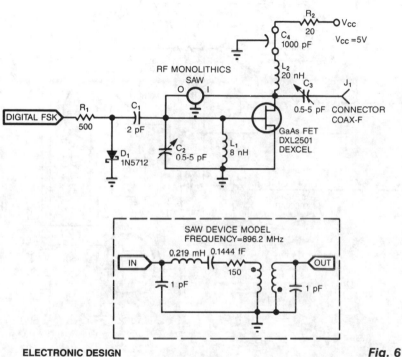

ELECTRONIC DESIGN

Fig. 61-3

Adding a diode, a resistor, and a capacitor to the SAW (surface-acoustic-wave) oscillator allows you to use the oscillator in FSK (frequency-shift-keying) applications. D1, R1, and C1 form a simple diode switch in which D1 shunts C1 to ground. When the digital FSK input to R1 is low, D1 is off, and the small junction capacitance of D1 couples C1 to ground. A high FSK signal causes current to flow through R1 and D1. D1's dynamic impedance is small when it is in forward conduction. Therefore, C1 sees a lower impedance path to ground. Thus, the FSK input effectively switches C1 in and out of the oscillator's circuit.

When C1 is in the circuit—digital FSK is high—it pulls the frequency of the circuit slightly lower because of the additional phase shift C1 introduces at the GaGs FET gate terminal (available from: Dexcel, Div. of Gould, Santa Clara, CA). The SAW device (available from: RF Monolithics, Dallas, TX) restricts the amount of frequency shifting—usually less than 20 ppm for a high-Q SAW device.

The oscillator shown produces a center frequency of 896.2 MHz with an FSK deviation of 17 kHz when you drive the FSK input with a 0 to 5 V signal. The frequency also depends on L1 and C2.

RF MODULATOR

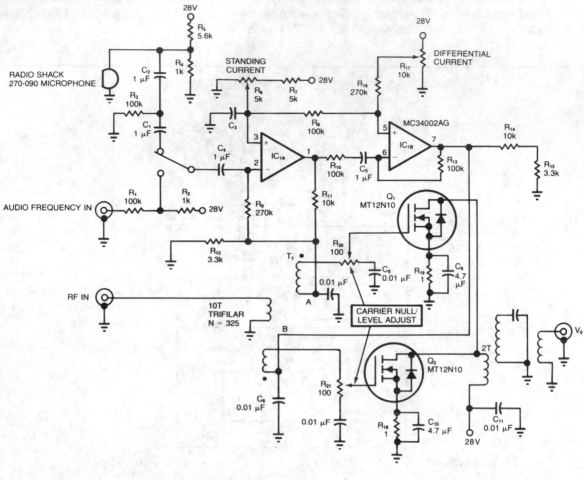

Fig. 61-4

Because power MOSFETs offer high power gain at both audio and radio frequencies, they are useful in many areas of radio-circuit design. For rf applications, a MOSFET's large safe operating area, V_{DS} vs I_D, protects it against damage from reflected rf energy. As a modulator, a MOSFET's transfer linearity aids fidelity. In the suppressed-carrier modulator, an rf signal is applied to the primary of transformer T1, whose secondaries provide equal-amplitude, opposite-phase rf drive signals to output FETs Q1 and Q2. Output V_o is zero when no audio-frequency signals are present, because the opposite-phase rf signals from Q1 and Q2 cancel. When audio-frequency signals appear at nodes A and B, you obtain a modulated rf output (V_o). Source resistors R18 and R19 improve the dc stability and low-frequency gain. A phase inverter, based on the dual op amp IC1A and IC1B, generates the out-of-phase, equal-amplitude, audio-frequency modulation signals.

MODULATION MONITOR

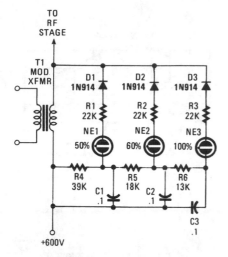

Switching diodes are used to fire the neon lamps when negative-peak modulation hits 50, 60, and 100%. To use the circuit, keep an eye on the lamps. You should attempt to keep the 50% lamp firing all the time, the 60% lamp should be on as much as possible, but try to prevent the 100% lamp from lighting.

POPULAR ELECTRONICS   *Fig. 61-5*

PULSE-POSITION MODULATOR

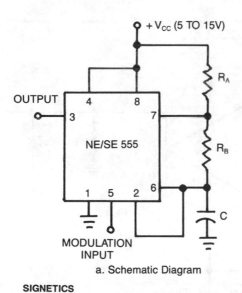

a. Schematic Diagram

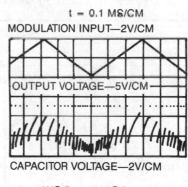

R_A—3KΩ R_B = 500ΩC = .01 μF

b. Expected Waveform

SIGNETICS   *Fig. 61-6*

This application uses the timer connected for astable (free-running) operation, with a modulating signal again applied to the control voltage terminal. The pulse position varies with the modulating signal, since the threshold voltage and the time delay is varied.

PULSE-WIDTH MODULATOR

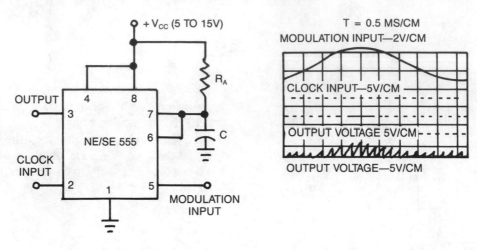

a. Device Schematic

b. Expected Waveforms

SIGNETICS

Fig. 61-7

In this application, the timer is connected in the monostable mode. The circuit is triggered with a continuous pulse train and the threshold voltage is modulated by the signal applied to the control voltage terminal at pin 5. This has the effect of modulating the pulse width as the control voltage varies. The figure shows the actual waveform generated with this circuit.

BALANCED MODULATOR

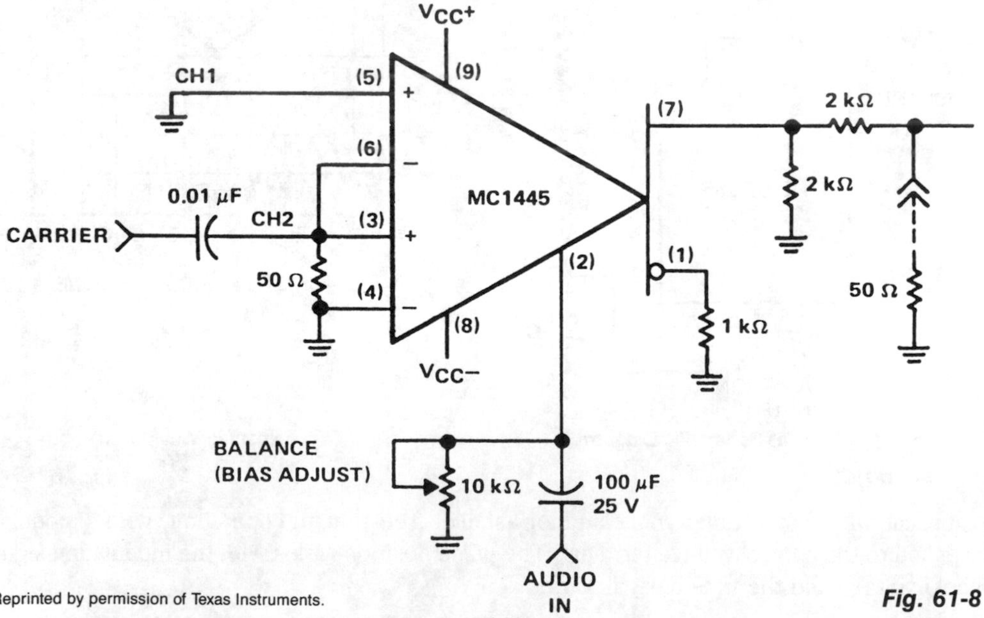

Fig. 61-8

BALANCED MODULATOR (*Cont.*)

When the carrier level is adequate to switch the cross-coupled pair of differential amplifiers, the modulation signal, which has been applied to the gate, will be switched at the carrier rate, between the collector loads. When switching occurs, it will result in the modulation being multiplied by a symmetrical switching function. If the modulation gate remains in the linear region, only the first harmonic will be present. To balance the MC1445 modulator, equal gain must be achieved in the two separate channels. To remain in the linear region, the modulation input must be restricted to approximately 200 mV pk-pk. Because the gate bias point is sensitive to the amount of carrier suppression, a high-resolution, 10-turn potentiometer should be used.

DOUBLE-SIDEBAND SUPPRESSED-CARRIER MODULATOR

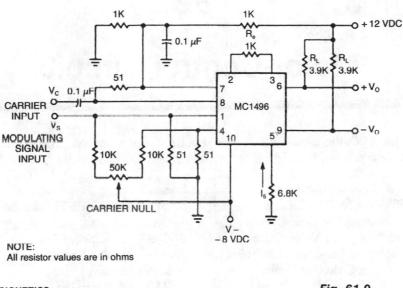

NOTE:
All resistor values are in ohms

SIGNETICS   *Fig. 61-9*

The basic current allows no carrier to be present in the output. By adding offsets to the carrier differential pairs, controlled amounts of carrier appear at the output. The amplitude becomes a function of the modulation signal—AM modulation.

62

Motor-Control Circuits

The sources of the following circuits are contained in the Sources section beginning on page 782. The figure number contained in the box of each circuit correlates to the sources entry in the Sources section.

PWM SERVO AMPLIFIER

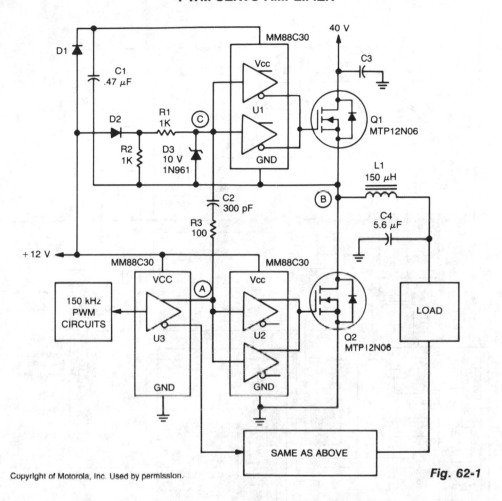

Fig. 62-1

A major feature of the PWM servo amplifier is elimination of a pulse transformer. A 150 kHz pulse-width modulated signal is applied to U3, with its complementary outputs applied to identical circuits to drive the load. When point A increases, Q2 is on and point B is at ground potential. The V_{CC} for U1 is maintained through D1, and Q1 is held off by D2. When point A decreases, Q2 turns off, point C is pulled low by C2, which turns Q1 on. The time constant for R1, R3, and C2 can hold Q1 on just long enough to allow the voltage at point B to start rising. As point B rises, it charges C2 by forward biasing D3, maintaining point C low with respect to U1, and keeping Q1 turned on.

With point B at 40 V, D2 is off and point C is held low by R1 and R2, and V_{CC} for U1 is maintained by the charge on C1. When point A increases again, Q2 again turns on, C2 pushes point C high, and turns Q1 off long enough to allow the voltage at point B to start falling. C2 is now discharged by reverse-biased D3, which keeps point C high with respect to U1, and keeps Q1 off. Once point B reaches ground potential, D1 again turns on, recharging C1, and maintaining V_{CC} to U1. D2 also turns on and keeps Q1 off.

PWM SPEED CONTROL AND ENERGY-RECOVERING BRAKE

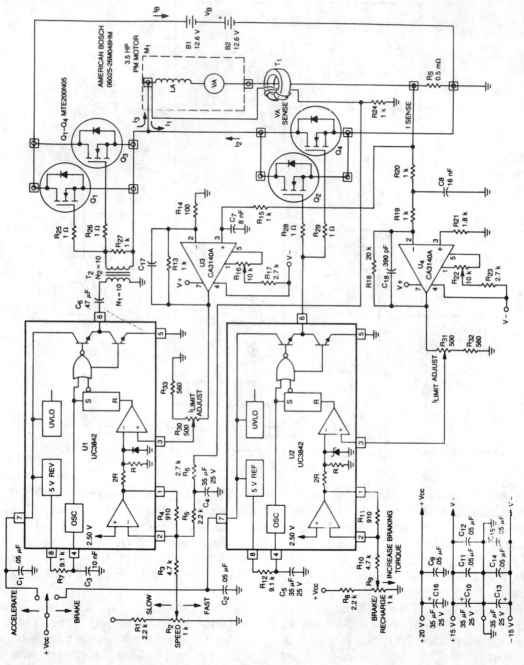

Fig. 62-2

This circuit uses the main drive motor as a generator/brake to recover the battery charge during vehicle braking. When this is done, it can increase the overall range and efficiency of an electric vehicle.

In the accelerate mode, Q1 through Q3 receive gate pulses from U1, an on-line, current-mode, PWM controller IC. Assuming negligible effects of R, when Q1 through Q3 turn on, current I_1 builds up through the motor at a rate of:

$$dI_1 = V_B - V_A$$

where: V_A = Battery voltage
V_B = Motor's back EMF
L_A = Motor inductance in henries

Motor current and torque continue to rise until the voltage on the I_{SENSE} line is greater than I_{RESET}, as determined by the speed potentiometer. At this time, Q1 through Q3 are switched off, current I_2 begins to flow and decreases at a rate of:

$$\frac{dI_2}{dt} = \frac{V_A}{L_A}$$

until the next clock period begins.

As vehicle braking occurs, accelerate PWM IC U1 is switched off and braking PWM IC U2 and Q2 through Q4 are switched on. During this time, the back EMF source voltage causes current I_3 to begin to flow at the rate of:

$$\frac{dI_3}{dt} = \frac{V_A}{L_A}$$

The current I_3 continues to rise until I_{SENSE} is greater than I_{RESET}. Now, Q2 through Q4 are switched off and I_B is forced to flow back into the storage battery, thus energy is recovered.

The braking torque produced by the motor is proportional to the average reverse current that flows through the motor on the duty cycle of Q2 through Q4. The braking force can continue until:

$$V_A = 0$$

For reliable performance, voltage supplies should be independent of the main battery voltage.

START-AND-RUN MOTOR CIRCUIT

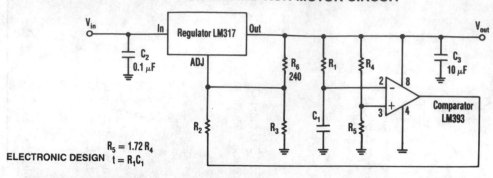

$R_5 = 1.72 R_4$
ELECTRONIC DESIGN $t = R_1 C_1$

Fig. 62-3

The timed two-voltage circuit can start and run a small dc motor or solenoid. The input voltage to the LM317 three-terminal regulator ranges from 5 to 40 V, and the output voltage can range from 2 to 36 V. With input voltage V_{IN} initially applied to the input, and capacitor C1 in a discharged state, the LM393 comparator's open-collector output circuit is open-circuited. Then the higher start-up output voltage is:

$$V_{OUT} = 1.25 \, [1 + (R_3/240)]$$

At a time t after start-up, when:

$$t = -R_1 C_1 \text{Ln}[R_4/(R_4 + R_5)]$$

or:

$$t = -R_1 C_1, \text{ if } R_5 = 1.72 R_4$$

the comparator output decreases. At that time, the output voltage switches to a lower value to run the device at its proper operating level.

AUTOMATIC FAN SPEED CONTROLLER

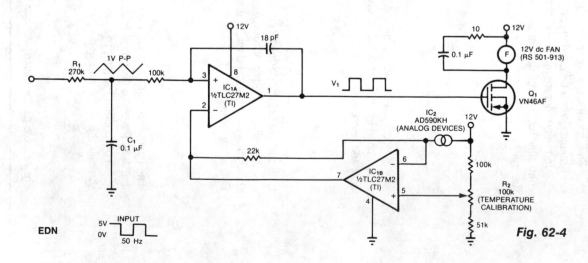

EDN

Fig. 62-4

AUTOMATIC FAN SPEED CONTROLLER (Cont.)

The controller circuit can reduce a fan's noise, power consumption, and wear, particularly in the presence of a low, fluctuating ambient temperature. Mount a temperature sensor in the fan's airstream, and the circuit will adjust the fan speed as necessary to maintain a relatively constant sensor temperature. Input components R1 and C1 integrate the input square wave, producing a quasitriangular wave at the non-inverting input of op amp IC1A. At this inverting input is a reference voltage that decreases as temperature increases. The two-terminal sensor produces 1 μA/°K. The result is a rectangular wave at the output of IC1A with a duty cycle proportional to absolute temperature. Thus, a rise in temperature triggers a counteracting cooling effect by delivering more power to the fan. To calibrate the system with the sensor at room temperature, simply adjust R2 for a 50% duty cycle at V_1. The fan will switch off at approximately 0°C and will be fully on at 44°C.

EFFICIENT SWITCHING CONTROLLER

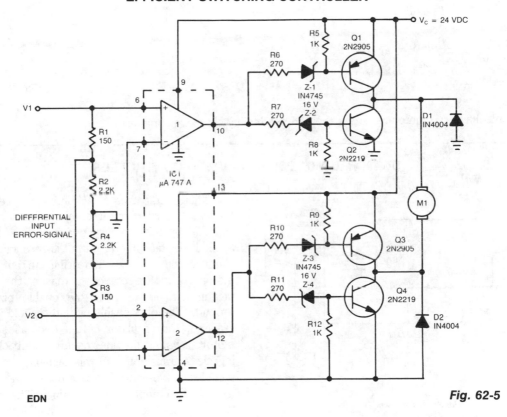

EDN

Fig. 62-5

This high-performance switching controller for a low-power dc servo motor uses a symmetrical complementary-transistor bridge. The bridge acts as a reversing switch between the motor and a single-ended power supply. Since the transistors operate either fully on or completely off, except during a very short transition period, much less heat is dissipated than in linear-amplifier circuits. Damping is provided by the circuit's inherent dynamic braking. Since either maximum or zero voltage is applied to the motor, the dynamic response is faster than that of linear servo drives.

SERVO SYSTEM CONTROLLER

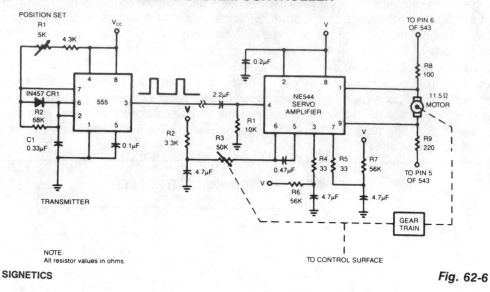

NOTE:
All resistor values in ohms.

SIGNETICS

TRANSMITTER

Fig. 62-6

To control a servo motor remotely, the 555 needs only six extra components.

SWITCHED-MODE MOTOR-SPEED CONTROLLER

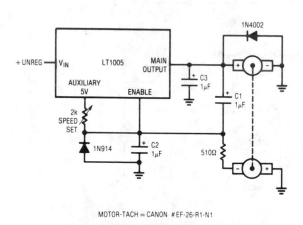

MOTOR-TACH = CANON # EF-26-R1-N1

LINEAR TECHNOLOGY CORP.   Fig. 62-7

This circuit uses a tachometer to generate a feedback signal which is compared to a reference supplied by the auxiliary output. When power is applied, the tachometer output is zero and the regulator output comes on, forcing current into the motor. As motor rotation increases, the negative tachometer output pulls the enable pin toward ground. When the enable pin's threshold voltage is reached, the regulator output decreases and the motor slows. C1 provides positive feedback, ensuring clean transitions. In this fashion, the motor's speed is servo-controlled at a point determined by the 2-KΩ potentiometer setting. The regulator free-runs at whatever frequency and duty cycle are required to maintain the enable pin at its threshold. The loop bandwidth and stability are set by C2 and C3. The 1N914 diode prevents the negative output tachometer from pulling the enable pin below ground, and the 1N4002 commutates the motor's negative flyback pulse.

CLOSED-LOOP MOTOR-SPEED CONTROL

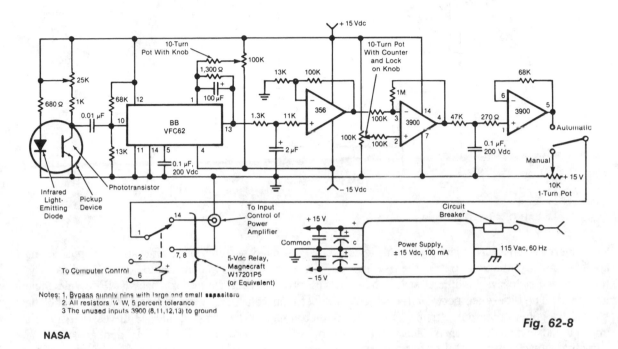

Notes: 1. Bypass supply pins with large and small capacitors
2. All resistors ¼ W, 5 percent tolerance
3 The unused inputs 3900 (8,11,12,13) to ground

NASA

Fig. 62-8

This electronic motor-speed control circuit is designed to operate in an electrically noisy environment. The circuit includes an optoelectronic pickup device, which is placed inside the motor housing to provide a speed feedback signal. The circuit automatically maintains the speed of the motor at the commanded value.

The pickup device contains an infrared LED and a phototransistor. The radiation from the diode is chopped into pulses by the motor fan blades, which are detected by the phototransistor. The train of pulses from the phototransistor is fed to a frequency-to-voltage converter, the output of which is a voltage proportional to the speed of the motor. This voltage is low-pass filtered, amplified, and compared with a manually-adjustable control voltage that represents the commanded speed.

The difference between the speed-measurement and speed-command signals is amplified and fed as a control voltage to an external power amplifier that drives the motor. A selector switch at the output of the final amplifier of this circuit also enables the operator to bypass the circuit and manually set the control voltage for the external amplifier.

TACHLESS MOTOR-SPEED CONTROLLER

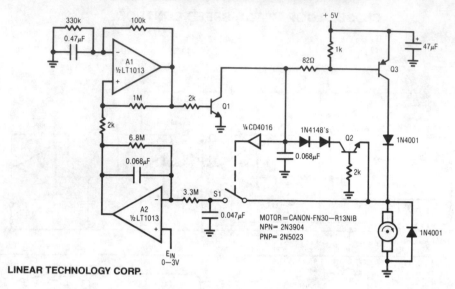

LINEAR TECHNOLOGY CORP.

Fig. 62-9

This circuit is particularly applicable to digitally-controlled systems in robotic and X-Y positioning applications. By functioning from the 5-V logic supply, it eliminates additional motor-drive supplies. The *tachless* feedback saves additional space and cost. The circuit senses the motor's back EMF to determine its speed. The difference between the speed and a set point is used to close a sampled loop around the motor. A1 generates a pulse train. When A1's output is high, Q1 is biased, and Q3 drives the motor's ungrounded terminal. When A1 decreases, Q3 turns off and the motor's back EMF appears after the inductive flyback ceases. During this period, S1's input is turned on, and the 0.047-μF capacitor acquires the back EMF's value. A2 compares this value with the set point and the amplified difference (trace D) changes A1's duty cycle, controlling the motor speed.

CONSTANT-SPEED MOTOR DRIVER

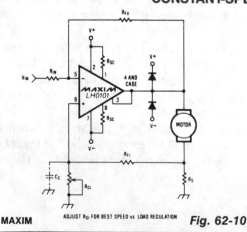

MAXIM

ADJUST R$_{GI}$ FOR BEST SPEED vs LOAD REGULATION   Fig. 62-10

When the torque load on the motor increases, its current increases. This current increase is sensed across R_S, and positive feedback is applied to the noninverting terminal of the LH0101, thereby increasing the motor voltage to compensate for the increased torque load. With the proper amount of positive feedback, the motor-speed variation can be kept below 1% from no load to full load.

DC MOTOR DRIVE WITH FIXED SPEED CONTROL

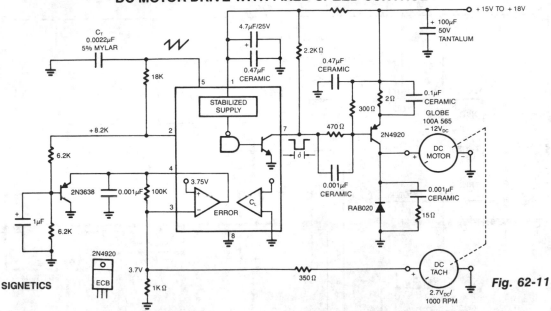

SIGNETICS

Fig. 62-11

The NE5561 provides pulse-proportional drive and speed control based on dc tachometer feedback. This simple switching circuit consists of transistor 2N4920 pnp with a commutation diode used to deliver programmed pulse energy to the motor. A frequency of approximately 20 kHz is used to eliminate audio noise. The dc tach delivers 2.7 V/1000 RPM. Negative feedback occurs when this voltage is applied to the error amplifier of the NE5561. The duty cycle varies directly with load torque demand. The no-load current is ≈ 0.3 A and full load is 0.6 A.

BRIDGE-TYPE AC SERVO AMPLIFIER

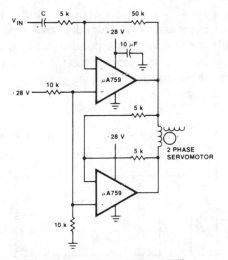

This motor driver circuit uses a μA759 power amplifier to drive a two-phase servomotor.

FAIRCHILD CAMERA AND INSTRUMENT CORP.   **Fig. 62-12**

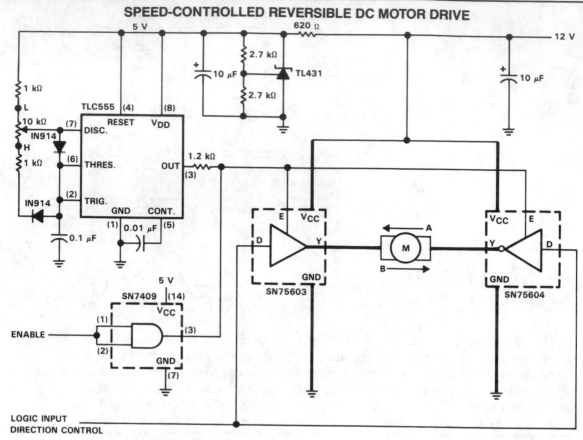

SPEED-CONTROLLED REVERSIBLE DC MOTOR DRIVE

LOGIC INPUT
DIRECTION CONTROL

Reprinted by permission of Texas Instruments.

Fig. 62-13

The figure illustrates a reversible dc motor drive application with adjustable speed control. The D inputs for these drivers are complementary and can be tied together and driven from the same logic control for bidirectional motor drive. The enables are tied together and driven by a pulse-width-modulated generator providing on duty cycles of 10 to 90% for speed control. A separate enable control is provided through an SN7409 logic gate. See the truth table for this motor controller application.

Definitions for the terms used in the truth table are as follows:

EN Enable
DC Direction control
SP.C Speed control
A Direction of current—right to left
B Direction of current—left to right
H Logic 1 voltage level
L Logic 0 voltage level
N Speed control set for narrow pulse width
W Speed control set for wide pulse width
X Irrelevant

Truth Table for Motor Control Circuit

EN	DC	SP.C	MOTOR DIRECTION	MOTOR SPEED
L	X	X	OFF	OFF
H	L	N	A	SLOW
H	L	W	A	FAST
H	H	N	B	SLOW
H	H	W	B	FAST

PWM MOTOR CONTROLLER

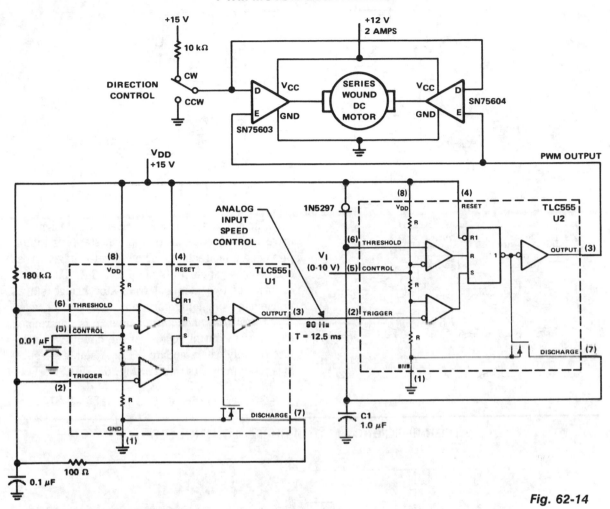

Fig. 62-14

Reprinted by permission of Texas Instruments.

The PWM controller uses complementary half-H peripheral drivers SN75603 and SN75604, with totem-pole outputs rated at 40 V and 2.0 A. These drivers effectively place the motor in a full-bridge configuration, which has the ability to provide bidirectional control.

Timer U1 operates in the astable mode at a frequency of 80 Hz. The 100-Ω discharge resistor results in an 8-μs trigger pulse which is coupled to the trigger input of timer U2. Timer U2 serves as the PWM generator. Capacitor C1 is charged linearly with a constant current of 1 mA from the 1N5297, which is an FET current-regulator diode.

Motor speed is controlled by feeding a dc voltage of 0 to 10 V to control input pin 5 of U2. As the control voltage increases, the width of the output pulse pin 3 also increases. These pulses control the on/off time of the two motor drivers. The trigger pulse width of timer U1 limits the minimum possible duty cycle from U2.

STEPPING MOTOR DRIVE

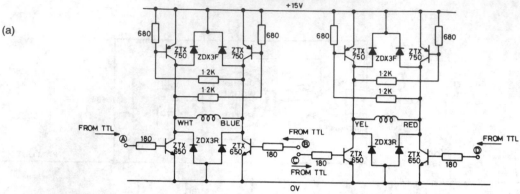

(a)

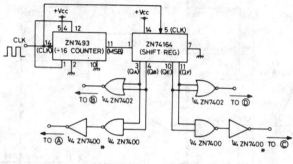

(b)

The circuit shown in Fig. 62-15A is designed to drive a 15-V, two-phase, bipolar stepping motor, providing a bidirectional single level voltage across each winding at currents of up to 9.6 A. The circuit consists of two identical transistor bridge stages employing complementary npn and pnp devices. The transistor conduction sequence is determined by external control logic, and the circuit will interface directly with standard TTL. A suitable control logic system is illustrated in Fig. 62-15B.

ZᴇTᴇX, formerly FERRANTI   *Fig. 62-15*

HIGH-EFFICIENCY MOTOR-SPEED CONTROLLER

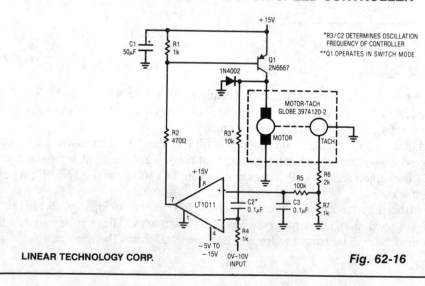

LINEAR TECHNOLOGY CORP.   *Fig. 62-16*

63

Multiplexers

The sources of the following circuits are contained in the Sources section beginning on page 782. The figure number contained in the box of each circuit correlates to the sources entry in the Sources section.

Two-Level Multiplexer
1-of-15 Cascaded Video MUX
Low-Cost Four-Channel Multiplexer
Demultiplexer
One-of-Eight Channel Transmission
    System

Three-Channel Multiplexer with
    Sample-and-Hold
Analog Multiplexer with
    Buffered Input and Output

TWO-LEVEL MULTIPLEXER

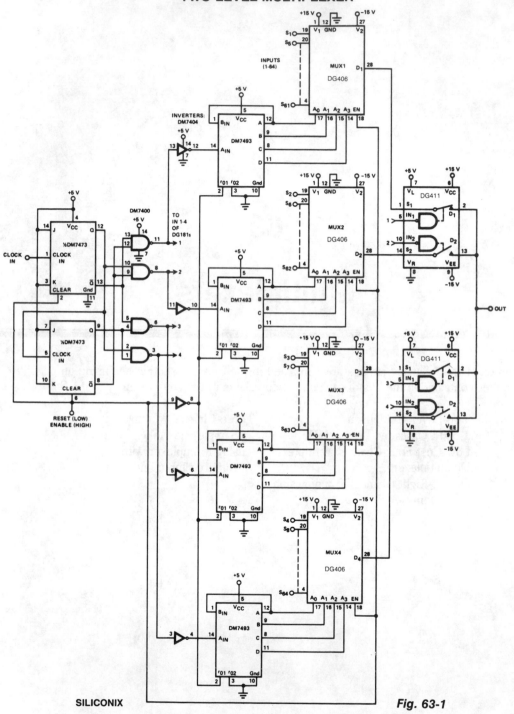

SILICONIX

Fig. 63-1

TWO-LEVEL MULTIPLEXER (*Cont.*)

When a large number of channels are multiplexed, the outputs of two or more multiplexers can be connected together and each multiplexer sequentially enabled. In the inhibit mode, the multiplexer draws less power and its output and inputs act as open circuits. Theoretically, an infinite number of channels can be accommodated in this way; in practice, the accumulated output capacitance and leakage of many paralleled multiplexers limits the speed and accuracy of the system. A much better method is the two-level multiplex system. The two-level system has a bank of high-speed switches at the output which sequentially switch between the four DG406s. Each DG406 is able to switch during the time the other three are being interrogated. The DG406s contribute leakage and capacitance at the output only, when they are switched on by the DG411—$1/4$ of the time. The two-level multiplex system is very useful in communications links, high-speed interfacing with comparators, or wherever a large number of channels must be multiplexed at high speeds.

1-OF-15 CASCADED VIDEO MUX

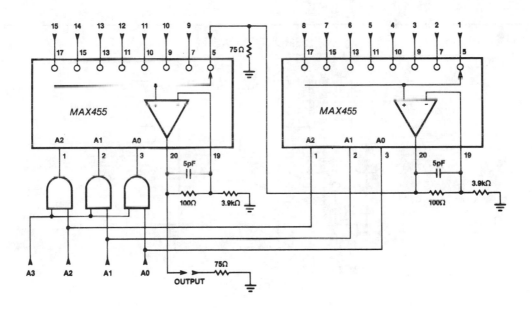

MAXIM

Fig. 63-2

Two MAX455s can be cascaded to form a 1 of 15 video MUX by connecting the output of one MUX to one of the input channels of a second MUX. Although the two devices are usually close to one another, the output of the first MUX should be terminated to preserve its bandwidth.

LOW-COST FOUR-CHANNEL MULTIPLEXER

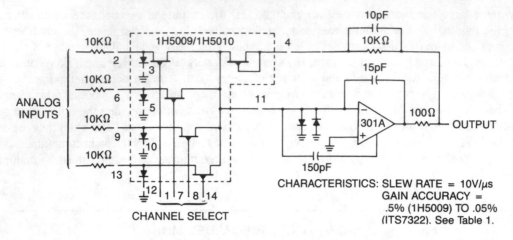

ANALOG INPUTS

CHANNEL SELECT

CHARACTERISTICS: SLEW RATE = 10V/μs
GAIN ACCURACY =
.5% (1H5009) TO .05%
(ITS7322). See Table 1.

INTERSIL

Fig. 63-3

DEMULTIPLEXER

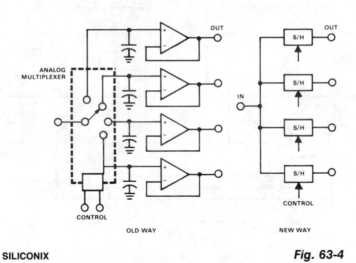

SILICONIX

Fig. 63-4

This circuit reconstructs and separates analog signals which have been time-division multiplexed. The conventional method, shown on the left, has several restrictions, particularly when a short dwell time and a long, accurate hold time is required. The capacitors must charge from a low-impedance source through the resistance and current-limiting characteristics of the multiplexer. When holding, the high-impedance lines are relatively long and subject to noise pickup and leakage. When FET input buffer amplifiers are used for low leakage applications, severe temperature offset errors are often introduced.

1-OF-8 CHANNEL TRANSMISSION SYSTEM

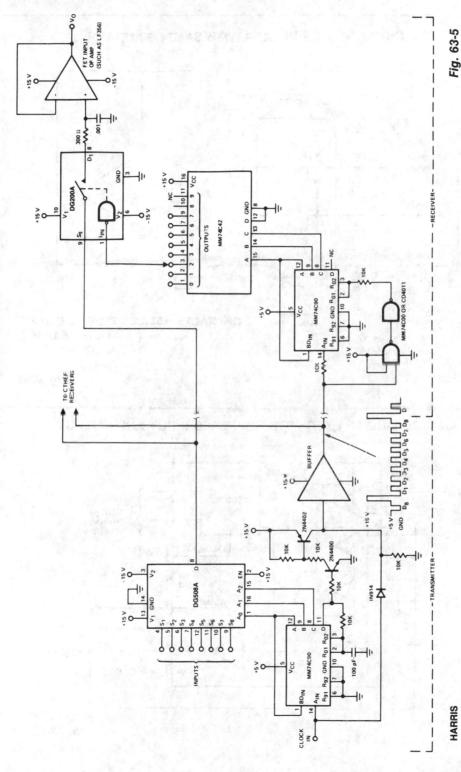

This circuit shows a typical multiplex system intended to carry one of 8 inputs into a remote location. A 5-V pulse train is sent down a separate channel to perform timing and synchronizing functions. A 15-V reset pulse is superimposed on the 5-V clock, which is detected by the MM74C00 in the receiver. Using this system, many remote points can be monitored, one at a time, at any of several locations.

HARRIS

Fig. 63-5

395

THREE-CHANNEL MULTIPLEXER WITH SAMPLE-AND-HOLD

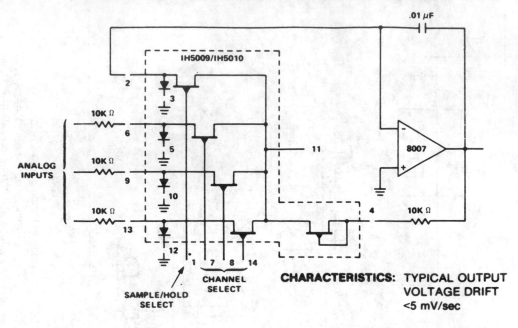

ANALOG INPUTS

10K Ω

10K Ω

10K Ω

IH5009/IH5010

SAMPLE/HOLD SELECT

CHANNEL SELECT

.01 µF

8007

10K Ω

CHARACTERISTICS: TYPICAL OUTPUT VOLTAGE DRIFT <5 mV/sec

INTERSIL

Fig. 63-6

ANALOG MULTIPLEXER WITH BUFFERED INPUT AND OUTPUT

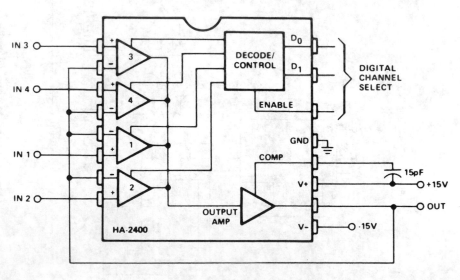

IN 3

IN 4

IN 1

IN 2

DECODE/ CONTROL

D_0

D_1

DIGITAL CHANNEL SELECT

ENABLE

GND

COMP

V+

15pF

+15V

OUT

OUTPUT AMP

V−

−15V

HA-2400

HARRIS

Fig. 63-7

396

ANALOG MULTIPLEXER WITH BUFFERED INPUT AND OUTPUT (*Cont.*)

This circuit is used for analog signal selection or time division multiplexing. As shown, the feedback signal places the selected amplifier channel in a voltage follower (noninverting unity gain) configuration, and provides very high input impedance and low output impedance. The single package replaces four input buffer amplifiers, four analog switches with decoding, and one output buffer amplifier. For low-level input signals, gain can be added to one or more channels by connecting the (−) inputs to a voltage divider between output and ground. The bandwidth is approximately 8 MHz, and the output will slew from one level to another at about 15.0 V per μs.

Expansion to multiplex 5 to 12 channels can be accomplished by connecting the compensation pins of two or three devices together, and using the output of only one of the devices. The enable input on the unselected devices must be low.

Expansion to 16 or more channels is accomplished easily by connecting outputs of four 4-channel multiplexers to the inputs of another 4-channel multiplexer. Differential signals can be handled by two identical multiplexers addressed in parallel. Inverting amplifier configurations can also be used, but the feedback resistors might cause crosstalk from the output to unselected inputs.

64

Noise Reduction Circuits

The sources of the following circuits are contained in the Sources section beginning on page 782. The figure number contained in the box of each circuit correlates to the sources entry in the Sources section.

Dolby B/C Noise Reduction System
Dolby B Noise Reduction Circuit in Encode Mode
Dolby B Noise Reduction Circuit in Decode Code

DOLBY B/C NOISE REDUCTION SYSTEM

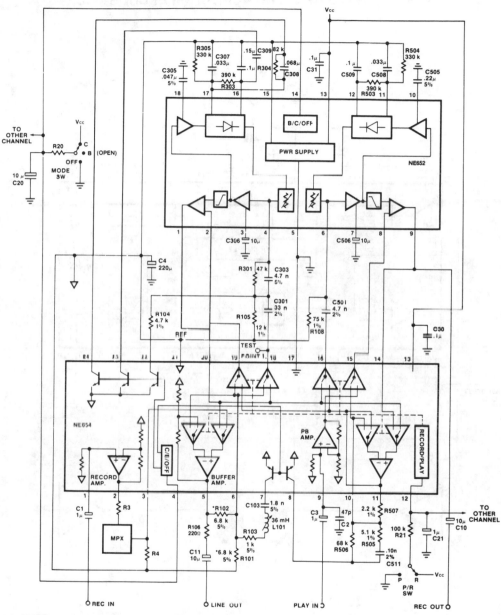

NOTES:
1. *Line output programming resistors.
2. Split supply operation and coupling capacitors are optional.
3. Time constant for mode switch is optional.
4. Applications info is for reference only. Final design configuration/values are found in relevant Dolby Labs Bulletins and Licensee manuals.
5. R20 value is equal to 6.8KΩ divided by "N" where "N" equals the number of switched channels.
6. R106 is recommended for large capacitive loads on line out.
7. Switches shown in REC position.

SIGNETICS

Fig. 64-1

DOLBY B NOISE REDUCTION CIRCUIT IN ENCODE MODE

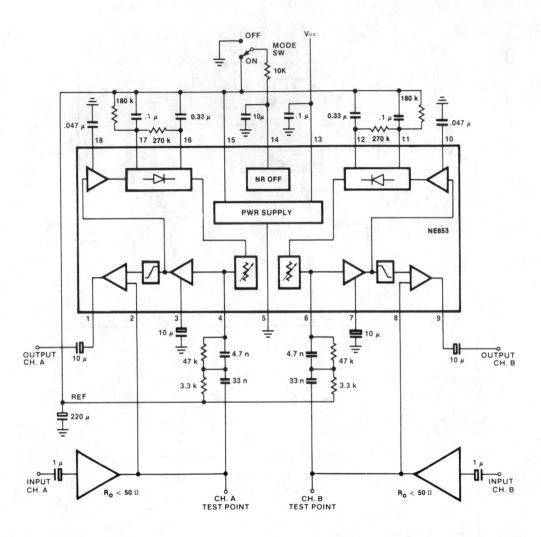

SIGNETICS

Fig. 64-2

DOLBY B NOISE REDUCTION CIRCUIT IN DECODE MODE

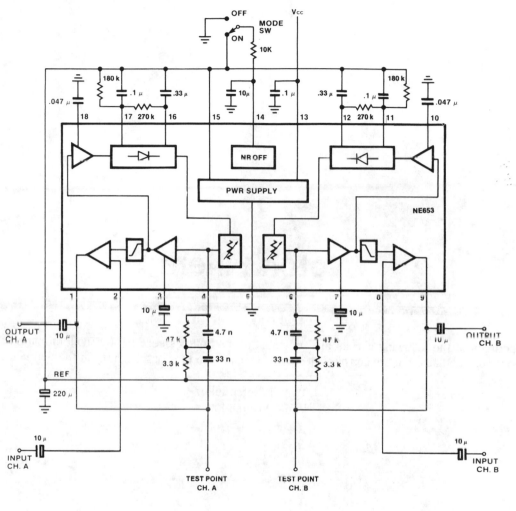

65

Notch Filters

The sources of the following circuits are contained in the Sources section beginning on page 782. The figure number contained in the box of each circuit correlates to the sources entry in the Sources section.

Twin-T Notch Filter
High-Q Notch Filter

TWIN-T NOTCH FILTER

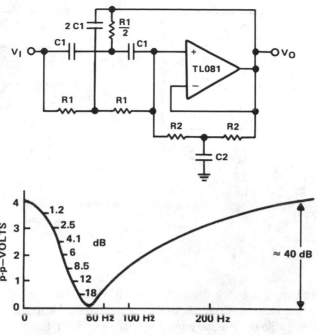

60-Hz Twin-T Notch Filter Response

Fig. 65-1

This filter is used to reject or block a frequency or band of frequencies. These filters are often designed into audio and instrumentation systems to eliminate a single frequency, such as 60 Hz. Commercial grade components with 5% – 10% tolerance produce a null depth of at least 30 to 40 dB. When a twin-T network is combined with a TL081 op amp in a circuit, an active filter can be implemented. The added resistor capacitor network, R2 and C2, work effectively in parallel with the original twin-T network, on the input of the filter. These networks set the Q of the filter. The op amp is basically connected as a unity-gain voltage follower. The Q is found from:

$$Q = \frac{R_2}{2R_1} = \frac{C_1}{C_2}$$

For a 60-Hz notch filter with a Q of 5, it is usually best to pick the C1 capacitor value and calculate the resistor R1. Let C1 = 0.22 μF. Then:

$$R_1 = 12 \text{ K}\Omega$$
$$R_2 = 120 \text{ K}\Omega$$
$$C_2 = 0.047 \ \mu\text{F}$$

Standard 5% resistors and 10% capacitors produce a notch depth of about 40 dB, as shown in the frequency response curve.

HIGH-Q NOTCH FILTER

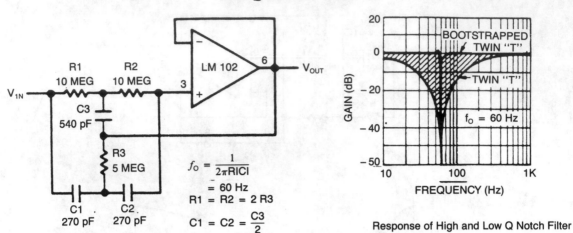

$$f_O = \frac{1}{2\pi RICI}$$
$$= 60 \text{ Hz}$$
$$R1 = R2 = 2\,R3$$
$$C1 = C2 = \frac{C3}{2}$$

Response of High and Low Q Notch Filter

NATIONAL SEMICONDUCTOR CORP.

Fig. 65-2

This circuit shows a twin-T network connected to an LM102 to form a high-Q, 60-Hz notch filter. The junction of R3 and C3 which is normally connected to ground, is bootstrapped to the output of the follower. Because the output of the follower is a very low impedance, neither the depth nor the frequency of the notch change; however, the Q is raised in proportion to the amount of signal fed back to R3 and C3. Shown is the response of a normal twin-T and the response with the follower added.

66

Operational Amplifiers

The sources of the following circuits are contained in the Sources section beginning on page 782. The figure number contained in the box of each circuit correlates to the sources entry in the Sources section.

Operational Amplifiers

OPERATIONAL AMPLIFIERS

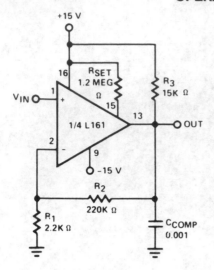

The L161 as a X100 Operational Amplifier

(A)

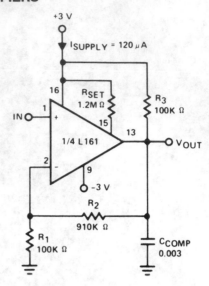

A Micropower X10 Op Amp

(B)

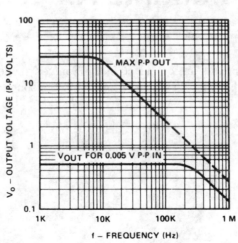

**Frequency Response and Maximum Output
for the X100 Op Amp**

SILICONIX

Fig. 66-1

This is a single gain-of-100 amplifier with a gain-bandwidth product of 20 MHz! The primary limitation in the performance is the low slew rate (0.3 V/μs) imposed by I_{OH} charging C_{COMP}. The effects of slew rate and compensation are shown. A lower gain amplifier requires a larger C_{COMP}, which in turn further reduces slew rate. For this reason, it might actually be advantageous in certain areas to lower the gain by placing a resistive divider at the input rather than raising R_l. Figure 66-1B shows a 700-μW, X10 op amp whose slew rate is 0.02 V/μs and is 3 dB down at 100 kHz.

67

Optically-Coupled Circuits

The sources of the following circuits are contained in the Sources section beginning on page 782. The figure number contained in the box of each circuit correlates to the sources entry in the Sources section.

High-Voltage Ac Switcher
Solid-State Zero-Voltage Switchers (ZVS)
Triggering SCR Series
Normally Closed Half-Wave ZVS Contact
   Circuit
Normally Open and Normally Closed
   Dc Solid-State Relays
Indicator Lamp Driver
Ambient-Light-Ignoring Optical Sensor

Optical CMOS Coupler
Line-Current Detector
Line-Operated Power Outage Light
Optical TTL Coupler
Zero-Voltage Switching, Solid-State Relay
   with Antiparallel SCR Output
Dc Latching Relay
Ac Relay
Photodiode Source Follower

HIGH-VOLTAGE AC SWITCHER

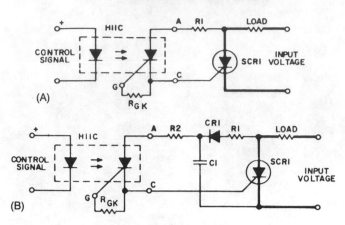

(A)

(B)

A basic circuit to trigger an SCR is shown in Fig. 67-1A. This circuit has the disadvantage that the blocking voltage of the photon-coupler output device determines the circuit-blocking voltage, irrespective of higher main SCR capability.

Adding capacitor C1 to the circuit, as shown in Fig. 67-1B, will reduce the *dV/dt* seen by the photon-coupler output device. The energy stored in C1, when discharged into the gate of SCR1, will improve the *di/dt* capability of the main SCR.

Using a separate power supply for the coupler adds flexibility to the trigger circuit; it removes the limitation of the blocking voltage capability of the photon-coupler output device. The flexibility adds cost and more than one power supply might be necessary for multiple SCRs if no common reference points are available.

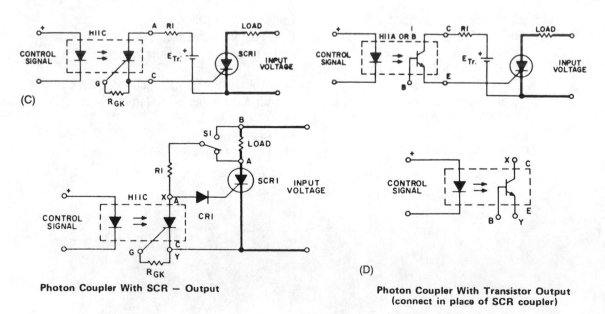

(C)

Photon Coupler With SCR — Output

(D)

Photon Coupler With Transistor Output
(connect in place of SCR coupler)

In Fig. 67-1C, R1 can be connected to Point A, which will remove the voltage from the coupler after SCR1 is triggered, or to Point B so that the coupler output will always be biased by input voltage. The former is preferred since it decreases the power dissipation in R1. A more practical form of SCR triggering is shown in Fig. 67-1F. Trigger energy is obtained from the anode supply and stored in C1. Coupler voltage is limited by the zener voltage. This approach permits switching of higher voltages than the blocking voltage capability of the output device of the photon coupler. To reduce the power losses in R1 and to obtain shorter time constants for charging C1, the zener diode is used instead of a resistor.

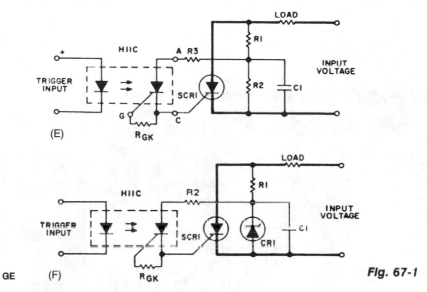

Fig. 67-1

A guide for selecting the component values would consist of the following steps:

- Choose C1 in a range of 0.05 to 1 μF. The maximum value might be limited by the recharging time constant $(R_L + R_1) C_1$ while the minimum value will be set by the minimum pulse width required to ensure SCR latching.
- R2 is determined from peak gate current limits, if applicable, and minimum pulse width requirements.
- Select a zener diode. A 25-V zener is a practical value, since this will meet the usual gate requirement of 20 V and 20 Ω. This diode will also eliminate spurious triggering because of voltage transients.
- Photon coupler triggering is ideal for the SCR's driving inductive loads. By ensuring that the LASCR latches on, it can supply gate current to SCR1 until it stays on.
- Component values for dc voltage are easily computed from the following formulae:

$$R_1 = \frac{E_{\text{IN}} - V_Z}{I_G}$$

where: V = zener voltage
$P_{(R1)} = I_G \cdot (E_{\text{IN}} - V_Z)$
$P_{(\text{ZENER})} = I_G \cdot V_Z$

SOLID-STATE ZERO-VOLTAGE SWITCHING (ZVS) CIRCUITS

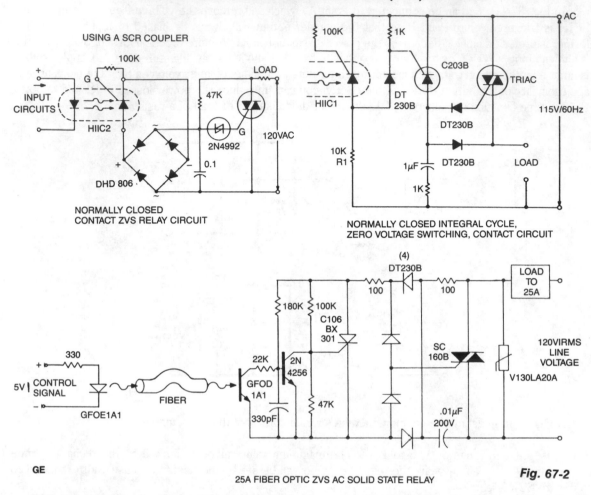

GE

25A FIBER OPTIC ZVS AC SOLID STATE RELAY

Fig. 67-2

This circuit is effective for lamp and heater loads. Some circuits driving reactive loads require integral cycling and zero-voltage switching—when an identical number of positive and negative half cycles of voltage are applied to the load during a power period. The circuit, although not strictly a relay because of the three-terminal power connection, performs the integral cycle ZVS function when interfaced with the previous coil circuits.

Fiber optics offers advantages in power control systems. Electrical signals do not flow along the nonconducting fiber, minimizing shock hazard to both operator and equipment. EMI/RFI pick up on the fiber is nonexistent—although high gain receiver circuits might require shielding, eliminating noise pick-up errors caused by sources along the cable route. Both ac and dc power systems can be controlled by fiber optics using techniques similar to the optoisolator solid-state relay. Triac triggering is accomplished through the C106BX301, a low gate trigger current SCR, switching line voltage derived current to the triac gate via the full-wave rectifier bridge. The primary difference between fiber optics solid-state relay circuits and optoisolator circuits is the gain; photo currents are much smaller.

TRIGGERING SCR SERIES

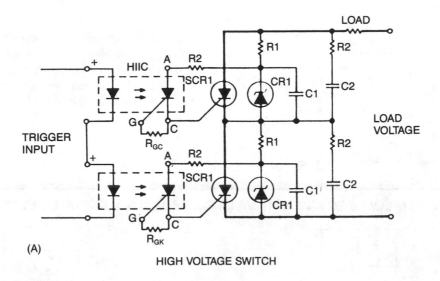

(A)

HIGH VOLTAGE SWITCH

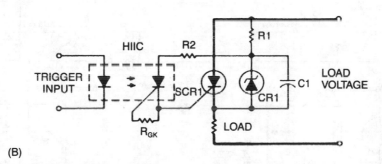

(B)

CONNECTION OF LOAD TO CATHODE OF MAIN SCR

GE

Fig. 67-3

Snubber circuit R2C2, as shown, might be necessary since R1 and C1 are tailored for optimized triggering and not for dV/dt protection. Fiber-optic pairs can be used with discrete SCRs to switch thousands of volts. A photon coupler with a transistor output will limit the trigger-pulse amplitude and rise time because of CTR and saturation effects. Using the H11C1, the rise time of the input pulse to the photon coupler is not critical, and its amplitude is limited only by the H11C1 turn-on sensitivity. The load can also be connected to the cathode as illustrated in Fig. 67-3B.

NORMALLY CLOSED HALF-WAVE ZVS CONTACT CIRCUIT

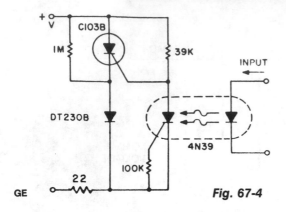

A normally closed contact circuit that provides zero-voltage switching is designed around the 4N39 SCR optocoupler. The circuit illustrates the method of modifying the normally open contact circuit by using the photo SCR to hold off the trigger SCR.

Fig. 67-4

NORMALLY OPEN AND NORMALLY CLOSED DC SOLID-STATE RELAYS

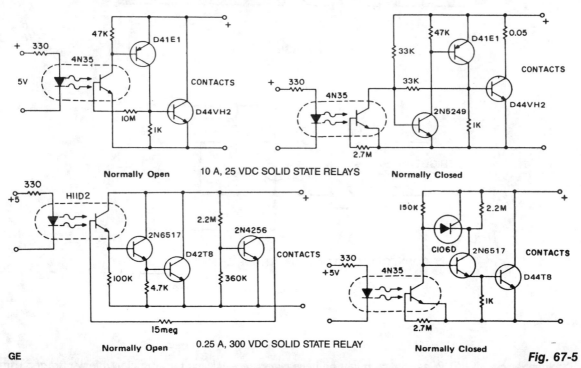

Fig. 67-5

The phototransistor and photodarlington couplers act as dc relays in saturated switching at currents up to 5 mA and 50 mA, respectively. When higher currents or higher voltage capabilities are required, additional devices are required to buffer or amplify the photocoupler output. The addition of hysteresis to provide fast switching and stable pick-up and drop-out points can be easily implemented simultaneously. These circuits provide several approaches to implement the dc relay function and serve as practical, cost-effective examples.

INDICATOR LAMP DRIVER

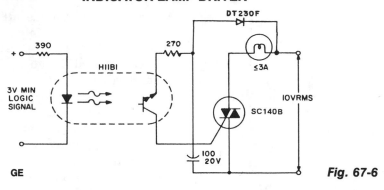

GE

Fig. 67-6

A simple solid-state relay circuit drives the 10-Vac telephone indicator lamps from logic circuitry, while maintaining complete isolation between the 10-V line and the logic circuit.

AMBIENT-LIGHT-IGNORING OPTICAL SENSOR

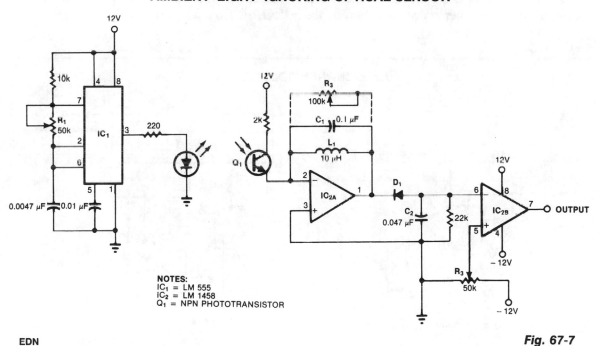

NOTES:
IC₁ = LM 555
IC₂ = LM 1458
Q₁ = NPN PHOTOTRANSISTOR

EDN

Fig. 67-7

A resonance-tuned narrow-band amplifier reduces this optical object detector's sensitivity to stray light. C1 and L1 in IC2A's feedback loop cause the op amp to pass only those frequencies at or near the LED's 5-kHz modulation rate. IC2B's output increases when the received signal is sufficient to drop the negative voltage across C2 below the reference set by R2.

OPTICAL CMOS COUPLER

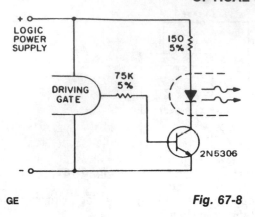

LOGIC POWER SUPPLY

DRIVING GATE

75K 5%

150 5%

2N5306

GE

Fig. 67-8

Since standard CMOS logic operates down to 3-V supply voltages and is specified as low as 30 μA maximum current sinking/sourcing capability, it is necessary to use a buffer transistor to provide the required current to the IRED if CMOS is to drive the optocoupler. As in the case of the low output TTL families, the H74A output can drive a multiplicity of CMOS gate inputs or a standard TTL input given the proper bias of the IRED. A one-logic stage drives the IRED on. This circuit will provide worst-case drive criteria to the IRED for logic supply voltages from 3 to 10 V, although lower power dissipation can be obtained by using higher value resistors for high supply voltages. If this is desired, the worst-case drive must be supplied to the IRED with minimum supply voltage, minimum temperature and maximum resistor tolerances, gate saturation resistance, and transistor saturation voltages applied. For the H74 devices, minimum IRED current at worst-case conditions, zero logic state output of the driving gate, is 6.5 mA and the H11L1 is 1.6 mA.

LINE-CURRENT DETECTOR

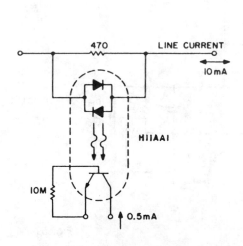

LINE CURRENT

470

10 mA

H11AA1

10M

0.5mA

POLARITY INSENSITIVE LINE CURRENT DETECTOR

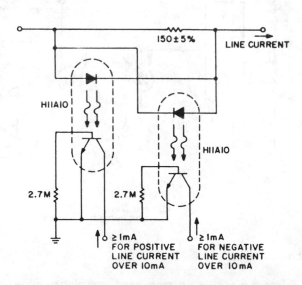

150 ± 5%

LINE CURRENT

H11A10

H11A10

2.7M

2.7M

≥ 1mA FOR POSITIVE LINE CURRENT OVER 10mA

≥ 1mA FOR NEGATIVE LINE CURRENT OVER 10mA

POLARITY INDICATING LINE CURRENT DETECTOR

GE

Fig. 67-9

Detection of line-current flow and indicating the flow to an electrically remote point is required in line status monitoring at a variety of points in the telephone system and auxiliary systems. The line should be minimally unbalanced or loaded by the monitor circuit, and relatively high levels of 60-Hz induced voltages must be ignored. The H11AA1 allows line currents of either polarity to be sensed without discrimination and will ignore noise up to approximately 2.5 mA. In applications where greater noise immunity or polarity-sensitive line-current detection is required, the H11A10 threshold coupler can be used. This phototransistor coupler is specified to provide a minimum 10% current transfer ratio at a defined input current, while leaking less than 50 μA at half that input current over the full $-55°C$ to $+100°C$ temperature range. The input current range, at which the coupler is on, is programmable by a single resistor from 5 to 10 mA.

LINE-OPERATED POWER OUTAGE LIGHT

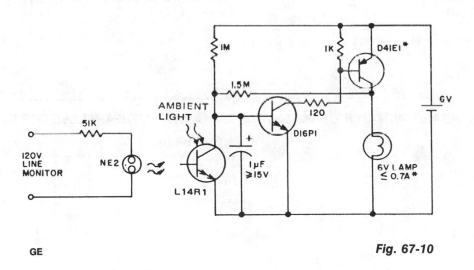

GE

Fig. 67-10

This circuit provides emergency lighting during a power outage. The phototransistor should be positioned to maximize coupling of both neon light and ambient light into the pellet, without allowing self-illumination from the 6-V lamp. Many circuits of this type also use line voltage to charge the battery.

OPTICAL TTL COUPLER

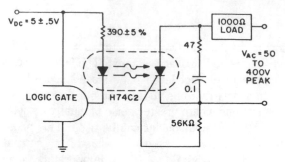

LOGIC TO POWER COUPLING H74 BIAS CIRCUIT

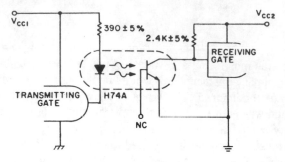

LOGIC TO LOGIC COUPLERS H74A1 BIAS CIRCUIT

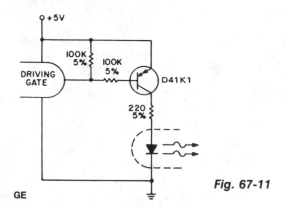

Fig. 67-11

IRED DRIVE FROM LOW POWER, MSI AND LSI TTL

For higher speed applications, up to 1-MHz NRZ, the Schmitt-trigger output H11L series optoisolator provides many features. The 1.6-mA drive current allows fan-in circuitry to drive the IRED, while the 5-V, 270-Ω sink capability and 100-ns transition times of the output add to the logic coupling flexibility.

ZERO-VOLTAGE SWITCHING, SOLID-STATE RELAY WITH ANTIPARALLEL SCR OUTPUT

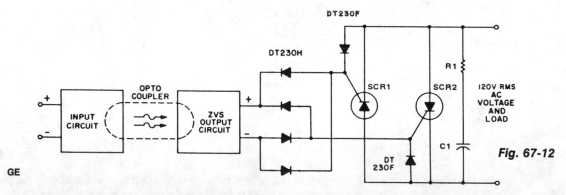

Fig. 67-12

A higher line voltage can be used if the diode, varistor, ZVS, and power thyristor settings are at compatible levels. For applications beyond triac current ratings, antiparallel SCRs might be triggered by the ZVS network.

DC LATCHING RELAY

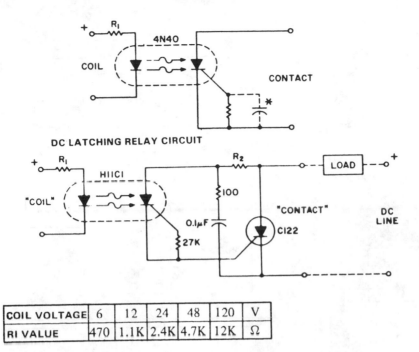

DC LATCHING RELAY CIRCUIT

COIL VOLTAGE	6	12	24	48	120	V
RI VALUE	470	1.1K	2.4K	4.7K	12K	Ω

LINE VOLTAGE	12	24	48	120	V
C122 PART	U	F	A	B	D
R2 VALUE	200	470	1K	2.2K	Ω

FOR HEAT SINK RATINGS
SEE C122 SPECIFICATION
SHEET NUMBER 150.35 AND
APPLICATION NOTE NUMBER
200.55

NO HEAT SINK RATINGS AT $T_A \leqslant 50°$

I CONTACT, MAX.	PULSE WIDTH	DUTY CYCLE
0.67 A	D.C.	100%
4.0 A	160 msec.	12%
8.0 A	160 msec.	3%
12 A	160 msec.	1%
15 A	160 msec.	0.3%

GE

Fig. 67-13

The H11C supplies the dc latching relay function and reverse polarity blocking, for currents up to 300 mA, depending on ambient temperature. For dc use, the gate cathode resistor can be supplemented by a capacitor to minimize transient and *dV/dt* sensitivity. For pulsating dc operation, the capacitor value must be designated to either retrigger the SCR at the application of the next pulse or prevent retriggering at the next power pulse. If not, random or undesired operation might occur. For higher current contacts, the H11C can be used to trigger an SCR capable of handling the current, as illustrated.

AC RELAY

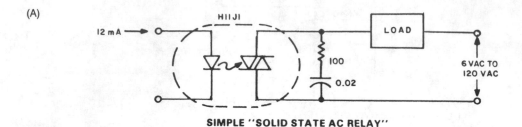

(A)

SIMPLE "SOLID STATE AC RELAY"

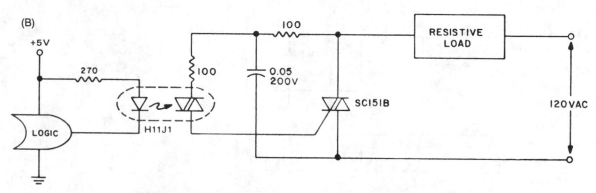

(B)

MINIMUM PARTS COUNT ISOLATED LOGIC TRIGGERED TRIAC

GE

Fig. 67-14

When zero voltage switching is not required, methods of providing this function are illustrated. The lowest parts count version of a solid-state relay is an optoisolator, the triac driver H11J. Unfortunately, the ability of the H11J to drive a load on a 60-Hz line is severely limited by its power dissipation and the dynamic characteristics of the detector. These factors limit applications to 30 – 50 mA resistive loads on 120 Vac, and slightly higher values at lower voltages. These values are compatible with neon lamp drive, pilot, and indicator incandescent bulbs; low voltage control circuits, such as furnace and bell circuits, if dV/dt are sufficient; but less than benign loads require a discrete triac.

The H11J1 triac trigger optocoupler potentially allows a simple power switching circuit utilizing only the triac, a resistor, and the optocoupler. This configuration will be sensitive to high values of dV/dt and noise on normal power-line voltages, leading to the need for the configuration shown in Fig. 67-14B, where the triac snubber acts as a filter for line voltage to the optocoupler.

Since the snubber is not usually used for resistive loads, the cost effectiveness of the circuit is compromised somewhat. Even with this disadvantage, the labor, board space, and inventory of parts savings of this circuit prove it cost-optimized for isolated logic control of power-line switching. In applications where transient voltages on the power line are prevalent, provisions should be made to protect the H11J1 from breakover triggering.

PHOTODIODE SOURCE FOLLOWER

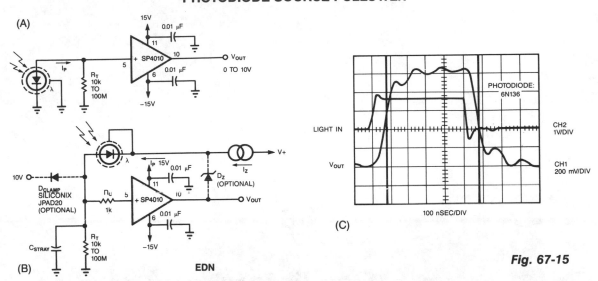

Fig. 67-15

EDN

A common method of transforming the output current of a photodiode into a voltage signal, paralleling the photodiode with a high-value load resistor, produces a nonlinear response. Also the combination of the load's transresistance, R_T, and the photodiode's junction capacitance, C_J, slows the circuit's response time. Figure 67-11B shows virtually the same components as Fig. 67-11A rearranged to maximize the inherent speed and linearity of the photodiode. The SP4010 (available from Hybrid Systems, Billerica, MA) is a unity voltage-gain buffer with a JFET input, 60 MHz 3 dB bandwidth, and 18-bit, 0.0004%, linearity over a ± 10 V input range.

In the circuit of Fig. 67-11B, the photodiode sees a constant voltage across its terminals, which is essential for linear photodiode outputs. The optional zener diode, D_Z, sets a reverse bias at the photodiode for lower junction capacitance and higher speed. If you don't use D_Z, be sure to connect the feedback loop. An optional diode, D_{CLAMP}, limits the output in case of unexpected light bursts, but results in increased dark-current leakage and lower speed. The buffered output of the circuit equals the photodiode current times the transresistance, R_T. Figure 67-11C shows the circuit's response to a fast light pulse.

419

68

Oscillators

The sources of the following circuits are contained in the Sources section beginning on page 782. The figure number contained in the box of each circuit correlates to the sources entry in the Sources section.

Discrete Sequence Oscillator
Fixed-Frequency Variable
   Duty-Cycle Oscillator
RLC Oscillator
HC-Based Oscillators
Programmable-Frequency
   Sine-Wave Oscillator
Variable Wien-Bridge Oscillator
Wide-Range Oscillator
HCV/HCT-Based Oscillator
50% Duty-Cycle Oscillator
High-Frequency Oscillator
Last-Cycle Completing Gated
   Oscillator

Audio Oscillator
Low-Frequency Oscillator
Quadrature Oscillator
Wien-Bridge Oscillator
CMOS Oscillator
XOR-Gate Oscillator
SCR Relaxation Oscillator
CMOS Oscillator
Code-Practice Oscillator
Precision Voltage-Controlled
   Oscillator
5-V Oscillator
Low-Voltage Wien-Bridge
   Oscillator

DISCRETE SEQUENCE OSCILLATOR

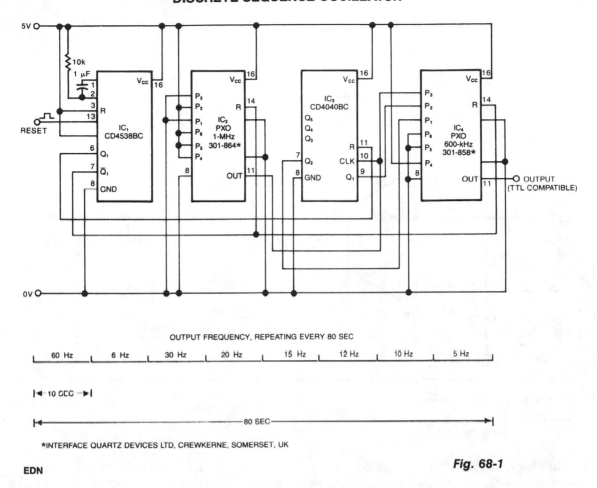

*INTERFACE QUARTZ DEVICES LTD, CREWKERNE, SOMERSET, UK

EDN

Fig. 68-1

The swept-frequency oscillator offers an inexpensive source of discrete frequencies for use in testing digital circuits. In this configuration, the circuit generates an 80-second sequence of eight frequencies, dwelling for 10 seconds on each frequency. You can change the dwell time or the number of frequencies. Frequencies can range from 0.005 Hz to 1 MHz.

The programmable crystal oscillators, PXOs, IC2 and IC4 can each generate 57 frequencies in response to an 8-bit external code. IC2 contains a 1-MHz crystal and produces a 0.05-Hz output. IC4 contains a 600-kHz crystal; its output changes in response to the combined outputs of the 12-stage binary counter IC3 (Q1 and Q2) and the PXO IC2.

To generate more frequencies, you can use one or more of IC3's outputs, (Q3, Q4, Q5) to drive one or more of IC4's inputs (P4, P5, P6). Similarly, you can rewire IC2 or drive it with other logic to control the duration of each frequency. IC1, a monostable multivibrator, provides a system reset. It initiates the sequence shown, beginning at 60 Hz, in response to a positive pulse.

FIXED-FREQUENCY VARIABLE DUTY-CYCLE OSCILLATOR

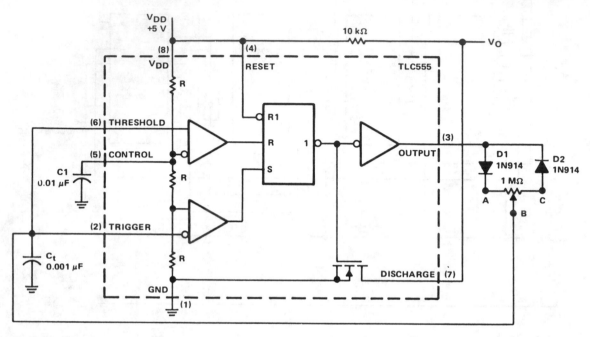

Fig. 68-2

In a basic astable timer, configuration timing periods t_1 and t_2 are not controlled independently. The lack of control makes it difficult to maintain a constant period, T, if either t_1 or t_2 is varied. In this circuit, charge R_{AB} and discharge R_{BC} resistances are determined by the position of common wiper arm B of the potentiometer. So, it is possible to adjust the duty-cycle by adjusting t_1 and t_2 proportionately, without changing period T.

At start-up, the voltage across C_t is less than the trigger level voltage ($1/2\ V_{DD}$), causing the timer to be triggered via pin 2. The output of the timer at pin 3 increases, turning off the discharge transistor at pin 7 and allowing C_t to charge through diode D1 and resistance R_{AB}. When capacitor C_t charges to upper threshold voltage $2/3\ V_{DD}$, the flip-flop is reset and the output at pin 3 decreases. Capacitor C_t then discharges through diode D2 and resistor R_{BC}. When the voltage at pin 2 reaches $1/3\ V_{DD}$, the lower threshold or trigger level, the timer triggers again and the cycle is repeated. In this circuit, the oscillator frequency remains fixed and the duty cycle is adjustable from less than 0.5% to greater than 99.5%.

RLC OSCILLATOR

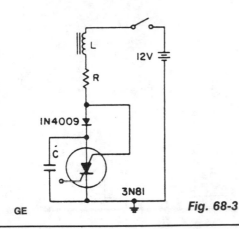

A positive transient, such as the power switch closing, charges C through L to a voltage above the supply voltage, if Q is sufficient. When the current reverses, the diode blocks and triggers the SCS. As the capacitor discharges, the anode gate approaches ground potential, depriving the anode of holding current. This turns off the SCS, and C charges to repeat the cycle.

GE

Fig. 68-3

HC-BASED OSCILLATORS

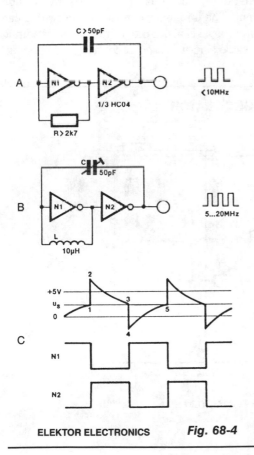

Two inverters, one resistor, and one capacitor are all that is required to make a HC(T)-based oscillator that gives reliable operation up to about 10 MHz. The use of two HC inverters produces a fairly symmetrical rectangular output signal. In the same circuit, HCT inverters give a duty factor of about 25%, rather than about 50%, since the toggle point of an HC and an HCT inverter is $1/2$ V_{CC}, and slightly less than 2 V, respectively. If the oscillator is to operate above 10 MHz, the resistor is replaced with a small inductor, as shown in Fig. 68-4B.

The output frequency of the circuit in Fig. 68-4A is given as about 1/1.8rc, and can be made variable by connecting a 100-KΩ potentiometer in series with R. The solution adopted for the oscillator in Fig. 68-4B is even simpler: C is a 50-pF trimmer capacitor.

ELEKTOR ELECTRONICS   **Fig. 68-4**

PROGRAMMABLE-FREQUENCY SINE-WAVE OSCILLATOR

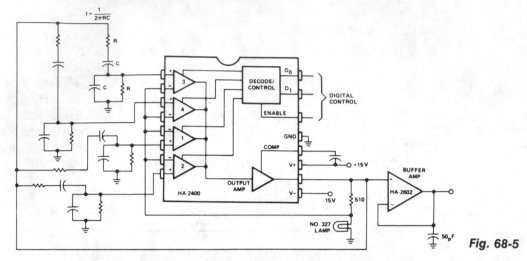

$$f = \frac{1}{2\pi RC}$$

HARRIS

Fig. 68-5

This Wien-bridge oscillator is very popular for signal generators, since it is easily turned over a wide frequency range, and has a very low distortion sine-wave output. The frequency determining networks can be designed from about 10 Hz to greater than 1 MHz; the output level is about 6.0-V rms. By substituting a programmable attenuator for the buffer amplifier, a very versatile sine-wave source for automatic testing, etc. can be constructed.

VARIABLE WIEN-BRIDGE OSCILLATOR

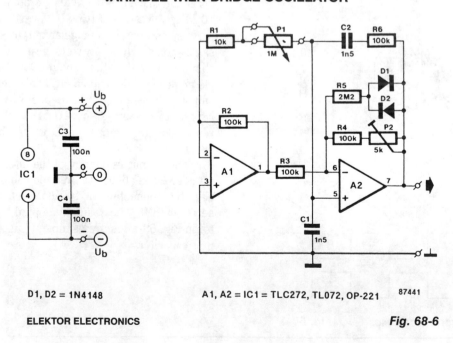

D1, D2 = 1N4148

A1, A2 = IC1 = TLC272, TL072, OP-221   87441

ELEKTOR ELECTRONICS

Fig. 68-6

VARIABLE WIEN-BRIDGE OSCILLATOR (*Cont.*)

A Wien-bridge oscillator can be made variable by using two frequency-determining parts that are varied simultaneously at high tracking accuracy. High-quality tracking potentiometers or variable capacitors are, however, expensive and difficult to obtain. To avoid having to use such a component, this oscillator was designed to operate with a single potentiometer. The output frequency, f_o, is calculated from:

$$f_o = 1/(2nRC \, v \, a)$$

where:

$$R = R_2 = R_3 = R_4 = R_6, \quad C = C_1 = C_2, \text{ and } a = (P_1 + R_1)R$$

With preset P2 you can adjust the overall amplification so that the output signal has a reasonably stable amplitude, 3.5 V_{PP} max., over the entire frequency range. The stated components allow the frequency to be adjusted between 350 Hz and 3.5 kHz.

WIDE-RANGE OSCILLATOR

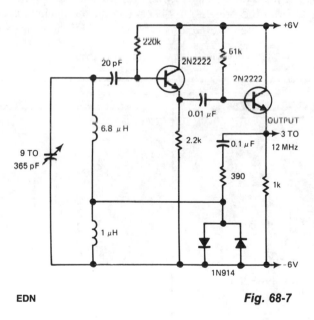

EDN

Fig. 68-7

The gain control allows the oscillator to maintain essentially constant output over its range. The circuit functions over 160 kHz to 12 MHz with essentially constant amplitude.

HCU/HCT-BASED OSCILLATOR

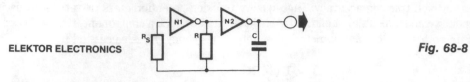

ELEKTOR ELECTRONICS

Fig. 68-8

N1, N2 = 1/3 IC1 = 74HCT04, 74HCU04

When frequency stability is not of prime importance, a simple, yet reliable, digital clock oscillator can be made with the aid of relatively few components. High-speed CMOS (HCU/HCT) inverters or gates with an inverter function are eminently suitable to make such oscillators, thanks to their low power consumption, good output signal definition, and extensive frequency range.

The circuit as shown uses two inverters in a 74HCT04 or 74HCU04. The basic design equations are:

$$\text{For HCU:} \quad f=1/T; \quad T=2.2\,RC; \quad V_{CC}\ 6\ \text{V}; \quad I_c=13\ \text{mA}$$
$$\text{For HCT:} \quad f=1/T; \quad T=2.4\,RC; \quad V_{CC}\ 5.5\ \text{V}; \quad I_c=2.2\ \text{mA}$$

With R_s and R calculated for a given frequency and value of C, both resistors can be realized as presets to enable precise setting of the output frequency and the duty factor. Do not forget, however, to fit small series resistors in series with the presets, in observance of the minimum values for R and R_s as given in the design equations. The values quoted for I_c are only valid if the inputs of the remaining gates are grounded.

50% DUTY-CYCLE OSCILLATOR

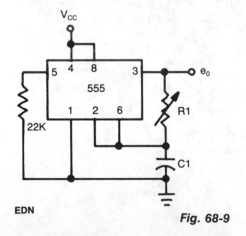

EDN

Fig. 68-9

Frequency of oscillation depends on the R1/C1 time constant and allows frequency adjustment by varying R1.

HIGH-FREQUENCY OSCILLATOR

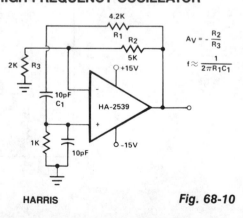

$$A_V = -\frac{R_2}{R_3}$$
$$f \approx \frac{1}{2\pi R_1 C_1}$$

HARRIS

Fig. 68-10

Intended primarily as a building block for a QRP transmitter, this 20-MHz oscillator delivered a clean 6-V, pk-pk signal into a 100-Ω load.

LAST-CYCLE COMPLETING GATED OSCILLATOR

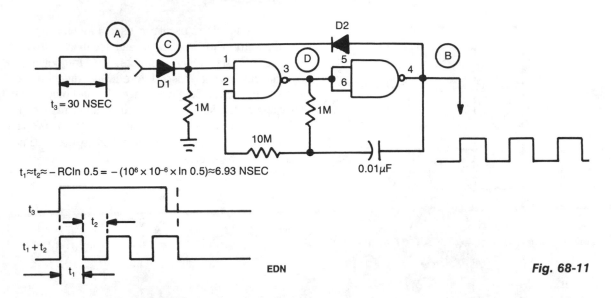

$t_1 \approx t_2 \approx - RC\ln 0.5 = -(10^6 \times 10^{-6} \times \ln 0.5) \approx 6.93$ NSEC

EDN

Fig. 68-11

Regenerative feedback at C enables the oscillator to complete its timing cycle, rather than immediately shutting it off. The IC used was a CD4011AE, although an equivalent will work.

AUDIO OSCILLATOR

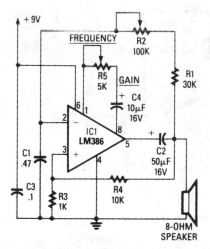

The circuit's frequency of oscillation is $f = 2.8/[C_1 \times (R_1 + R_2)]$. Using the values shown, the output frequency can be varied from 60 Hz to 20 kHz by rotating potentiometer R2.

A portion of IC1's output voltage is fed to its noninverting input at pin 3. The voltage serves as a reference for capacitor C1, which is connected to the noninverting input at pin 2 of the IC. That capacitor continually charges and discharges around the reference voltage, and the result is a square-wave output. Capacitor C2 decouples the output.

Fig. 68-12

LOW-FREQUENCY OSCILLATOR

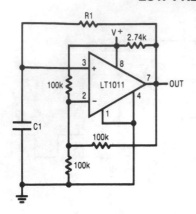

This simple rc oscillator uses a medium-speed comparator with hysteresis and feedback through R1 and C1 as timing elements. The frequency of oscillation is, at least theoretically, independent from the power supply voltage. If the comparator swings to the supply rails, if the pull-up resistor is much smaller than the resistor R_h, and if the propagation delay is negligible compared to the rc time constant, the oscillation frequency is:

$$f_{OSC} = \frac{0.72}{R_1 C_1}$$

LINEAR TECHNOLOGY CORP.   *Fig. 68-13*

QUADRATURE OSCILLATOR

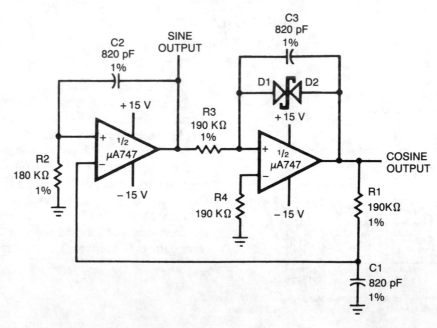

FAIRCHILD CAMERA AND INSTRUMENT   *Fig. 68-14*

WIEN-BRIDGE OSCILLATOR

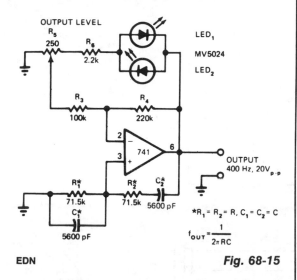

EDN

Fig. 68-15

LEDs function as both pilot lamps and as an AGC (automatic gain control) in this unconventional amplitude-stabilized oscillator.

CMOS OSCILLATOR

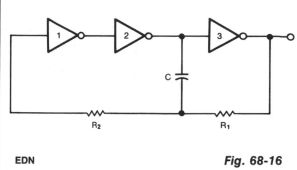

EDN

Fig. 68-16

This circuit is guaranteed to oscillate at a frequency of about $2.2/(R_1 \times C)$ if R_2 is greater than R_1. You can reduce the number of gates further if you replace gates 1 and 2 with a noninverting gate.

XOR-GATE OSCILLATOR

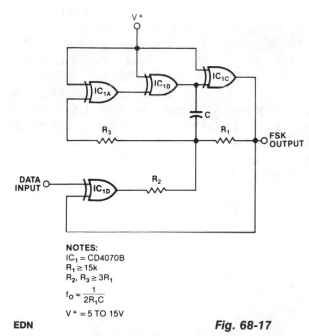

EDN

Fig. 68-17

An exclusive-OR gate, IC1D, turns a simple CMOS oscillator into an FSK generator. When the data input increases, IC1D inverts, and negative feedback through R2 lowers the circuit's output frequency. A low input results in positive feedback and a higher output frequency. R1 and C set the oscillator's frequency range, and R2 determines the circuit's frequency shift. To ensure frequency stability, make R3 much greater than R1 and use a high-quality feedback capacitor. The three gates constituting the oscillator itself need not be exclusive-OR types; use any CMOS inverter.

SCR RELAXATION OSCILLATOR

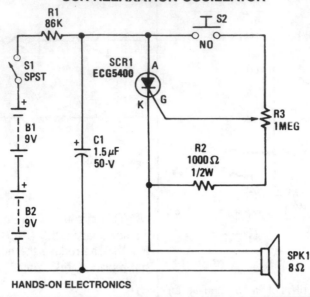

HANDS-ON ELECTRONICS

Fig. 68-18

CMOS OSCILLATOR

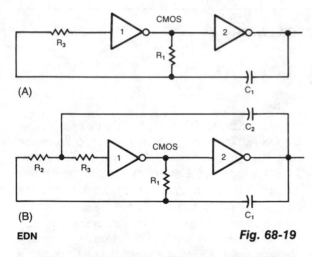

(A)

(B)

EDN

Fig. 68-19

The common clock oscillator in Fig. 68-19A has two small problems: It might not, in fact, oscillate if the transition regions of its two gates differ. If it does oscillate, it might sometimes oscillate at a slightly lower frequency than its equation predicts because of the finite gain of the first gate. If the cir- cuit does work, oscillation occurs usually because both gates are in the package and, therefore, have logic thresholds only a few millivolts apart.

The circuit in Fig. 68-19B resolves both prob- lems by adding a resistor and a capacitor. The R2/ C2 network provides hysteresis, thus delaying the onset of gate 1's transition until C1 has enough voltage to move gate 1 securely through its transi- tion region. When gate 1 is finally in its transition region, C2 provides positive feedback, thus rapidly moving gate 1 out of its transition region.

The equations for the oscillator in Fig. 68-19B are:

$$R_2 = 10R_1$$
$$R_3 = 10R_2$$
$$C_1 = 100C_2$$

$$f = \cong \frac{1}{1.2R_1C_1}$$

CODE-PRACTICE OSCILLATOR

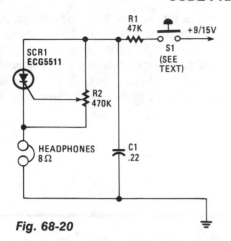

Capacitor C1 charges through resistor R1, and when the gate level established by potentiometer R2 is high enough, the SCR is triggered. Current flows through the SCR and earphones, discharging C1. The anode voltage and current drop to a low level, so the SCR stops conducting and the cycle is repeated. Resistor R2 lets the gate potential across C1 be adjusted, which charges the frequency or tone. Use a pair of 8-Ω headphones. The telegraph key goes right into the B+ line, 9-V battery.

Fig. 68-20

HANDS-ON ELECTRONICS/
POPULAR ELECTRONICS

PRECISION VOLTAGE-CONTROLLED OSCILLATOR

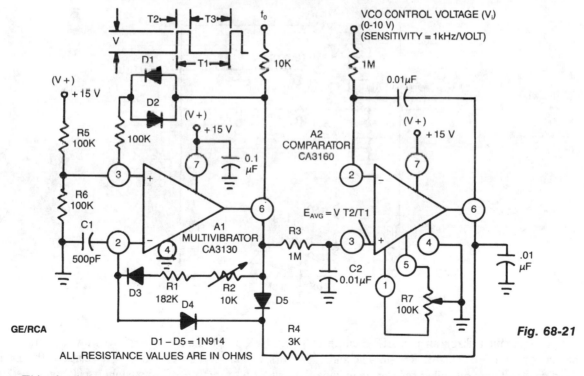

Fig. 68-21

This circuit uses a CA3130 BiMOS op amp as a multivibrator and CA3160 BiMOS op amp as a comparator. The oscillator has a sensitivity of 1 kHz/V, with a tracking error in the order of 0.02%, and a temperature coefficient of 0.01%/°C.

5-V OSCILLATOR

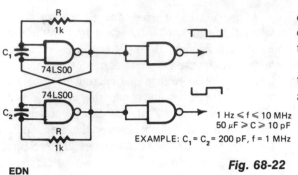

Consistently self-starting and yet capable of operating from over 1 Hz to 10 MHz, this low-cost oscillator requires only five components. Calculate the period of oscillation by using this relationship: $P = 5 \times 10^3 \, C$ sec when $C = C_1 = C_2$. By changing the ratio of C1 to C2, the duty cycle can be as low as 20%.

1 Hz ≤ f ≤ 10 MHz
50 μF ≥ C ≥ 10 pF
EXAMPLE: $C_1 = C_2 = 200$ pF, f = 1 MHz

EDN

Fig. 68-22

LOW-VOLTAGE WIEN-BRIDGE OSCILLATOR

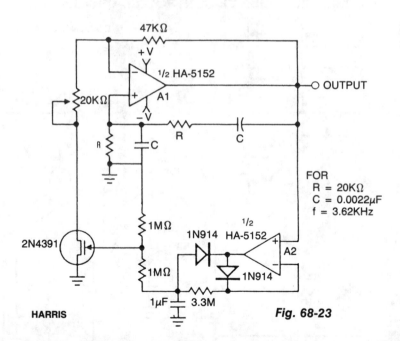

FOR
R = 20KΩ
C = 0.0022μF
f = 3.62KHz

HARRIS

Fig. 68-23

This circuit utilizes an HA-5152 dual op amp and FET to produce a low-voltage, low-power, Wein-bridge sine-wave oscillator. Resistors R and capacitors C control the frequency of oscillation; the FET, used as a voltage-controlled resistor, maintains the gain of A1 exactly 3 dB to sustain oscillation. The 20-KΩ pot can be used to vary the signal amplitude. The HA-5152 has the capability to operate from ±1.5-V supplies. This circuit will produce a low-distortion sine-wave output while drawing only 400 uA of supply current.

69

Oscilloscope Circuits

The sources of the following circuits are contained in the Sources section beginning on page 782. The figure number contained in the box of each circuit correlates to the sources entry in the Sources section.

Scope Extender
Eight-Channel Voltage Display
Oscilloscope Calibrator
Scope Sensitivity Amplifier
Add-On Scope Multiplexer

Oscilloscope Preamplifier
Oscilloscope/Counter Preamplifier
Oscilloscope-Triggered Sweep
CRO Doubler

SCOPE EXTENDER

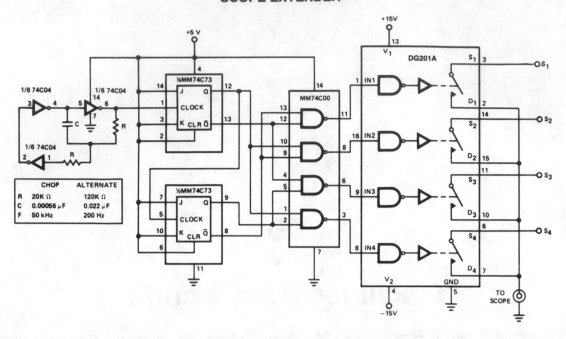

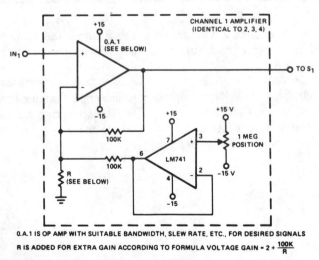

0.A.1 IS OP AMP WITH SUITABLE BANDWIDTH, SLEW RATE, ETC., FOR DESIRED SIGNALS

R IS ADDED FOR EXTRA GAIN ACCORDING TO FORMULA VOLTAGE GAIN = $2 + \frac{100K}{R}$

SILICONIX

Fig. 69-1

The adapter allows four inputs to be displayed simultaneously on a single trace scope. For low-frequency signals, less than 500 Hz, the adapter is used in the *chop* mode at a frequency of 50 kHz. The clock can be run faster, but switching glitches and the actual switching time of the DG201A limit the maximum frequency to 200 kHz. High frequencies are best viewed in the alternate mode, with a clock frequency of 200 Hz. When the clock is below 100 Hz, trace flicker becomes objectionable. One of the four inputs is used to trigger the horizontal trace of the scope.

EIGHT-CHANNEL VOLTAGE DISPLAY

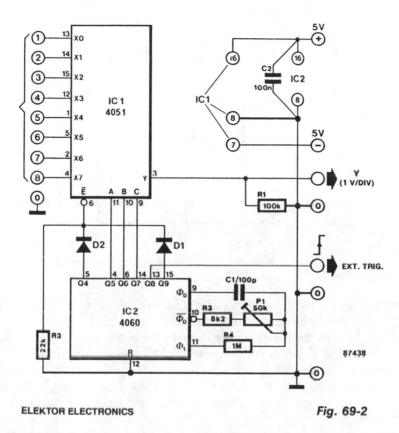

ELEKTOR ELECTRONICS

Fig. 69-2

This circuit turns a common oscilloscope into a versatile eight-channel display for direct voltages. The trend of each of the eight input levels is readily observed, albeit that the attainable resolution is not very high.

The circuit diagram shows the use of an eight-channel analog multiplexer IC1, which is the electronic version of an eight-way rotary switch with contacts X0 through X7 and pole Y. The relevant channel is selected by applying a binary code to the A-B-C inputs. For example, binary code 011 (A-B-C) enables channel 7 (X6 Y). The A-B-C inputs of IC1 are driven from three successive outputs of binary counter IC2, which is set to oscillate at about 50 kHz with the aid of P1. Since the counter is not reset, the binary state of outputs Q5, Q6, and Q7 steps from 0 to 7 in a cyclic manner. Each of the direct voltages at input terminals 1 to 8 is therefore briefly connected to the Y input of the oscilloscope. All eight input levels can be seen simultaneously by setting the timebase of the scope, in accordance with the time it takes the counter to output states 0 through 7, on outputs Q5, Q6, and Q7.

The timebase on the scope should be set to 0.5 ms/div, and triggering should occur on the positive edge of the external signal. Set the vertical sensitivity to 1 V/div. The input range of this circuit is from − 4 V to +4 V; connected channels are terminated in about 100 KΩ.

OSCILLOSCOPE CALIBRATOR

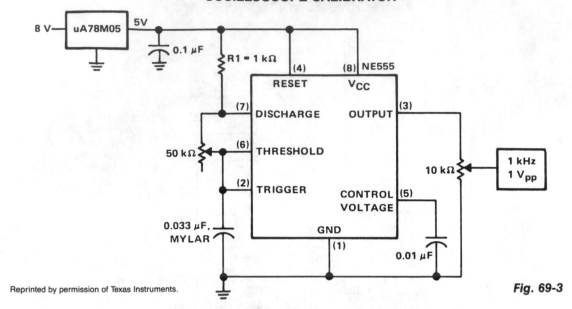

Reprinted by permission of Texas Instruments.

Fig. 69-3

The calibrator can be used to check the accuracy of a time-base generator, as well as to calibrate the input level of amplifiers. The calibrator consists of an NE555 connected in the astable mode. The oscillator is set to exactly 1 kHz by adjusting potentiometer P1 while the output at pin 3 is being monitored against a known frequency standard or frequency counter. The output level, likewise, is monitored from potentiometer P2's center arm to ground with a standard instrument. P2 is adjusted for 1 V pk-pk at the calibrator output terminal. During operation, the calibrator output terminal will produce a 1-kHz, square-wave signal at 1 V pk-pk with about 50% duty cycle. For long-term oscillator frequency stability, C1 should be a low-leakage mylar capacitor.

SCOPE SENSITIVITY AMPLIFIER

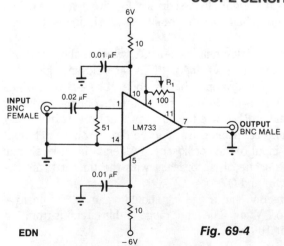

EDN

Fig. 69-4

This circuit provides 20 ±0.1 dB voltage gain from 0.5 to 25 MHz and ±3 dB from 70 kHz to 55 MHz. An LM733 video amplifier furnishes a low input-noise spec, 10-μV typical, measured over a 15.7-MHz bandwidth. The scale factor of the instrument can be preserved by using a trimmer R1 or a selected precision resistor, to set the circuit's voltage gain to exactly 100.

ADD-ON SCOPE MULTIPLEXER

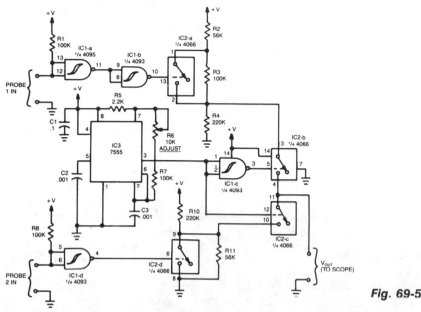

Fig. 69-5

The operation of the unit revolves around three ICs: a 4093 quad NAND Schmitt trigger, a 4066 quad analog switch, and a 7555 timer. When a high is fed to probe 1 in, it is inverted to IC1a and once again by IC1b, so that the input to IC2a is high. That high causes the *switch contacts* in IC2a to close. With the *contacts* closed, a high-level output is presented to the input of IC2b. The high output is fed to probe 2 in. That signal is then inverted by IC1d and routed to IC2d, causing its *contacts* to open, and the unit to output a logic-level high. The output of IC2d is then fed to IC2c.

OSCILLOSCOPE PREAMPLIFIER

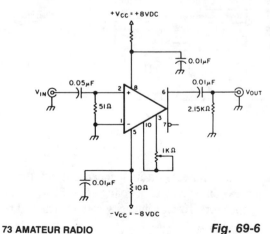

This circuit provides about 20 dB voltage gain with a frequency range from 0.5 to 50 MHz. You can extend the low-frequency response of this circuit by increasing the value of the 0.05-μF capacitor—or try removing the capacitor. This circuit delivers a particularly small level of input noise, measured at approximately 20 μV over a bandwidth range of 15 MHz.

Calibrate the gain by adjusting the gain potentiometer connected between pins 3 and 10, then adjust the 1-KΩ trimmer potentiometer for an exact voltage gain of 10; this helps preserve the scale factor of the oscilloscope.

73 AMATEUR RADIO   *Fig. 69-6*

OSCILLOSCOPE/COUNTER PREAMPLIFIER

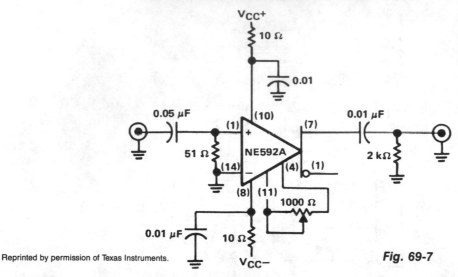

Reprinted by permission of Texas Instruments.

Fig. 69-7

The circuit will provide a 20 $\pm$ 0.1 dB voltage gain from 500 kHz to 50 MHz. The low-frequency response of the amplifier can be extended by increasing the value of the 0.05-μF capacitor connected in series with the input terminal. This circuit will yield an input-noise level of approximately 10 μV over a 15.7-MHz bandwidth. The gain can be calibrated by adjusting the potentiometer connected between pins 4 and 11. The 1-KΩ potentiometer can be adjusted for an exact voltage gain of 10. This preserves the scale factor of the instrument.

OSCILLOSCOPE-TRIGGERED SWEEP

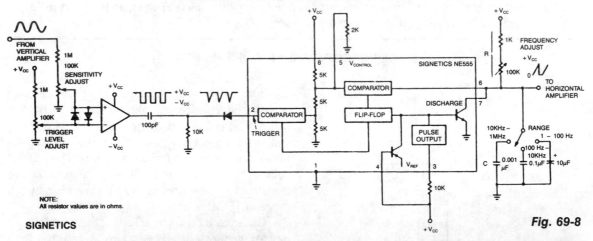

NOTE:
All resistor values are in ohms.

SIGNETICS

Fig. 69-8

The circuit's input op amp triggers the timer, sets its flip-flop and cuts off its discharge transistor so that capacitor C can charge. When capacitor voltage reaches the timer's control voltage of 0.33 V_{CC}, the flip-flop resets and the transistor conducts, discharging the capacitor. Greater linearity can be achieved by substituting a constant-current source for frequency adjust resistor R.

CRO DOUBLER

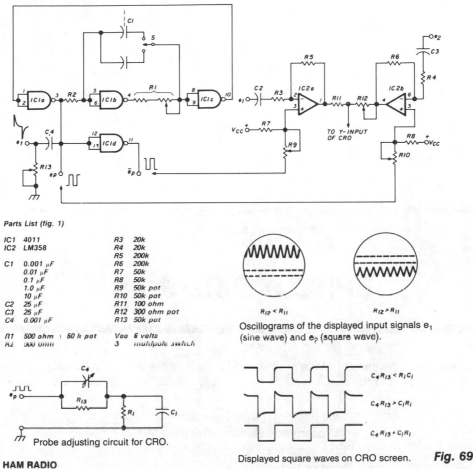

Parts List (fig. 1)

IC1	4011	R3	20k
IC2	LM358	R4	20k
		R5	200k
C1	0.001 μF	R6	200k
	0.01 μF	R7	50k
	0.1 μF	R8	50k
	1.0 μF	R9	50k pot
	10 μF	R10	50k pot
C2	25 μF	R11	100 ohm
C3	25 μF	R12	300 ohm pot
C4	0.001 μF	R13	50k pot
R1	500 ohm - 50 k pot	Vcc	6 volts
R2	900 ohm	S	multipole switch

Oscillograms of the displayed input signals e_1 (sine wave) and e_p (square wave).

Probe adjusting circuit for CRO.

HAM RADIO

Displayed square waves on CRO screen.

Fig. 69-9

IC1a, IC1b, and IC1c of the quad two-input NAND gate 4011 are connected as an astable multivibrator; IC1d is connected as an inverter. Terminals 3 and 11 of the 4011 produce square waves with opposite phases. The square waves, e_p, at the output of IC1a, passing through differentiator C4R13, then form positive and negative pulses, e_t. The dual op amps of the LM358 are used as two gated amplifiers for singles e_1 and e_2 and fed through terminals 2 and 6, to be displayed simultaneously on the CRO screen.

The two opposite-phase square waves $\bar{e}_p$ and e_p are used to gate IC2a and IC2b at terminals 3 and 5 of the LM358, respectively. Resistances R9 and R10 are preadjusted so that one op amp is driven to saturation while the other works normally as an amplifier. Thus, they will amplify signals e_1 and e_2 alternately, and two separate traces will be displayed on the screen. Resistance R12 can be varied to adjust the vertical separation of the two traces.

Select a suitable value for C1 with switch S, and adjust the pot of R1. The frequency of square waves can be varied from 1 to 10^6 cps. This process is necessary for stabilizing the waveforms displayed on the screen. A common supply of 6 V is used in the circuit.

70

Phase Detectors

The sources of the following circuits are contained in the Sources section beginning on page 782. The figure number contained in the box of each circuit correlates to the sources entry in the Sources section.

Phase Selector/Phase Detector/
    Synchronous Rectifier/Balanced Modulator
Phase Sequence Detector
Phase Detector

PHASE SELECTOR/PHASE DETECTOR/
SYNCHRONOUS RECTIFIER/BALANCED MODULATOR

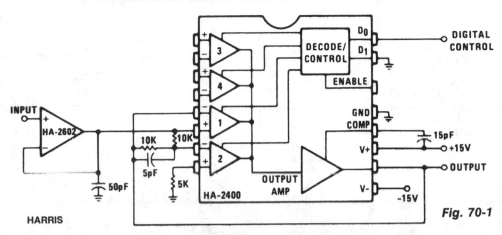

HARRIS

Fig. 70-1

This circuit passes the input signal at unity gain, either unchanged or inverted, depending on the digital control input. A buffered input is shown, since low-source impedance is essential. Gain can be added by modifications to the feedback networks. Signals up to 100 kHz can be handled with 20.0-V pk-pk, output. The circuit becomes a phase detector when driving the digital control input with a reference phase at the same frequency as the input signal; the average dc output is proportional to the phase difference, with 0 V at ±90°. By connecting the output to a comparator, which in turn drives the digital control, a synchronous full-wave rectifier is formed. With a low-frequency input signal and a high-frequency digital control signal, a balanced (suppressed carrier) modulator is formed.

PHASE-SEQUENCE DETECTOR

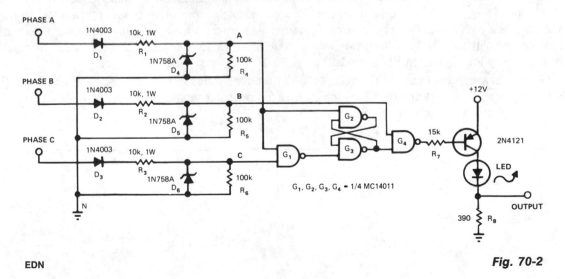

EDN

Fig. 70-2

441

PHASE DETECTOR

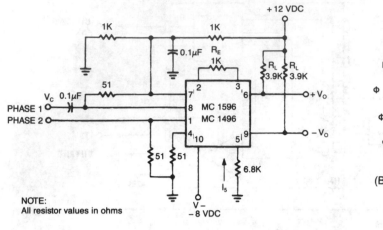

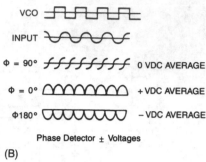

Phase Detector ± Voltages

(B)

NOTE:
All resistor values in ohms

(A)

Fig. 70-3

SIGNETICS

The output of the detector contains a term related to the cosine of the phase angle. Two signals of equal frequency are applied to the inputs. The frequencies are multiplied together, producing the sum and difference frequencies. Equal frequencies cause the difference component to become dc, while the undesired sum component is filtered out. The dc component is related to the phase angle by the graph of Fig. 70-2B. At 90°, the cosine becomes zero, while being at maximum positive or maximum negative at 0° and 180°, respectively. The advantage of using the balanced modulator over other types of phase comparators is the excellent conversion linearity. This configuration also provides a conversion gain, rather than a loss for greater resolution. Used in conjunction with a phase-locked loop, for instance, the balanced modulator provides a very low-distortion FM demodulator.

Correct phase sequences (ABC, BCA, or CAB) produce trains of output pulses and illuminate the LED. The output stays low and the LED remains dark for incorrect sequences (BAC, ACB, or CBA) or for phase loss (phase A, B, or C missing).

71

Photography-Related Circuits

The sources of the following circuits are contained in the Sources section beginning on page 782. The figure number contained in the box of each circuit correlates to the sources entry in the Sources section.

Slide-Show Timer
Camera Alarm Trigger
Darkroom Enlarger Timer
Flash Meter
Slave Photographic Xenon Flash Trigger
Slide Timer
Electronic Photoflash

SLIDE-SHOW TIMER

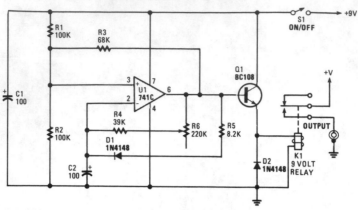

POPULAR ELECTRONICS/HANDS-ON ELECTRONICS   *Fig. 71-1*

You can set the interval from about 5 – 30 seconds. A relay operates the slide-change mechanism. Op amp U1 forms a sort of Schmitt trigger. Resistors R1 and R2 bias the noninverting input at pin 3 of U1 to half the supply voltage. Feedback resistor R3 increases or reduces the bias to pin 3, depending on whether the output of U1 is high or low.

When power is first applied to the circuit, C2 has a zero charge and the inverting input of the op amp is at a lower voltage than its noninverting input. When the output of U1 is high, C2 begins to charge through R5 and D1. It takes about one second for the charge on C2 to reach the same voltage as that at the noninverting input of U1. At that time, the output of U1 begins a negative swing.

Because of the positive feedback through R3, the voltage at the noninverting input is reduced and the output becomes more negative. The voltage at the noninverting input is about $1/4$ of the supply voltage, and C2 begins to discharge through the resistor bank. The timing is controlled by R6.

The resulting pulses are fed to the base of Q1, configured as an emitter-following buffer stage, which is used to activate relay K1. Transistor Q1 is necessary because op amps usually have an output current in the 20-mA range, which is too low to activate the relay.

CAMERA ALARM TRIGGER

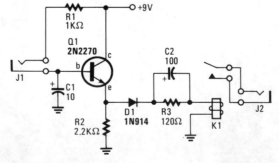

HANDS-ON ELECTRONICS   *Fig. 71-2*

444

CAMERA ALARM TRIGGER (*Cont.*)

Transistor Q1 remains off until the magnetic switch connected to J1 closes. When that happens, the 9 V is connected through R1 to Q1'S base. Q1 turns on, thereby charging C2 through relay K1, which causes K1's contacts to close. Since the contacts connect via J2 to the remote control jack in the camera, which in turn connects to the camera's shutter release, the closure of K1's contacts will cause the camera's shutter to trigger. After C2 charges, K1 opens because current through its coil ceases; the camera won't take another picture. If Q1 turns off because the magnetic switch on the window or gate opens, C2 discharges and the circuit is ready for another cycle. As long as Q1 remains on, C2 stays charged and prevents K1 from triggering more photos. Capacitor C1 bypasses spurious magnetic switch noises from physical phenomena, such as a rattling window or a gate shaking in the wind, thus reducing the likelihood of an unwanted picture. Resistor R2 biases Q1 and dampens K1/C2 oscillations which might cause contact bounce. Diode D1 prevents C2 from discharging through K1; the relay coil isn't polarity conscious and C2 discharging through it would trigger an unwanted picture.

DARKROOM ENLARGER TIMER

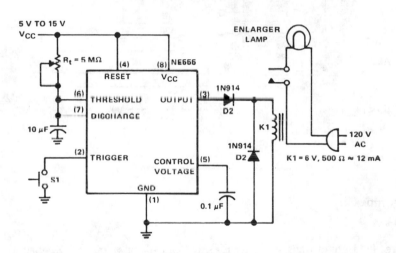

Reprinted by permission of Texas Instruments.

Fig. 71-3

The NE555 circuit is a basic one-shot timer with a relay connected between the output and ground. It is triggered with the normally open momentary contact switch, which when operated, grounds the trigger input at pin 2. This causes a high output to energize K1 which closes the normally open contacts in the lamp circuit. They remain closed during the timing interval, then open at time out. Timing is controlled by a 5-MΩ potentiometer, R_t. All timer-driven relay circuits should use a reverse clamping diode, such as D1, across the coil. The purpose of diode D2 is to prevent a timer output latch-up condition in the presence of reverse spikes across the relay.

With the rc time constant shown, the full-scale time is about 1 minute. A scale for the 5-MΩ potentiometer shaft position can be made and calibrated in seconds. Longer or shorter full-scale times can be achieved by changing the values of the rc timing components.

FLASH METER

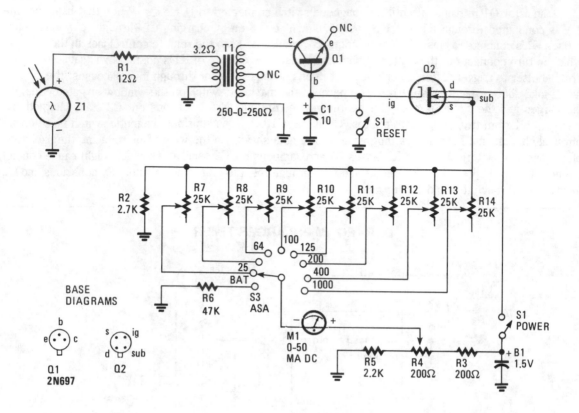

Fig. 71-4

Insulated-gate, field-effect transistor (IGFET), Q2 and silicon photo cell Z1 form the heart of this circuit. Transformer T1 is an audio-output type, but it's reversed in the circuit. A sudden flash from a photoflash unit detected by Z1 sends a voltage pulse through the low-impedance winding of T1 via R1. That voltage pulse is stepped-up in T1's 500-Ω, primary winding before being rectified by Q1. Transistor Q1 is used as a diode; its emitter lead was snipped off close to the case. Q1 then charges C1 to a value proportional to the amplitude of the electrical pulse generated by the light from a flash unit.

Capacitor C1 controls the current flowing through Q2, which has a very high-input impedance. The current through Q2 is read by meter M1, a 0–50 μA dc unit, which has been calibrated in f-stops. The extremely high internal resistances of Q1 and Q2 will allow C1 to retain its charge for several minutes; this is more than enough time for you to take your reading of M1. The charge on C1 is shorted to ground and returned to 0 V by depressing reset button S1. The flashmate is ready to read the next photoflash. Trim potentiometers, R7 through R14, are adjusted to values which will yield correct readings for corresponding film sensitivities, or exposure indexes.

SLAVE PHOTOGRAPHIC XENON FLASH TRIGGER

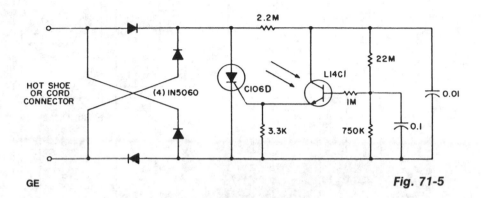

GE   *Fig. 71-5*

This circuit is used for remote photographic flash units that flash at the same time as the flash attached to the camera. This circuit is designed to the trigger cord or *hot shoe* connection of a commercial portable flash unit and triggers the unit from the light produced by the light of the flash unit attached to the camera. This provides remote operation without the need for wires or cables between the various units. The flash trigger unit should be connected to the slave flash before turning the flash on to prevent a *dV/dt* triggered flash on connection. The L14C1 phototransistor has a wide, almost cosine viewing angle, so alignment is not critical. If a very sensitive, more directional remote trigger unit is desired, the circuit can be modified using an L14G2 lensed phototransistor as the sensor.

The lens on this transistor provides a viewing angle of approximately 10° and gives over a 10 to 1 improvement in light sensitivity (3 to 1 range improvement). Note that the phototransistor is connected in a self-biasing circuit which is relatively insensitive to slow-changing ambient light, and yet discharges the 0.01-μF capacitor into the C106D gate when illuminated by a photo flash. For a physically smaller size, the C106D can be replaced by a C205D, if the duty cycle is reduced appropriately.

SLIDE TIMER

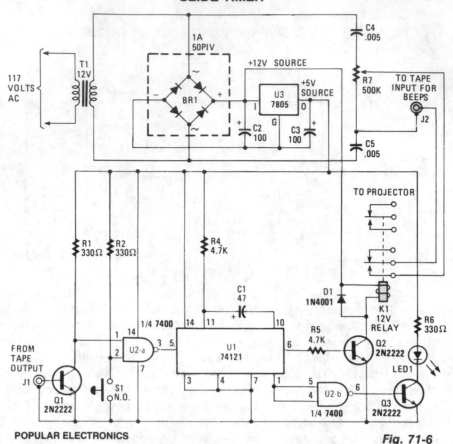

POPULAR ELECTRONICS

Fig. 71-6

This circuit will record commentary and/or music on one track of a tape and put the beeps that change the slides on another track. Gate U2a is used to trigger U1, which is configured as a timer when a pulse is received from either the tape input or via pushbutton switch S1. Timer U2 outputs one pulse for every input pulse received, no matter how long S1 is depressed.

The Q output of U2 at pin 1 is fed to U2b, which is set up as an inverter. When pin 1 of U1 becomes low, Q3 is activated, lighting LED1. The Q output of U1 at pin 6 is tied to the base of Q2, through R5, so that when pin 6 becomes high, Q2 is turned on. When Q2 is turned on, relay K1 is energized, and a signal is fed to the tape input through J2. The second set of contacts of K1 are used to trigger the projector.

Power for the circuit is provided by a 7805 regulator. The unregulated 12 V output of BR1 is used to power the relay. The 12-V relay needs to have two sets of contacts as shown: to advance the projector, and to supply the beeps when recording. The LED indicates projector advance.

To record the beeps, connect beeper jack J2 to the input of the tape recorder and connect the controller to the projector-advance plug. The 60-Hz line frequency is used to produce beeps that are recorded on half of the stereo tape. The other track is used for commentary. The beeper output is controlled via 500-KΩ potentiometer R6.

SLIDE TIMER *(Cont.)*

Use pushbutton switch S1 to put the beeps on the tape, where required, to advance the projector. The beep length is automatic, and the projector will advance once for every push of S1.

When presenting your program, disconnect one speaker from the recorder, and connect the recorder to the jack on the controller and plug into the earphone jack. Connect the controller to the projector. The beeps will not be heard and the projector will advance at precisely the correct time.

ELECTRONIC PHOTOFLASH

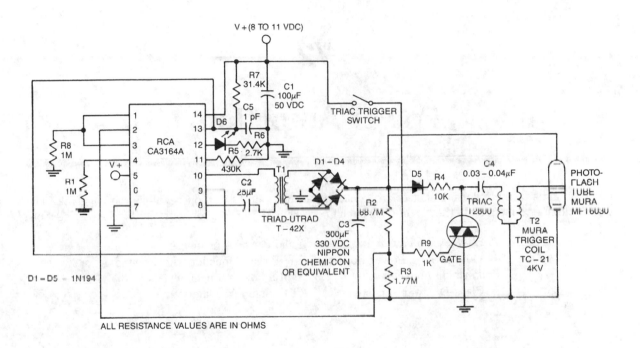

ALL RESISTANCE VALUES ARE IN OHMS

GE/RCA

Fig. 71-7

Using the CA3164A BiMOS control chip consumes less than 15 μA during standby, yet it can provide 100 mA of chopped current to the dc-to-dc converter during the energy-reservoir charging cycle. The CA3164A drives the primary of T1 with symmetrically chopped current at a 400 to 2000 Hz rate. The CA3164A's chopper frequency is about 500 Hz; the duty cycle, 50% when R7 is 3.1 4 KΩ.

72

Power Amplifiers

The sources of the following circuits are contained in the Sources section beginning on page 782. The figure number contained in the box of each circuit correlates to the sources entry in the Sources section.

50-W Audio Power Amplifier
Output-Stage Power Booster
Portable Amplifier
Class-D Power Amplifier

Audio Power Amplifier
6-W Power Amplifier with Preamp
Hybrid Power Amplifier
20-W Audio Amplifier

50-W AUDIO POWER AMPLIFIER

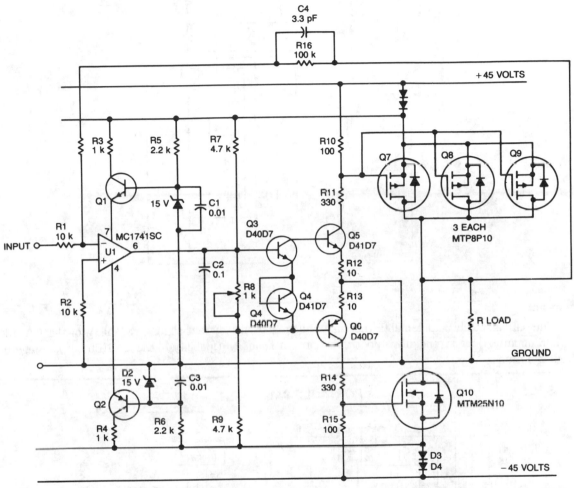

Fig. 72-1

This audio amplifier design approach employs TMOS Power FETs operating in a complementary common-source configuration. They are biased to cutoff, then turn on very quickly when a signal is applied. The advantage of this approach is that the output stage is very stable from a thermal point of view.

U1 is a high slew-rate amp that drives Q3, Q4, and Q6 (operating class AB) providing level transition for the output stage consisting of Q7, Q8, Q9, and Q10. The positive temperature coefficient of the TMOS device enables parallel operation of Q7, Q8, and Q9 and provides a higher power *complementary* device for Q10. These TMOS Power FETs must be driven from a low-source impedance of 100 Ω, in order to actually obtain high turn-on speeds.

OUTPUT-STAGE POWER BOOSTER

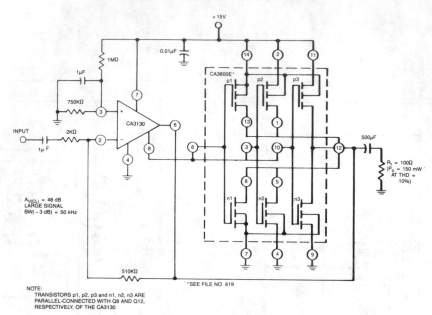

NOTE:
TRANSISTORS p1, p2, p3 and n1, n2, n3 ARE
PARALLEL-CONNECTED WITH Q8 AND Q12,
RESPECTIVELY, OF THE CA3130

GE/RCA

Fig. 72-2

This circuit easily supplements the current-sourcing and current-sinking capability of the CA3130 BiMOS op amp. This arrangement boosts the current-handling capability of the CA3130 output stage by about 2.5 times.

PORTABLE AMPLIFIER

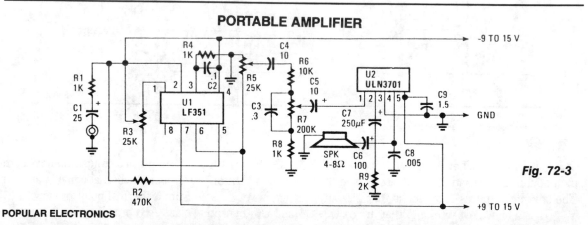

POPULAR ELECTRONICS

Fig. 72-3

U1, an FET op amp needs a bipolar voltage at pins 4 and 7 with a common ground for optimum gain. You can calculate the gain by dividing R2 by R1. Zero-set balance can be had through pins 1 and 5 through R3. Put a voltmeter between pin 6 and ground and adjust R3 for zero voltage. Once you've established that, you can measure the ohmic resistance at each side of R3's center tap and replace the potentiometer with fixed resistors. R6, R7, R8, and C3 form a tone control that will give you added bass boost, if needed.

CLASS-D POWER AMPLIFIER

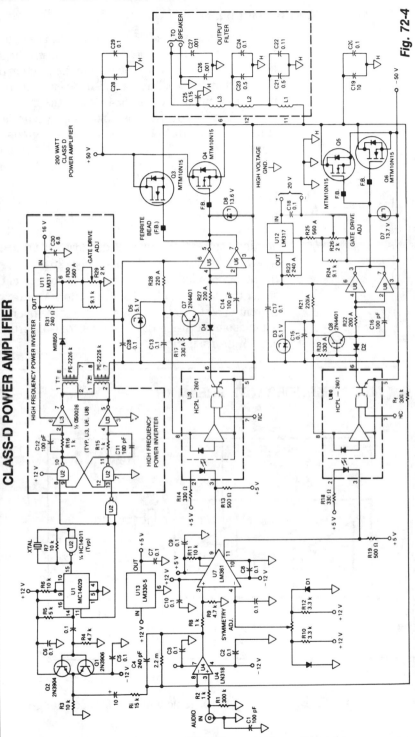

Fig. 72-4

In this circuit, a 2-MHz clock is divided by eight in U1, providing a stable 250-kHz carrier. Q1 and Q2 buffer the clock and provide a low-impedance drive for op amp U4, which is a high-gain amplifier and integrator. U4 accepts audio inputs and converts the 250-kHz square wave into a triangular wave. The summed audio and triangular-wave signal is applied to the input of comparator U7, where it is compared with a dc reference to produce a pulse-width modulated signal at the output of U7.

The output devices switch between the +50 V and −50 V rails in a complementary fashion, driving the output filter that is a sixth-order Butterworth low-pass type, which demodulates the audio and attenuates the carrier and high frequency components. Feedback is provided R_f; amplifier gain is $R_f R_i$.

Specifications: 200 W continuous power into a 4-Ω load; 20 to 20 kHz frequency response +0.5, −1.0 dB at 200 W; THD, IMD 0.5% at 200 W; 1.5-V rms input for rated output; 69 dB S/N ratio, A weighting; 6.6-μs ms slew rate.

453

AUDIO POWER AMPLIFIER

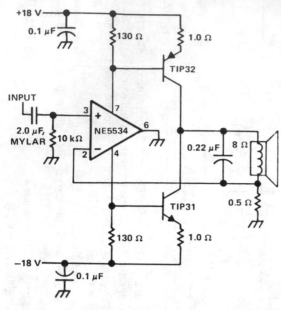

Reprinted by permission of Texas Instruments.

Fig. 72-5

The single speaker amplifier circuit uses current feedback, rather than the more popular voltage feedback. The feedback loop is from the junction of the speaker terminal and a 0.5-Ω resistor, to the inverting input of the NE5534. When the input to the amplifier is positive, the power supply supplies current through the TIP32 and the load to ground. Conversely, with a negative input, the TIP31 supplies current through the load to ground. The gain is set to about 15 (gain = SPKR 8 Ω/0.5 Ω feedback). The 0.22-μF capacitor across the speaker rolls off its response beyond the frequencies of interest. Using the 0.22-μF capacitor specified, the amplifier current output is 3 dB down at 90 kHz where the speaker impedance is about 20 Ω. To set the recommended class A output collector current, adjust the value of either 130-Ω resistor. An output current of 50 to 100 mA will provide a good operating midpoint between the best crossover distortion and power dissipation.

6-W AUDIO AMPLIFIER WITH PREAMP

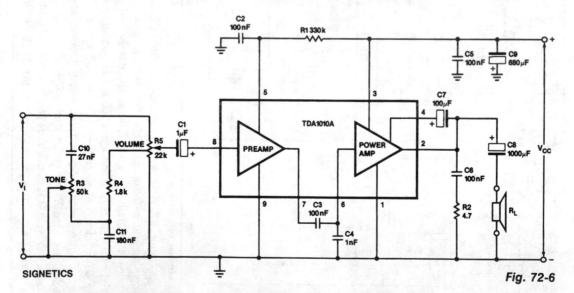

Fig. 72-6

This monolithic IC, class-B, audio amplifier circuit is a 6-W car radio amplifier for use with 4-Ω and 2-Ω load impedances.

HYBRID POWER AMPLIFIER

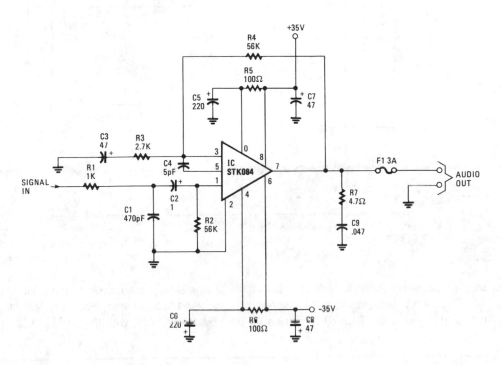

Fig. 72-7

The input is ac coupled to the amplifier through C2, which blocks dc signals that might also be present at the input. The R1/C1 combination forms a low-pass filter, which eliminates unwanted high-frequency signals by bypassing them to ground when they appear at the circuit input, which has an impedance of about 52 Ω. The gain of the amplifier is set at about 26 dB by resistors R3 and R4. The R5/C5/C7 combination on the positive supply and its counterpart R6/C6/C8 on the negative supply provides power-supply decoupling. R7 and C9 together prevent oscillation at the output of the amplifier. From that point, the amplifier's output signal is direct coupled to the speaker through a 3-A fuse, F1. The dc output of the amplifier at pin 7 is 0 V, so no dc current flows through the speaker. Should there be a catastrophic failure of the output stage, fuse F1, which should be a fast-acting type, prevents dc from flowing through the speaker.

20-W AUDIO AMPLIFIER

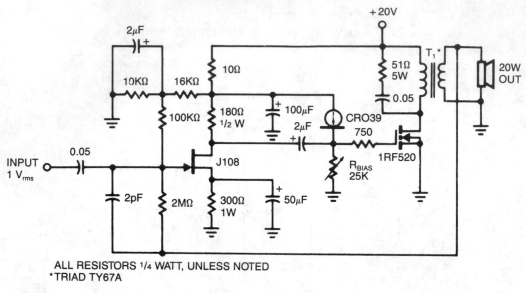

ALL RESISTORS ¼ WATT, UNLESS NOTED
*TRIAD TY67A

SILICONIX

Fig. 72-8

This amplifier delivers 20 W into an 8-Ω load using a single IRF520 driving a transformer coupled output stage. This circuit is similar to the audio output stage used in many inexpensive radios and phonographs. Distortion is less than 5% at 10 W, using very little feedback (3%), with the IRF520 biased at 3 A.

73

Fixed Power Supplies

The sources of the following circuits are contained in the Sources section beginning on page 782. The figure number contained in the box of each circuit correlates to the sources entry in the Sources section.

SWITCHING POWER SUPPLY

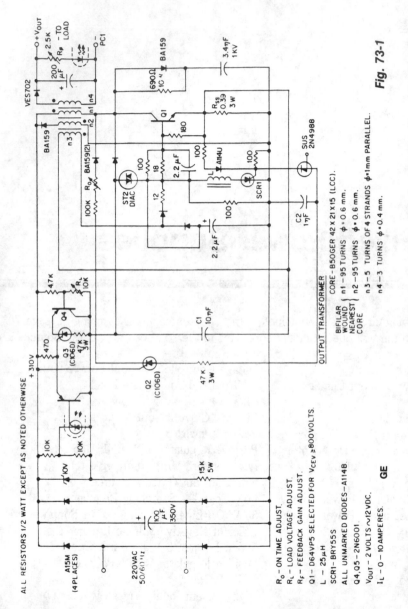

Fig. 73-1

This low-voltage high-current output, switching dc power supply is running off the 220-Vac input. In this circuit, an ST2 diac relaxation oscillator, Q3, C1, and the diac, initiates conduction of the output switching transistor Q1, the on-time of which is maintained constant by a separate timing/commutation network consisting of Q2, C2, SUS, and SCR 1. The output voltage, consequently, is dependent on the duty cycle. To compensate for unwanted variations of output voltage because of input voltage or load resistance fluctuations, an H11C wired as a linear-model unilateral PNP transistor in a stable differential amplifier configuration is connected into the galvanically isolated negative-feedback loop. The loop determines the duty cycle and hence the output voltage. Of further interest in this circuit is the use of several low-current, high-voltage, 400V V_{DRM} thyristors (Q2, Q3,) which are also used as pnp remote-base transistors. Short-circuit protection is assured by coupling Q1 collector-current feedback into the turn-off circuitry via R_{SS}.

LOW COST, LOW DROPOUT LINEAR REGULATOR

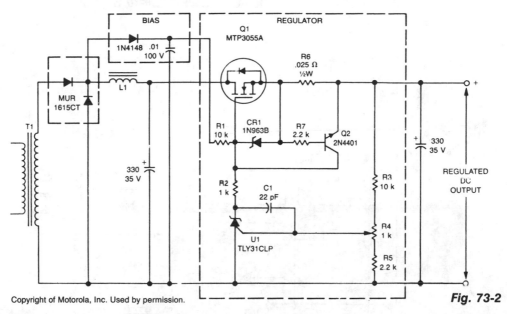

Fig. 73-2

This linear post regulator provides 12 V at 3 A. It employs TL431 reference U1 which, without additional amplification, drives TMOS MTP3055A gate Q1 series pass regulator. Bias voltage is applied through R1 to Q1's gate, which is protected against overvoltage by diode CR1. Frequency compensation for closed-loop stability is provided by C1.

Key performance features are:

Dropout voltage:	0.6 V	Load regulation:	10 mV
Line regulation:	± 5 mV	Output ripple:	10 mV pk-pk

VOLTAGE DOUBLER

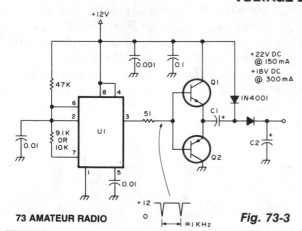

This circuit drives relays of 24 and 18 Vdc from a 12-V power supply. Use this circuit with almost any pnp or npn power transistor.

Parts: U1: NE 555 timer. C1 and C2: 50 µF/ 25 Vdc. Q1: TIP 29, TIP120, 2N4922, TIP61, TIP110, or 2N4921. Q2: TIP30, TIP125, 2N4919, TIP62, TIP115, or 2N4918.

73 AMATEUR RADIO   **Fig. 73-3**

ISOLATED FEEDBACK POWER SUPPLY

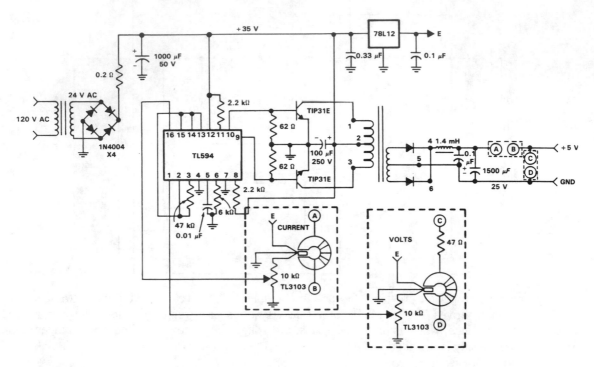

Reprinted by permission of Texas Instruments.

Fig. 73-4

Figure 73-4 is a power supply circuit using the isolated feedback capabilities of the TL3103 for both current and voltage sensing. This supply is powered from the ac power line and has an output of 5 V at 1.5 A. Both output voltage and current are sensed and the error voltages are applied to the error amplifiers of the TL594 PWM control IC. The 24-V transformer produces about 35 V at the 1000-µF filter capacitor. The 20-kHz switching frequency is set by the 6-KΩ resistor and the 0.01-µF capacitor on pins 6 and 5, respectively. The TL594 is set for push-pull operation by typing pin 13 high. The 5-V reference on pin 14 is tied to pin 15, which is the reference or the current error amplifier. The 5-V reference is also tied to pin 2 which is the reference for the output voltage error amplifier. The output voltage and current limit are set by adjustment of the 10-KΩ pots in the TL3103 error-sensing circuits. A pair of TIP31E npn transistors are used as switching transistors in a push-pull circuit.

HAND-HELD TRANSCEIVER DC ADAPTER

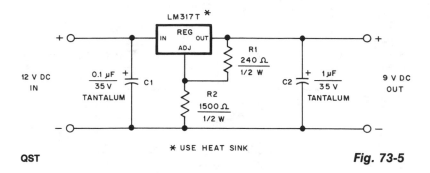

Fig. 73-5

This dc adapter provides a regulated 9-V source for operating a Kenwood TR-2500 hand-held transceiver in the car. The LM317T's mounting tab is electrically connected to its output pin, so take this into account as you construct your version of the adapter. The LM317T regulator dissipates 2 or 3 W in this application, so mount it on a 1-×-2-inch piece of $^{1}/_{8}$-inch-thick aluminum heatsink.

LOW-DROPOUT 5 V REGULATOR

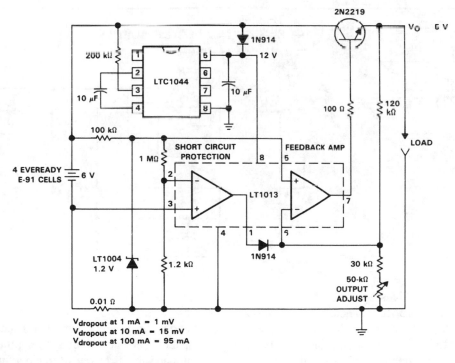

Reprinted by permission of Texas Instruments.

Fig. 73-6

DUAL-TRACKING REGULATOR

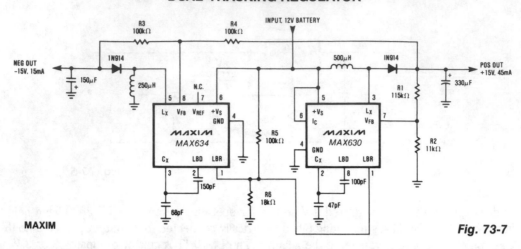

MAXIM

Fig. 73-7

A MAX634 inverting regulator is combined with a MAX630 to provide a dual tracking ±15 output from a 12-V battery. The reference for the −15 V output is derived from the positive output via R3 and R4. Both regulators are set to maximize output power at low battery voltages by reducing the oscillator frequency, via LBR, when V_{BATT} falls to 8.5 V.

+15 V 1-A REGULATED POWER SUPPLY

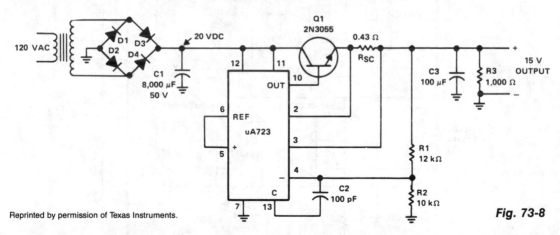

Fig. 73-8

The supply receives +20 Vdc from the rectifier/filter section. This is applied to pins 11 and 12 of the uA723, as well as to the collector of the 2N3055 series-pass transistor. The output voltage is sampled through R1 and R2, providing about 7 V with respect to ground at pin 4. The reference terminal at pin 6 is tied directly to pin 5, the noninverting input of the error amplifier. For fine trimming the output voltage, a potentiometer can be installed between R1 and R2. A 100-pF capacitor from pin 13 to pin 4 furnishes gain compensation for the amplifier.

462

Base drive to the 2N3055 pass transistor is furnished by pin 10 of the uA723. Since the desired output of the supply is 1 A, maximum current limit is set to 1.5 A by resistor R_{SC} whose value is 0.433 Ω.

A 100-μF electrolytic capacitor is used for ripple voltage reduction at the output. A 1-KΩ output resistor provides stability for the power supply under no-load conditions. The 2N3055 pass transistor must be mounted on an adequate heatsink.

–15 V 1-A REGULATED POWER SUPPLY

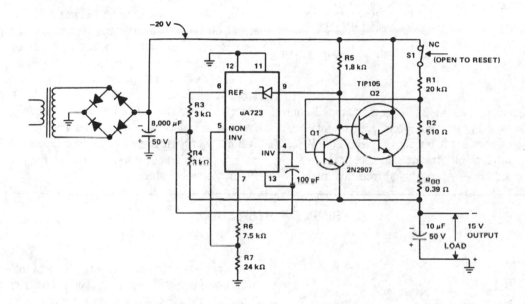

Fig. 73-9

The supply receives –20 V from the rectifier/filter which is fed to the collector of the Darlington pnp pass transistor, a TIP105. The base drive to the TIP105 is supplied through resistor R5. The base of the TIP105 is driven from V_Z terminal at pin 9, which is the anode of a 6.2-V zener diode that connects to the emitter of the uA723 output control transistor. The method of providing the positive feedback required for foldback action is shown. This technique introduces positive feedback by increased current flow through resistors R1 and R2 under short-circuit conditions. This forward biases the base-emitter junction of the 2N2907 sensing transistor, which reduces base drive to the TIP105.

12-VDC BATTERY-OPERATED 120-VAC POWER SOURCE

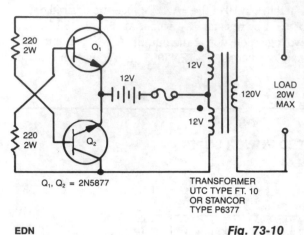

Q₁, Q₂ = 2N5877

TRANSFORMER
UTC TYPE FT. 10
OR STANCOR
TYPE P6377

EDN

Fig. 73-10

A simple 120 V: 24 V, center-tapped control transformer and four additional components can do the job. This circuit outputs a clean 200 V pk-pk square wave at 60 Hz and can supply up to 20 W. The circuit is self-starting and free-running.

If Q1 is faster and has a higher gain than Q2, it will turn on first when you apply the input power and will hold Q2 off. Load current and transformer magnetizing current then flows in the upper half of the primary winding, and auto transformer action supplies the base drive until the transformer saturates. When that action occurs, Q1 loses its base drive. As it turns off, the transformer voltages reverse, turning Q2 on and repeating the cycle. The output frequency depends on the transformer iron and input voltage, but not on the load. The frequency will generally range between 50 to 60 Hz with a 60-Hz transformer and car battery or equivalent source. The output voltage depends on turns ratio and the difference between input voltage and transistor saturation voltage. For higher power, use larger transformers and transistors. This type of inverter normally is used in radios, phonographs, hand tools, shavers, and small fluorescent lamps. It will not work with reactive loads (motors) or loads with high inrush currents, such as coffee pots, frying pans, and heaters.

SIMPLE POWER SUPPLY

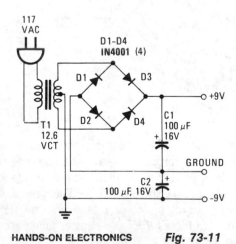

HANDS-ON ELECTRONICS

Fig. 73-11

This power supply delivers plus and minus 9 V to replace two 9-V batteries. The rectifier circuit is actually two separate full-wave rectifiers fed from the secondary of the transformer. One full-wave rectifier is composed of diodes D1 and D2, which develop +9 V, and the other is composed of D3 and D4, which develop −9 V.

Each diode from every pair rectifies 6.3 Vac, half the secondary voltage, and charges the associated filter capacitor to the peak value of the ac waveform, $6.3 \times 1.414 = 8.9$ V. Each diode should have a PIV, Peak Inverse Voltage, rating that is at least twice the peak voltage from the transformer, $2 \times 8.9 = 18$ V. The 1N4001 has a PIV of 50 V.

GENERAL-PURPOSE POWER SUPPLY

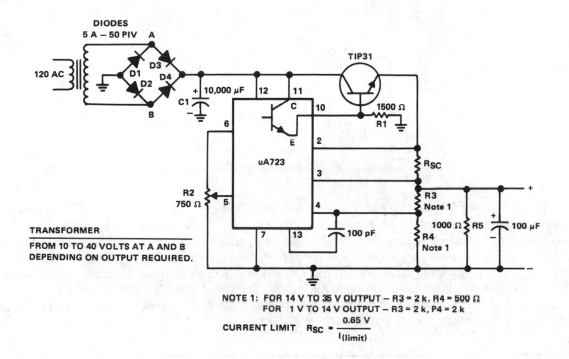

NOTE 1: FOR 14 V TO 35 V OUTPUT – R3 = 2 k, R4 = 500 Ω
         FOR 1 V TO 14 V OUTPUT – R3 = 2 k, P4 = 2 k

CURRENT LIMIT $R_{SC} = \dfrac{0.65\ V}{I_{(limit)}}$

Reprinted by permission of Texas Instruments.

Fig. 73-12

The supply 6-66 can be used for supply output voltages from 1 to 35 V. The line transformer should be selected to give about 1.4 times the desired output voltage from the positive side of filter capacitor C1 to ground. Potentiometer R2 sets the output voltage to the desired value by adjusting the reference input. R_{SC} is the current limit set resistor. Its value is calculated as:

$$R_{SC} = \frac{0.65\ V}{I_L}$$

For example, if the maximum current output is to be 1 A, $R_{SC} = 0.65/1.0 = 0.65\ Ω$. The 1-KΩ resistor, R_S, is a light-loaded resistor designed to improve the no-load stability of the supply.

LOW-POWER INVERTER

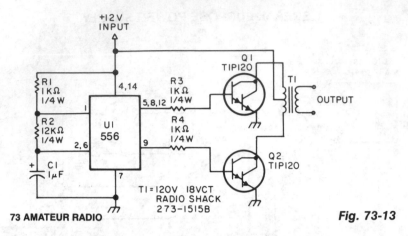

73 AMATEUR RADIO

Fig. 73-13

This low-power inverter uses only 9 parts and turns 10 to 16 Vdc into 60-Hz, 115-V square-wave power to operate ac equipment up to 25 W. The first section of the 556 timer chip is wired as an astable oscillator with R2 and C1 setting the frequency. The output is available at pin 5. The second section is wired as a phase inverter. That output is available at pin 9. Resistors R3 and R4 keep output transistors Q1 and Q2 from loading down the oscillator. The two transistors drive the transformer push-pull fashion. When one transistor is biased-on, the other is cut-off. The transformer is a 120 V/18 VCT unit that is connected backwards, so that it steps the voltage up rather than down. Oscillator circuit U1, R1, R2, and C1 operates from about 4 to 16 V with a very stable output.

THREE-RAIL POWER SUPPLY

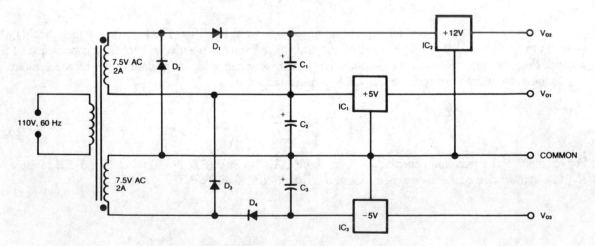

EDN

Fig. 73-14

THREE-RAIL POWER SUPPLY (Cont.)

This circuit generates three supply voltages using a minimum of components. Diodes D2 and D3 perform full-wave rectification, alternately charging capacitor C2 on both halves of the ac cycle. On the other hand, diode D1 with capacitor C1, and diode D4 with capacitor C3 each perform half-wave rectification. The full- and half-wave rectification arrangement is satisfactory for modest supply currents drawn from -5 and $+12$-V regulators IC3 and IC2. You can use this circuit as an auxiliary supply in an up-based instrument, for example, and avoid the less attractive alternatives of buying a custom-wound transformer, building a more complex supply, or using a secondary winding, say 18 Vac, and wasting power in the 5-V regulators.

PROGRAMMABLE POWER SUPPLY

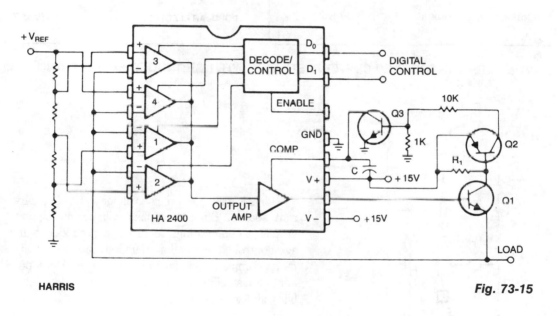

HARRIS

Fig. 73-15

Many systems require one or more relatively low-current voltage sources which can be programmed to a few predetermined levels. The circuit shown above produces positive output levels, but could be modified for negative or bipolar outputs. Q1 is the series regulator transistor, selected for the required current and power capability. R1, Q2, and Q3 form an optional short circuit protection circuit, with R1 chosen to drop about 0.7 V at the maximum output current. The compensation capacitor, C, should be chosen to keep the overshoot, when switching, to an acceptable level.

TRIAC-CONTROLLED VOLTAGE DOUBLER

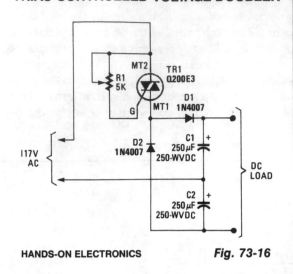

Fig. 73-16

HIGH STABILITY 10-V REGULATOR

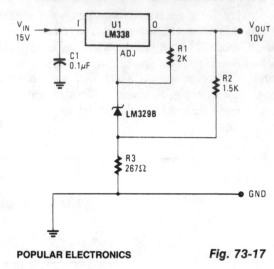

Fig. 73-17

VOLTAGE-CONTROLLED CURRENT SOURCE WITH GROUNDED SOURCE AND LOAD

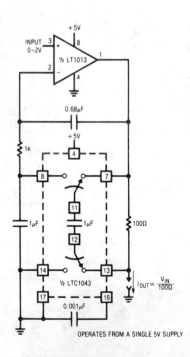

This is a simple, precise voltage-controlled current source. Bipolar supplies will permit bipolar output. Configurations featuring a grounded voltage-control source and a grounded load are usually more complex and depend upon several components for stability. In this circuit, accuracy and stability almost entirely depend upon the 100-Ω shunt.

Fig. 73-18

CHARGE POOL POWER SUPPLY

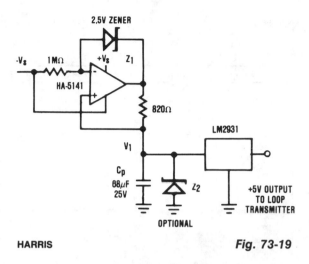

HARRIS   ***Fig. 73-19***

It is usually desirable to have the remote transmitter of a 4 to 20 mA current loop system powered directly from the transmission line. In some cases, this is not possible because of the high-power requirements set by the remote sensor/transmitter

system. In these cases, an alternative to the separate power supply is still possible. If the remote transmitter can be operated in a pulsed mode where it is active only long enough to perform its function, then a charge pool power supply can still allow the transmitter to be powered directly by the current loop. In this circuit, constant current $I1$ is supplied to the charge pool capacitor, CP, by the HA-5141 (where $I1 = 3$ mA). The voltage $V1$ continues to rise until the output of the HA-5141 approaches $+V_s$ or the optional voltage limiting provided by $Z2$. The LM2931 voltage regulator supplies the transmitter with a stable $+5$ V supply from the charge collected by CP. Available power supply current is determined by the duration, allowable voltage droop on CP, and required repetition rate. For example, if $V1$ is allowed to droop 4.4 V and the duration of operation is 1 ms, the available power supply current is approximately:

$$= CP \frac{dV_1}{dt} = 68\mu F \times \frac{4.4 \text{ V}}{1 \text{ ms}} = 30 \text{ mA}$$

BILATERAL CURRENT SOURCE

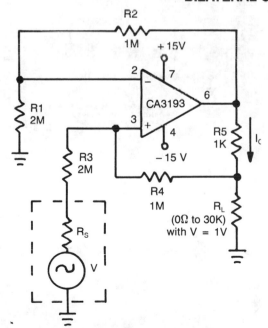

ALL RESISTORS ARE 1%
ALL RESISTANCE VALUES ARE IN OHMS
IF R1 = R3 AND R2 ≈ R4 + R5 THEN

I_L IS INDEPENDENT OF VARIATIONS IN R_L
FOR R_L VALUES OF 0Ω to 3KΩ WITH V = 1V

$$I_L = \frac{V \text{ R4}}{R3 \text{ R5}} = \frac{V \text{ 1M}}{(2M)(1K)} = \frac{V}{2K} = 500\mu A$$

GE/RCA   ***Fig. 73-20***

This circuit uses a CA3193 precision op amp to deliver a current independent of variations in R_L. With $R1$ set equal to $R3$, and $R2$ approximately equal to $R4 + R5$, the output current, I_L, is: V_{IN} $(R4)/(R3)$ $(R5)$. 500-μA load current is constant for load values from 0 to 3 Ω.

POWER CONVERTER

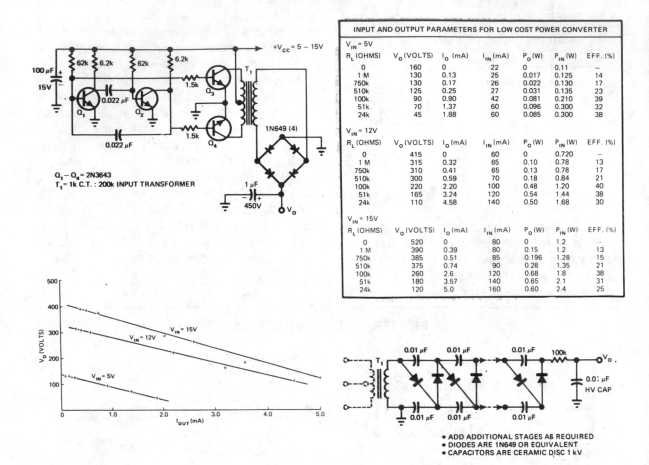

INPUT AND OUTPUT PARAMETERS FOR LOW COST POWER CONVERTER						
V_{IN} = 5V						
R_L (OHMS)	V_O (VOLTS)	I_O (mA)	I_{IN} (mA)	P_O (W)	P_{IN} (W)	EFF. (%)
0	160	0	22	0	0.11	–
1 M	130	0.13	25	0.017	0.125	14
750k	130	0.17	26	0.022	0.130	17
510k	125	0.25	27	0.031	0.135	23
100k	90	0.90	42	0.081	0.210	39
51k	70	1.37	60	0.096	0.300	32
24k	45	1.88	60	0.085	0.300	38
V_{IN} = 12V						
R_L (OHMS)	V_O (VOLTS)	I_O (mA)	I_{IN} (mA)	P_O (W)	P_{IN} (W)	EFF. (%)
0	415	0	60	0	0.720	–
1 M	315	0.32	65	0.10	0.78	13
750k	310	0.41	65	0.13	0.78	17
510k	300	0.59	70	0.18	0.84	21
100k	220	2.20	100	0.48	1.20	40
51k	165	3.24	120	0.54	1.44	38
24k	110	4.58	140	0.50	1.68	30
V_{IN} = 15V						
R_L (OHMS)	V_O (VOLTS)	I_O (mA)	I_{IN} (mA)	P_O (W)	P_{IN} (W)	EFF. (%)
0	520	0	80	0	1.2	–
1 M	390	0.39	80	0.15	1.2	13
750k	385	0.51	85	0.196	1.28	15
510k	375	0.74	90	0.28	1.35	21
100k	260	2.6	120	0.68	1.8	38
51k	180	3.57	140	0.65	2.1	31
24k	120	5.0	160	0.60	2.4	25

$Q_1 - Q_4$ = 2N3643
T_1 = 1k C.T. : 200k INPUT TRANSFORMER

- ADD ADDITIONAL STAGES AS REQUIRED
- DIODES ARE 1N649 OR EQUIVALENT
- CAPACITORS ARE CERAMIC DISC 1 kV

EDN

Fig. 73-21

This circuit consists of an astable multivibrator driving a push-pull pair of transistors into the transformer primary. The multivibrator frequency should equal around 1 or 2 kHz. For higher dc voltages, voltage multipliers on the secondary circuit have been used successfully to generate 10 kV from a 40-stage multiplier like the one shown.

POSITIVE REGULATOR WITH PNP BOOST

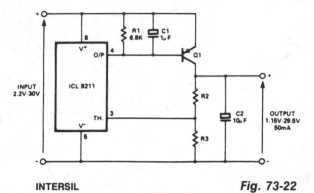

Fig. 73-22

The IC8211 provides the voltage reference and regulator amplifier, while Q1 is the series pass transistor. R1 defines the output current of the IC8211, while C1 and C2 provide loop stability and also act to suppress feedthrough of input transients to the output supply. R2 and R3 determine the output voltage as follows:

output supply. R2 and R3 determine the output voltage as follows:

$$V_{OUT} = 1.5 \times \frac{R_2 + R_3}{R_3}$$

In addition, the values of R2 and R3 are chosen to provide a small amount of standing current in Q1, which gives additional stability margin to the circuit. Where accurate setting of the output voltage is required, either R2 or R3 can be made adjustable. If R2 is made adjustable, the output voltage will vary linearly with the shaft angle; however, if the potentiometer wiper was to open the circuit, the output voltage would rise. In general, therefore, it is better to make R3 adjustable, since this gives fail-safe operation.

LOW FORWARD-DROP RECTIFIER CIRCUIT

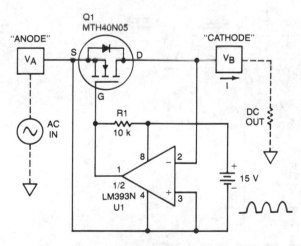

Fig. 73-23

A TMOS power FET, Q1, and an LM393 comparator provide a high-efficiency rectifier circuit. When V_A exceeds V_B, U1's output becomes high and Q1 conducts. Conversely, when V_B exceeds V_A, the comparator output becomes low and Q1 does not conduct.

The forward drop is determined by Q1's on resistance and current I. The MTH40N05 has an on resistance of $0.028 \, \Omega$; for $I = 10$ A, the forward drop is less than 0.3 V. Typically, the best Schottky diodes do not even begin conducting below a few hundred mV.

SAFE CONSTANT-CURRENT SOURCE

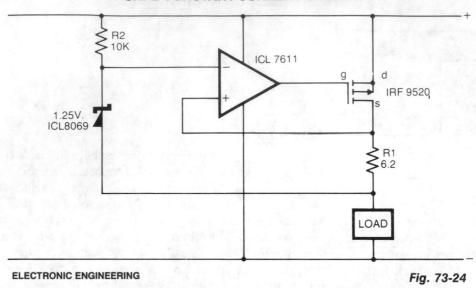

ELECTRONIC ENGINEERING

Fig. 73-24

In the circuit shown, a CMOS op amp controls the current through a p-channel HEXFET power transistor to maintain a constant voltage across R1. The current is given by: $1 = V_{REF}/R1$. The advantages of this configuration are: (a) in the event of a component failure, the load current is limited by R1; and (b) the overhead voltage needed by the op amp and the HEXFET is extremely low.

LOW-COST 3-A SWITCHING REGULATOR

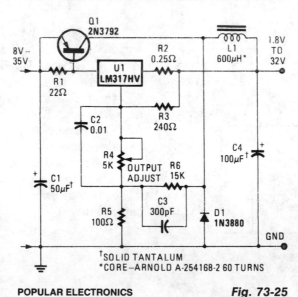

POPULAR ELECTRONICS

Fig. 73-25

50-W OFF-LINE SWITCHING POWER SUPPLY

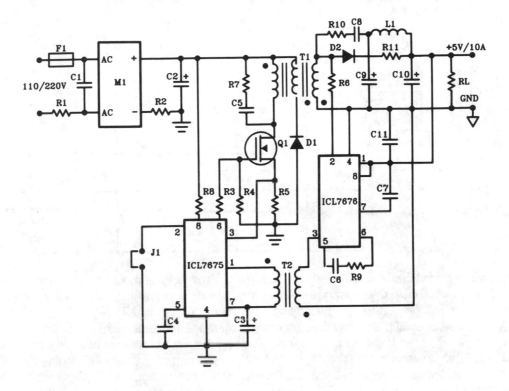

Component Values Table

C1	0.022 μF/400V	R1	100Ω at 25°C	L1	25 μH
C2	470 μF/250V	R2	1Ω/1W	D1	1N4937
C3	470 μF/16V	R3	10Ω/0.25W	D2	MBR1035
C4	220 pF/100V	R4	100 kΩ/0.25W	T1	Lp = 9 mH, n = 1:15
C5	470 pF/500V	R5	0.33Ω/1W	T2	50 μH, n = 1:3
C6	2200 pF/500V	R6	10 kΩ/0.25W	F1	Fuse 1 A/SB
C7	270 pF/500V	R7	390Ω/2W	M1	Diode Bridge
C8	39 pF/500V	R8	22 kΩ/10W	Q1	BUZ80A/IXTP4N80
C9	11,000 μF/6.3V	R9	68Ω/0.25W		(220VAC)
C10	10 μF/16V	R10	10Ω/0.5W	Q1	GE IRF823
C11	0.047 μF/10V	R11	3.3Ω/0.5W		(110VAC)
		RL	5Ω/10W		

INTERSIL

Fig. 73-26

The schematic shows a 50-W power supply with a 5-V 10-A output. It is a flyback converter operating in the continuous mode. The circuit features a primary side and secondary side controller will full-protection from fault conditions such as overcurrent. After the fault condition has been removed, the power supply will enter the soft-start cycle before recommencing normal operation.

EFFICIENT NEGATIVE VOLTAGE REGULATOR

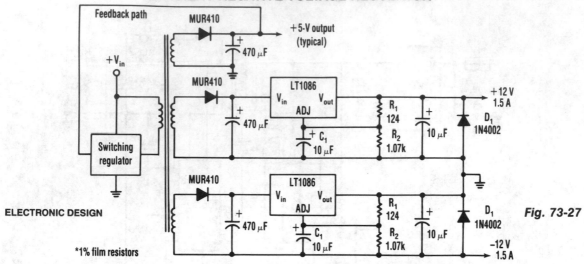

ELECTRONIC DESIGN

Fig. 73-27

*1% film resistors

One way to provide good negative-voltage regulation is with a low-dropout positive-voltage regulator operating from a well-isolated secondary winding of switch-mode circuit transformer. The technique works with any positive-voltage regulator, although highest efficiency occurs with low-dropout types.

Under all loading conditions, the minimum voltage difference between the regulator V_{IN} and V_{OUT} pins must be at least 1.5 V, the LT1086's low-dropout voltage. If this requirement isn't met, the output falls out of regulation. Two programming resistors, R1 and R2, set the output voltage to 12 V, and the LT1086's servo the voltage between the output and its adjusting (ADJ) terminals to 1.25 V. Capacitor C1 improves ripple rejection, and protection diode D1 eliminates common-load problems.

Since a secondary winding is galvanically isolated, a regulator's 12 V output can be referenced to ground. Therefore, in the case of a negative-voltage output, the positive-voltage terminal of the regulator connects to ground, and the -12 V output comes off the anode of D1. The V_{IN} terminal floats at 1.5 V or more above ground.

5 V-TO-ISOLATED 5 V AT 20 MA CONVERTER

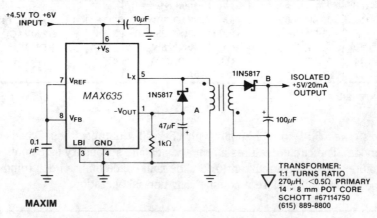

MAXIM

Fig. 73-28

In this circuit, a negative output voltage dc-dc converter generates a −5 V output at pin A. In order to generate −5 V at point A, the primary of the transformer must fly back to a diode drop more negative than −5 V. If the transformer has a tightly coupled 1/1 turns ratio, there will be a 5 V plus a diode drop across the secondary. The 1N5817 rectifies this secondary voltage to generate an isolated 5-V output. The isolated output is not fully regulated since only the −5 V at point A is sensed by the MAX635.

POSITIVE REGULATOR WITH NPN AND PNP BOOST

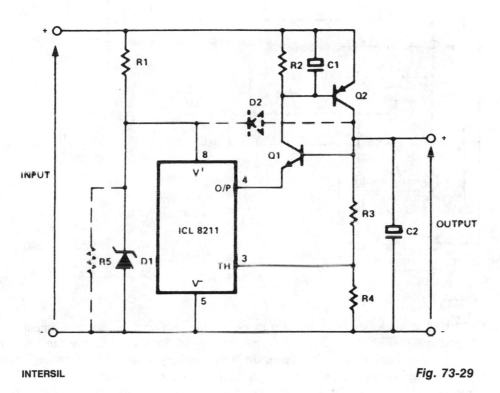

INTERSIL

Fig. 73-29

In the circuit, Q1 and Q2 are connected in the classic SCR or thyristor configuration. Where higher input voltages or minimum component count are required, the circuit for thyristor boost can be used. The thyristor is running in a linear mode with its cathode as the control terminal and its gate as the output terminal. This is known as the remote base configuration.

HIGH-CURRENT INDUCTORLESS, SWITCHING REGULATOR

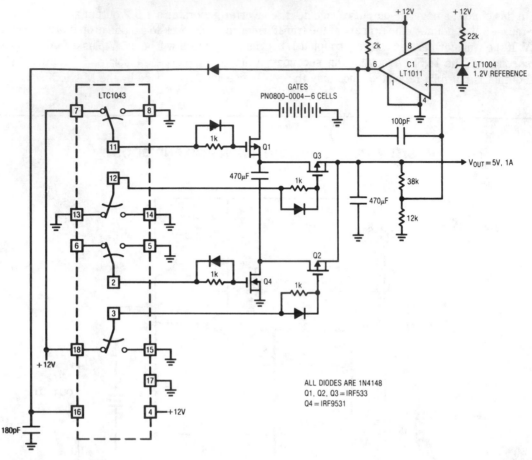

LINEAR TECHNOLOGY CORP.

Fig. 73-30

The LTC10432 switched-capacitor building block provides nonoverlapping complementary drive to the Q1 to Q4 power MOSFETs. The MOSFETs are arranged so that C1 and C2 are alternately placed in series and then parallel. During the series phase, the +12 V battery's current flows through both capacitors, charging them, and furnishing load current. During the parallel phase, both capacitors deliver current to the load. Q1 and Q2 receive similar drive from pins 3 and 11. The diode-resistor networks provide additional nonoverlapping drive characteristics, preventing simultaneous drive to the series-parallel phase switches. Normally, the output would be one-half of the supply voltage, but C1 and its associated components close a feedback loop, forcing the output to 5 V. With the circuit in the series phase, the output heads rapidly positive. When the output exceeds 5 V, C1 trips, forcing the LTC1043 oscillator pin, trace D, high; this truncates the LTC1043's triangular-wave oscillator cycle. The circuit is forced into the parallel phase and the output coasts down slowly, until the next LTC1043 clock cycle begins. C1's output diode prevents the triangle down-slope from being affected and the 100-pF capacitor provides sharp transitions. The loop regulates the output to 5 V by feedback controlling the turn-off point of the series phase.

SLOW TURN-ON 15 V REGULATOR

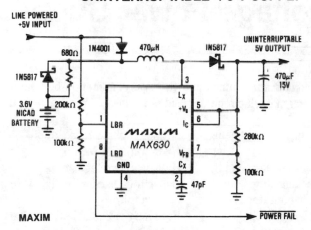

V_{IN} → I U1 LM338 O V_{OUT} 15V

ADJ

C2 0.1

R1* 120Ω

R3 50K

D1 1N4002

R2 2.7K

Q1 2N2905

+ C1 100µF

*R1 = 240Ω FOR LM138 AND LM238

POPULAR ELECTRONICS **Fig. 73-31**

AC VOLTAGE REGULATOR

U1 LM317 O

ADJ

R3 120Ω

12 V_{PP}

R1 480Ω

R4 480Ω

6 V $_{PP}$ 1A

R2 120Ω

ADJ

I U2 LM317 O

POPULAR ELECTRONICS **Fig. 73-32**

UNINTERRUPTABLE +5 V SUPPLY

LINE POWERED +5V INPUT

680Ω

1N4001

470µH

1N5817

UNINTERRUPTABLE 5V OUTPUT

1N5817

3.6V NICAD BATTERY

200kΩ

470µF 15V

100kΩ

1 LBR

MAXIM MAX630

Lx

+V$_S$ 5

I$_C$ 6

280kΩ

8 LBD

GND Cx

V$_{FB}$ 7

100kΩ

4

2 47pF

MAXIM

POWER FAIL

Fig. 73-33

This circuit provides a continuous supply of regulated +5 V, with automatic switch-over between line power and battery backup. When the line-powered input voltage is a +5 V, it provides 4.4 V to the MAX630 and trickle charges the battery. If the line-powered input falls below the battery voltage, the 3.6 V battery supplies power to the MAX630, which boosts the battery voltage up to +5 V, thus maintaining a continuous supply to the uninterruptable +5 V bus. Since the +5 V output is always supplied through the MAX630, there are no power spikes or glitches during power transfer. The MAX630's low-battery detector monitors the line-powered +5 V, and the LBD output can be used to shut down unnecessary sections of the system during power failures. Alternatively, the low-battery detector could monitor the NiCad battery voltage and provide warning of power loss when the battery is nearly discharged. Unlike battery backup systems that use 9-V batteries, this circuit does not need +12 or +15 V to recharge the battery. Consequently, it can be used to provide +5 V backup on modules or circuit cards which only have 5 V available.

74

High-Voltage Power Supplies

The sources of the following circuits are contained in the Sources section beginning on page 782. The figure number contained in the box of each circuit correlates to the sources entry in the Sources section.

ARC-JET POWER SUPPLY AND STARTING CIRCUIT

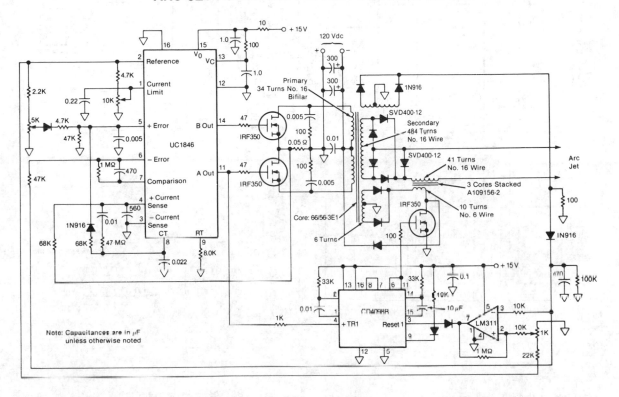

Fig. 74-1

This circuit for starting arc jets and controlling them in steady operation is capable of high power efficiency and can be constructed in a lightweight form. The design comprises a pulse-width-modulated power converter, which is configured in a closed control loop for fast current control. The series averaging inductor maintains nearly constant current during rapid voltage changes, and thereby allows time for the fast-response regulator to adjust its pulse width to accommodate load-voltage changes. The output averaging inductor doubles as the high-voltage pulse transformer for ignition. The starting circuit operates according to the same principle as that of an automobile ignition coil. When the current is interrupted by a transistor switch, the inductor magnetic field collapses, and a high-voltage pulse is produced. The pulse is initiated every 0.25 second until arc current is detected, then the pulser is automatically turned off.

479

PREREGULATED HIGH-VOLTAGE SUPPLY

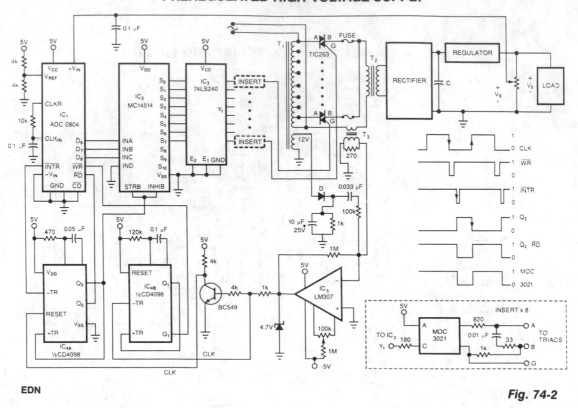

EDN

Fig. 74-2

One of the control circuit's triacs selects the tap on main transformer T1, which provides the proper, preregulated voltage to the secondary regulator. T2 and its associated components comprise the secondary regulator.

The ADC 0804, IC1, digitizes a voltage-feedback signal from the secondary regulator's output. The MC1415 demultiplexer, IC2, decodes the digitizer's output. IC2, in turn, drives T1's optoisolated triacs via the 74LS240 driver chip, IC3, and associated optoisolators.

Transformer T3 samples the circuit's current output. The auxiliary, 12 V winding on T1 ensures no-load starting. The combination of op amp IC5 and the inverting transistor, Q1, square this current signal. The output of Q1 is the CLK signal, which triggers one-half of the one shot, IC4A, to begin the circuit's A/D conversion. The one shots' periods are set to time out within 1/2 cycle of the ac input.

Upon completion of its A/D conversion, IC1's INTR output triggers the other half of the one shot, IC4B, which enables the converter's data outputs. The rising edge of the CLK signal resets the one shot and latches the new conversion value into IC2. The latch, associated driver, and optoisolator trigger a selected triac according to the latest value of the voltage-feedback signal, V_o.

HIGH-VOLTAGE BUCKING REGULATOR

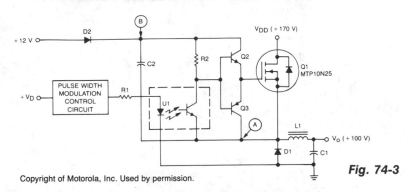

Copyright of Motorola, Inc. Used by permission.

Fig. 74-3

This circuit is basically the classic bucking regulator, except it uses a TMOS N-channel power FET for the chopper and creates its own supply for the gate control.

The unique aspect of this circuit is how it generates a separate supply for the gate circuit, which must be greater than V_{DD}. When power is applied, C2 charges, through D2, to +12 V. At this time, Q1 is off and the voltage at point A is just below zero. When the pulse-modulated signal is applied, the optoisolator transistors, Q2 and Q3, supply a signal to Q1 that turns it on. The voltage at point A then goes to V_{DD}, C2 back-biases D2, and the voltage at point B becomes 12 V above V_{DD}.

After Q1 is turned on, current starts to flow through L1 into C1, increasing until Q1 turns off. The current still wants to flow through L1, so the voltage at point A moves toward negative infinity, but is clamped by D1 to just below zero. Current flows less and less into C1, until Q1 turns on again. Q2 and Q3 drive Q1's gate between the voltages at point A and B, which is always a 12 V swing, so V_{GS} max. is never exceeded. For proper operation, the 12-V supply has to be established before the pulse-width modulator signal is applied.

HIGH-VOLTAGE DC GENERATOR

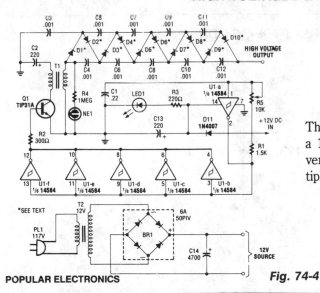

This circuit is fed from a 12-Vdc power supply. The input to the circuit is then amplified to provide a 10,000-Vdc output. The output of the up-converter is then fed into a 10 stage, high-voltage multiplier to produce an output of 10,000 Vdc.

Fig. 74-4

BATTERY-POWERED HIGH-VOLTAGE GENERATOR

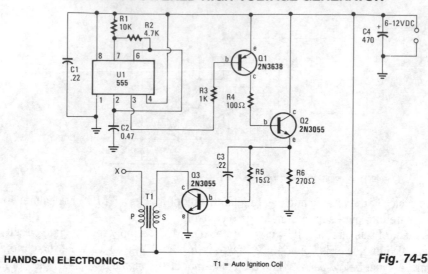

HANDS-ON ELECTRONICS

T1 = Auto Ignition Coil

Fig. 74-5

Output voltage great enough to jump a 1-inch gap can be obtained from a 12-V power source. A 555 timer IC is connected as an astable multivibrator that produces a narrow negative pulse at pin 3. The pulse turns Q1 on for the duration of the time period. The collector of Q1 is direct-coupled to the base of the power transistor Q2, turning it on during the same time period. The emitter of Q2 is direct-coupled through current limiting resistor R5 to the base of the power transistor. Q3 switches on, producing a minimum resistance between the collector and emitter. The high-current pulse going through the primary of high-voltage transformer T1 generates a very high pulse voltage at its secondary output terminal (labeled X). The pulse frequency is determined by the values of R1, R2, and C2. The values given in the parts list were chosen to give the best possible performance when an auto-ignition coil is used for T1.

OPTOISOLATED HIGH-VOLTAGE DRIVER

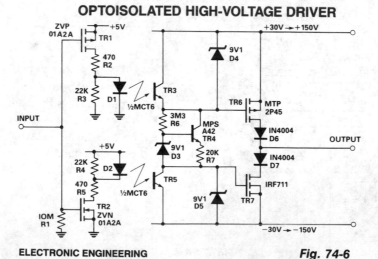

ELECTRONIC ENGINEERING

Fig. 74-6

OPTOISOLATED HIGH-VOLTAGE DRIVER (*Cont.*)

This circuit takes as an input a signal from a 5-V CMOS logic circuit and outputs a high voltage of the same polarity. The high-voltage supply can be varied from ± 30 V to ± 150 V without the need to change circuit components. The input voltage is applied to the gates of transistors TR1 and TR2.

TR3 is optically coupled to D1 as is TR5 to D2. R5 limits the current through D2, while R3 and R4 reduce the affects of leakage current. The light transmitted by D1 turns TR3 on and discharges the gate-source capacitance of TR6, which turns TR6 off. At the same time, TR5 is off and a constant current produced by R6, R7, D3, and TR4 charges the gate-sourced capacitance of TR7, thus turning TR7 on. With TR7 on and TR6 off, the output is pulled close to the lower supply rail. When the input is high, TR1 is off and TR2 is on. Therefore, D2 conducts, which turns on TR5. With TR3 off and TR5 on, TR6 turns on and TR7 off. The output is pulled towards the higher supply rail.

SIMPLE HIGH-VOLTAGE SUPPLY

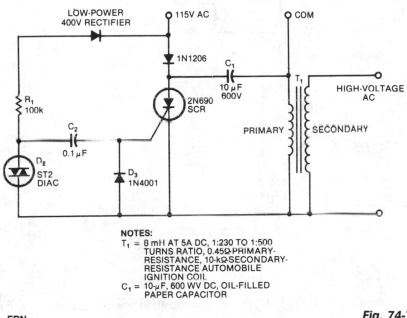

NOTES:
T_1 = 8 mH AT 5A DC, 1:230 TO 1:500
TURNS RATIO, 0.45Ω-PRIMARY-
RESISTANCE, 10-kΩ-SECONDARY-
RESISTANCE AUTOMOBILE
IGNITION COIL
C_1 = 10-μF, 600 WV DC, OIL-FILLED
PAPER CAPACITOR

EDN

Fig. 74-7

This circuit can generate high-voltage pulses with an inexpensive auto ignition coil. Add a rectifier on the output and the circuit produces high-voltage dc. The circuit's input is 115 Vac. During the input's positive half cycle, energy is stored in capacitor C1, which is charged via diode D1 and the primary winding of transformer T1, the coil. The SCR and its trigger circuitry are inactive during this period. During the input's negative half cycle, energy is stored in capacitor C2 until diac D2 reaches its trigger voltage, whereupon D2 conducts abruptly and C2 releases its energy into the SCR's gate. The SCR then discharges C1 into the transformer's primary and ceases to conduct. This store-and-release cycle repeats on the line's positive and negative half cycles, producing high-voltage pulses at the transformer's secondary.

HIGH-VOLTAGE INVERTER

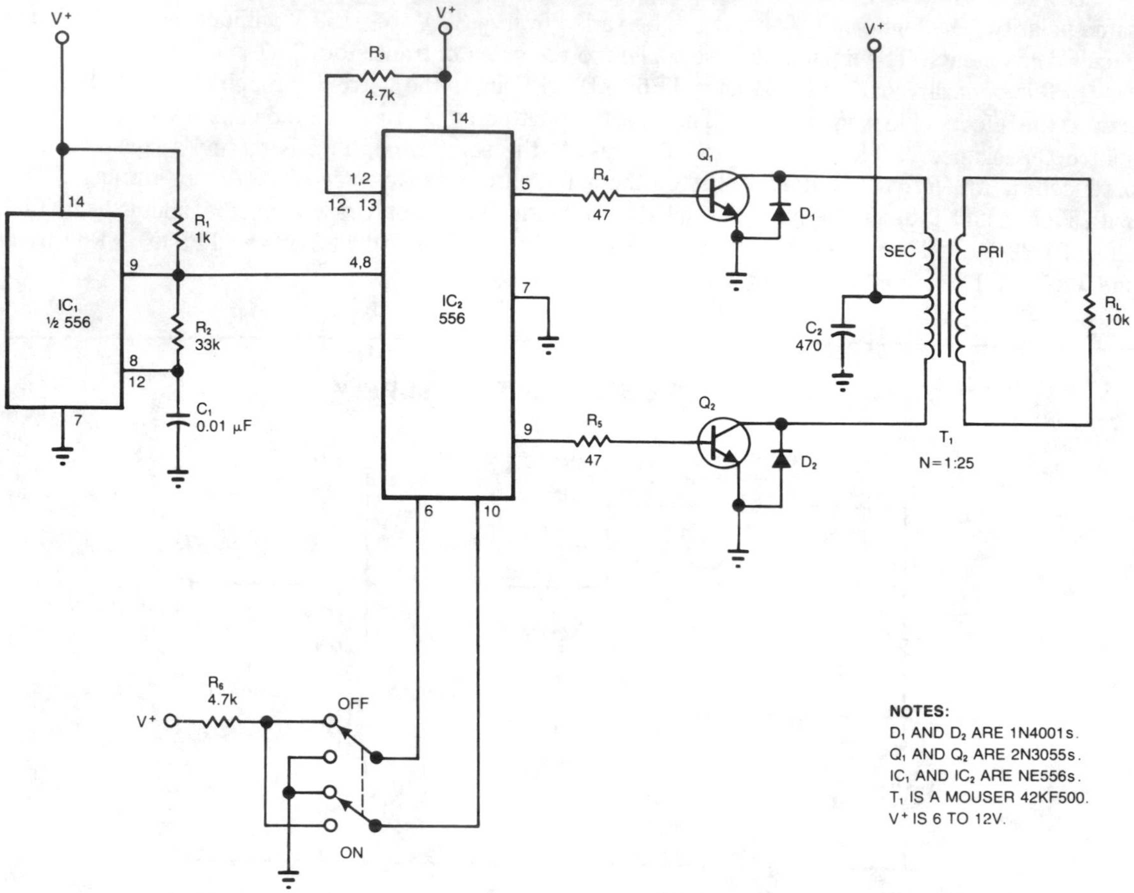

NOTES:
D₁ AND D₂ ARE 1N4001s.
Q₁ AND Q₂ ARE 2N3055s.
IC₁ AND IC₂ ARE NE556s.
T₁ IS A MOUSER 42KF500.
V⁺ IS 6 TO 12V.

EDN

Fig. 74-8

The circuit converts a dc voltage (V^+) to a high-amplitude square wave in the audio-frequency range. The dual timer, IC2, provides an inexpensive alternative to the traditional transformer for providing complementary base drive to the power transistors, Q1 and Q2. You can convert a 6 to 12 V battery output, for example, to an ac amplitude, which is limited primarily by the power rating of transformer T1. Connect timer IC1 as an oscillator to provide a symmetrical square-wave drive to both inputs of IC2. The timing components, R2 and C1, produce a 2.2-kHz output frequency. By connecting half of IC2 in the inverting mode and the other half in noninverting mode, the timer's outputs alternately drive the two transistors. You can operate the audio-output transformer, T1, as a step-up transformer by connecting it backwards— using the output winding as an input. The transformer delivers an output voltage across R_L of $4 \times N \times V^+$V pk-pk, where N is the transformer turns ratio. For the circuit shown, the output swing is $100 \times V^+$V pk-pk.

HIGH-VOLTAGE REGULATOR

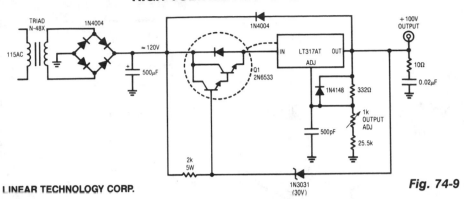

LINEAR TECHNOLOGY CORP.

Fig. 74-9

The regulator delivers 100-V at 100 mA and withstands shorts to ground. Even at 100 V output, the LT317A functions in the normal mode, maintaining 1.2 V between its output and adjustment pin. Under these conditions, the 30-V zener is off and Q1 conducts. When an output short occurs, the zener conducts, forcing Q1's base to 30 V. This causes Q1's emitter to clamp 2 V_{BE}s below V_Z, well within the V_{IN}-V_{OUT} rating of the regulator. Under these conditions, Q1, a high-voltage device, sustains 90 V-V_{CE} at whatever current the transformer specified saturates at 130 mA, while Q1 safely dissipates 12 W. If Q1 and the LT317A are thermally coupled, the regulator will soon go into thermal shutdown and oscillation will commence. This action will continue, protecting the load and the regulator as long as the output remains shorted. The 500-pF capacitor and the 10 Ω/0.02 μF damper aid transient response and the diodes provide safe discharge paths for the capacitors.

CAPACITOR-DISCHARGE HIGH-VOLTAGE GENERATOR

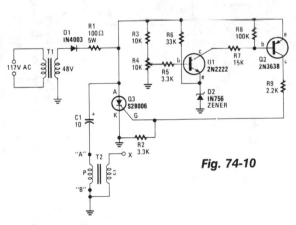

Fig. 74-10

HANDS-ON ELECTRONICS

Stepdown transformer T1 drops the incoming line voltage to approximately 48 Vac which is rectified by diode D1; the resultant dc charges capacitor C1—through current limiting resistor R1—to a voltage level preset by R4. When the voltage on R4's wiper reaches about 8.6 V, Q1 begins to turn on, drawing current through R7 and the base-emitter junction of Q2. Q2 turns on and supplies a positive voltage to the gate of silicon-controlled rectifier Q3. The positive gate voltage causes Q3 to conduct, thereby discharging C1 through the primary winding of step-up transformer T2, which results in a high-voltage arc at output terminal X. The voltage developed at T2's output is determined by the value of C1, the voltage across C1, and the turns ratio of transformer T2. The frequency or pulse rate of the high voltage is determined by the resistance of T1's primary and secondary windings, the value of R1, and the value of C1. The lower the value of each item, the higher the output pulse rate; the peak output voltage will only remain unchanged if C1's value remains unchanged.

485

REMOTELY ADJUSTABLE SOLID-STATE HIGH-VOLTAGE SUPPLY

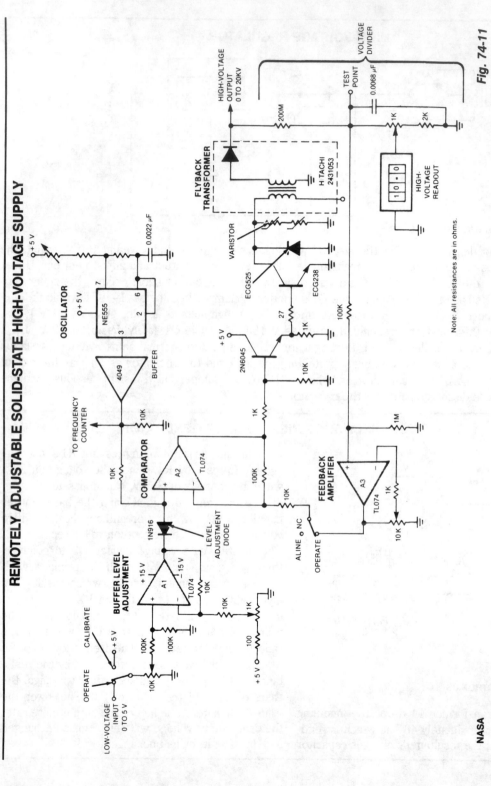

NASA

Fig. 74-11

The output voltage changes approximately linearly up to 20 KV as the input voltage is varied from 0 to 5 V. The oscillator is tuned by a 5-Ω potentiometer to peak the output voltage at the frequency of maximum transformer response between 45 and 55 kHz. The feedback voltage is applied through a 100-KΩ resistor, an op amp, and a comparator to a high-voltage amplifier. A diode and varistors on the primary side of the transformer protect the output transistor. The transformer is a flyback-type used in color-television sets. A feedback loop balances between the high-voltage output and the low-voltage input.

75

Variable Power Supplies

The sources of the following circuits are contained in the Sources section beginning on page 782. The figure number contained in the box of each circuit correlates to the sources entry in the Sources section.

100-kHz MULTIPLE-OUTPUT SWITCHING POWER SUPPLY

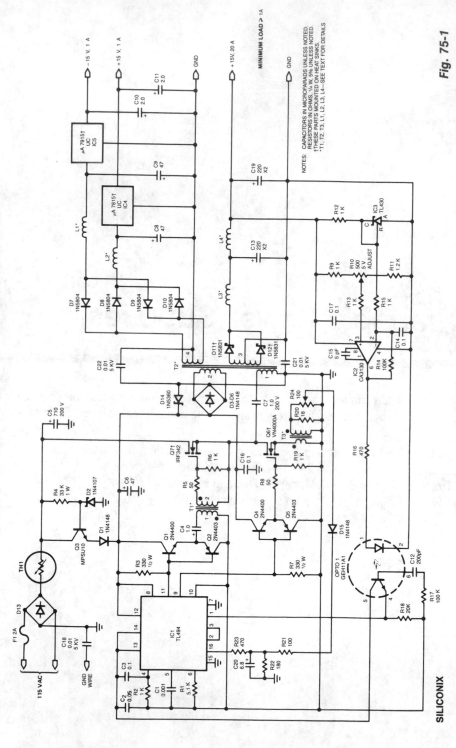

SILICONIX

Fig. 75-1

The power supply uses two VN4000A 400-V MOSPOWER FETs in a half-bridge power switch configuration. Outputs available are +5 V at 20 A and ±15 V (or ±12 V) at 1 A. Since linear three-terminal regulators are used for the low-current outputs, either ±12 V or ±15 V can be made available with a simple change in the transformer secondary windings. A TL494 switching regulator IC provides pulse-width modulation control and drive signals for the power supply. The upper MOSPOWER FET, Q7, in the power switch stage is driven by a simple transformer drive circuit. The lower MOS, Q6, since it is ground referenced, is directly driven from the control IC.

3 – 30 V UNIVERSAL POWER SUPPLY MODULE

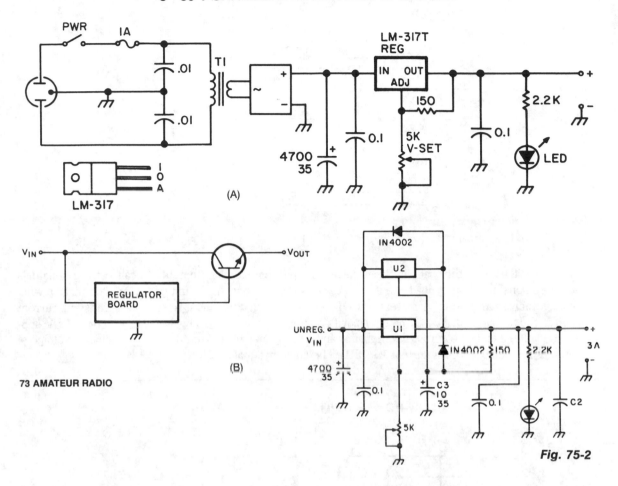

Fig. 75-2

U1, an LM317 adjustable regulator provides short-circuit protection and automatic current limiting at 1.5 A. The input voltage to the regulator is supplied by DB1, a 4-A 100 PIV full-wave bridge rectifier. Capacitor C1 provides initial filtering. U1 provides additional electronic filtering as part of the regulating function. The output level of the regulator is set by trim-pot R1. Bypass capacitors on the input and output of U1 prevent high-frequency oscillation. The current rating of the transformer must be at least 1.8 times the rated continuous-duty output of the supply. This means that a 1.5-A supply should use a 2.7-A transformer. For light or intermittent loads, a smaller 2.0-A transformer should suffice.

Wiring a second LM317, U2, in parallel with U1 is a quick and clean way to increase the current-limiting threshold to 3 A without sacrificing short-circuit protection. When more than 3 A is required, the regulator module can be used to drive the base of one or more pass-transistors (see Fig. 75-2B).

REGULATOR/CURRENT SOURCE

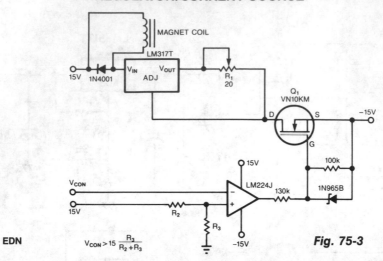

EDN

$$V_{CON} > 15 \frac{R_3}{R_2 + R_3}$$

Fig. 75-3

The circuit powers the load via the regulator's input instead of its output. Because the regulator's output sees constant dummy load R1, it tries to consume a constant amount of current, no matter what the voltage across the actual load really is. Hence, the regulator's input serves as a constant-current source for the actual load. Power the circuit with any one of the commonly available ±15 or ±12 V supplies. The voltage dropped across the regulator and dummy load decreased the total compliance voltage of the circuit. You set the load's current with R1. The current equals 1.25 A/Ω × R1.

LOW-POWER SWITCHING REGULATOR

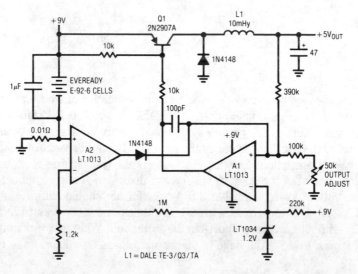

LINEAR TECHNOLOGY CORP.

Fig. 75-4

LOW-POWER SWITCHING REGULATOR (*Cont.*

A simple battery-powered switching regulator provides 5 V out from a 9-V source with 80% efficiency and 50 mA output capability. When Q1 is on, its collector voltage rises, forcing current through the inductor. The output voltage rises, causing A1's output to rise. Q1 cuts off and the output decays through the load. The 100-pF capacitor ensures clean switching. The cycle repeats when the output drops low enough for A1 to turn on Q1. The 1-μF capacitor ensures low battery impedance at high frequencies, preventing *sag* during switching.

VARIABLE VOLTAGE REGULATOR

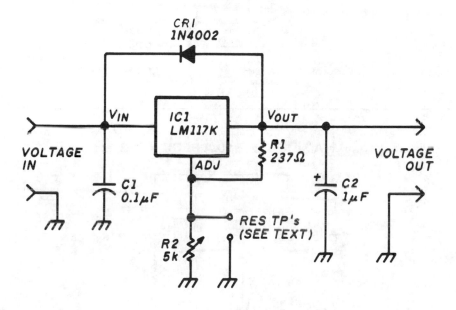

HAM RADIO   *Fig. 75-5*

The variable voltage regulator lets you adjust the output voltage of a fixed dc power supply between 1.2 and 37 Vdc, and will supply the output current in excess of 1.5 A. The circuit incorporates an LM117K three-terminal adjustable output positive voltage regulator in a TO-3 can. Thermal overload protection and short-circuit current-limiting constant with temperature are included in the package. Capacitor C1 reduces sensitivity to input line impedance, and C2 reduces excessive ringing. Diode CR1 prevents C2 from discharging through the IC during an output short.

TRACKING PREREGULATOR

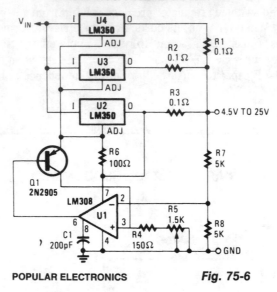

POPULAR ELECTRONICS   *Fig. 75-6*

ADJUSTABLE 10-A REGULATOR

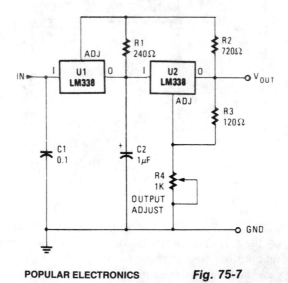

POPULAR ELECTRONICS   *Fig. 75-7*

76

Power Supply Monitors

The sources of the following circuits are contained in the Sources section beginning on page 782. The figure number contained in the box of each circuit correlates to the sources entry in the Sources section.

Power-Supply Balance Indicator
Single-Supply Fault Monitor

POWER-SUPPLY BALANCE INDICATOR

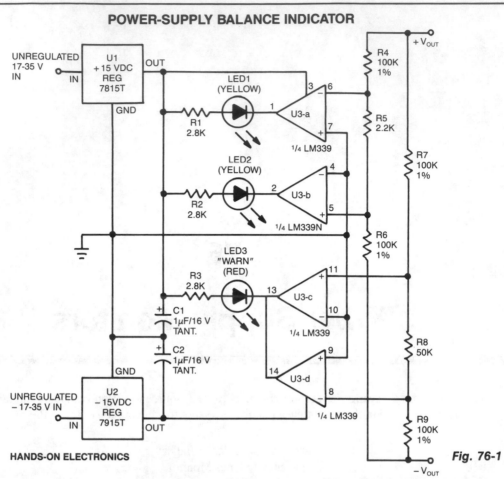

HANDS-ON ELECTRONICS

Fig. 76-1

This circuit uses two comparator pairs from an LM339N quad comparator; one pair drives the yellow positive (+) and negative (−) indicators, the other jointly drives the red warn LED3. The circuit draws its power from the unregulated portion of the power supply. The four comparators get their switching inputs from two parallel resistor-divider strings. Both strings have their ends tied between the power supply's positive and negative output terminals. The first string, consisting of R4, R5, and R6, divides the input voltage in half, with output taps at 0.5%. The other string, made up of R7, R8, and R9, also divides the input voltage in half, with taps at +10%. The 0.5% R4/R5/R6 string drives the two comparators controlling the positive and negative indicators (LED1 and LED2). Their inputs are crossed so that LED2 does not fire until the positive supply is at least 0.5% higher than the negative; the positive indicator does not go off until the negative supply is at least 0.5% higher than the positive—in relative levels. That overlap permits both LEDs to be on when the two supplies are in 1% or better balance. The +10T R7/R8/R9 string drives the other two comparators, which control the warn indicator. If either side of the supply is 10% or more higher than the other, one of the two comparators will switch its output low and light the red LED3—the LM339N has opened-collector outputs, allowing such wired OR connections. The inputs are not crossed, as with the other comparator pair, so there is a band in the middle where neither comparators output is low and the LED remains off.

SINGLE-SUPPLY FAULT MONITOR

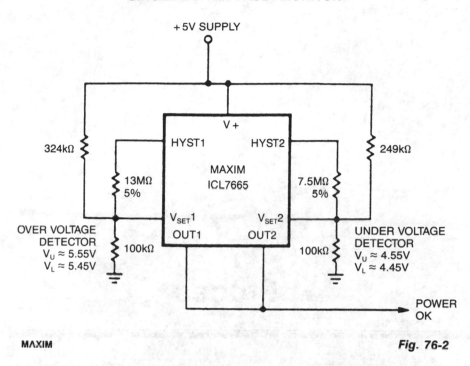

MAXIM

Fig. 76-2

This circuit shows a typical over/under-voltage fault monitor for a single supply. The upper trip points, controlling OUT 1, are centered on 5.5 V with 100 mV of hysteresis (V_U = 5.55 V, V_L − 5.45 V); and the lower trip points, controlling OUT 2, are centered on 4.5 V, also with 100 mV of hysteresis. OUT 1 and OUT 2 are connected together in a wired OR configuration to generate a *power OK* signal.

77

Probes

The sources of the following circuits are contained in the Sources section beginning on page 782. The figure number contained in the box of each circuit correlates to the sources entry in the Sources section.

DIGITAL LOGIC PROBE

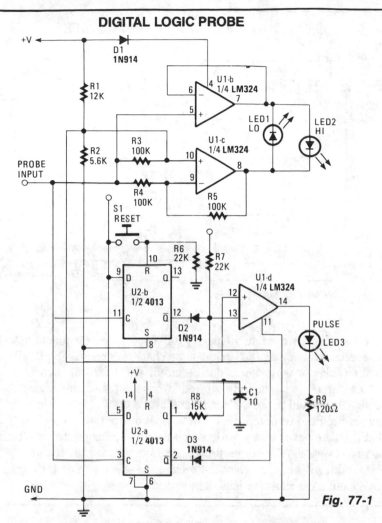

Fig. 77-1

The probe relies on the power supply of the CUT (circuit-under-test). The input to the probe, at probe tip, is fed along two paths. One path flows to the clock inputs of U2a and U2b. The other path feeds both the inverting input of U1c, which is set up as an inverting-mode integrator, and the noninverting input of U1b, which is configured as a noninverting unity-gain amplifier, in a logic-low state.

That low, below the reference set at pin 10, causes U1b's output at pin 7 to become high. With Ub1 outputting low and U1c outputting high, LED1 is forward-biased, and lights. LED2, reverse-biased, remains dark. Suppose that the logic level on the same pin becomes high. That high is applied to pin 5 of U1b, causing its output to be high. LED2 is now forward-biased and lights, while LED1 is reverse-biased and becomes dark.

Assume that a clock frequency is sensed at the probe input; LED1 and LED2 alternately light, and depending on the frequency of the signal, can appear constantly lit. That frequency, which is also applied to the clock input of both flip-flops, causes the Q outputs of U2a and U2b to simultaneously alternate between high and low. Each time that the Q outputs of the two flip-flops decrease, the output of U1d increases, lighting LED3, indicating that a pulse stream has been detected.

LOW INPUT CAPACITANCE BUFFER

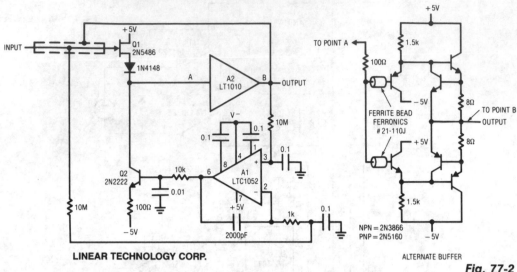

LINEAR TECHNOLOGY CORP.

ALTERNATE BUFFER

Fig. 77-2

Q1 and Q2 constitute a simple, high-speed FET input buffer. Q1 functions as a source follower, with the Q2 current-source load setting the drain-source channel current. The LT1010 buffer provides output drive capability for cables or whatever load is required. The LTC1052 stabilizes the circuit by comparing the filtered circuit output to a similarly filtered version of the input signal. The amplified difference between these signals is used to set Q2's bias, and hence Q1's channel current. This forces Q1's V_{GS} to whatever voltage is required to match the circuit's input and output potentials. The diode in Q1's source line ensures that the gate never forward biases and the 2000-pF capacitor at A1 provides stable loop compensation. The rc network in A1's output prevents it from seeing high-speed edges coupled through Q2's collector-base junction. A2's output is also fed back to the shield around Q1's gate lead, bootstrapping the circuit's effective input capacitance to less than 1 pF.

RF PROBE

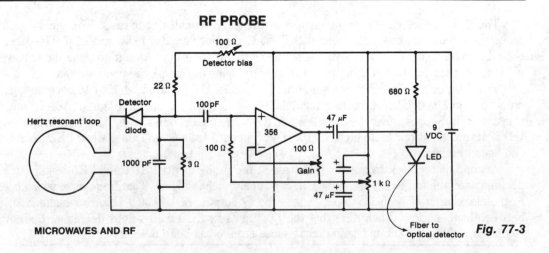

MICROWAVES AND RF

Fig. 77-3

RF PROBE (*Cont.*)

This rf probe is coupled with a fiber-optic cable to the test equipment. It utilizes inexpensive components to improve probe performance at UHF frequencies. The receiving antenna in this probe feeds an envelope-detector diode. After amplification by the LF356 op amp, the low-frequency output modulates the LED, which in turn feeds the optical fiber. The design facilitates the use of a single battery for the op amp, with voltage splitting by means of the 1-KΩ potentiometer, and miniature 47-μF tantalum capacitors to provide decoupling. The gain control is easily adjusted to give the best dynamic range for a specific LED.

CMOS UNIVERSAL LOGIC PROBE

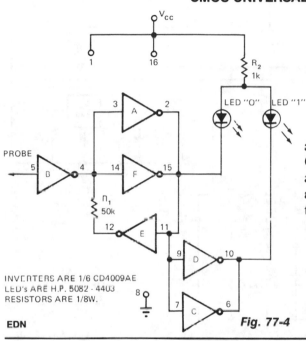

Only the CD4009AE hex buffer, two resistors, and two LEDs are required for a logic probe. CMOS logic probe features $10^{12}\Omega$ input impedance and covers 3 to 15 V range. While LEDs are visible at all voltages, a 1-KΩ pot in place of R2 will allow the user to increase brightness at lower voltages.

INVERTERS ARE 1/6 CD4009AE
LED's ARE H.P. 5082 - 4403
RESISTORS ARE 1/8W.

EDN

Fig. 77-4

4 – 220 V TEST PROBE

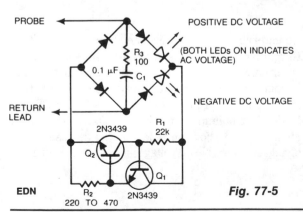

EDN

Fig. 77-5

Using inexpensive components, you can fit a simple probe circuit into a pencil-sized enclosure. When both LEDs are on, the probe indicates the presence of an ac voltage; either LED alone indicates the presence and polarity of a dc voltage. The diode-bridge arrangement allows one-way current source R1, R2, Q1, and Q2 to light either LED (or both) when the probe is activated by a test voltage. Diodes, provide the necessary peak-inverse voltage rating; R3 and C1 provide a spike-suppression network to protect the current-source transistors.

BATTERY-POWERED GROUND-NOISE PROBE

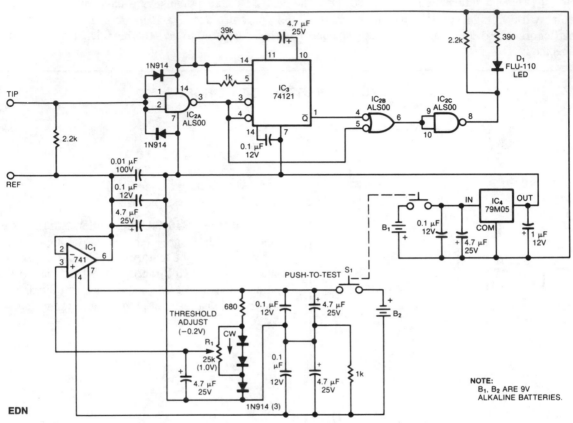

Fig. 77-6

EDN

NOTE:
B₁, B₂ ARE 9V
ALKALINE BATTERIES.

Oscilloscope measurements of ground noise can be unreliable because noise can enter your circuit via the scope's three-pronged power plug. You can avoid this problem by using the ground-noise tester shown. Powered by two 9-V batteries, the circuit dissipates power only while push-to-test switch S1 is depressed. Noise pulses that reach IC2A's switching threshold of about 1.5 to 1.8 V create a logic transition that triggers the monostable multivibrator IC3, which stretches the pulse to produce a visible blink from LED D1. You set the noise reference level by adjusting threshold-adjust potentiometer R1, which lets the circuit respond to minimum pulse amplitudes ranging from about 0 to 1 V. For convenience, you can use a one-turn potentiometer for R1 and calibrate the dial by applying an adjustable dc voltage, monitored by an accurate voltmeter.

FET PROBE

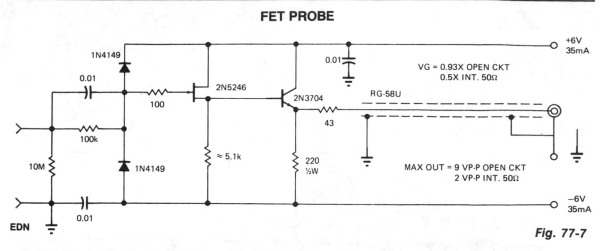

VG = 0.93X OPEN CKT
0.5X INT. 50Ω

RG-58U

MAX OUT = 9 VP-P OPEN CKT
2 VP-P INT. 50Ω

Fig. 77-7

This FET probe has an input impedance of 10 MΩ shunted by 8 pF. Eliminating the protective diodes reduces this impedance to about 4 pF. The frequency response of the probe extends from dc to 20 MHz (−1 dB), although higher frequency operation is possible through optimized construction and use of a UHF-type transistor. Zero dc offset at the output is achieved by selecting a combination of a 2N5246 and source resistor that yields a gate-source bias equal to the V_{BE} of the 2N3704 at approximately 0 V. At medium frequencies, the probe can be used unterminated for near-unity gain; for optimum impedance converter probe high-frequency response, the cable must be terminated into 50 Ω. The voltage gain, when properly terminated, is precisely 0.5 X.

pH PROBE AND DETECTOR

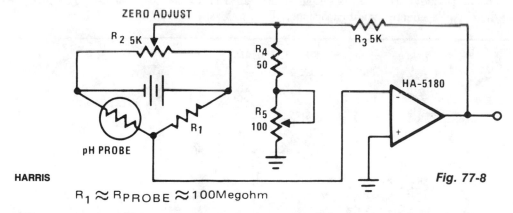

$R_1 \approx R_{PROBE} \approx 100 Megohm$

Fig. 77-8

The greatest sensitivity is achieved if R1 is approximately equal to the probe resistance. The circuit can be *zeroed* with R2, while the full-scale voltage is controlled by R5. The correlation between pH and output voltage might not be linear, which would necessitate a shaping circuit. A calibration scheme, using solutions of known pH, might prove adequate and more reliable over a period of time because of probe variance.

STABILIZED LOW-INPUT CAPACITANCE BUFFER

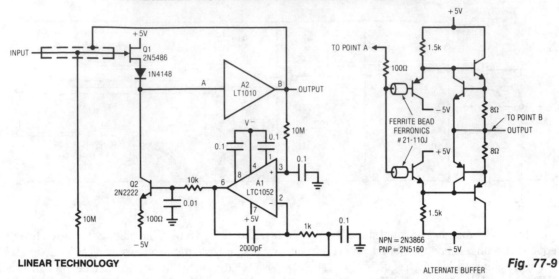

LINEAR TECHNOLOGY

ALTERNATE BUFFER

Fig. 77-9

Q1 and Q2 constitute a simple, high-speed FET input buffer. Q1 functions as a source follower, with the Q2 current source load setting the drain-source channel current. The LT1010 buffer provides output drive capability for cables or whatever load is required. Normally, this open-loop configuration would be quite drifty because there is no dc feedback. The LTC1052 contributes this function to stabilize the circuit. It does this by comparing the filtered circuit output to a similarly filtered version of the input signal. The amplified difference between these signals is used to set Q2's bias, and hence Q1's channel current. Q1's source line ensures that the gate never forward biases, and the 2000 pF capacitor at A1 provides stable loop compensation. The rc network in A1's output prevents it from seeing high-speed edges coupled through Q2's collector-base junction. A2's output is also fed back to the shield around Q1's gate lead, bootstrapping the circuit's effective input capacitance down to less than 1 pF.

RF PROBE

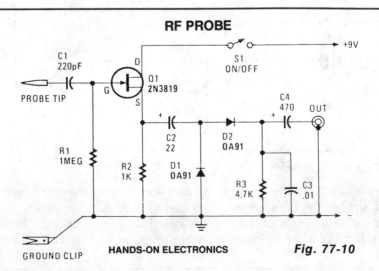

PROBE TIP

GROUND CLIP

HANDS-ON ELECTRONICS

Fig. 77-10

RF PROBE (Cont.)

Transistor Q1-configured as a source-follower buffer stage, offering a bit under unity voltage gain—gives the unit a high-impedance input of about 1 MΩ shunted by about 10 pF, which keeps only minimal loading on the equipment being tested. C1 serves as input dc blocking capacitor. The Q1 output is coupled by C2 to a simple AM detector circuit made up of D1, D2, R3 and C3. Capacitor C4 provides output dc blocking. Total current consumption should be somewhere around 1 mA. The circuit responds to frequencies from 100 kHz to well over 50 MHz.

78

Programmable Amplifiers

The sources of the following circuits are contained in the Sources section beginning on page 782. The figure number contained in the box of each circuit correlates to the sources entry in the Sources section.

Inverting Programmable-Gain Amplifier
Noninverting Programmable Gain Amplifier
Wide Range Digitally Controlled Variable-Gain
   Amplifier

Digitally Programmable Precision Amplifier
Programmable-Gain Differential-Input Amplifier
Programmable Amplifier

INVERTING PROGRAMMABLE-GAIN AMPLIFIER

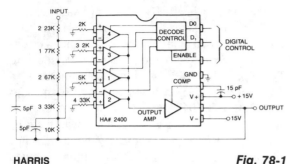

This circuit can be programmed for a gain of 0, −1, −2, −4, or −8. This could also be accomplished with one input resistor and one feedback resistor per channel in the conventional manner, but this would require eight resistors, rather than five.

HARRIS

Fig. 78-1

NONINVERTING PROGRAMMABLE GAIN AMPLIFIER

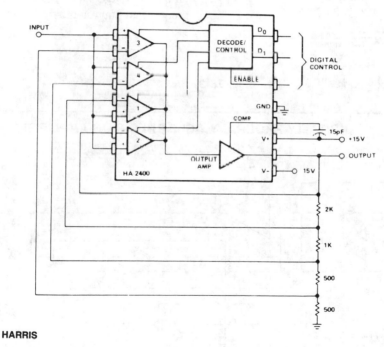

HARRIS

Fig. 78-2

This is a noninverting amplifier configuration with feedback resistors chosen to produce a gain of 0, 1, 2, 4, or 8, depending on the digital control inputs. Comparators at the output could be used for automatic gain selection for auto-ranging meters, etc.

WIDE RANGE DIGITALLY CONTROLLED VARIABLE-GAIN AMPLIFIER

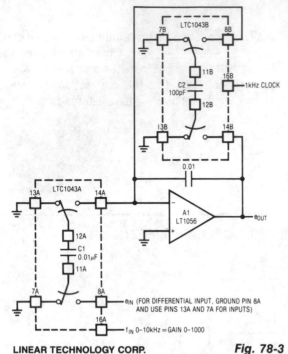

The circuit uses the LTC1043 in a variable gain amplifier which features continuously variable gain, gain stability of 20 ppm/°C, and single-ended or differential inputs. The circuit uses two separate LTC1043s. LTC1043B is continuously clocked by a 1-kHz source, which could also be processor supplied. Both LTC1043s function as the sampled data equivalent of a resistor within the bandwidth set by A1's 0.01-μF value and the switched-capacitor equivalent feedback resistor. The time-averaged current delivered to the summing point by LTC-1043A is a function of the 0.01-μF capacitor's input-derived voltage and the commutation frequency at pin 16. Low-commutation frequencies result in small time-averaged current values, and require a large input resistor. Higher frequencies require an equivalent small input resistor.

LINEAR TECHNOLOGY CORP.   *Fig. 78-3*

DIGITALLY PROGRAMMABLE PRECISION AMPLIFIER

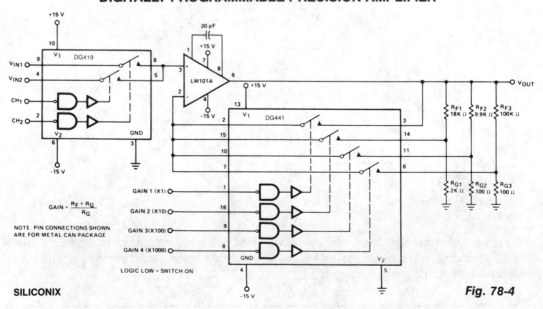

SILICONIX   *Fig. 78-4*

The DG419 *looks* into the high input impedance of the op amp, so the effects of $R_{DS(on)}$ are negligible. The DG441 is also connected in series with R_{IN} and is not included in the feedback dividers, thus contributing negligible error to the overall gain. Because the DG419 and DG441 can handle ± 15 V, the unity gain follower connection, X1, is capable of the full op-amp output range of ± 12 V.

PROGRAMMABLE-GAIN DIFFERENTIAL-INPUT AMPLIFIER

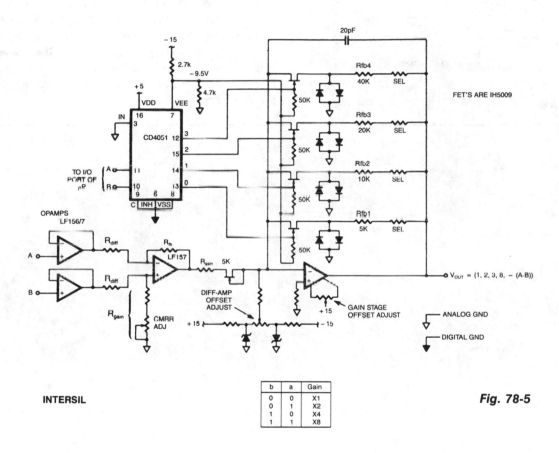

INTERSIL

b	a	Gain
0	0	X1
0	1	X2
1	0	X4
1	1	X8

Fig. 78-5

This programmable gain circuit employs a CD4051 CMOS Analog Multiplexer as a two to four line decoder, with appropriate FET drive for switching between feedback resistors to program the gain to any one of four values.

PROGRAMMABLE AMPLIFIER

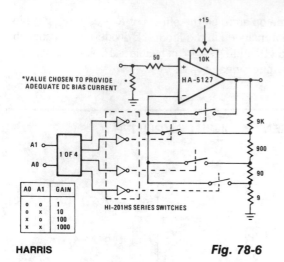

*VALUE CHOSEN TO PROVIDE
ADEQUATE DC BIAS CURRENT

HI-201HS SERIES SWITCHES

A0	A1	GAIN
o	o	1
o	x	10
x	o	100
x	x	1000

HARRIS   *Fig. 78-6*

Often a circuit will be called upon to perform several functions. In these situations, the variable gain configuration of this circuit could be quite useful. This programmable gain stage depends on CMOS analog switches to alter the amount of feedback, and thereby, the gain of the stage. Placement of the switching elements inside the relatively low-current area of the feedback loop, minimizes the effects of bias currents and switch resistance on the calculated gain of the stage. Voltage spikes can occur during the switching process, resulting in temporarily reduced gain because of the make-before-break operation of the switches. This gain loss can be minimized by providing a separate voltage divider network for each level of gain.

79

Protection Circuits

The sources of the following circuits are contained in the Sources section beginning on page 782. The figure number contained in the box of each circuit correlates to the sources entry in the Sources section.

Electric Crowbars
Ac Power-Line Connections Monitor
Line-Voltage Monitor
Power-Failure Alarm
Ac Circuit Breaker
Fast Overvoltage Protector

ELECTRIC CROWBARS

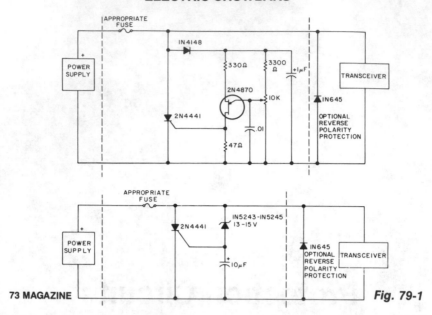

73 MAGAZINE

Fig. 79-1

To avoid grief when using 12-V power supplies with mobile transceivers, especially when there is a short-circuit failure of the series pass transistor, crowbar circuits provide protection by clamping the power line and blowing the fuse within microseconds of an overvoltage condition. It is a good idea to incorporate the crowbar directly into the transceiver. The main difference between the two circuits is that less complex circuit B depends on component tolerances for the exact trigger level, while the circuit A includes a unijunction trigger to permit precise setting of the operating point.

AC POWER-LINE CONNECTIONS MONITOR

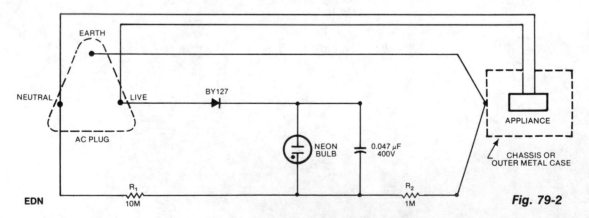

EDN

Fig. 79-2

A continuous glow signifies that everything is normal; a blinking or extinguished neon bulb indicates a broken earth-ground connection, or interchanged neutral and live wires.

LINE-VOLTAGE MONITOR

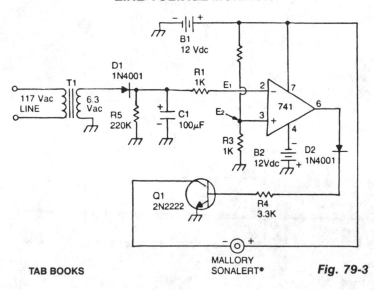

TAB BOOKS

MALLORY
SONALERT®

Fig. 79-3

This circuit uses a type 741 op amp as a voltage comparator. One input of the 741 is connected to a reference voltage (a 12-V battery) through a resistor voltage divider. The potential at the noninverting input of the 741 is approximately 3 V. The inverting input of the op amp comparator is connected to the output of a line-operated 8-V power supply. When the ac power main fails, T1 will no longer be energized, so the charge stored in capacitor C1 will begin to discharge through resistor R5. When the capacitor voltage drops below the reference voltage of 3 V, the output of the comparator becomes high. This output condition will forward bias transistor Q1, causing the Sonalert to sound the alarm. The time constant of the R5/C1 combination is 22 seconds—long enough to prevent noise from triggering the alarm.

POWER-FAILURE ALARM

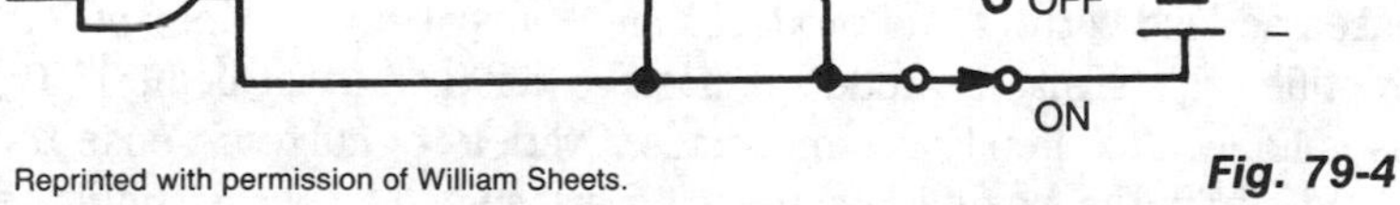

SC 628
SONALERT

IN4007

1.8K

22 μF
150V

1.8K

IN4007

OFF

ON

12V

Fig. 79-4

With power ac off, the alarm sounds when S1 is closed on. The 12-V battery is kept charged when the circuit is plugged in and the switch is left on.

AC CIRCUIT BREAKER

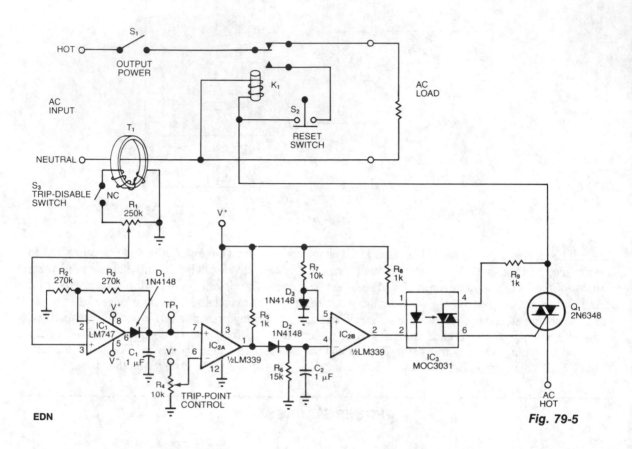

EDN

Fig. 79-5

The adjustable circuit breaker responds in 0.02 s under all conditions—provided you select a fast relay for K1. For moderate overload conditions, it's preferable to use the fuse or the fast-acting breaker. Toroid transformer T1 senses ac load current and produces an ac signal at the wiper of R1, when switch S3 is closed. Diode D1 rectifies this signal to produce a positive voltage at test point TP1. Because R1 allows you to calibrate this voltage, the circuit accommodates a variety of current-sense transformers. To calibrate the trip threshold, apply the maximum expected overload and adjust R1 until the TP1 voltage is 0.7 V below the positive saturation level for IC1. Then adjust R4 for the desired trip point. To reset the circuit breaker after it has tripped, open S1 or S2.

FAST OVERVOLTAGE PROTECTOR

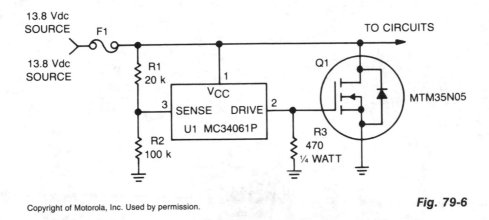

Fig. 79-6

This circuit protects expensive portable equipment against all types of improper hookups and environmental hazards that could cause an overvoltage condition. It operates very quickly and does not latchup, that is, it recovers when the overvoltage condition is removed. In contrast, SCR overvoltage circuits can latch and do not recover, unless the power is removed.

Here, U1 senses an overvoltage condition when the drop across R1 exceeds 2.5 V. This causes U1 to apply a positive signal to the gate of Q1, turning it on and shorting the line going to the external circuits. Fuse 1 opens if the transient condition lasts long enough to exceed the i^2t rating.

80

Proximity Sensors

The sources of the following circuits are contained in the Sources section beginning on page 782. The figure number contained in the box of each circuit correlates to the sources entry in the Sources section.

Capacitive Sensor Alarm
UHF Movement Detector
Proximity Switch
SCR Proximity Alarm
Proximity Sensor

CAPACITIVE SENSOR ALARM

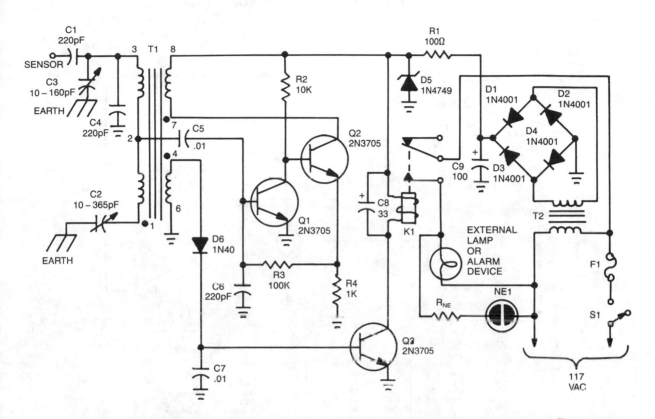

Fig. 80-1

The unit is constructed around a balanced-bridge circuit, using both capacitance and inductance. The bridge consists of capacitors C2 and C3, and the center-tapped winding of T1. One end of the bridge is coupled to ground by C4, while capacitance changes are introduced through C1. A small capacitance change unbalances the bridge and produces an ac signal at the base of Q1. Transistors Q1 and Q2 are connected to form a modified-Darlington amplifier. The collector load for Q2 is a separate winding of T1 that is connected out-of-phase with the incoming ac signal. That produces a large, distorted signal each time the bridge is unbalanced.

The distorted signal is taken from the bridge circuit by a third winding of transformer T1. That signal is then rectified by D6 and applied as a dc signal to the base of Q3. The applied signal energizes the relay, K1, as soon as the unbalanced condition occurs, and the relay drops out as soon as the circuit balance is restored. Of course, for normal alarm use, the relay should be made self-latching, so that the alarm condition remains in effect until the system is reset.

An audible alarm, such as a bell or klaxon horn, can be operated from the relay. If a silent alarm is needed, a light bulb can be used. Transformer T1 can be purchased as part #6182 from: Pulse Engineering, P.O. Box 12235, San Diego, CA 92112.

UHF MOVEMENT DETECTOR

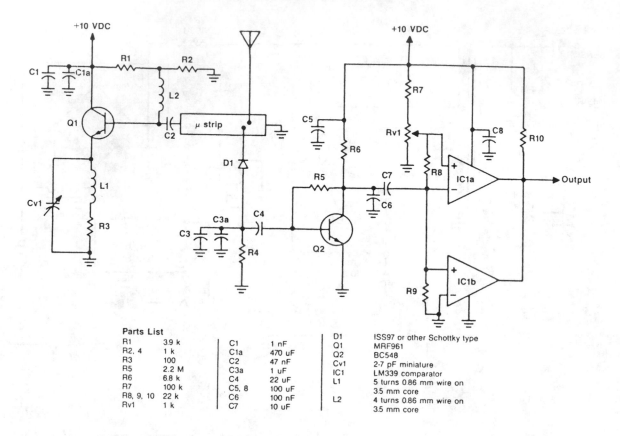

Parts List

R1	3.9 k	C1	1 nF	D1	ISS97 or other Schottky type	
R2, 4	1 k	C1a	470 uF	Q1	MRF961	
R3	100	C2	47 nF	Q2	BC548	
R5	2.2 M	C3a	1 uF	Cv1	2-7 pF miniature	
R6	6.8 k	C4	22 uF	IC1	LM339 comparator	
R7	100 k	C5, 8	100 uF	L1	5 turns 0.86 mm wire on 3.5 mm core	
R8, 9, 10	22 k	C6	100 nF			
Rv1	1 k	C7	10 uF	L2	4 turns 0.86 mm wire on 3.5 mm core	

D. Huisman, RF Design, December 1986, p. 41.

Fig. 80-2

The oscillator is a standard UHF design which delivers about 10 mW at 1.2 GHz. R1 and R2 bias the base of Q1 to 1.2 V via L2. Collector current is set by R3 to about 30 mA. C2 couples the base of Q1 to the stripline circuit. Tuning is provided by CV1, and C1 plus C1a decouple the collector. R2 and R3 are not decoupled, since this could cause instability.

Q2 is a simple one-transistor amplifier. C4 and C7 reduce gain below 1.5 and above 100 Hz; the remaining band of frequencies is amplified and passed on to the level detector. Two comparators of IC1 provide level detection. The trigger voltage is set by R7, Rv1, R8, and R9; it is adjustable from 8 to 60 mV by Rv1.

Positive voltage swings above the trigger level cause the IC1a output to become low, while negative swings cause IC1b to become low. C8 decouples IC1 from the power supply, and R10 is a pull-up resistor for the open collector output of IC1.

PROXIMITY SWITCH

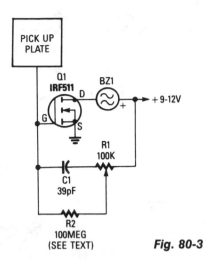

PICK UP PLATE

Q1
IRF511

BZ1

+ 9-12V

R1
100K

C1
39pF

R2
100MEG
(SEE TEXT)

Fig. 80-3

POPULAR ELECTRONICS/HANDS-ON ELECTRONICS

A 3-×-3-inch piece of circuit board, or similar size metal object which functions as the pick-up sensor, is connected to the gate of Q1. A 100-MΩ resistor, R2, isolates Q1's gate from R1, allowing the input impedance to remain very high. If a 100-MΩ resistor cannot be located, just tie five 22-MΩ resistors in series and use that combination for R2. In fact, R2 can be made even higher in value for added sensitivity.

Potentiometer R1 is adjusted to where the piezo buzzer just begins to sound off and then carefully backs off to where the sound ceases. Experimenting with the setting of R1 will help in obtaining the best sensitivity adjustment for the circuit. Resistor R1 can be set to where the pick-up must be contacted to set off the alarm sounder. A relay or other current-hungry component can take the place of the piezo sounder to control most any external circuit.

SCR PROXIMITY ALARM

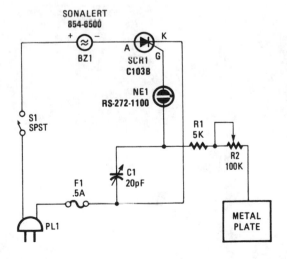

SONALERT
854-6500

BZ1

SCR1
C103B

NE1
RS-272-1100

S1
SPST

R1
5K

R2
100K

F1
.5A

C1
20pF

PL1

METAL PLATE

HANDS-ON ELECTRONICS

Fig. 80-4

517

PROXIMITY SENSOR

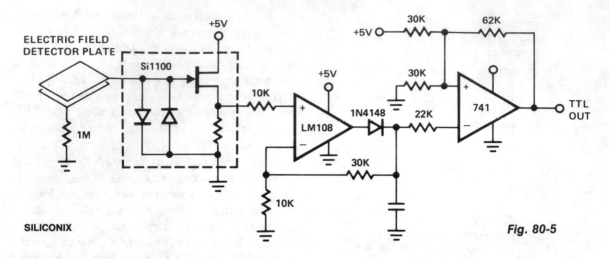

SILICONIX

Fig. 80-5

The Si1100 series circuit input is connected to a capacitive field sensor—possibly a piece of double-sided circuit board. Any induced voltage change on the plate is fed to the input of the peak detector section of the op-amp circuit. The Schmitt trigger monitors the voltage across the capacitor and changes its output state when the capacitor voltage crossed the 2.5-trigger point. The output from the Schmitt trigger switches between 0 and 5 V and is microprocessor compatible for sensor applications, such as computer-controlled intruder alarms.

82

Ramp Generators

The sources of the following circuits are contained in the Sources section beginning on page 782. The figure number contained in the box of each circuit correlates to the sources entry in the Sources section.

LOGIC PULSER

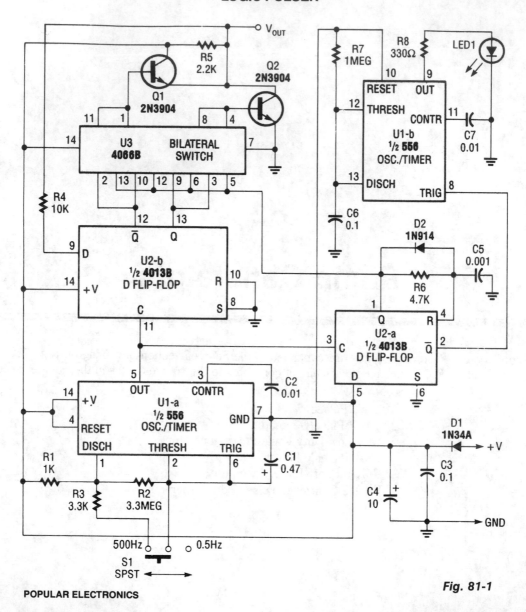

POPULAR ELECTRONICS

Fig. 81-1

The pulser generates pulses at a user-selected frequency of 0.5 or 500 Hz, with a pulse width of about 5 ms. If the input to be pulsed is already being driven high or low by another output, the pulser automatically pulses the input to the opposite logic state. The pulser is powered by the circuit under test, and operates from supplies of from +5 to +15 Vdc.

300-V PULSE GENERATOR

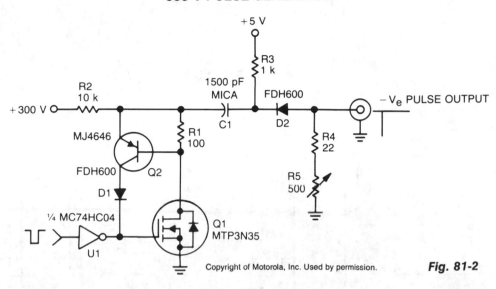

Copyright of Motorola, Inc. Used by permission.

Fig. 81-2

In this TMOS pulser, a negative-going pulse is applied to U1, a high-speed CMOS buffer, which directly drives the gate of Q1, an MTP3N35. If only a 100-V pulse is required, the MTA6N10 can be used. The pulse output across R2 is differentiated by R3/C1 and appears as a negative-going spike at the output terminal.

VERY LOW DUTY-CYCLE PULSE GENERATOR

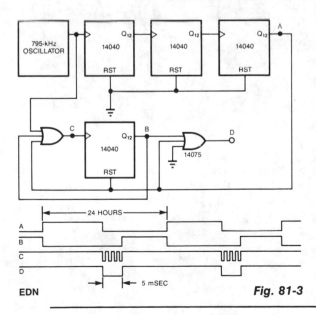

EDN

Fig. 81-3

Using a precision oscillator and a few CMOS counters, you can build a precise, very low duty-cycle pulse generator. You can add as many counters as you desire to make the period as long as you wish. The circuit will generate a pulse about 5 ms long every 24 hours—one 5.9 −6% duty cycle.

WIDE-RANGING PULSER

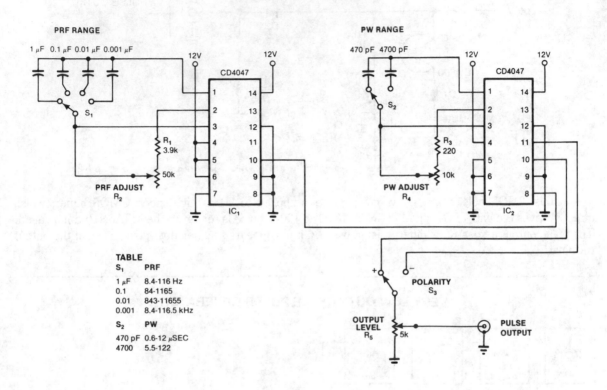

TABLE

S₁	PRF
1 μF	8.4-116 Hz
0.1	84-1165
0.01	843-11655
0.001	8.4-116.5 kHz

S₂	PW
470 pF	0.6-12 μSEC
4700	5.5-122

Fig. 81-4

An output pulse's characteristics depend upon two multivibrator's timing components. IC1's free-running astable-mode frequency sets the pulse's prf, whereas the pulse's width comes from IC2's monostable operation.

CMOS SHORT - PULSE GENERATOR

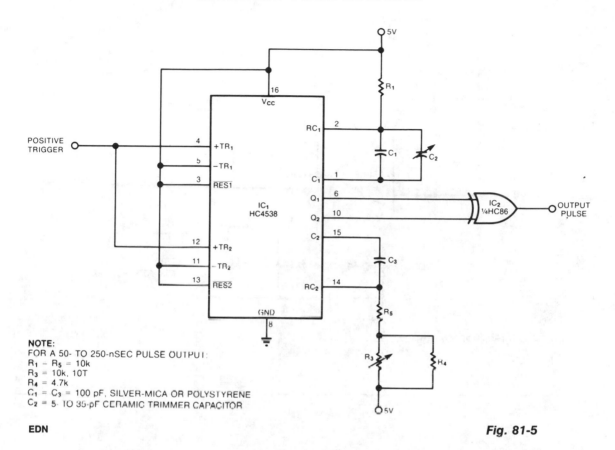

NOTE:
FOR A 50- TO 250-nSEC PULSE OUTPUT:
$R_1 - R_5 = 10k$
$R_3 = 10k, 10T$
$R_4 = 4.7k$
$C_1 = C_3 = 100 pF,$ SILVER-MICA OR POLYSTYRENE
$C_2 = 5-$ TO 35-pF CERAMIC TRIMMER CAPACITOR

EDN

Fig. 81-5

Comprising two low-power, CMOS chips, the pulse generator produces a precise pulse width in the 50 to 500 ns range. IC1 is a dual monostable multivibrator (one shot) in which each positive trigger pulse initiates simultaneous positive output pulses at pins 6 and 10. In response, XOR gate IC2 produces a positive pulse whose duration is equal to the difference between the two input-pulse durations. Section 1 of the one shot generates an approximate 1-μs reference pulse—shorter pulses are more susceptible to manufacturing variations caused by parasitic layout capacitance. Variable capacitor C2 lets you adjust this pulse width. Section 2 of the one shot generates a variable-length pulse; you adjust its width by using potentiometer R3. Resistors R4 and R5 set the output pulse's maximum and minimum width, respectively. Because the XOR gate's rise and fall times are about 20 ns for reasonable values of load capacitance, you should calibrate the circuit using C2 for a minimum output-width of 50 ns.

VOLTAGE CONTROLLER/PULSE-WIDTH GENERATOR

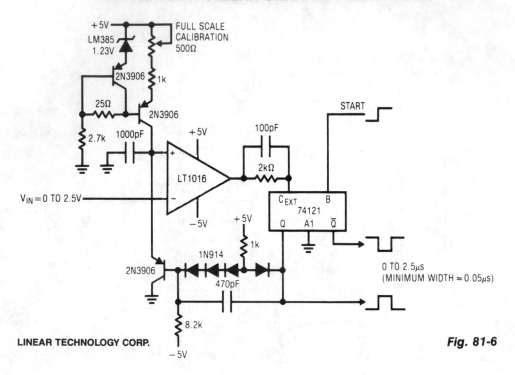

LINEAR TECHNOLOGY CORP.

Fig. 81-6

82

Ramp Generators

The sources of the following circuits are contained in the Sources section beginning on page 782. The figure number contained in the box of each circuit correlates to the sources entry in the Sources section.

Accurate Ramp Generator
Integrator/Ramp Generator with Initial Condition Reset

ACCURATE RAMP GENERATOR

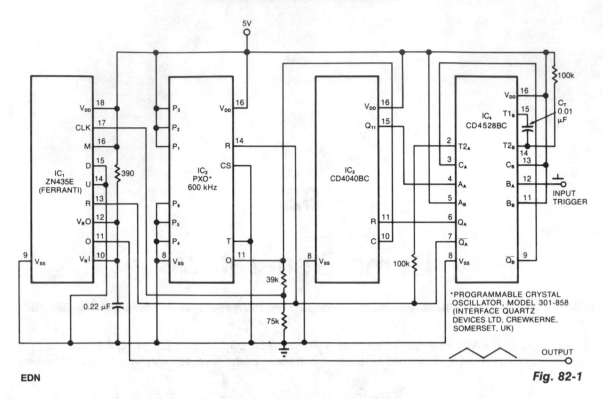

EDN

Fig. 82-1

The ramp generator, an inexpensive alternative to commercial function generators, provides a more linear and repeatable output than conventional analog integrators. The circuit provides a triangle waveform in burst mode; in this case, two cycles of 10.24 ms each per input trigger pulse. IC4 is a dual monostable multivibrator (one shot) in which the A side is configured as a latch (see *Multivibrator IC performs extra tasks*, EDN, September 6, 1984, p. 232). The rising edge of each input pulse triggers the B side, producing at pin 9 an output pulse whose duration depends on the timing capacitor's, C_T, value—A 0.01-μF value gives a 500 μs pulse. This output provides a reset to the A side latch. While the latch is reset with Q_A high, Q_A low, the other three ICs are active. The P1 through P6 connections, as shown, set oscillator IC2's frequency to 50 kHz at pin 11.

Counter IC3 counts upward. The output at pin 11 of multifunction converter IC1 ramps up to full-scale, reverses, ramps down to zero, and then repeats this sequence of events. As this output completes its second cycle, IC3 reaches a count of 1024, causing the Q11 output to become high and toggle the IC4 latch. The resulting change of state on Q_A and Q_A resets the other three ICs, terminating further activity until the arrival of the next input trigger pulse. IC2 is included for its synchronous-reset capability, and it therefore drives the internal clock of IC1, which cannot be synchronously reset. Still IC2 can be omitted in some applications. The circuit operates from a 5-V supply. You can modify the output by changing IC2's frequency and IC3's output connection.

INTEGRATOR/RAMP GENERATOR WITH INITIAL CONDITION RESET

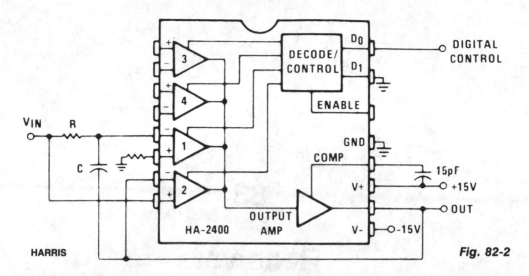

HARRIS

Fig. 82-2

Channel 1 is wired as a conventional integrator, and channel 2 as a voltage follower. When channel 2 is switched on, the output will follow V_{IN} and C will discharge to maintain 0 V across it. When channel 1 is then switched on, the output will initially be at the instantaneous value of V_{IN}, and then will commence integrating towards the opposite polarity. This circuit is particularly suitable for timing ramp generation using a fixed dc input. Many variations, such as building programmable time constant integrators, are possible.

83

Receivers

The sources of the following circuits are contained in the Sources section beginning on page 782. The figure number contained in the box of each circuit correlates to the sources entry in the Sources section.

AM Radio
FM Tuner
FM MPX/SCA Receiver
Narrow-Band FM Receiver
Low-Cost Line Receiver

FSK Data Receiver
Simple Ham-Band Receiver
Digital Data Line Receiver
Integrated AM Receiver

AM RADIO

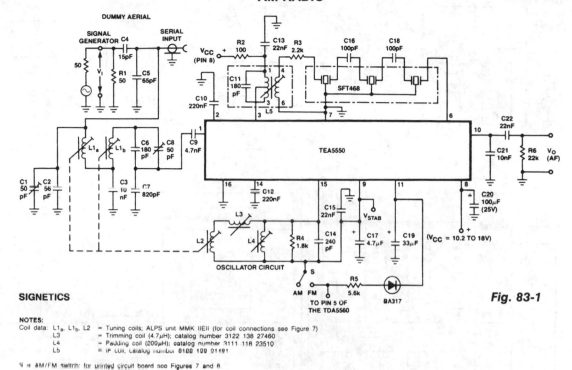

SIGNETICS

NOTES:
Coil data: L1$_a$, L1$_b$, L2 = Tuning coils; ALPS unit MMK IIEII (for coil connections see Figure 7)
L3 = Trimming coil (4.7µH); catalog number 3122 138 27460
L4 = Padding coil (200µH); catalog number 3111 118 23510
L5 = IF coil, catalog number 0108 109 01191

S is AM/FM switch; for printed circuit board see Figures 7 and 8.

Fig. 83-1

This circuit diagram is for a double-tuned, AM-channel, in-car radio receiver using the TEA5550.

FM TUNER

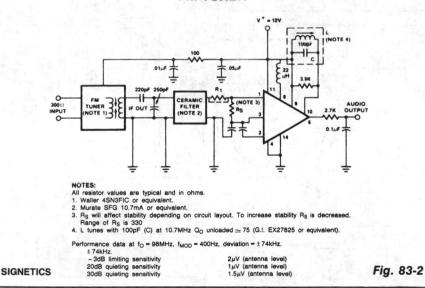

NOTES:
All resistor values are typical and in ohms.
1. Waller 4SN3FIC or equivalent.
2. Murate SFG 10.7mA or equivalent.
3. R$_S$ will affect stability depending on circuit layout. To increase stability R$_S$ is decreased.
   Range of R$_S$ is 330
4. L tunes with 100pF (C) at 10.7MHz Q$_O$ unloaded ≃ 75 (G.I. EX27825 or equivalent).

Performance data at f$_O$ = 98MHz, f$_{MOD}$ = 400Hz, deviation = ± 74kHz.
    ± 74kHz.
    – 3dB limiting sensitivity   2µV (antenna level)
    20dB quieting sensitivity   1µV (antenna level)
    30dB quieting sensitivity   1.5µV (antenna level)

SIGNETICS

Fig. 83-2

FM MPX/SCA RECEIVER

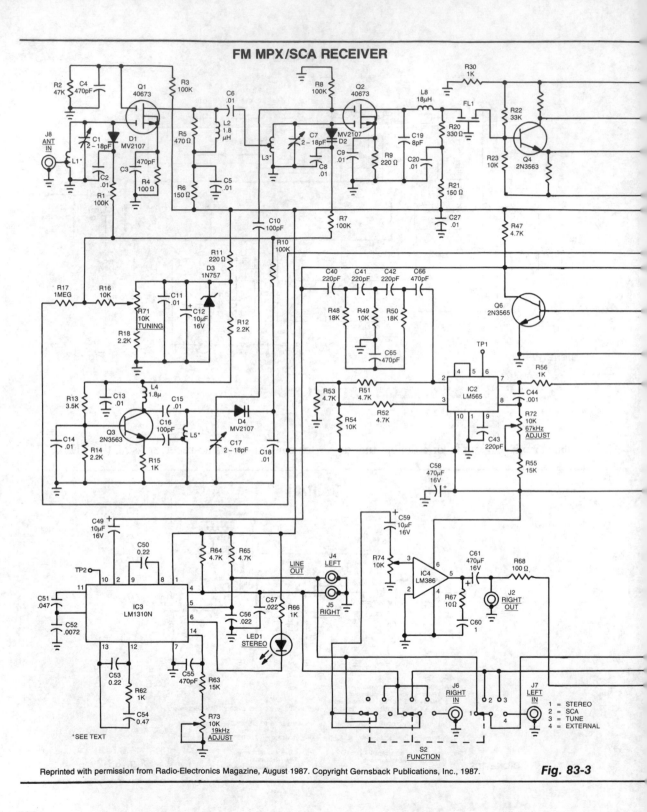

Reprinted with permission from Radio-Electronics Magazine, August 1987. Copyright Gernsback Publications, Inc., 1987.

Fig. 83-3

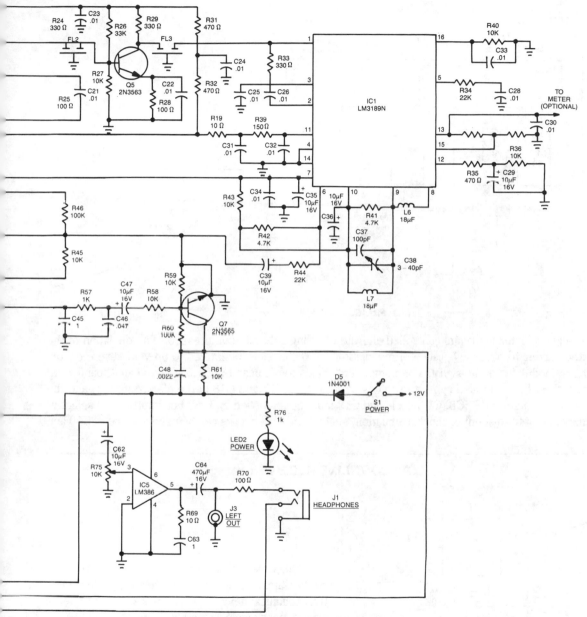

This receiver is capable of better than 1.5 μ VIHF sensitivity and uses MOSFET front-end circuitry with varactors to eliminate conventional bulky tuning capacitors. It also features high dynamic range, ceramic i-f filters requiring no alignment, and a quadrature-type detector with excellent limiting and AM rejection capability. The receiver operates from nominal 12-V supply. The kit is available from North Country Radio, P.O. Box 53, Wykagyl Station, NY 10804.

NARROW-BAND FM RECEIVER

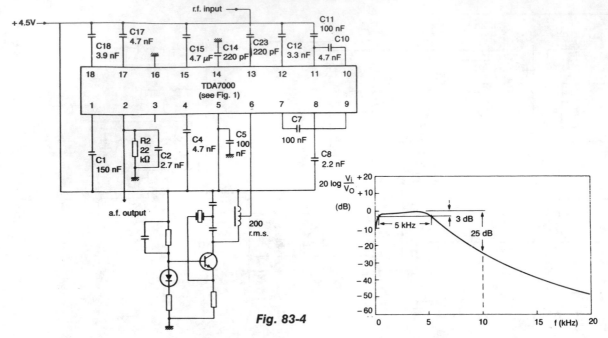

Fig. 83-4

The local oscillator is crystal-controlled and the i-f swing is hardly compressed. The deviation of the transmitted carrier frequency, because of modulation, must therefore be limited to prevent severe distortion of the demodulated audio signal. The component values result in an i-f of 4.5 kHz and an i-f bandwidth of 5 kHz. If the i-f is multiplied by N, the values of capacitors C17 and C18 in the all-pass filters, and the values of filter capacitors C7, C8, C10, C11, and C12 must be multiplied by $1/N$. For improved i-f selectivity to achieve greater adjacent channel attenuation, second-order networks can be used in place of C10 and C11.

LOW-COST LINE RECEIVER

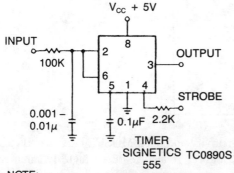

NOTE:
ALL RESISTOR VALUES ARE IN OHMS.

SIGNETICS   **Fig. 83-5**

This timer makes an excellent line receiver for control applications involving relatively slow electromechanical devices. It can work without special drivers over single, unshielded lines.

FSK DATA RECEIVER

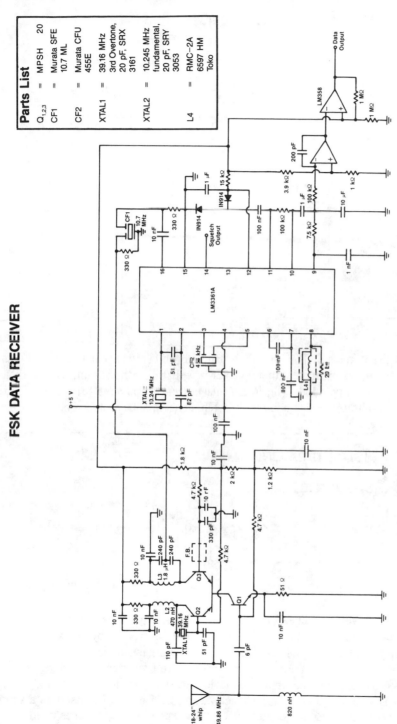

Fig. 83-6

M. Lee, "A Simple FSK Data Receiver," RF Design, March 1987, p. 46.

The various signal frequencies are obtained for an incoming carrier centered at 49.86 MHz. The receiver employs double conversion, with i-fs at 10.7 MHz and 455 kHz. Ceramic filters are used in both i-fs for selectivity and reduced-coil count. A quadrature detector is used to recover the baseband data, and an integrator and Schmitt trigger are used to filter the demodulated output. Also included is a squelch circuit that functions as a status line, and the open-collector output switches high when a signal is received. The LM3361A functions as the 2nd LO, 2nd mixer, limiting i-f, quadrature detector, and squelch; yet, it consumes less than 4 mA from a 5-V logic supply. The entire receiver requires approximately 10 mA.

533

SIMPLE HAM-BAND RECEIVER

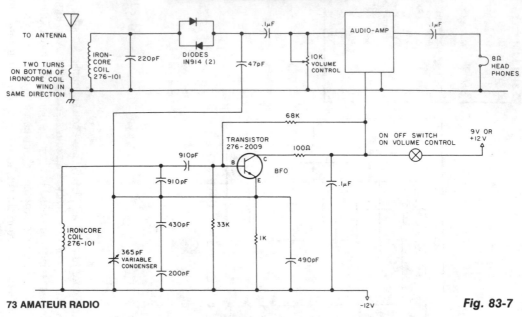

73 AMATEUR RADIO

Fig. 83-7

This circuit is configured for the 80m band. The 365-pF, broadcast-band variable capacitor should have a vernier drive with a six-to-one ratio, which makes tuning easier by separating the stations on the dial. A good antenna and ground are also recommended. The Radio Shack iron-core chokes (276-101), used in the bfo part of the circuit, can be calibrated by listening for the bfo signal in a calibrated receiver.

DIGITAL DATA LINE RECEIVER

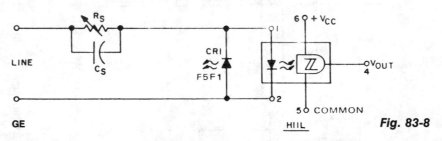

GE   H11L   **Fig. 83-8**

When digital data is transmitted over long lines (longer than 1 meter), proper transfer is often disturbed by the parasitic effects of ground level shifts and ground loops, as well as by extraneous noise picked up along the way. An optocoupler, such as the H11L, combining galvanic isolation to minimize ground loop currents and their concomitant common-mode voltages, with predictable switching levels to enhance noise immunity, can significantly reduce erratic behavior. Resistor R_S is programmed for the desired switching threshold, C_S is an optional speed-up capacitor, and CR1 is an LED used as a simple diode to provide perfect line balance and a discharge path for C_S if the speed-up capacitor is used.

INTEGRATED AM RECEIVER

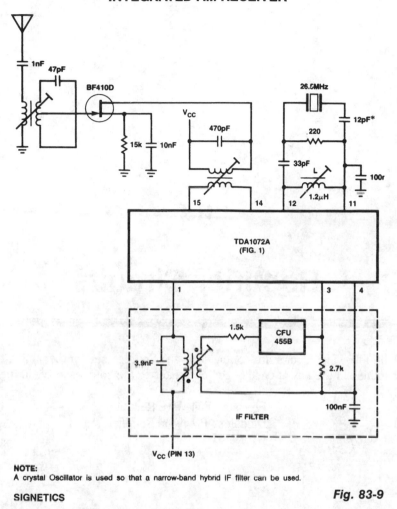

NOTE:
A crystal Oscillator is used so that a narrow-band hybrid IF filter can be used.

SIGNETICS

Fig. 83-9

This circuit has aerial and local oscillator circuits for a 27-MHz receiver for remote control of garage doors, projectors, curtains, etc.

84

Rectifier Circuits

The sources of the following circuits are contained in the Sources section beginning on page 782. The figure number contained in the box of each circuit correlates to the sources entry in the Sources section.

Precision Full-Wave Rectifier
Diodeless Precision Rectifier

PRECISION FULL-WAVE RECTIFIER

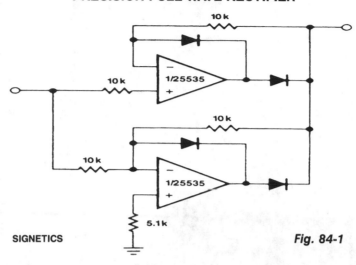

SIGNETICS

Fig. 84-1

This circuit provides accurate full-wave rectification. The output impedance is low for both input polarities, and the errors are small at all signal levels. Note that the output will not sink heavy currents, except a small amount through the 10-KΩ resistors. Therefore, the load applied should be referenced to ground or a negative voltage. The reversal of all diode polarities will reverse the polarity of the output. Since the outputs of the amplifiers must slew through two diode drops when the input polarity changes, the 741-type devices give 5% distortion at about 300 Hz.

DIODELESS PRECISION RECTIFIER

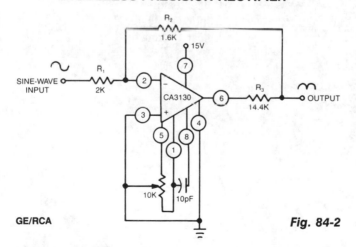

GE/RCA

Fig. 84-2

A CA3130 BiMOS op amp, acts as an attenuator for positive inputs and as a conventional op amp for negative signals. With 1-V rms input and a circuit gain of 0.8, its frequency response is −1% at 60 kHz and −1 dB at 300 kHz.

85

Resistance/Continuity Meters

The sources of the following circuits are contained in the Sources section beginning on page 782. The figure number contained in the box of each circuit correlates to the sources entry in the Sources section.

Cable Tester
Continuity Tester
Linear Ohmmeter

CABLE TESTER

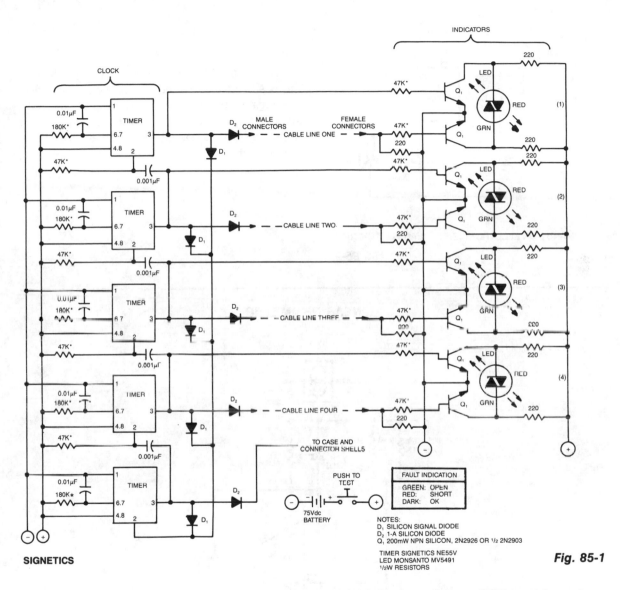

Fig. 85-1

This compact tester checks cables for open-circuit or short-circuit conditions. A differential transistor pair at one end of each cable line remains balanced as long as the same clock pulse generated by timer IC appears at both ends of the line. A clock pulse, just at the clock end of the line, lights a green LED, and a clock pulse, only at the other end, lights a red LED.

CONTINUITY TESTER

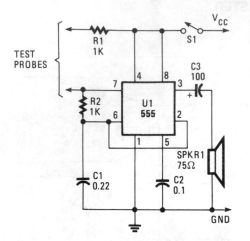

The continuity tester feeds a voltage through the positive probe to the circuit-under-test, while the negative probe serves as the return line. Voltage that returns to the tester through the negative probe triggers the circuit, giving an audible indication of continuity.

HANDS-ON ELECTRONICS/POPULAR ELECTRONICS

Fig. 85-2

LINEAR OHMMETER

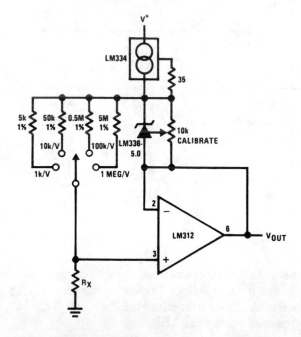

NATIONAL SEMICONDUCTOR CORP.   **Fig. 85-3**

86

Rf Amplifiers

The sources of the following circuits are contained in the Sources section beginning on page 782. The figure number contained in the box of each circuit correlates to the sources entry in the Sources section.

1296-MHz Solid-State Power Amplifier
10 dB-Gain Amplifier
2 – 30 MHz Amplifier
450-MHz Common-Gate Amplifier
Rf Wideband Adjustable AGC Amplifier

1-MHz Meter-Driver Amplifier
5-W 150-MHz Amplifier
UHF-TV Preamplifier
60-W 225 – 400 MHz Amplifier

1296-MHz SOLID-STATE POWER AMPLIFIER

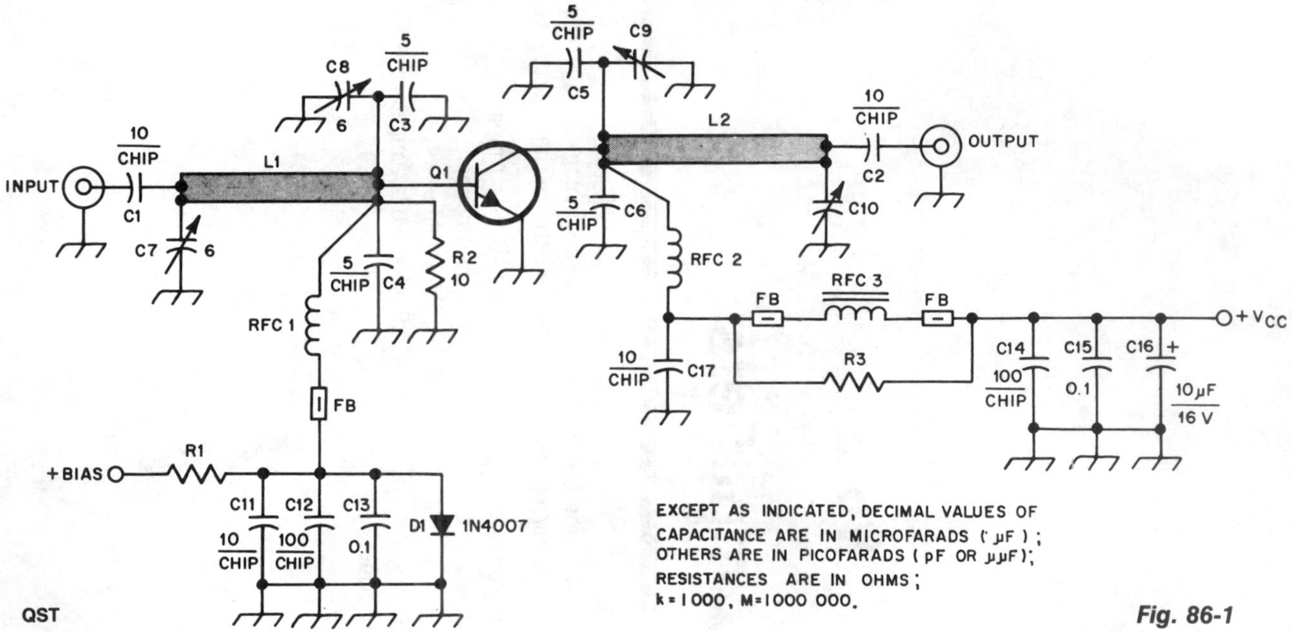

Fig. 1—Schematic diagram of the NEL1306 and NEL1320 1296-MHz solid-state power amplifiers. The schematic is identical for both versions. Component values are the same except as noted.

C1, C2, C11, C17—10-pF chip capacitor.
C3, C4, C5, C6—3.6- to 5.0-pF chip capacitor.
C7, C8—1.8- to 6.0-pF miniature trimmer capacitor (Mouser 24AA070 or equiv. See text).
C9, C10—Same as C7 and C8 for the NEL1306 amplifier. For the NEL1320 version, 0.8- to 10-pF piston trimmers are used (Johanson 5200 series or equiv.).
C12, C14—100-pF chip capacitor.

C13, C15—0.1-μF disc ceramic capacitor.
C16—10-μF electrolytic capacitor.
D1—1N4007 diode.
L1, L2—30-ohm microstripline, ¼-wavelength long (see text).
Q1—NEC NEL130681-12 (6 W) or NEL132081-12 (18 W) transistor.
R1—82- to 100-Ω resistor, 2-W minimum. Vary for specified idling current.

R2—10-Ω, ¼-W carbon-composition resistor with "zero" lead length. See text.
R3—15-Ω, 1-W carbon-composition resistor.
RFC1—3t no. 24 wire, 0.125 inch ID, spaced 1 wire diam.
RFC2—1t no. 24 wire, 0.125 inch ID, spaced 1 wire diam.
RFC3—1-μH RF choke: 18t no. 24 enam. close-spaced on a T50-10 toroid core.

EXCEPT AS INDICATED, DECIMAL VALUES OF CAPACITANCE ARE IN MICROFARADS (μF); OTHERS ARE IN PICOFARADS (pF OR μμF); RESISTANCES ARE IN OHMS ; k = 1000, M = 1000 000.

Fig. 86-1

QST

The design incorporates 30-Ω, ¼λ microstrip lines on the input and output. C3, C4, C7, and C8, along with L1, form a pi network that matches the low-input impedance of the device to 50 Ω. C5, C6, C9, C10, and 30-Ω transmission line L2 form an output pi network that maximizes power transfer to 50 Ω. C10 is not always necessary, depending on variations among devices and circuit-board material. Bias is provided by R1, R2, and D1. R1 can be optimized, if desired, to adjust the collector idling current.

10 dB-GAIN AMPLIFIER

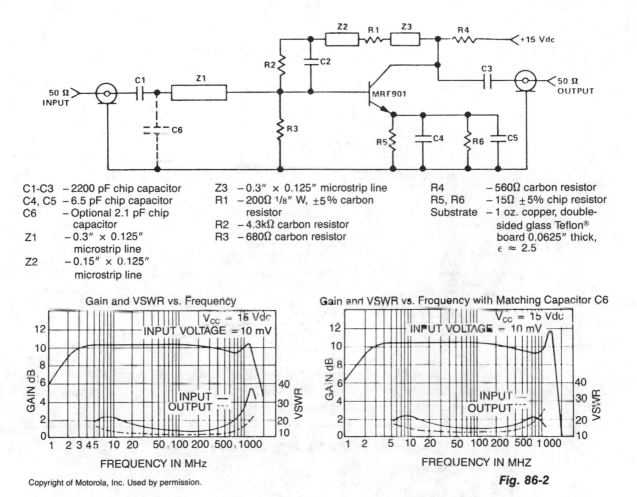

C1-C3	– 2200 pF chip capacitor	Z3	– 0.3″ × 0.125″ microstrip line
C4, C5	– 6.5 pF chip capacitor	R1	– 200Ω ¹/₈″ W, ±5% carbon
C6	– Optional 2.1 pF chip		resistor
	capacitor	R2	– 4.3kΩ carbon resistor
Z1	– 0.3″ × 0.125″	R3	– 680Ω carbon resistor
	microstrip line		
Z2	– 0.15″ × 0.125″		
	microstrip line		

R4	– 560Ω carbon resistor
R5, R6	– 15Ω ±5% chip resistor
Substrate	– 1 oz. copper, double-sided glass Teflon® board 0.0625″ thick, $\epsilon \approx 2.5$

Fig. 86-2

This circuit design is a class A amplifier employing both ac and dc feedback. Bias is stabilized at 15 mA of the collector current using dc feedback from the collector. The ac feedback, from collector to base, and in each of the partially bypassed emitter circuits, compensates for the increase in device gain with decreasing frequency, yielding a flat response over a maximum bandwidth. The amplifier shows a nominal 10-dB power gain from 3 MHz to 1.4 GHz. With only a minimum matching network used at the amplifier input, the input VSWR remains less than 2.5:1 to approximately 1 GHz, while the output VSWR stays under 2:1. Note that a slight degradation in gain flatness and output VSWR occurs with the addition of C6. A more elaborate network design would probably optimize impedance matching, while maintaining gain flatness.

2 – 30 MHz AMPLIFIER

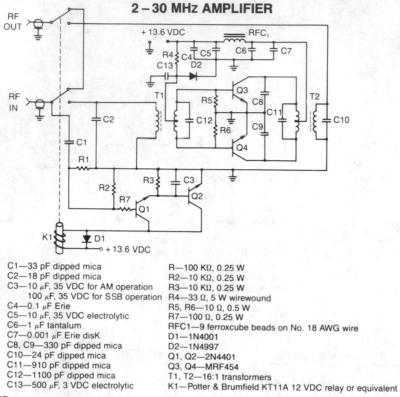

C1—33 pF dipped mica	R—100 KΩ, 0.25 W
C2—18 pF dipped mica	R2—10 KΩ, 0.25 W
C3—10 µF, 35 VDC for AM operation	R3—10 KΩ, 0.25 W
100 µF, 35 VDC for SSB operation	R4—33 Ω, 5 W wirewound
C4—0.1 µF Erie	R5, R6—10 Ω, 0.5 W
C5—10 µF, 35 VDC electrolytic	R7—100 Ω, 0.25 W
C6—1 µF tantalum	RFC1—9 ferroxcube beads on No. 18 AWG wire
C7—0.001 µF Erie disK	D1—1N4001
C8, C9—330 pF dipped mica	D2—1N4997
C10—24 pF dipped mica	Q1, Q2—2N4401
C11—910 pF dipped mica	Q3, Q4—MRF454
C12—1100 pF dipped mica	T1, T2—16:1 transformers
C13—500 µF, 3 VDC electrolytic	K1—Potter & Brumfield KT11A 12 VDC relay or equivalent

MICROWAVES AND RF

Fig. 86-3

This amplifier provides 140-W PEP nominal output power when supplied with input levels as low as 3 W. Both input and output transformers have a 4:1 turn ratio and a 16:1 impedance ratio to achieve low input VSWR across the band with high-saturation capability.

450-MHz COMMON-GATE AMPLIFIER

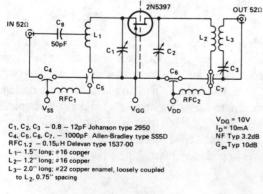

C_1, C_2, C_3 — 0.8 – 12pF Johanson type 2950
C_4, C_5, C_6, C_7 — 1000pF Allen-Bradley type SS5D
$RFC_{1,2}$ — 0.15µH Delevan type 1537-00
L_1 — 1.5" long; #16 copper
L_2 — 1.2" long; #16 copper
L_3 — 2.0" long; #22 copper enamel, loosely coupled
  to L_2, 0.75" spacing

V_{DG} = 10V
I_D = 10mA
NF Typ 3.2dB
G_{ps} Typ 10dB

This is a low noise, 3-dB typical NF, amplifier with about 10-dB gain at 450 – 470 MHz for VHF two-way applications.

Fig. 86-4

RF WIDEBAND ADJUSTABLE AGC AMPLIFIER

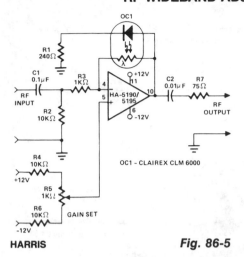

HARRIS

Fig. 86-5

fied output signal drive the OC1 LED into a conducting state. Since the resistance of the OC1 photosensitive element is inversely proportional to light intensity, the higher the signal level, the lower the feedback resistance to the op amp inverting input. The greater negative feedback lowers stage gain. Any changes in gain occur smoothly because the inherent memory characteristic of the photoresistor acts to integrate the peak signal inputs. In practice, the stage gain is adjusted automatically to where the output signal positive peaks are approximately one diode drop above ground.

Gain set control R5 applies a fixed dc bias to the op amp noninverting input, thus establishing the steady state-zero input signal current through the OC1 LED and determining the signal level at which AGC action begins.

The effective AGC range depends on a number of factors, including individual device characteristics, the nature of the rf drive signal, the initial setting for R5, et al. Theoretically, the AGC range can be as high as 4000:1 for a perfect op amp because the OC1 photoresistor can vary in value from 1 MΩ with the LED dark to 250 Ω with the LED fully on.

This circuit functions as a wideband adjustable AGC amplifier. With an effective bandwidth of approximately 10 MHz, it is capable of handling rf input signal frequencies from 3.2 to 10 MHz at levels ranging from 40 mV up to 3 V pk-pk.

AGC action is achieved by using optocoupler/isolater OC1 as part of the gain control-feedback loop. In operation, the positive peaks of the ampli-

---

1-MHz METER-DRIVER AMPLIFIER

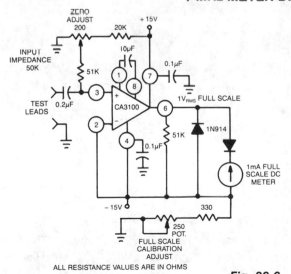

ALL RESISTANCE VALUES ARE IN OHMS

GE/RCA

Fig. 86-6

This circuit uses the CA3100 BiMOS op amp to drive a 1-mA meter movement to full scale with 1-V rms input.

545

5-W 150-MHz AMPLIFIER

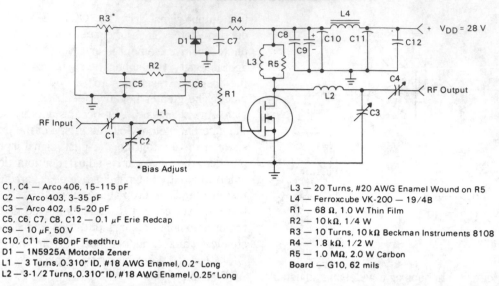

C1, C4 — Arco 406, 15–115 pF
C2 — Arco 403, 3–35 pF
C3 — Arco 402, 1.5–20 pF
C5, C6, C7, C8, C12 — 0.1 µF Erie Redcap
C9 — 10 µF, 50 V
C10, C11 — 680 pF Feedthru
D1 — 1N5925A Motorola Zener
L1 — 3 Turns, 0.310" ID, #18 AWG Enamel, 0.2" Long
L2 — 3-1/2 Turns, 0.310" ID, #18 AWG Enamel, 0.25" Long

L3 — 20 Turns, #20 AWG Enamel Wound on R5
L4 — Ferroxcube VK-200 — 19/4B
R1 — 68 Ω, 1.0 W Thin Film
R2 — 10 kΩ, 1/4 W
R3 — 10 Turns, 10 kΩ Beckman Instruments 8108
R4 — 1.8 kΩ, 1/2 W
R5 — 1.0 MΩ, 2.0 W Carbon
Board — G10, 62 mils

Fig. 86-7

This circuit utilizes the MRF123 TMOS power FET. The MRF134 is a very high gain FET that is potentially unstable at both VHF and UHF frequencies. Note that a 68-Ω input loading resistor has been utilized to enhance stability. This amplifier has a gain of 14 dB and a drain efficiency of 55%.

UHF-TV PREAMPLIFIER

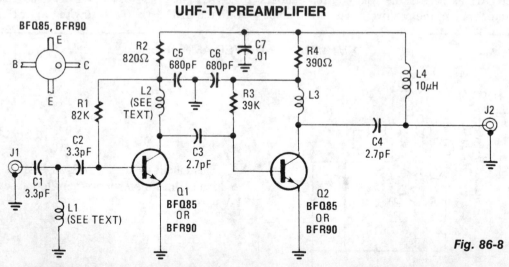

Fig. 86-8

An inexpensive, antenna-mounted, UHF-TV preamplifier can add more than 25 dB of gain. The first stage of the preamp is biased for optimum noise, the second stage for optimum gain. L1, L2 strip line ≈ λ/8 part of PC board.

60-W 225 – 400 MHz AMPLIFIER

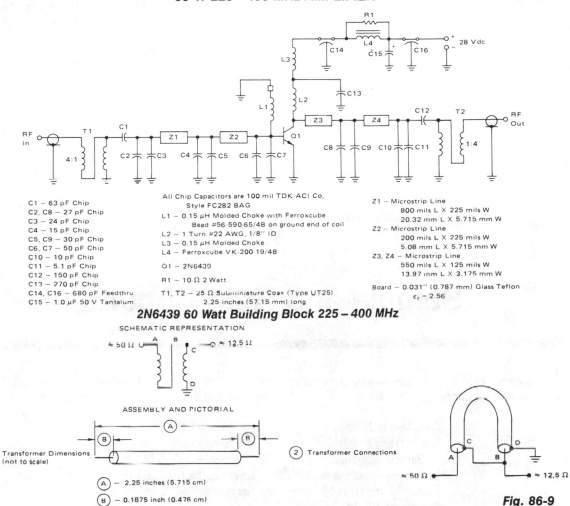

C1 – 63 pF Chip
C2, C8 – 27 pF Chip
C3 – 24 pF Chip
C4 – 15 pF Chip
C5, C9 – 30 pF Chip
C6, C7 – 50 pF Chip
C10 – 10 pF Chip
C11 – 5.1 pF Chip
C12 – 150 pF Chip
C13 – 270 pF Chip
C14, C16 – 680 pF Feedthru
C15 – 1.0 µF 50 V Tantalum

All Chip Capacitors are 100 mil TDK-ACI Co,
    Style FC282 BAG
L1 – 0.15 µH Molded Choke with Ferroxcube
    Bead #56-590-65/4B on ground end of coil
L2 – 1 Turn #22 AWG, 1/8'' ID
L3 – 0.15 µH Molded Choke
L4 – Ferroxcube VK-200-19/4B
Q1 – 2N6439
R1 – 10 Ω 2 Watt
T1, T2 – 25 Ω Subminiature Coax (Type UT25)
    2.25 inches (57.15 mm) long

Z1 – Microstrip Line
    800 mils L X 225 mils W
    20.32 mm L X 5.715 mm W
Z2 – Microstrip Line
    200 mils L X 225 mils W
    5.08 mm L X 5.715 mm W
Z3, Z4 – Microstrip Line
    550 mils L X 125 mils W
    13.97 mm L X 3.175 mm W
Board – 0.031'' (0.787 mm) Glass Teflon
    ϵ_r = 2.56

2N6439 60 Watt Building Block 225 – 400 MHz

SCHEMATIC REPRESENTATION

ASSEMBLY AND PICTORIAL

① Transformer Dimensions (not to scale)

Ⓐ – 2.25 inches (5.715 cm)
Ⓑ – 0.1875 inch (0.476 cm)

② Transformer Connections

Fig. 86-9

Construction Details of the 4:1 Unbalanced to Unbalanced Transformers

This 60-W, 28-V broadband amplifier covers the 225 – 400 MHz military communications band. The amplifier may be used singly as a 60-W output stage in a 225 – 400 MHz transmitter, or by using two of these amplifiers combined with quadrature couplers, a 100-W output amplifier stage can be constructed. The circuit is designed to be driven from a 50-Ω source and work into a nominal 50-Ω load. The input network consists of two microstrip L-sections composed of Z1, Z2, and C2 through C6. C1 serves as a dc-blocking capacitor. A 4:1 impedance ratio coaxial transformer T1 completes the input matching network. L1 and a ferrite bead serve as a base decoupling choke. The output circuit consists of shunt inductor L2 at the collector, followed by two microstrip L-sections composed of Z3, Z4, and C8 through C11. C12 serves as a dc blocking capacitor, and is followed by another 4:1 impedance ratio coaxial transformer. Collector decoupling is accomplished through the use of L3, L4, C14, C15, C16, and R1.

87

Sample-and-Hold Circuits

The sources of the following circuits are contained in the Sources section beginning on page 782. The figure number contained in the box of each circuit correlates to the sources entry in the Sources section.

SAMPLE-AND-HOLD

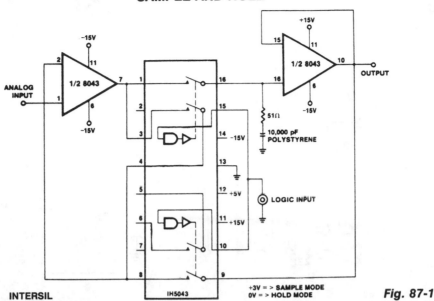

INTERSIL   IH5043   +3V = > SAMPLE MODE
0V = > HOLD MODE   **Fig. 87-1**

Two important properties of the 8043 are used to advantage in this circuit. The low input bias currents give rise to slow output decay rates (droop) in the hold mode, while the high slew rate at 6 V/μs improves the tracking speed and the response time of the circuit. The upper waveform is the input 10 V/div, the lower waveform the output 5 V/div. The logic input is high.

The center waveform is the analog input, a ramp moving at about 67 V/ms, the lower waveform is the logic input to the sample-and-hold; a logic 1 initiates the sample mode. The upper waveform is the output, displaced by about one scope division 2 V from the input to avoid superimposing traces. The hold mode, during which the output remains constant, is clearly visible. At the beginning of a sample period, the output takes about 8 μs to catch up with the input, after which it tracks, until the next hold period.

BASIC TRACK-AND-HOLD/SAMPLE-AND-HOLD

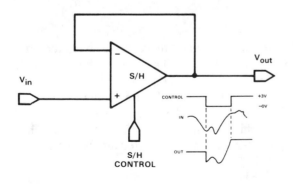

HARRIS   **Fig. 87-2**

Feedback is the same as a conventional op-amp voltage follower which yields a unity-gain, noninverting output. This hookup also has a very high input impedance. The only difference between a track-and-hold and a sample-and-hold is the time period during which the switch is closed. In track-and-hold operation, the switch is closed for a relatively long period; the output signal might change appreciably and would hold the level present at the instant the switch is opened. In sample-and-hold operation, the switch is closed only for the time necessary to fully charge the holding capacitor.

HIGH-SPEED SAMPLE-AND-HOLD

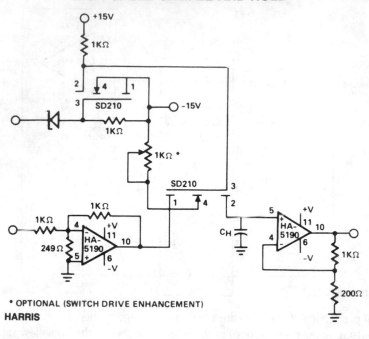

* OPTIONAL (SWITCH DRIVE ENHANCEMENT)

HARRIS

Fig. 87-3

This circuit uses the speed and drive capability of the HA-5190 coupled with two high-speed DMOS FET switches. The input amplifier is allowed to operate at a gain of −5, although the overall circuit gain is unity. Acquisition times of less than 100 ns to 0.1% of a 1-V input step are possible. Drift current can be appreciably reduced by using FET input buffers on the output stage of the sample-and-hold.

FILTERED SAMPLE-AND-HOLD

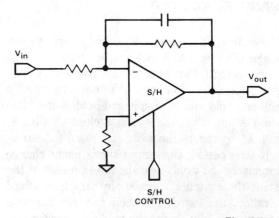

HARRIS

Fig. 87-4

It is often required that a signal be filtered prior to sampling. This can be accomplished with only one device. Use any of the inverting and noninverting filters that can be built with op amps. However, it is necessary that the sampling switch be closed for a sufficient time for the filter to settle when active filter types are connected around the device.

SAMPLE-AND-HOLD

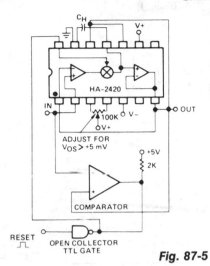

ADJUST FOR $V_{OS} > +5$ mV

COMPARATOR

RESET

OPEN COLLECTOR
TTL GATE

HARRIS

Fig. 87-5

The sample-and-hold function has often been accomplished with separate analog switches and op amps. These designs always involve performance tradeoffs between acquisition time, charge injection, and droop rate. The HA-242-/2425 monolithic sample-and-hold, has many better tradeoffs, and usually a lower total cost than the other approaches. The switching element is a complementary bipolar circuit with feedback, which allows high charging currents of 30 mA, a low charge injection of 10 pC, and an ultra-low off leakage current of 5 pA; a combination that is not approached in any other electronic switch. These factors make it also superior as an integrator reset switch, or as a precision peak detector.

SAMPLE-AND-HOLD

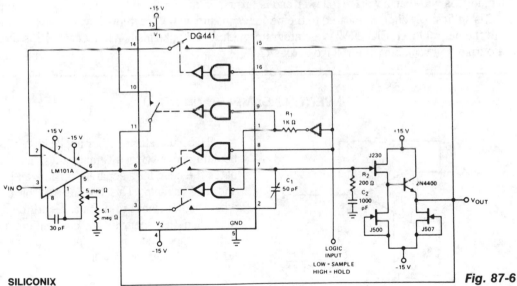

SILICONIX

Fig. 87-6

The LM101A provides gain and buffers the input from storage capacitor C2. R2 adds a zero in the open loop response to compensate for the pole caused by the switch resistance and C2, improving the closed-loop stability. R1 provides a slight delay in the digital drive to pins 1 and 9. C1 provides cancellation of coupled charge, keeping the sample-and-hold offset below 5 mV over the analog signal range of −10 through +10 V. Aperture time is typically 1 μs, the switching time of the DG441. Acquisition time is 25 μs, but this can be improved by using a faster slewing op amp. Droop rate is typically less than 5 mV/s at 25°C.

TRACK-AND-HOLD/SAMPLE-AND-HOLD

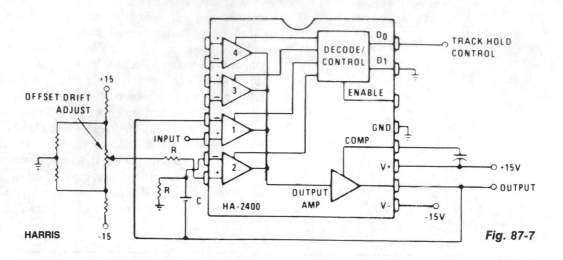

Fig. 87-7

Channel 1 is wired as a voltage follower and is turned on during the track/sample time. If the product of $R \times C$ is sufficiently short compared to the period of maximum output frequency, or sample time; C will charge to the output level. Channel 2 is an integrator with zero input signal. When channel 2 is then turned on, the output will remain at the voltage across C.

INVERTING SAMPLE-AND-HOLD

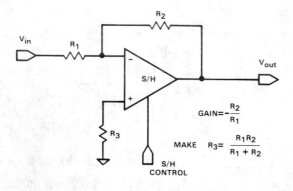

$$\text{GAIN} = -\frac{R_2}{R_1}$$

$$\text{MAKE} \quad R_3 = \frac{R_1 R_2}{R_1 + R_2}$$

HARRIS   Fig. 87-8

This illustrates another application in which the hookup versatility of a sample-and-hold often eliminates the need for a separate op amp and a sample-and-hold module. This hookup will have a somewhat higher input-to-output feedthrough during hold than the noninverting connection, since the output impedance is an open-loop value during hold. The feedthrough will

$$\frac{V_{IN} R_0}{R_1 + R_2 + R_0}$$

SAMPLE-AND-HOLD

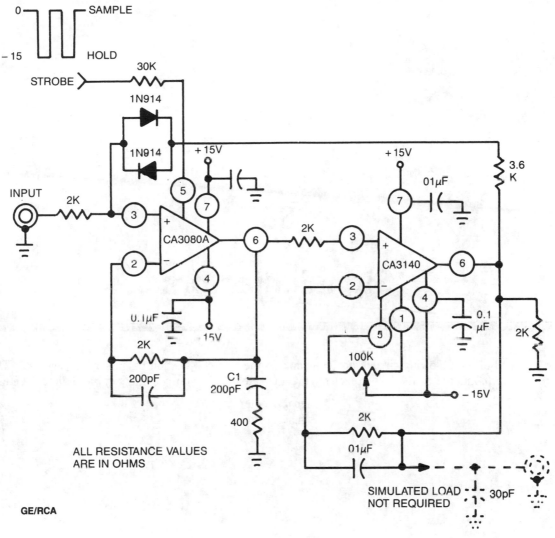

Fig. 87-9

This circuit uses a CA3140 BiMOS op amp as the readout amplifier for the storage capacitor C1, and a CA3080A variable op amp as input buffer amplifier and low feedthrough transmission switch. Offset nulling is accomplished with the CA3140.

88

Signal Injectors

The sources of the following circuits are contained in the Sources section beginning on page 782. The figure number contained in the box of each circuit correlates to the sources entry in the Sources section.

Signal Injector
Signal Injector

SIGNAL INJECTOR

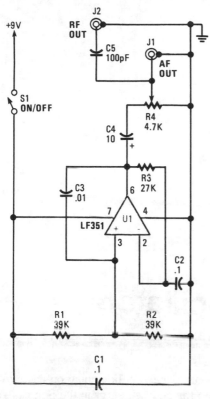

This unit is a single oscillator built around an LF351 JFET-input op amp. Resistors R1 and R2 bias the noninverting input while R3 biases the inverting input from the output. This layout provides 100% negative feedback, but the decoupling caused by C2 gives reduced feedback and high-voltage gain when dealing with audio frequencies. The fundamental operating frequency is about 800 Hz. Potentiometer R4 is the output-level control. To use it start at the speaker. If no tone is heard, move back to the amplifer input, and listen for the tone. Still if no tone is heard, continue backtracking from the output to the input, covering all stages in between. The stage where the signal is lost is the one that is not operating.

POPULAR ELECTRONICS/HANDS-ON ELECTRONICS   *Fig. 88-1*

SIGNAL INJECTOR

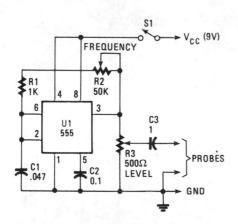

The unit provides a square-wave output that is rich in harmonic content. The circuit's output frequency can be varied from 50 Hz to 15 kHz. The heart of the circuit is a 555 astable connected in its equal mark/space mode. The frequency is controlled by potentiometer R2 and capacitor C1. Resistor R3 controls the output level with the output ac-coupled through C3.

POPULAR ELECTRONICS/HANDS-ON ELECTRONICS   *Fig. 88-2*

89

Sine-Wave Oscillators

The sources of the following circuits are contained in the Sources section beginning on page 782. The figure number contained in the box of each circuit correlates to the sources entry in the Sources section.

LOW-DISTORTION THERMALLY STABILIZED WIEN-BRIDGE OSCILLATOR

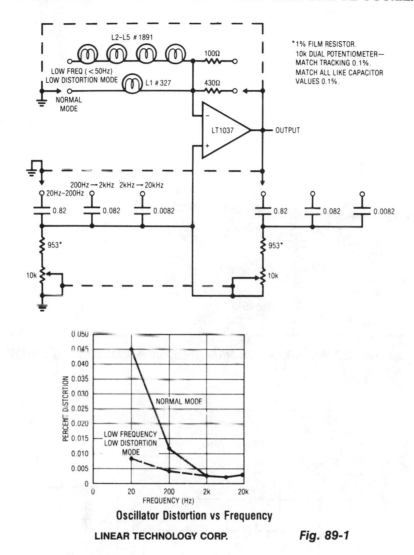

Oscillator Distortion vs Frequency

LINEAR TECHNOLOGY CORP.   *Fig. 89-1*

A variable Wien bridge provides frequency tuning from 20 Hz to 20 kHz. Gain control comes from the positive temperature coefficient of the lamp. When power is applied, the lamp is at a low resistance value, the gain is high, and oscillation amplitude builds. The lamp's gain-regulating behavior is flat within 0.25 dB over the 20 Hz – 20 kHz range of the circuit. Distortion is below 0.003%. At low frequencies, the thermal time constant of the small normal-mode lamp begins to introduce distortion levels about 0.01%. This is because of *hunting* when the oscillator's frequency approaches the lamp's thermal time constant. This effect can be eliminated, at the expense of reduced output amplitude and longer amplitude settling time, by switching to the low-frequency, low-distortion mode. The four large lamps give a longer thermal time constant, and distortion is reduced.

SINGLE-SUPPLY WIEN-BRIDGE OSCILLATOR

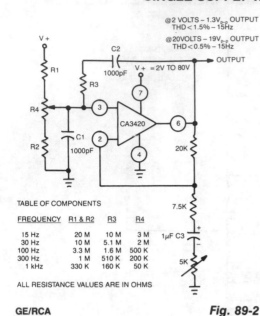

The adjustment of R4 contributes to the comparatively symmetrical output transfer characteristic of the CA3420 BiMOS op amp. To extend the lower operating frequency, remove C3 and use a dual supply.

@2 VOLTS – 1.3V$_{p-p}$ OUTPUT
THD < 1.5% – 15Hz

@20VOLTS – 19V$_{p-p}$ OUTPUT
THD < 0.5% – 15Hz

V+ = 2V TO 80V

TABLE OF COMPONENTS

FREQUENCY	R1 & R2	R3	R4
15 Hz	20 M	10 M	3 M
30 Hz	10 M	5.1 M	2 M
100 Hz	3.3 M	1.6 M	500 K
300 Hz	1 M	510 K	200 K
1 kHz	330 K	160 K	50 K

ALL RESISTANCE VALUES ARE IN OHMS

GE/RCA   **Fig. 89-2**

SUPER-LOW-DISTORTION VARIABLE SINE-WAVE OSCILLATOR

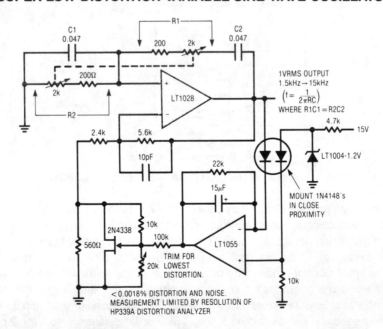

1VRMS OUTPUT
1.5kHz → 15kHz

$$\left(f = \frac{1}{2\pi RC}\right)$$

WHERE R1C1 = R2C2

MOUNT 1N4148's
IN CLOSE
PROXIMITY

TRIM FOR
LOWEST
DISTORTION.

< 0.0018% DISTORTION AND NOISE.
MEASUREMENT LIMITED BY RESOLUTION OF
HP339A DISTORTION ANALYZER

LINEAR TECHNOLOGY CORP.   **Fig. 89-3**

AUDIO GENERATOR

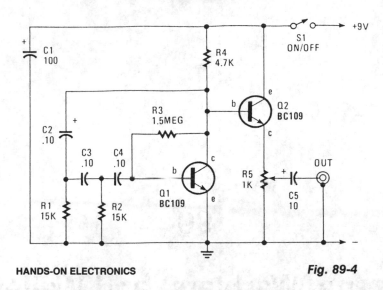

Fig. 89-4

This circuit produces a sinusoidal output of about 8 V pk-pk, which can be varied down to zero, at about 500 Hz. The signal is generated by a phase-shift oscillator.

90

Sirens, Warblers, and Wailers

The sources of the following circuits are contained in the Sources section beginning on page 782. The figure number contained in the box of each circuit correlates to the sources entry in the Sources section.

Electronic Bagpipe
Two-Tone Siren
Yelping Siren
Programmable-Frequency
    Adjustable-Rate Siren
The Wailing Siren
Linear IC Siren

Super Sound Generator
Hee-Haw Siren
Electronic Siren
555 Beep Transformer
Siren
Two-State Siren
Steam Train with Whistle

ELECTRONIC BAGPIPE

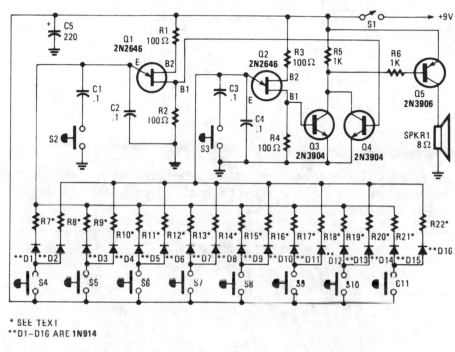

* SEE TEXT
**D1–D16 ARE 1N914

POPULAR ELECTRONICS

Fig. 90-1

This circuit mimics the dual-tone drone sound that's produced by the unusual wind instrument. Unijunction transistors Q1 and Q2 are connected in similar audio-oscillator circuits. Each of the oscillator frequencies is determined by one of the two resistors selected by one of the pushbutton switches, S4 through S11. The odd-numbered resistors in R7 to R21, determine the frequency for the Q1 oscillator circuit and the even-numbered resistors in R8 through R22, determine the frequency for Q2's circuit.

When S4 is pressed, the positive supply is connected to both R7 and R8 through isolation diodes D1 and D2, causing both oscillators to operate. A narrow, fast-rising positive pulse is developed at B1 of both Q1 and Q2 for each cycle of operation. Transistors Q3 and Q4 serve as a simple audio mixer, which is used to combine the pulses from each oscillator. The mixed signal at the collectors of Q3 and Q4 is coupled through R6 to the base of Q5, which amplifies and drives an 8-Ω speaker, SPKR1. Switches, S2 and S3 are used to reduce the oscillator's frequency by about 50% when closed, to produce a new group of tones.

TWO-TONE SIREN

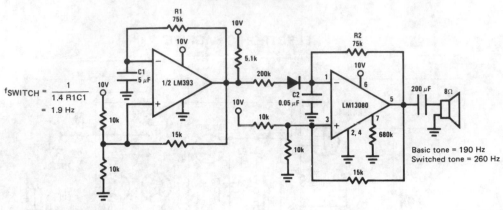

$$f_{SWITCH} = \frac{1}{1.4\ R1C1}$$
$$= 1.9\ Hz$$

Basic tone = 190 Hz
Switched tone = 260 Hz

NATIONAL SEMICONDUCTOR

Fig. 90-2

This siren provides a constant audio output, but alternates between two separate tones. The LM13080 is set to oscillate at one basic frequency; this frequency is changed by adding a 200-KΩ charging resistor in parallel with the feedback resistor, R2.

YELPING SIREN

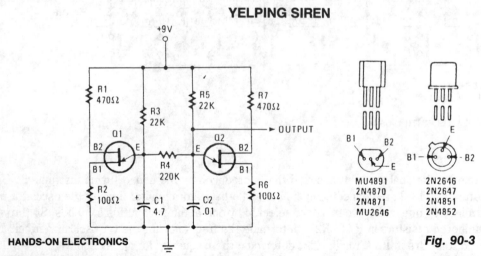

MU4891
2N4870
2N4871
MU2646

2N2646
2N2647
2N4851
2N4852

HANDS-ON ELECTRONICS

Fig. 90-3

Unijunction transistors Q1 and Q2 are both connected as relaxation-type, sawtooth oscillators. Transistor Q1 is the low-frequency control oscillator and Q2 is the tone generator. Sawtooth waveforms are produced at the emitter terminals. Without R4 connecting the two emitters, each oscillator operates independently, with its frequency determined mainly by the rc time constant. With the values shown, Q1 operates from 1. to 1.5 Hz and Q2 operates from 400 to 500 Hz. When R4 is connected between the two emitters, it couples the low-frequency sawtooth from Q1 directly across capacitor C2. That coupling causes the frequency of the tone generator to increase, along with the rise in sawtooth voltage from Q1. The tone generator's frequency drops to its lower design value when C1 discharges and produces the falling edge of the sawtooth.

PROGRAMMABLE-FREQUENCY ADJUSTABLE-RATE SIREN

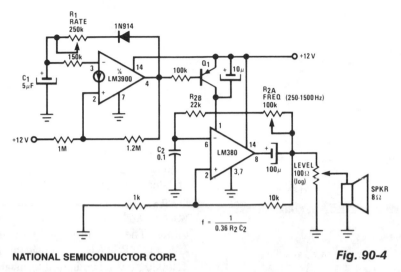

NATIONAL SEMICONDUCTOR CORP.

Fig. 90-4

$$f = \frac{1}{0.36\, R_2\, C_2}$$

The LM380 operates as an astable oscillator with the frequency determined by R2/C2. Adding Q1 and driving its base, with the output of an LM3900 wired as a second astable oscillator, acts to gate the output of the LM380 on and off, at a rate fixed by R1/C1.

THE WAILING SIREN

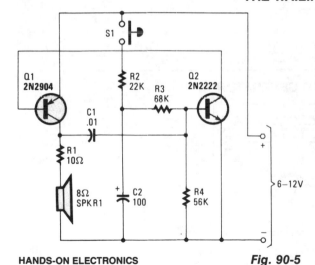

HANDS-ON ELECTRONICS

Fig. 90-5

Transistors Q1 and Q2, with feedback provided via C1 from the collector of Q1 to the base of Q2, forms a voltage-controlled oscillator (VCO). De-

pending on the voltage applied to Q2's base, the VCO frequency ranges from around 60 Hz to 7.5 kHz. The instantaneous voltage applied to the base of Q2 is determined by the values of C2, R2, R3, and R4. When pushbutton switch S1 is closed, C2 charges fairly rapidly to the maximum supply voltage through R2, a 22-KΩ fixed resistor. That causes the siren sound to rise rapidly to its highest frequency. When the button is released, the capacitor discharges through R3 and R4 with a combined resistance of 124 KΩ, causing the siren sound to decay from a high-pitched wail to a low growl. If you want to experiment with the pitch of the sound at its highest frequency, try different values for C1. Increase its value for lower notes, and decrease it for higher ones. Different values for R2 will change the attack time. A 100-KΩ resistor provides equal attack and decay times. The way you handle the pushbutton varies the effect.

LINEAR IC SIREN

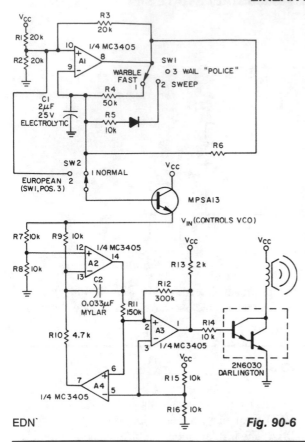

A low-frequency, op-amp oscillator and a VCO, both configured from a single MC3405 dual op amp and dual comparator, are the major components in a siren circuit that can be made to produce various warbles and wails, or serve as an audio sweep generator. The only other active components needed are an MPS A13 small-signal transistor and a 2N6030 power Darlington transistor.

EDN`

Fig. 90-6

SUPER SOUND GENERATOR

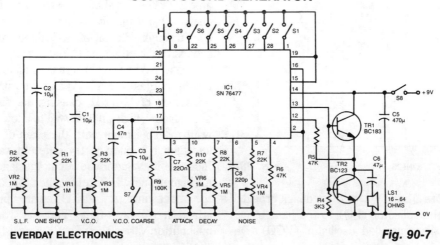

EVERDAY ELECTRONICS

Fig. 90-7

SUPER SOUND GENERATOR (*Cont.*)

Six preset controls and seven selector switches enable a vast range of different sounds to be produced and altered at will. Such sounds as steam trains chuffing, helicopters flying, bird chirping, and machine guns firing are possible, as well as the usual police sirens, phaser guns, and bomb explosions. The circuit incorporates an amplifier giving 150-mW output into a small loudspeaker. Alternatively, a separate amplifier system can be used for disco effects, car alarms, etc. Continuous or one-shot sounds are possible. For one-shot sounds, a push-button switch is provided, which can also be used to turn continuous sounds on and off. A single IC, SN76477, provides all of the sound generation circuits.

HEE-HAW SIREN

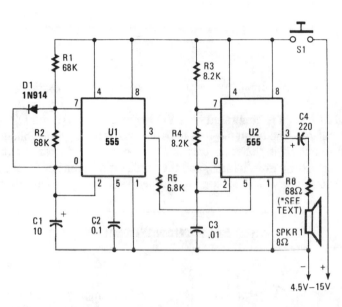

Fig. 90-8

A pair of timer IC's are the heart of a circuit that simulates the warbling hee-haw of a British police siren. One of the 555 timers, U2, is wired as an astable multivibrator operating at about 900 Hz. The other, U1, operates at approximately 1 Hz. Its output at pin 3 is a square wave with a 50% duty cycle—on and off cycles of about 0.5 second each. The output of U1 is applied to pin 5, the control-voltage terminal of U2. The frequency of the 555 timer IC is relatively independent of supply voltage, but can be varied over a fairly wide range by applying a variable voltage between pin 5 and ground. When U1's output becomes high, U2 operates at about 800 kHz. That switching between two frequencies produces the warbling hee-haw signal.

ELECTRONIC SIREN

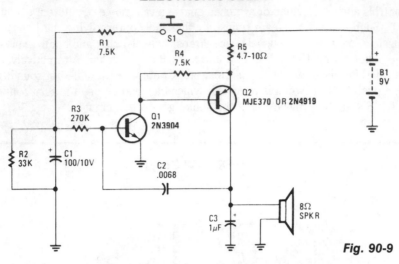

Fig. 90-9

The wailing sound of a siren is generated by a VFO consisting of Q1 and Q2. Capacitor C2 provides the feedback for the oscillator. The frequency of the oscillator is varied by the voltage applied to the base of Q1 through R3. When switch S1 is closed, capacitor C1 charges, thus increasing the oscillator frequency. When S1 is released, capacitor C1 discharges, and the oscillator frequency decreases. Capacitor C3 limits the maximum oscillator frequency. The average battery current drain is about 15 mA.

555 BEEP TRANSFORMER

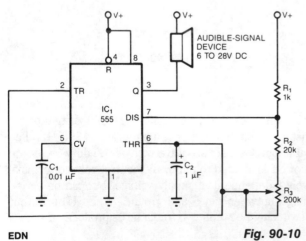

The simple circuit transforms the steady beep of an audible-signal device, such as a Mallory son-alert, into a distinctive warble or chirp. The value of C2 determines just what tone color you'll get. With the 1-μF value shown, the circuit produces a warble similar to the ring tone of an inexpensive phone. A 10-μF value produces a chirp similar to a truck's back-up alarm. One elaboration of this circuit would be to use the second section of a 555 timer to drive a piezoelectric transducer instead of a sonalert; that modification would vary the tone's pitch, as well as the chirp rate.

EDN

Fig. 90-10

SIREN

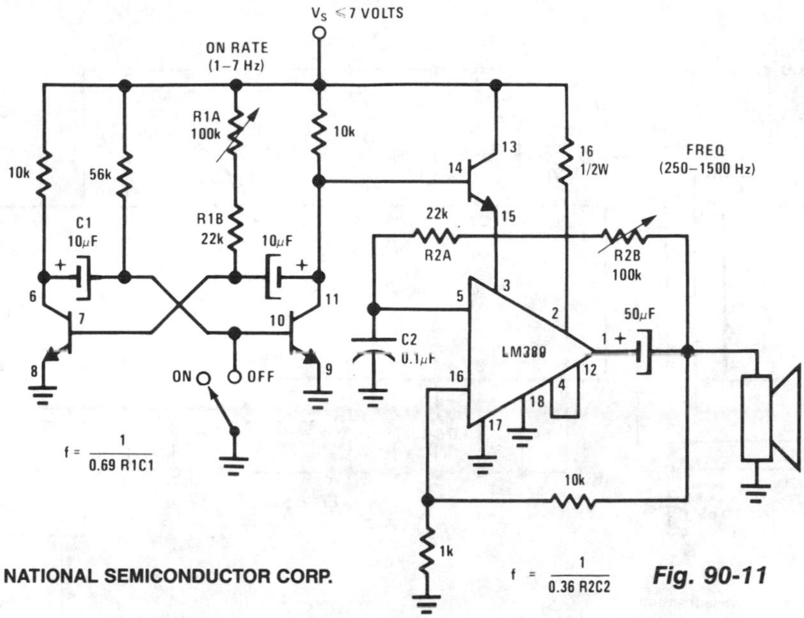

NATIONAL SEMICONDUCTOR CORP.

$$f = \frac{1}{0.69 \; R1C1}$$

$$f = \frac{1}{0.36 \; R2C2}$$

Fig. 90-11

This circuit uses one of the LM389 transistors to gate the power amplifier on and off by applying the muting technique. The other transistors form a cross-coupled multivibrator circuit that controls the rate of the square-wave oscillator. The power amplifier is used as the square-wave oscillator with individual frequency adjust provided by potentiometer R2B.

TWO-STATE SIREN

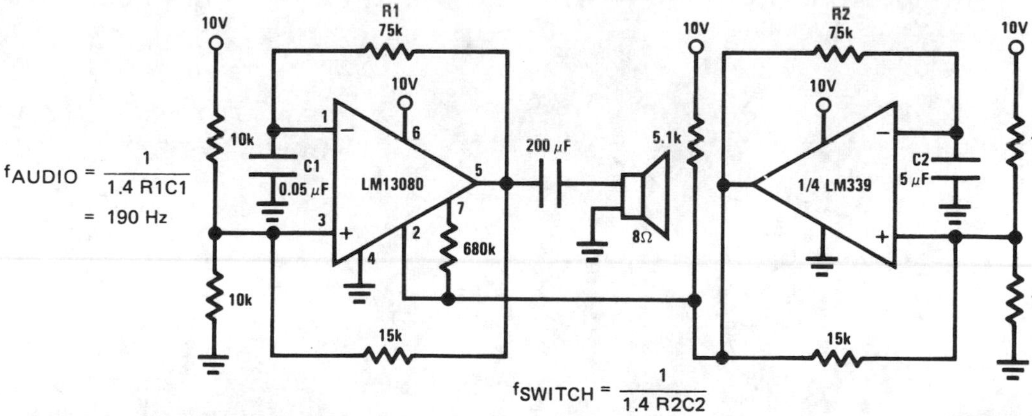

$$f_{AUDIO} = \frac{1}{1.4 \; R1C1}$$
$$= 190 \; Hz$$

$$f_{SWITCH} = \frac{1}{1.4 \; R2C2}$$
$$= 1.9 \; Hz$$

NATIONAL SEMICONDUCTOR CORP.

Fig. 90-12

This is a two-state or on/off-type siren where the LM13080 oscillates at an audio frequency and drives an 8-Ω speaker. The LM339 acts as a switch which controls the audio burst rate.

STEAM TRAIN WITH WHISTLE

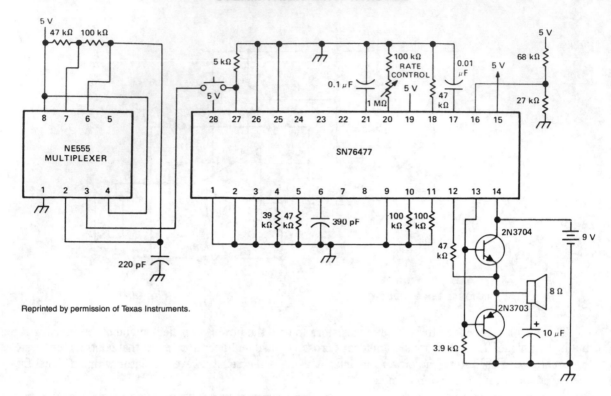

Reprinted by permission of Texas Instruments.

Fig. 90-13

91

Solid-State Relay Circuits

The sources of the following circuits are contained in the Sources section beginning on page 782. The figure number contained in the box of each circuit correlates to the sources entry in the Sources section.

Ac Solid-State Relays

AC SOLID-STATE RELAYS

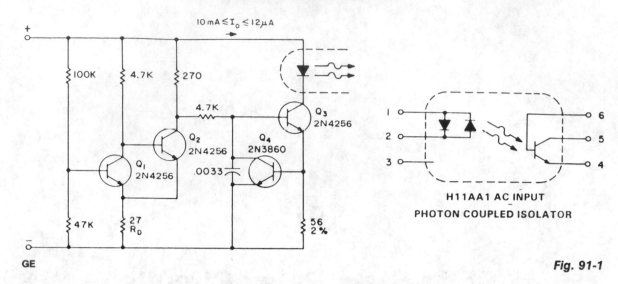

GE

Fig. 91-1

In the case where analog signals are being used as the logic control, hysteresis from a Schmitt-trigger input can be used to prevent half-wave power output. The circuit operation is as follows: at low input voltages, Q1 is biased in the off state. Q2 conducts and biases Q3, and the IRED turns off. When the base of Q1 reaches the biasing voltage of 0.6 V, plus the drop across R_D, Q1 turns on. Q3 is then supplied base drive, and the solid-state relay input will be activated. The combination of Q3 and Q4 acts as a constant-current source to the IRED. In order to turn-off Q3, the base drive must be reduced to pull it out of saturation. Because Q2 is in the off-state as the signal is reduced, Q1 will now stay on to a base bias-voltage lowered by the change in the drop across R_D. With these values, the highest turn-off voltage is 1.0 V, while turn-on will be at less than the 4.1 V supplied to the circuit.

For ac or bipolar input signals, there are several possible connections. If only positive signals are set to activate the relay, a diode, such as the A14, can be connected in parallel to protect the IRED from reverse voltage damage, since its specified peak reverse voltage capability is approximately 3 V. If ac signals are being used, or if activation is to be polarity insensitive, a H11AA coupler, which contains two LEDs in antiparallel connection, can be used. For high-input voltage designs, or for any easy means of converting a dc input relay to ac, a full-wave diode bridge can be used to bias the IRED.

92

Solenoid Drivers

The sources of the following circuits are contained in the Sources section beginning on page 782. The figure number contained in the box of each circuit correlates to the sources entry in the Sources section.

Power-Consumption Limiter
12-V Latch
Hold-Current Limiter

POWER-CONSUMPTION LIMITER

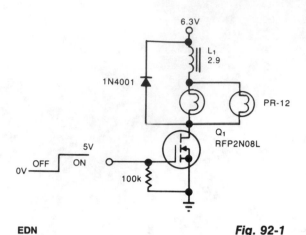

EDN

Fig. 92-1

since surge to on-current ratio is typically 5:1. The cold filament allows a surge of coil-activation current to pass through; as the filament heats up, it throttles the current to a more reasonable hold value. The solenoid driver circuit offers these features:

- 5-V logic swings turn the power-MOSFET switch, Q1, fully on and off.
- Two low-cost flashlight lamps, in parallel, handle the peak current. Because their dc current is only 50% of peak and because they operate at 60% of their rated voltage, the lamps have an operating life of 12,000 hours. Further, the lamp filaments' positive temperature coefficients raise each filament's resistance. This rise in resistance eliminates current-hogging problems and provides short-circuit protection.
- The steady-state on-current is 700 mA, vs. 1700 mA without the lamps.
- A 4.6-V min supply rating allows battery operation.

A simple solenoid driver uses incandescent lamp filaments as on-indicators to limit power consumption. High magnetic reluctance (opposition to flux) in the coil of an armature-driven device, such as a solenoid or relay, calls for a surge of activation current, followed by a lower dc level to remain on,

12-V LATCH

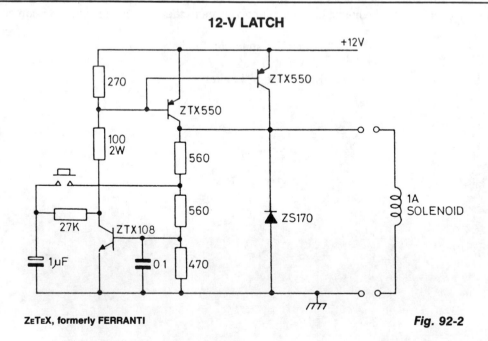

ZeTeX, formerly FERRANTI

Fig. 92-2

12-V LATCH (*Cont.*)

This circuit controls a solenoid by the operation of a single push-button switch. The circuit will supply loads of over 1 A and can be operated up to a maximum speed of once every 0.6 second. When power is first applied to the circuit, the solenoid will always start in its off position. Other features of the circuit are its automatic turn-off, if the load is shorted, and its virtually zero-power consumption when off.

HOLD-CURRENT LIMITER

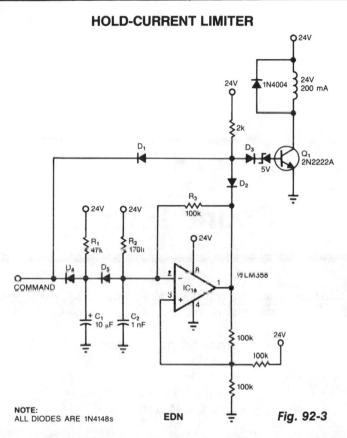

NOTE:
ALL DIODES ARE 1N4148s

EDN

Fig. 92-3

In many applications, a solenoid driver must first briefly supply a large amount of pull-in current, which quickly actuates the solenoid. Thereafter, the driver must supply a much lower holding current to avoid burning the solenoid out. To avoid using the customary, cumbersome, large capacitors or power-wasting resistors, you can use the switch technique.

As long as the input to the circuit is low, diode D1 holds Q1 off; a low input also prevents the op-amp circuit from oscillating. When the input reaches 24 V, Q1 switches on and pulls in the solenoid. Concurrently, D4 is back-biased, and C1 begins charging up. When C1 charges up, the op-amp circuit begins to oscillate, switching Q1 on and off.

The time constant defined by R1 and C1 determines the length of the period during which the solenoid receives full power. R3 and C2 set the oscillator's frequency, and R2 sets the oscillator's duty cycle. The hold current is directly proportional to the duty cycle. For the components shown, the full-power period is 300 ms, the oscillator's frequency is 3 kHz, and its duty cycle is 50%.

93

Sound Effects

The sources of the following circuits are contained in the Sources section beginning on page 782. The figure number contained in the box of each circuit correlates to the sources entry in the Sources section.

Sound-Effects Generator
Fuzz Box
Chug-Chug
Electronic Bird Chirper
Race-Car Motor/Crash

SOUND-EFFECTS GENERATOR

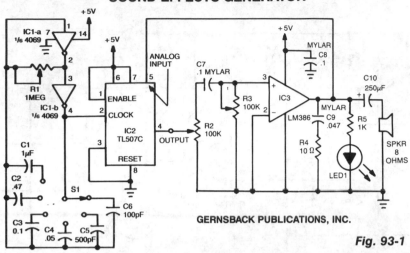

GERNSBACK PUBLICATIONS, INC.

Fig. 93-1

A variable clock-pulse generator is made up of two sections of IC1 (a 4069 CMOS hex inverter), R1, S1, and capacitors C1 through C6. By adjusting R1 and switching one of the capacitors into the circuit, the clock's pulse rate can be varied over a wide range.

The TL507C converts analog signals—in this case the output of IC3, an LM386 audio amplifier—into digital signals. The conversion is accomplished using the single-slope method; it involves comparing an internally generated ramp signal to the analog input signal and a 200 mV reference voltage.

The square-wave output from the a/d converter is fed to IC3 through a network consisting of R2, R3, and C7. Resistor R2 controls the amplitude of the pulses. Resistor R3 and capacitor C7 form a variable tone-control filter and a differentiator circuit that converts a square wave into a spiked waveform. That waveform is amplified by IC3, and the resulting output is fed back into the analog input of IC2, as well as to an 8-Ω speaker. By adjusting R1 and selecting one of the six capacitors with S1—thus varying the clock frequency—and by varying R2 and R3, you can produce many sounds.

FUZZ BOX

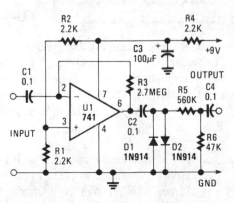

HANDS-ON ELECTRONICS   *Fig. 93-2*

The 741's maximum gain of 20,000 is pushed to nearly 3 million dB, and therefore distorts the output. That distortion provides the fuzz sound. The level is dropped by clipping the two diodes.

CHUG-CHUG

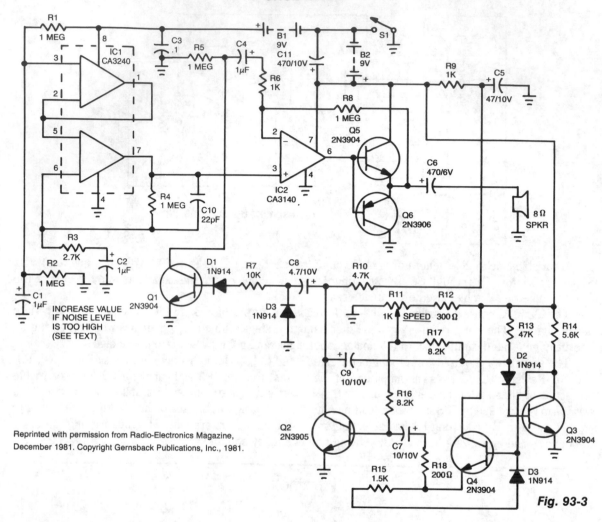

Reprinted with permission from Radio-Electronics Magazine, December 1981. Copyright Gernsback Publications, Inc., 1981.

Fig. 93-3

A CA3240 dual MOSFET-input device is used as a white-noise source. Op amp IC2 is used as a driver stage for the push-pull output stage formed by Q5 and Q6. Transistors Q2, Q3, and Q4 form a variable-frequency multivibrator. R11, the speed control, is used to control the multivibrator's frequency. The output is differentiated by C8 and applied to modulator transistor Q1, through D1 and R7. Transistor Q1 modulates the gain of the output amplifier stage by changing the impedance to ground, through R6 and C4. When the multivibrator's frequency is reduced using R11, C8 discharges slowly, creating a sound similar to escaping steam from a stopped locomotive.

To find the proper value for R3, short Q1's collector to ground. Then, increase the value of R3 until the current drain from the power supply is less than 60 mA. Then remove the short from Q1. To see if the device is operating properly, close switch S1 and reduce the resistance of R11. Wait 10 seconds, then rotate R11 slowly. You should hear a sound similar to a steam locomotive picking up speed.

ELECTRONIC BIRD CHIRPER

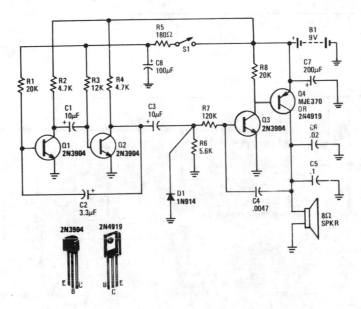

Fig. 93-4

Transistors Q1 and Q2 form the two halves of a free-running multivibrator whose frequency is determined by the voltage across C8. That capacitor is charged and discharged by closing and opening switch S1. Transistors Q3 and Q4 make up a VFO. The output of the free-running multivibrator frequency modulates the Q3/Q4 oscillator, causing the chirping bird sound. The number of chirps per second is determined by the frequency of the Q1/Q2 multivibrator, which also varies. The pitch of the chirps is determined by C5 and C6.

RACE-CAR MOTOR/CRASH

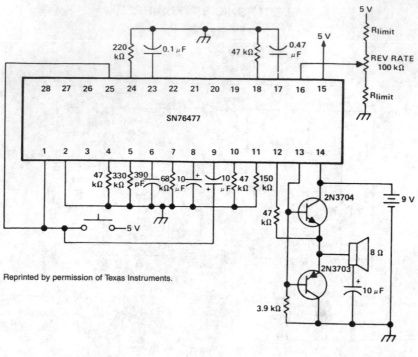

Reprinted by permission of Texas Instruments.

Fig. 93-5

For two simultaneous race-car sounds, the mixer can be multiplexed between the SLF and VCO functions.

94

Sound-Operated Circuits

The sources of the following circuits are contained in the Sources section beginning on page 782. The figure number contained in the box of each circuit correlates to the sources entry in the Sources section.

Sound-Activated Switch
Voice-Operated Switch

SOUND-ACTIVATED SWITCH

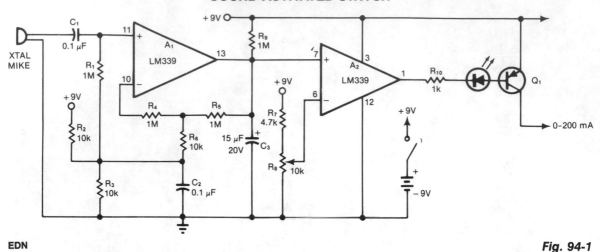

EDN

Fig. 94-1

A1 and A2 are two sections of a quad comparator. The first, A1, functions as an amplifier and detector. Resistors R5 and R6 set the gain at 100; the output of A1 is an open collector to negative-peak-rectify the output with a decay time constant determined by R9 and C3. This dc output is then compared with the reference level selected by R8. A2 triggers switch Q1, and an LED inserted in the base drive of Q1 gives visual indication of switch closure. The standby battery drain is 2 mA. Use potentiometer R8 to select the desired sensitivity.

VOICE-OPERATED SWITCH

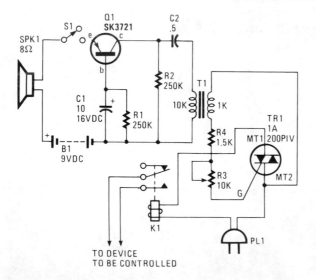

The sound picked up by SPKR1, which acts as a microphone, is fed to transistor amplifier Q1. The output of Q1 is applied across coupling transformer T1 and is used to drive the gate circuit of Triac TR1. TR1 is used to lend a latching effect to the action of the relay.

HANDS-ON ELECTRONICS

Fig. 94-2

95

Splitters

The sources of the following circuits are contained in the Sources section beginning on page 782. The figure number contained in the box of each circuit correlates to the source entry in the Sources section.

Wideband Signal Splitter
Precision Phase Splitter

WIDEBAND SIGNAL SPLITTER

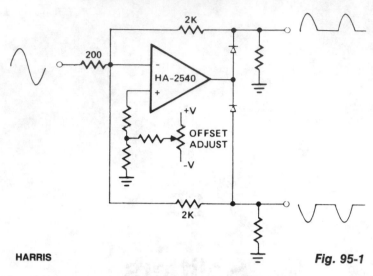

HARRIS

Fig. 95-1

With one HA-2539 or HA-2540 and two low-capacitance switching diodes, signals exceeding 10 MHz can be separated. This circuit is most useful for full-wave rectification, AM detection, or sync generation.

PRECISION PHASE SPLITTER

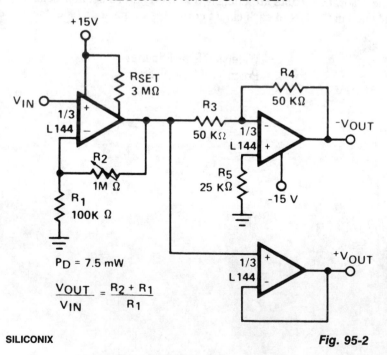

$$\frac{V_{OUT}}{V_{IN}} = \frac{R_2 + R_1}{R_1}$$

SILICONIX

Fig. 95-2

96

Square-Wave Generators

The sources of the following circuits are contained in the Sources section beginning on page 782. The figure number contained in the box of each circuit correlates to the sources entry in the Sources section.

Square-Wave Pulse Extractor
Nearly 50% Duty-Cycle Multivibrator
High-Current Oscillator
Quadrature-Output Oscillator

SQUARE-WAVE PULSE EXTRACTOR

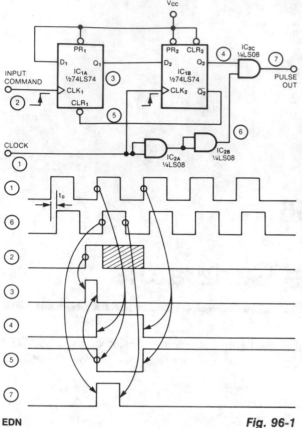

This circuit traps a single positive pulse from a square-wave train. Following the rising edge of an input command, the pulse-out signal emits a replica of one positive pulse of the clock signal simultaneous with the clock signal's next rising edge. The input command signal sets the Q1 output of flip-flop IC1A. Consequently, the next rising edge of the clock signal sets the Q2 output of IC1B, which allows AND gate IC2C to pass the clock signal's next positive pulse. AND gates IC2A and IC2B prevent the generation of brief output glitches by delaying the clock signal by t_D seconds (two propagation delays).

EDN

Fig. 96-1

NEARLY 50% DUTY-CYCLE MULTIVIBRATOR

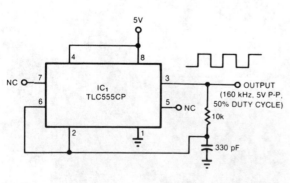

Three factors contribute to the output symmetry. The capacitor charges and discharges through the same external resistor. An internal resistive divider sets accurate switching thresholds within the chip, the bipolar types use dividers, as well. Most importantly, IC1's CMOS output stage switches fully between ground and V_{CC}, avoiding the errors from asymmetry that are often found in a TTL timer's output. The IC's internal switching-threshold tolerances can cause a deviation of several percent from the desired 50% duty cycle. To meet a tighter specification, you might have to select from a group of ICs.

EDN

Fig. 96-2

HIGH-CURRENT OSCILLATOR

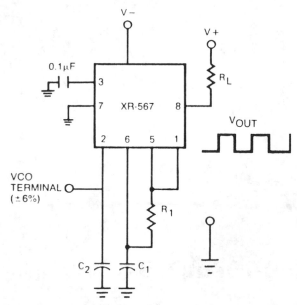

The oscillator output of the XR-567 can be amplified using the output amplifier and high-current logic output available at pin 8. In this manner, the circuit can switch 100-mA load currents without sacrificing oscillator stability. The oscillator frequency can be modulated over $\pm 6\%$ in frequency by applying a control voltage of pin 2.

EXAR   *Fig. 96-3*

QUADRATURE-OUTPUTS OSCILLATOR

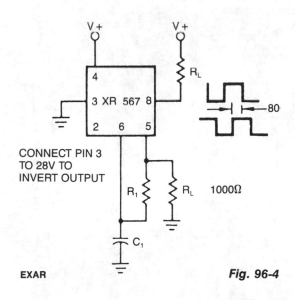

The XR-567 functions as a precision oscillator with two separate square-wave outputs at pins 5 and 8, that are at nearly quadrature phase with each other. Because of the internal biasing arrangement, the actual phase shift between the two outputs is typically 80%.

EXAR   *Fig. 96-4*

97

Staircase Generators

The sources of the following circuits are contained in the Sources section beginning on page 782. The figure number contained in the box of each circuit correlates to the sources entry in the Sources section.

μA2240 Staircase Generators
Staircase Generator

UA2240 STAIRCASE GENERATOR

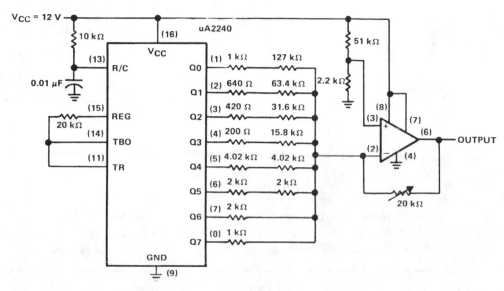

Reprinted by permission of Texas Instruments.

Fig. 97-1

The uA2240 timer/counter, combined with a precision resistor ladder network and an op amp, form the staircase generator. In the astable mode, once a trigger pulse is applied, the uA2240 operates continuously until it receives a reset pulse. The trigger input at pin 11 is tied to the time base output at pin 14, resulting in automatic starting and continuous operation. The frequency of the time-base oscillator, TBO, is set by the time constants R1 and C1 ($f = 1/R1C1$). For this example, a 10-KΩ resistor and a 0.01-μF capacitor form the timing network.

The counter outputs are connected to a precision resistor ladder network with binary-weighted resistors. The current sink through the resistors connected to the counter outputs correspond to the count number. For example, the current sink at Q7, the most significant bit, is 128 times the current sink at Q0, the least significant bit. As the count is generated by the uA2240 eight-bit counter, the current sink through each active binary-weighted resistor decreases the positive output of the op amp in discrete steps. The feedback potentiometer is set at a nominal 10 KΩ to supply a maximum output voltage range. An input of 12 V allows a 10-V output swing. With a 0.5-V input reference on pin 3 of the TCL271, the output will change from 10.46 V maximum, in 256 steps of 38.9 mV per step, to a 0.5 V minimum. Each step has a pulse duration of 100 μs and an amplitude decrease of 38.9 mV. The waveform output is repeated until a reset is applied to the uA2240.

STAIRCASE GENERATOR

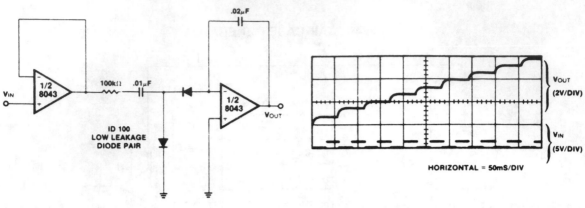

INTERSIL

Fig. 97-2

This circuit is a high-input impedance version of the so-called *diode pump* or *staircase generator*. Note that charge transfer takes place at the negative-going edge of the input signal. The most common application for staircase generators is in low-cost counters. By resetting the capacitor when the output reaches a predetermined level, the circuit can be made to count reliably up to a maximum of about 10.

98

Strobe Circuit

The sources of the following circuits are contained in the Sources section beginning on page 782. The figure number contained in the box of each circuit correlates to the sources entry in the Sources section.

Variable Strobe Light

VARIABLE STROBE LIGHT

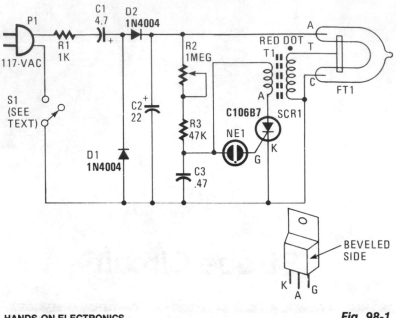

Fig. 98-1

In this strobe-light, two circuits are needed; one circuit charges a capacitor placing 320 Vdc between the cathode and anode of the flashtube. The other circuit provides bursts of approximately 4000 V to trigger the flashtube into conduction. The voltage-doubler works by summing two equal voltages in series, which results in a doubling of the voltage. The 4000 V needed to trigger the flashtube is provided by transformer T1—a voltage step-up transformer that develops 4000 V across its secondary coil when current flows in the primary coil. Silicon-controlled rectifier SCR1 controls the current flow in the primary coil of T1. When SCR1 conducts, current flows suddenly in the primary coil and 4000 Vac spikes appear across the secondary coil. For conduction, SCR1 needs a negative and positive voltage on the cathode and anode, respectively, and a positive voltage on the gate. It is the function of components R2, R3, C3, and NE1 to provide that positive gate voltage and turn on SCR1. Potentiometer R2, resistor R3, and capacitor C3 form an rc timing circuit. Control of charging time of C3 is accomplished by varying that resistance in the circuit. When the voltage on C3 reaches the firing voltage of the neon bulb, it causes NE1 to conduct, thus placing a positive voltage, from C3, on the gate of SCR1. The SCR now turns on and C3 discharges through SCR1 and the primary coil of T1. The 4000 V that is developed across the secondary coil of T1 fires the xenon tube, causing a bright flash. The whole process then repeats itself with C3 charging up, NE1 firing to short out SCR1, and T1 developing 4000 V to trigger the xenon flashtube.

99

Switching Circuits

The sources of the following circuits are contained in the Sources section beginning on page 782. The figure number contained in the box of each circuit correlates to the sources entry in the Sources section.

Rf Power Switch
Switch Debouncer
SCR-Replacing Latching Switch
One-MOSPOWER FET Analog Switch
On/Off Inverters

RF POWER SWITCH

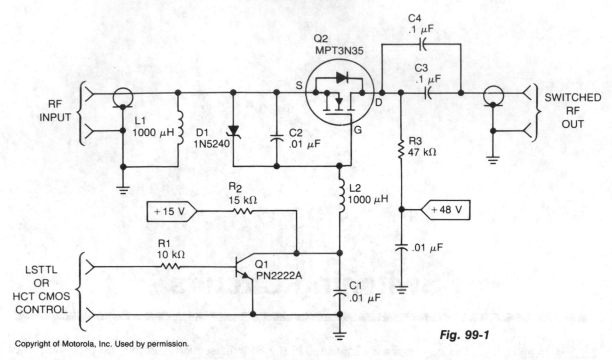

Fig. 99-1

This rf power switch operates at 1.7 MHz with a 50-V source and load. Its on loss is 0.2 dB and its off isolation is 30 dB. It provides 40-W PEP, 45 V_{PEAK} and 0.9 A_{PEAK}. The control input can come from CMOS, TTL, LS, etc., to turn on Q1, which turns on Q2, a TMOS MTP3N35.

SWITCH DEBOUNCER

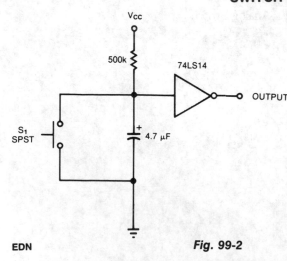

Fig. 99-2

TTL inverter 74LS14 has an internal 16-KΩ, pull-up resistor that pulls the gate input high when the switch is open. As you close the switch, the 4.7-μF capacitor discharges on the first contact. If the switch contacts bounce open, the internal resistor limits the capacitor's recharge to a rate sufficiently slow to prevent an undesired gate transition before the contacts again close. Note that the circuit correctly debounces the switch for both opening and closing. If you add an external pull-up resistor, you can use a CMOS Schmitt-trigger gate, 74HC14, and a smaller, 0.1-μF, capacitor.

EDN

SCR-REPLACING LATCHING SWITCH

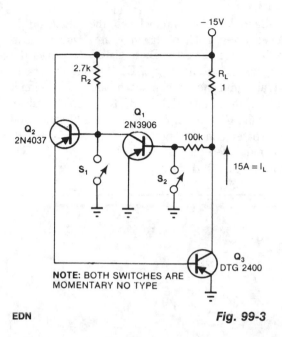

NOTE: BOTH SWITCHES ARE
MOMENTARY NO TYPE

EDN

Fig. 99-3

This circuit provides the turn-on characteristics of an SCR, but turns off with ease. The switch is comprised of three transistors with descending current ratings: Q3 has a high-current rating and Q2 has a medium rating. The current, I1, to be switched is 15 A. Momentarily depressing S2 removes Q1's base drive, turning Q1 off and allowing Q2 to turn on. Q2 then drives the base-emitter junction of Q3, turning Q3 on. Q3's collector-emitter voltage, which serves as Q1's base drive, is essentially zero, keeping Q1 off. To turn Q3 off, depress S1; this action momentarily shunts Q2's base current to ground, reversing the chain of events that turned Q3 on.

ONE-MOSPOWER FET ANALOG SWITCH

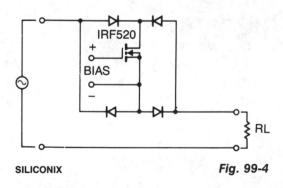

SILICONIX

Fig. 99-4

Using four diode in an array allows using only one MOSPOWER transistor for analog switching. The current flow is controlled by keeping the source-base connection of the MOSFET towards the load. Be sure to use diodes capable of handling the load current and a transistor whose breakdown voltage specification exceeds the peak analog voltage anticipated. Operationally, by increasing the gate-to-source bias voltage, the MOSFET turns on. For applications other than either full-on or full-off, care must be taken not to exceed the dissipation of the MOSPOWER transistor. A suitable heatsink cannot be overstressed in such applications.

ON/OFF INVERTERS

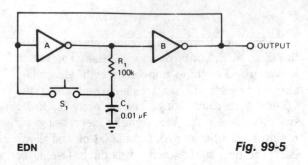

Each time the switch closes, the voltage on C1 causes inverter A to change state, with positive feedback from inverter B. Resistor R1 delays the charging and discharging of C1, making the circuit virtually immune to contact bounce. The circuit works with either CMOS or TTL gates. The values of R1 and C1 are not critical and can be increased for greater contact bounce protection, if needed. Recommended ranges are 10 K to 1 MΩ for R1, and 0.01 to 1.0 μF for C1.

EDN

Fig. 99-5

100

Tachometer Circuits

The sources of the following circuits are contained in the Sources section beginning on page 782. The figure number contained in the box of each circuit correlates to the sources entry in the Sources section.

LOW-FREQUENCY TACHOMETER

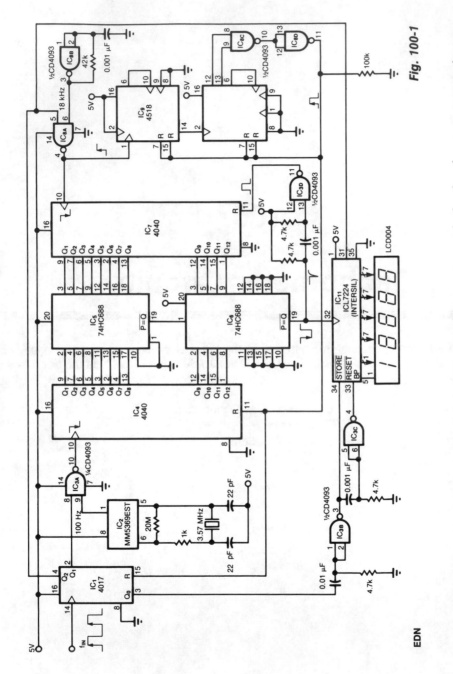

Fig. 100-1

EDN

This tachometer lets you measure heartbeats, respiratory rates, and other low-frequency events that recur at intervals of 0.33 to 40.96 seconds. The circuit senses the period of f_{IN}, computes the equivalent pulses per minute, and updates the LCD accordingly. Although the decimal readout equals 60 f_{IN}, the circuit doesn't actually produce a frequency of 60 f_{IN}. The computation involves counting and comparison techniques and takes 0.33 seconds.

TACHOMETER

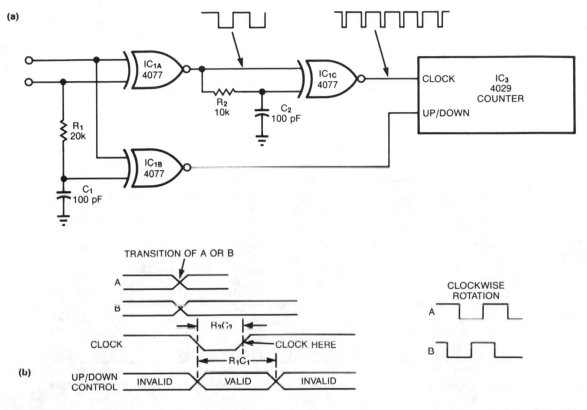

Fig. 100-2

A standard shaft encoder's A and B ports generate square waves with the same frequency as the shaft turns. The phase of A will lead or lag that of B by 90°, depending on the direction of rotation. To obtain maximum resolution, the tachometer circuit must count every change of the state for the A and B signals. Each such change causes a change of state at IC1A's output, followed by a 1-μs negative pulse at the output of IC1C. These clock pulses' positive (trailing) edges cause the counter to count up or down, according to the direction of shaft rotation.

You should set the R1C1 time constant, so that it is approximately twice that of the R2C2 product, to ensure adequate setup and hold times for the up/down signal with respect to the positive clock edges. IC1C supports this timing requirement by producing clock pulses of similar duration for either positive or negative transitions or IC1A.

The exclusive-NOR logic of IC1B generates the correct polarity of the up/down signal when necessary, at the positive clock edges, by combining the A value with the B value just prior to a transition of A or B. C1 provides memory by sorting the B value voltage for about 2 μs. The maximum frequency for A or B is approximately $(4R1C1)^{-1}$.

CALIBRATED TACHOMETER

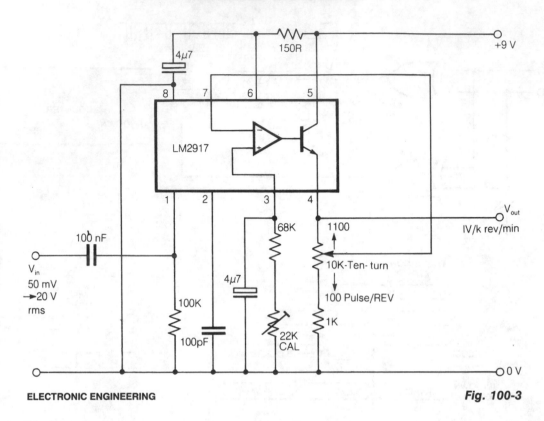

ELECTRONIC ENGINEERING

Fig. 100-3

Here is a simple tachometer circuit for use with a hand-held DVM or portable chart recorder. A novel feature is that the source frequency pulse/rev rate can be directly set on a ten-turn potentiometer to provide a convenient calibration of one V per 1000 rev/min. This is particularly useful when measuring a shaft or engine speed by sensing gear teeth.

The circuit uses an LM2917 IC which is specifically designed for tachometer applications. The ten-turn potentiometer, which provides the pulse/rev setting, is suitably configured in the output amplifier feedback path. The pulse/rev range is 100 to 1100, so the potentiometer dial mechanism should be set to start at 100 to provide direct calibration.

The IC's internal 7.5-V zener provides stable operation from a 9-V battery. The tachometer accepts an input signal between 50 mv and 20 V rms and has an upper speed limit of 6000 rev/min with the component values shown.

101

Tape-Recorder Circuits

The sources of the following circuits are contained in the Sources section beginning on page 782. The figure number contained in the box of each circuit correlates to the sources entry in the Sources section.

EXTENDED-PLAY CIRCUIT

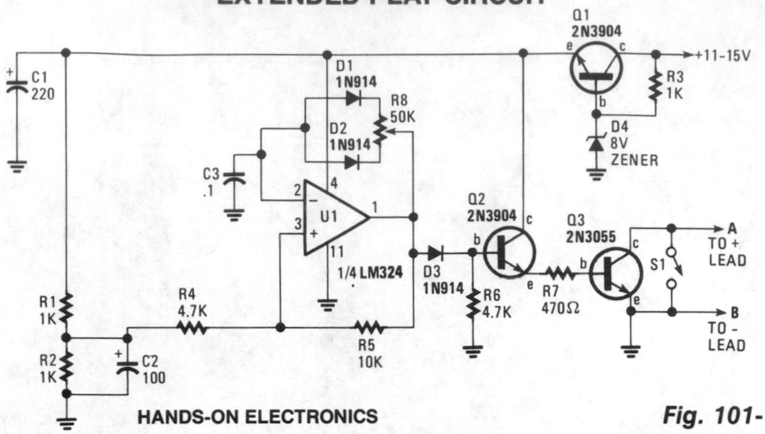

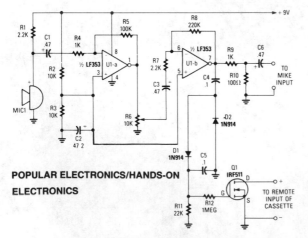

HANDS-ON ELECTRONICS

Fig. 101-1

A single op amp—one of four contained in the popular LM324—is operating in a variable pulsewidth, free-running squarewave oscillator circuit, with its timed output driving two transistors that control the on/off cycle of the tape-drive motor.

The oscillator's positive feedback path holds the secret to the successful operation of the variable on/off timing signal. The two diodes and pulsewidth potentiometer R8 allows the setting of the on and off time, without affecting the oscillator's operating frequency. One diode allows only the discharge current to flow through it and the section of R8 that it's connected to. The other diode, and its portion of R8, sets the charge time for the timing capacitor, C3. Since the recorder's speed is controlled by the precise off/on timing of the oscillator, a simple voltage-regulator circuit (Q1, R3, and D4) is included.

Connecting the speed control to most cassette recorders is a simple matter of digging into the recorder and disconnecting either of the motor's power leads, the ground or common side might be best, and connecting the recorder through a length of small, shielded cable to the control circuit. In some recorders, a remote input jack is furnished to remotely turn on and off the recorder. Before going in and modifying a recorder with a remote jack, try connecting the circuit to the external remote input.

SOUND-ACTIVATED SWITCH

POPULAR ELECTRONICS/HANDS-ON ELECTRONICS

A sensitive electret microphone picks up the sound and feeds the signal to a two-stage amplifier circuit, consisting of U1a and U1b. The amplified output of U1b is fed to a voltage-doubler circuit (comprised of D1, D2, C4, and C5). The output of the doubler is input to the gate of Q1. When the dc voltage reaches the gate's threshold level, Q1 switches on, starting the recorder. Resistor R6 sets the circuit's sensitivity and should be experimented with to obtain the optimum adjustment.

Fig. 101-2

SOUND-ACTIVATED TAPE SWITCH

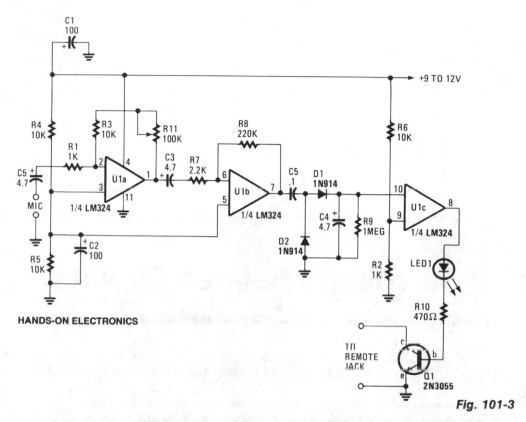

HANDS-ON ELECTRONICS

Fig. 101-3

This circuit can cause a cassette recorder to automatically turn on and record when a sound or noise is present. Another use, is when the sound-activated switch is used to turn on a cassette player so that it operates as a burglar-alarm detector and sounder. Op amps U1a and U1b are connected in tandem to amplify the sounds picked up by the detector's mike. The amplified audio voltage, output at pin 7 of U1b, is fed to a voltage-doubler circuit, consisting of D1 and D2. The elevated voltage from the doubler circuit is input to the positive input of op amp U1c, which is operating as a simple comparator circuit. The other input of U1c is connected to a voltage divider that sets the switching point for the dc signal voltage, to turn on when the signal level is greater than about 1.5 V. As the comparator switches on, its output at pin 8 becomes positive and supplies a forward bias to turn on D3 and Q1, which in turn, starts the recorder. The rc combination of C4/R9 sets the cassette's run time after the input sound has ceased, preventing the recorder from chopping-up or turning-off between closely spaced sounds or words picked up by the mike. The delay time is roughly 6 to 8 seconds. R11 sets the circuit's gain. Connect a low-impedance cassette mike to the amplifier's input, and connect the output of Q1 to the cassette's remote input or to the internal input and set the recorder to the record position. Talk and adjust the amplifier's gain with R11 for the desired sensitivity.

102

Telephone-Related Circuits

The sources of the following circuits are contained in the Sources section beginning on page 782. The figure number contained in the box of each circuit correlates to the sources entry in the Sources section.

Single-Chip Pulse/Tone Dialer
Telephone-Controlled Night Light
Hands-Free Telephone
Electronic Telephone Set with Redial
Ringer Relay
Tone-Dialing Telephone
Telephone Repeater
Speakerphone
Series Telephone Connection
Simple Touchtone™ Generator
Pulse-Dialing Telephone
Optically Interfaced Ring Detector
Parallel Telephone Connection
Add-On Telephone Hold Button

Telephone Handset Encoder
Dial Pulse Indicator
Telephone Sound Level Meter Monitor
Remote Telephone Ringer
Telephone Speech Activity Detector
Duplex Line Amplifier
Phone Recorder
Line-Activated Solid-State Switch
Cassette Interface
Ring Detector
Wireless Telephone Eavesdropper
Telephone Amplifier
Telephone Tap

SINGLE-CHIP PULSE/TONE DIALER

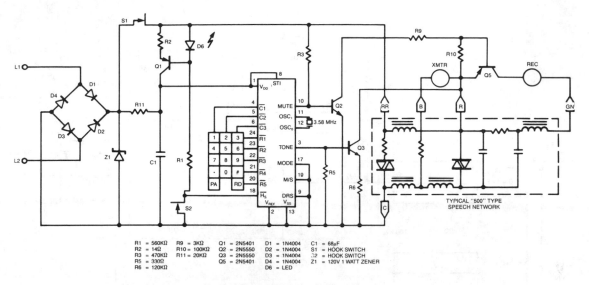

Typical Tone Dialing Application Circuit

R1 = 560KΩ	R9 = 3KΩ	Q1 = 2N5401	D1 = 1N4004	C1 = 68μF			
R2 = 14Ω	R10 = 100KΩ	Q2 = 2N5550	D2 = 1N4004	S1 = HOOK SWITCH			
R3 = 470KΩ	R11 = 20KΩ	Q3 = 2N5550	D3 = 1N4004	S2 = HOOK SWITCH			
R5 = 330Ω		Q5 = 2N5401	D4 = 1N4004	Z1 = 120V 1 WATT ZENER			
R6 = 120KΩ			D6 = LED				

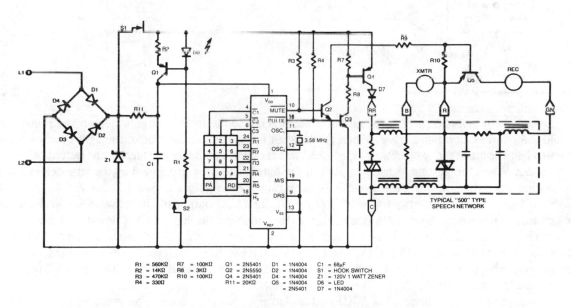

Typical Pulse Dialing Application Circuit

R1 = 560KΩ	R7 = 100KΩ	Q1 = 2N5401	D1 = 1N4004	C1 = 68μF
R2 = 14KΩ	R8 = 3KΩ	Q2 = 2N5550	D2 = 1N4004	S1 = HOOK SWITCH
R3 = 470KΩ	R10 = 100KΩ	Q4 = 2N5401	D4 = 1N4004	Z1 = 120V 1 WATT ZENER
R4 = 330Ω		R11 = 20KΩ	Q5 = 1N4004	D6 = LED
			= 2N5401	D7 = 1N4004

Fig. 102-1

EXAR

The XR-T5990 single-chip pulse/tone Dialer is a silicon gate CMOS circuit which performs both pulse and tone functions. It is designed to operate directly from the telephone line or on a separate small power supply. A 17-digit buffer is provided for redial feature.

TELEPHONE-CONTROLLED NIGHT LIGHT

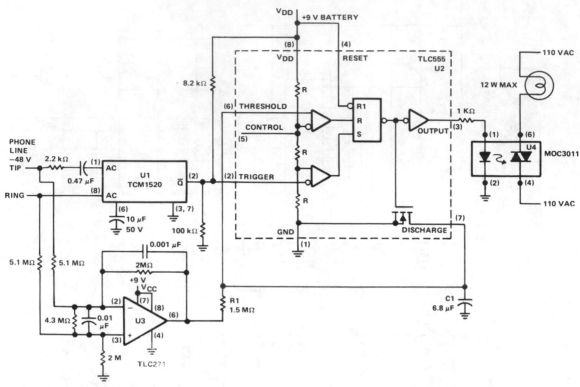

Reprinted by permission of Texas Instruments.

Fig. 102-2

When the telephone rings, or when the handset is lifted, the night light is turned on and remains on while the conversation takes place. When the handset is replaced in the cradle, the light remains on for about 11 s. During standby conditions, the -28 Vdc bias on the phone line maintains the output of U3 in a high state. When the ac ring signal is applied to the phone line, it is processed by the ring detector U1, producing a negative output pulse at pin 2 for each ring. These pulses trigger U2, causing its output to become high and the discharge transistor to turn off. The high output of U2 activates optoisolator U4, which turns on the night light. Each ring retriggers the timer and discharges C1, preventing it from reaching the $^2/_3\ V_{DD}$ threshold level. Thus, the night light will remain on while the phone is ringing and for about 11 s after the last ring. After 11 s, C1 will be charged to the U2 threshold level ($^2/_3\ V_{DD}$) resulting in the U2 output returning to a low level and its discharge output turning on, discharging C1. The lamp will turn off if the phone is not answered.

When the phone is answered, a 1-KΩ load is placed across the phone. This removes the differential input to op amp U3, causing its output to become low, and capacitor C1 starts discharging through R1. As long as the voltage across C1 remains low, timer U2 cannot start its cycle and the lamp will remain on. When the phone is hung up, the low impedance is removed from the phone line and the differential voltage across the line causes the U3 output to become high. This allows C1 to start charging, initiating the timing that will turn off the night light.

HANDS-FREE TELEPHONE

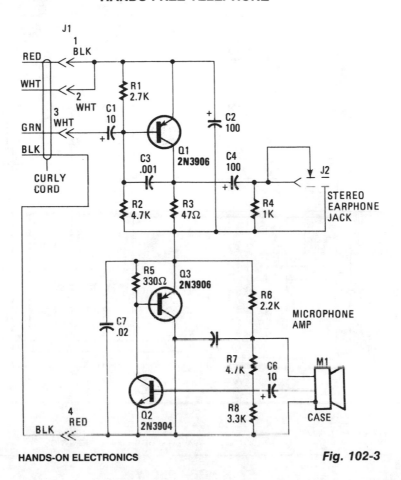

Fig. 102-3

Transistor Q1 of the headset amplifier circuit amplifies the 30 mV signal, that would have gone to the earphones, to .5 V, which sufficiently drives the stereo earphones. Capacitor C1 blocks any dc current from shorting back into the telephone base. Capacitor C2 provides the very important ac signal short around the amplifier. Capacitor C3 provides high-frequency rolloff characteristics and prevents the amplifier from oscillating. Capacitor C4 is a dc block to the 35-Ω impedance of the stereo earphones, and resistor R4 bleeds off any charge build up to prevent a popping sound when the stereo earphones are plugged into the mini-earphone jack J2. The headset amplifier has only about 2 Vdc across it. The microphone amplifier circuit is composed of transistors Q2 and Q3 in an inverted-Darlington configuration.

Another, and perhaps easier, way to understand the operation of this circuit is to consider Q3 as an emitter-follower stage. The electret microphone has a built-in FET IC amplifier that needs at least 3 V at 0.4 mA of clean supply power in order to provide an output impedance of 200 to 800 Ω. Resistors R6 and C5 provide that clean dc power to the FET IC and also provide the bias to Q2 without an ac feedback, which would have reduced Q2's gain. Capacitor C6 blocks the output dc bias from the FET IC.

ELECTRONIC TELEPHONE SET WITH REDIAL

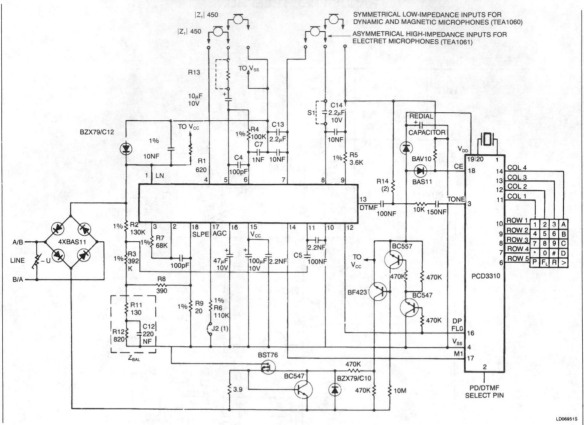

NOTES:
1. Automatic line compenstion obtained by connecting R6 to V$_{SS}$.
2. The value of resistor R14 is determined by the required level at LN and the DTMF gain of the TEA1060.   **SIGNETICS**

Fig. 102-4

RINGER RELAY

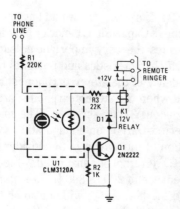

When the phone rings, the ring signal from the telephone company lights a neon lamp within a CLM3120 optocoupler. That causes a drop in the resistance of the CdS cell output of the device, turning on transistor Q1. When Q1 turns on, relay K1 is energized. The circuit should be connected in series with the lamp that is to be activated.

POPULAR ELECTRONICS

Fig. 102-5

TONE-DIALING TELEPHONE

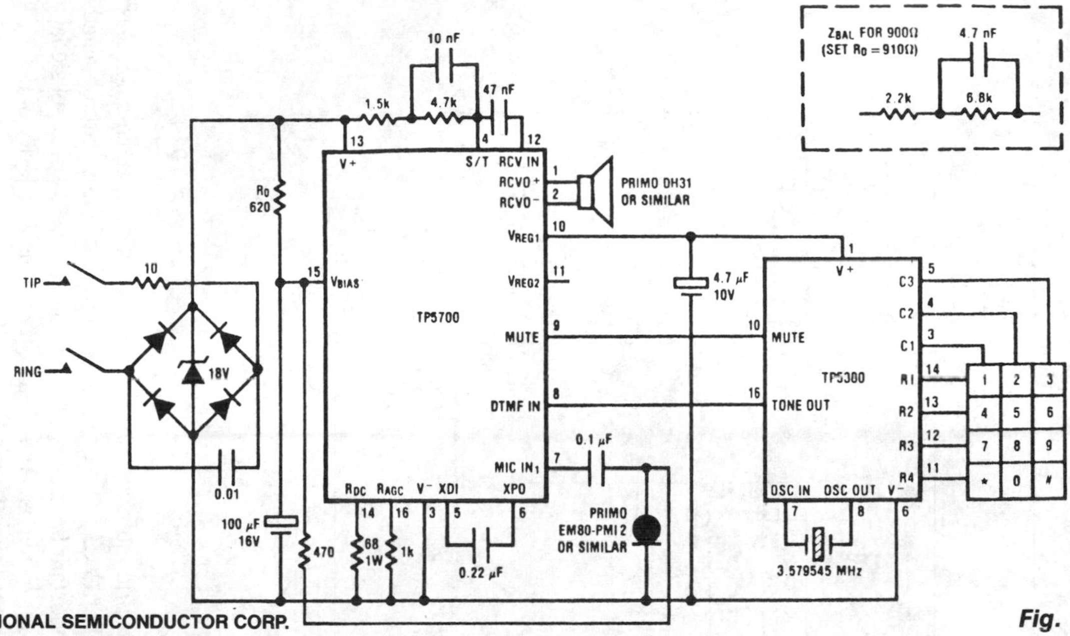

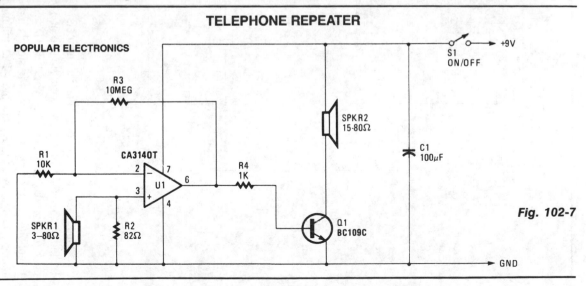

NATIONAL SEMICONDUCTOR CORP.

Fig. 102-6

This circuit shows the TP5700 directly interfacing to a low-voltage DTMF generator. V_{REG1} supplies the necessary 2 V minimum bias to enable the TP5380 to sense key closures and pull its mute output high. V_{REG1} then switches to a 3-V regulated output to sustain the tone dialer during tone generation. The TP5700 DTMF input incorporates the necessary load resistor to V− and provides gain, plus AGC action, to compensate for loop length. A muted tone level is heard in the receiver. For DTMF generators with a higher output level than the TP5380, a resistive potentiometer should be added to reduce the level at the speech circuit DTMF input.

TELEPHONE REPEATER

POPULAR ELECTRONICS

Fig. 102-7

SPEAKERPHONE

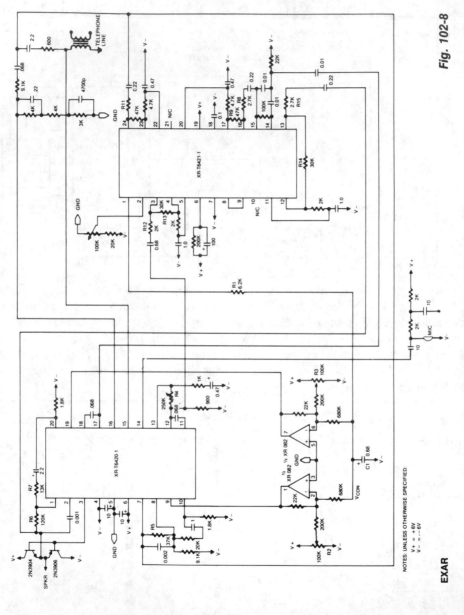

Fig. 102-8

EXAR

This circuit consists of two audio channels, a control circuit, and a hybrid interface circuit. The gain of each audio channel is controlled by the control circuitry, with the use of a voltage controlled amplifier (VCA). The inputs to the control circuit are obtained from each of the audio channels. The hybrid interface circuit performs three important functions. First, it couples the T_X channel signal to the telephone line. Second, it couples the signal on the telephone line to the R_X channels. And, finally, it cancels a majority of the T_X signal that can couple into the R_X channel. The amount of T_X signal that appears on the R_X channel is called sidetone.

SERIES TELEPHONE CONNECTION

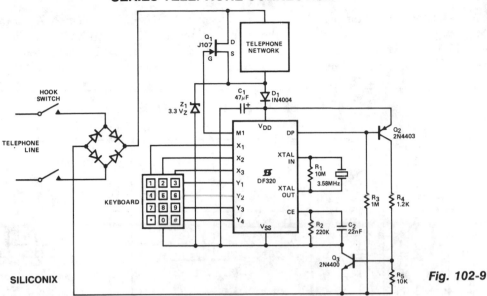

SILICONIX

Fig. 102-9

Here is a simple method of series connection into the telephone set suitable for PABX or short line applications. When the telephone handset is lifted, C1 is charged via D1 to $(V_{Z1} - 0.7)$ V and DF320 power on reset occurs. When the first keyed digit is recognized, M1 goes to logic *1*, muting the telephone network by switching on the low on resistance JFET Q1, and maximizing the line-loop current for impulsing. Impulsing occurs through DP switching Q2, and hence Q3 turns off. Rapid discharge of C1 through Z1 is prevented during line break by blocking diode D1. When dialing is complete, the circuit returns to the static standby condition, and Q1 is switched off. The circuit reset, during a line interruption by the cradle switch, is for the parallel connection mode.

SIMPLE TOUCHTONE™ GENERATOR

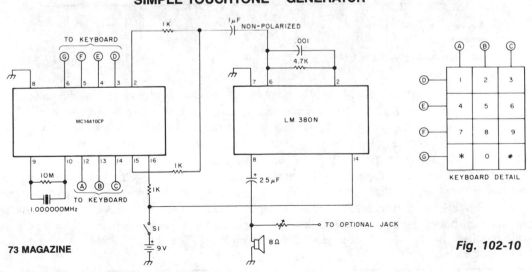

73 MAGAZINE

Fig. 102-10

The oscillator is a Motorola MC14410CP chip using a 1-MHz crystal. The chip generates both the high and low tones, feeding the energy to the amplifier through 1-K resistors and the 1-μF capacitor. Values for the output resistors can vary from a few hundred Ω to about 60 KΩ. The value of the resistor shunting the crystal can vary from about 3 to 15 MΩ. The amplifier consists of an LM-380N.

PULSE-DIALING TELEPHONE

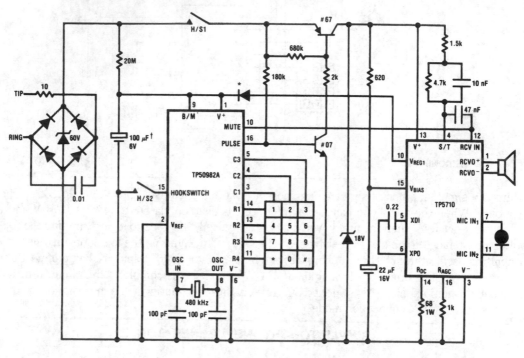

* Select as necessary to suit mic sensitivity

† Low leakage type

NATIONAL SEMICONDUCTOR CORP.

Fig. 102-11

The TP5700 or TP5710 can reduce the number of components required to build a pulse-dialing telephone, as shown. The usual current source can be eliminated by using the V_{REG1} output to power a TP50982A low-voltage (1.7 V) pulse dialer via a blocking diode. A low forward-voltage drop diode such as a Schottky type is necessary because V_{REG1} is used in its nonregulated mode and its output voltage might fall to 2 V on a 20-mA loop. A 100-μF decoupling capacitor is required to hold up the pulse dialer supply voltage during dialing. This capacitor will take about one second to charge up when the telephone is first connected to the line, but thereafter, the 20-MΩ resistor, required to retain the last-number dialed memory, will keep this capacitor charged. Partial muting is obtained by directly connecting the N-channel open-drain mute output of the pulse dialer to the RCV in pin on the speech circuit. A fully muted pulse dialer design requires the use of a shunt-mode dialer, such as the TP50981A or TP50985A.

OPTICALLY INTERFACED RING DETECTOR

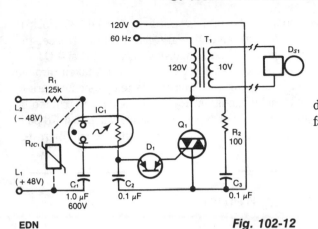

This ring detector, utilizing a neon-LDR (light-dependent resistor) optocoupler, simplifies interfacing with telephone lines.

EDN

Fig. 102-12

PARALLEL TELEPHONE CONNECTION

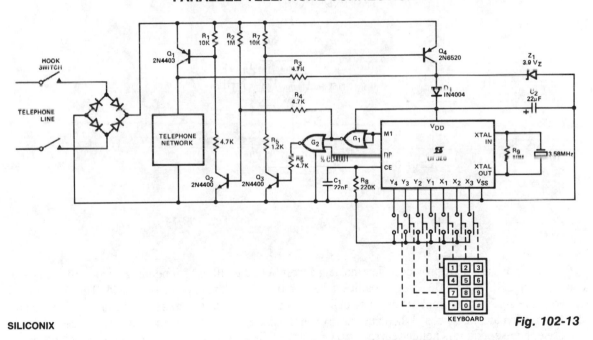

SILICONIX

Fig. 102-13

When the handset is lifted and power is applied to the circuit, Q2 is fed base current through R2, which in turn drives Q1. C2 is charged via R3 in series with D1 to $(V_{Z1} - 0.7)$ V. When the minimum operating V_{DD} voltage is reached, power on reset occurs via the rc network of C1 and R8. Q2 is maintained in the on condition by G1, while Q3, and hence Q4, are held off by G2. The DF320 network appears in parallel with the telephone as an impedance more than 10 KΩ in the standby condition with the telephone network connected in circuit through Q1. On recognition of the first keyed digit, the DF320 clock is started. M1 then goes to logic 1 causing Q2 and Q1 to turn off, and Q3 and Q4 to turn on. Hence, the majority of the line loop current now flows through Q4 and Z1. When impulsing occurs, Q3 and Q4 are turned off by DP acting on G2. Line loop current is then reduced to approximately 50 μA taken through R2, R4, and G2 in series. When dialing in, complete M1 goes to logic O, causing the telephone network to be reconnected. The DF320 then returns to the static standby condition. If the line loop is interrupted by the cradle switch during dialing, impulsing will continue until C2 discharges to a voltage, such that R8 pulls CE to logic O, causing the DF320 to reset. The diode bridge protects the network from line polarity reversal.

ADD-ON TELEPHONE HOLD BUTTON

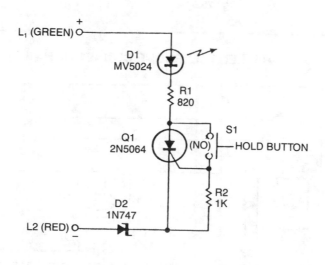

EDN

Fig. 102-14

A sensitive-gate SCR provides a line-holding current of 20 to 40 mA, depending on loop resistance. It also lights an LED to give the user a positive indication that the telephone line is on hold. The 20 to 40 mA should prove sufficient to hold the majority of lines, but it might require increasing—by decreasing the size of R1—in individual instances. When any receiver in the same loop is lifted, the low impedance of the off-hook telephone set shunts holding current away from the SCR, thereby releasing the line and extinguishing the LED. Zener diode D2 ensures that the line-holding current drops below the SCR's minimum conduction current. If the calling party tires of waiting on hold and hangs up, the release of the central-office relays from the calling side also releases the line from the hold mode.

TELEPHONE HANDSET ENCODER

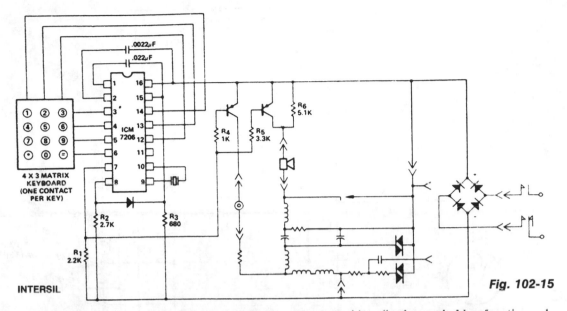

INTERSIL

Fig. 102-15

This encoder uses a single contact per key keyboard and provides all other switching functions electronically. The diode connected between terminals 8 and 15 prevents the output from going more than 1 V negative with respect to the negative supply V_{SS}. The circuit operates over the supply voltage range from 3.5 V to 15 V on the device side of the bridge rectifier. Transients as high as 100 V will not cause system failure, although the encoder will not operate correctly under these conditions. Correct operation will resume immediately after the transient is removed. The output voltage of the synthesized sine wave is almost directly proportional to the supply voltage ($V_{DD} - V_{SS}$) and will increase with the increase of supply voltage between terminals 8 and 16, after which the output voltage remains constant.

DIAL PULSE INDICATOR

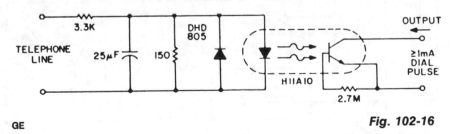

GE

Fig. 102-16

This indicator senses the switching on and off of the 48-Vdc line voltage and transmits the pulses to logic circuitry. An H11A10 threshold coupler, with capacitor filtering, gives a simple circuit which can provide dial pulse indication, and yet reject high levels of induced 60-Hz noise. The DHD805 provides reverse bias protection for the LED during transient over-voltage situations. The capacitive filtering removes less than 10 ms of the leading edge of a 40-V dial pulse, while providing rejection of up to 25-V rms at 60 Hz.

TELEPHONE SOUND LEVEL METER MONITOR

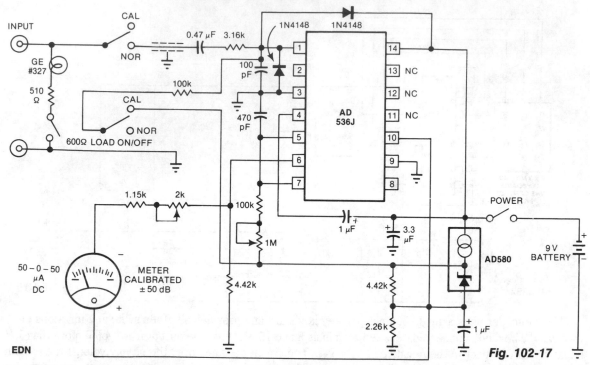

EDN

Fig. 102-17

The telephone-line decibel meter and line-voltage sensor shown lets you accurately monitor and adjust telephone sound levels. The 600-Ω resistor properly terminates the line. Power drain from the 9-V battery is 2 mA, and the meter provides ±30 dB range.

REMOTE TELEPHONE RINGER

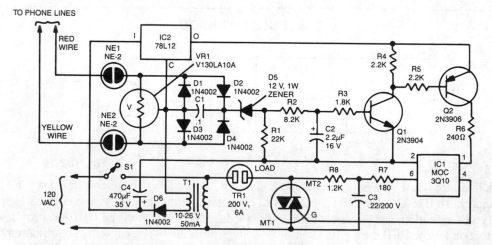

Reprinted with permission of Radio-Electronics Magazine, October 1983. Copyright Gernsback Publications, Inc., 1983.

Fig. 102-18

REMOTE TELEPHONE RINGER (*Cont.*)

The two neon bulbs will light when more than 100 V is across the ringing circuit. The bulbs provide line isolation between the unit and the telephone line. Finally, they act as a voltage divider for the bridge rectifier made up of D1 through D4. That voltage divider creates a positive voltage that is then applied through D5, is filtered by R2, R3, and C2, and causes Q1 and Q2 to conduct. When that happens, triac TR1 is fired through the optical coupler IC1; this turns on the triac, which applies 110 Vac to the load.

TELEPHONE SPEECH ACTIVITY DETECTOR

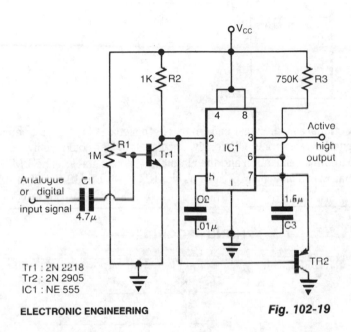

Tr1 : 2N 2218
Tr2 : 2N 2905
IC1 : NE 555

ELECTRONIC ENGINEERING

Fig. 102-19

This circuit can be used in telephone lines for speech activity detection purposes. This detection is very useful in the case of half-duplex conversation between two stations—in the case of simultaneous transmission of voice and data over the same pair of cables by the method of interspersion data on voice traffic, and also in echo suppressor devices. The circuit consists of a class-A amplifier to amplify the weak analog signals (25 – 400 mV). The IC1 which follows, is connected as a retriggerable monostable multivibrator with the TR2 discharging the timing capacitor C3, if the pulse train reaches the trigger input 2 of IC1 with period less than the time: $T_{HIGH} = 1.1\ R3C3$. The output 3 of IC1 is active on when an analog or digital signal is presented at the output, and it drops to a low level, T_{HIGH}, seconds after the input signal has ceased to exist.

DUPLEX LINE AMPLIFIER

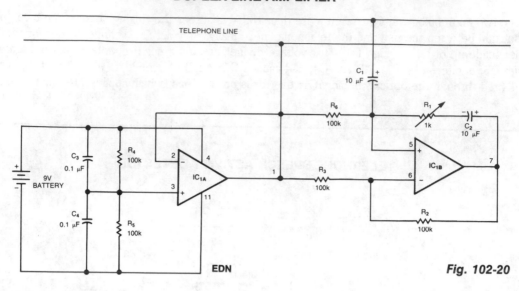

EDN

Fig. 102-20

This circuit is a bidirectional amplifier that can amplify both signals of a duplex telephone conversation. It uses the principle of negative resistance. Obviously, such an amplifier could easily be unstable; however, you can adjust R1 for maximum amplification and the circuit will remain stable. The LM324 op amps can be replaced with op amps that would distort less, such as the LM1558, LF412, LF353, or LF442.

PHONE RECORDER

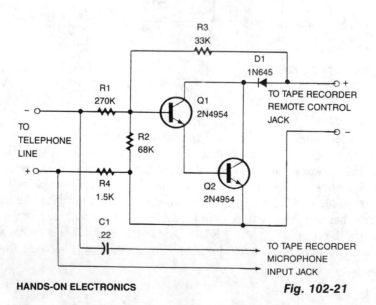

HANDS-ON ELECTRONICS

Fig. 102-21

PHONE RECORDER (*Cont.*)

This recorder can be connected to the telephone lines just about any place, and no external power source is needed. The tape recorder's switch terminals are applied to a pair of transistors, connected as Darlingtons, that are used to turn the recorder on and off. When the telephone is off-hook there's usually about 50 Vdc across the phone that's divided over R1, R2, and R4, so that Q1's base is negative enough to keep the recorder off. Pick up the receiver, and the voltage drops to 5 V. That leaves not quite-enough voltage on Q1's base to keep that transistor at cutoff, so the recorder begins. Remember to keep your recorder's switch in the on position, and depending on how many people use the telephone, remember to rewind or change tapes occasionally!

LINE-ACTIVATED SOLID-STATE SWITCH

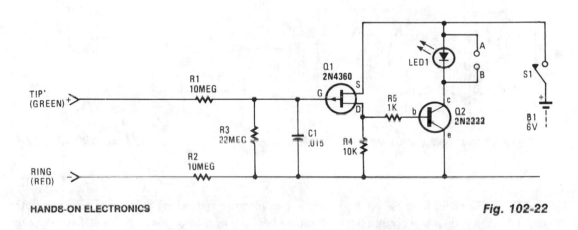

Fig. 102-22

Each and every time a phone on the same line or calling number is taken off-hook, the circuit will be activated to control an external electronic circuit. If several extension telephones are used on one phone line, the circuit can be useful as a *busy* indicator. LED1 contains a special flashing red LED that makes an excellent indicator for a *busy* circuit condition.

The solid-state switch can be used for several other phone-activated applications, such as automatically turning on a cassette recorder, starting a phone-use timer or counter, etc. A small relay can be connected at points A and B, in place of LED1, to control external circuits. A 117-Vac-to-6-Vdc plug-in power supply can be substituted for the battery to keep the operating cost at a minimum.

The 48-Vdc, on-hook, phone-line voltage keeps Q1 in the cut-off condition, allowing no current to flow through resistor R4, hence Q2 remains off. Resistors R1 and R2 keep the solid-state switch circuit from causing any problems with the telephone's central-office equipment. When a phone is taken off-hook, the line voltage (tip to ring) drops to 10 V or less, which forces Q1 to turn on; this, in turn, causes Q2 to trigger LED1, or a relay which might be used in lieu of LED1.

CASSETTE INTERFACE

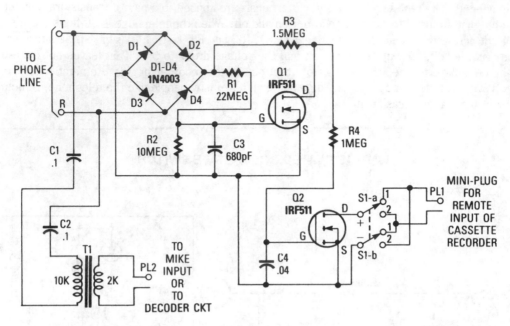

POPULAR ELECTRONICS/HANDS-ON ELECTRONICS

Fig. 102-23

Q1 and Q2 are used to form the basis of an interface circuit for attaching a cassette recorder to the phone line. The circuit does not require a power supply because operating power is drawn from the telephone line itself. The incoming signal is fed across a bridge-rectifier circuit, consisting of diodes D1 through D4.

When the phone is on hook, the voltage at the output of the bridge at the R1/R3 junction is near 48 V. That voltage is fed across a voltage divider consisting of R1 and R2. The voltage at the junction formed by R1 and R2 is fed to the gate of Q1, turning it on. That pulls the drain of Q1 low. Since the gate of Q2 is connected to the drain of Q1, the bias applied to the gate of Q2 is low, holding it in the OFF state.

When the answering machine responds to a call or a phone is taken off hook, the voltage across the phone lines drops below 10 V, causing Q1 to turn off. At that point, the voltage at Q1's drain rises, turning Q2 on. The remote input of the cassette is connected to Q2's drain and source through S1, and a miniature plug is connected to the remote input jack.

Switch S1 must be in a position so that the positive lead of the recorder's remote input connects, through switch position 1, to Q2's drain and the negative input to Q2's source. Switch S1 provides a convenient way to reverse the circuit's trigger output without having to unsolder and resolder leads. The phone's audio is coupled through C1, C2, and T1 to the microphone input of the cassette recorder.

RING DETECTOR

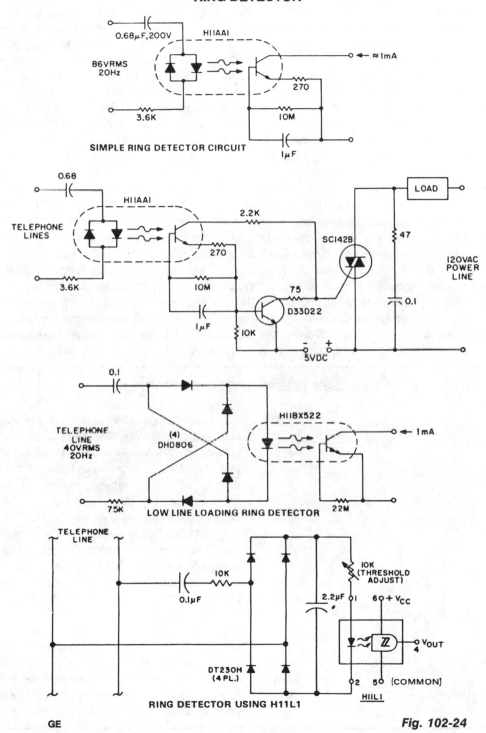

SIMPLE RING DETECTOR CIRCUIT

LOW LINE LOADING RING DETECTOR

RING DETECTOR USING H11L1

GE

Fig. 102-24

This circuit detects the 20 Hz, ≈ 86-V rms ring signal on telephone lines and initiates action in an electrically isolated circuit. Typical applications would include automatic answering equipment, and interconnect/interface and key systems. The circuits illustrated are *bare bones* circuits designed to illustrate concepts. They might not eliminate the ac/dc ring differentiation, 60-Hz noise rejection, dial tap rejection, and other effects that must be considered in field application. The first ring detector is the simplest and provides about 1-mA signal for a 7-mA line loading for $^1/_{10}$ sec after the start of the ring signal. The time delay capacitor provides a degree of dial tap and click suppression, as well as filtering out the zero crossing of the 20-Hz wave. This circuit provides the basis for a simple example, a ring extender that operates lamps and buzzers from the 120-V, 60-Hz power line, while maintaining positive isolation between the telephone line and the power line. Use of the isolated tab triac simplifies heat sinking by removing the constraint of isolating the triac heatsink from the chassis. Lower line current loading is required in many ring detector applications. This can be provided by using the H11BX522 photo-Darlington optocoupler, which is specified to provide a 1-mA output from a 0.5-mA input through the −25°C to +50°C temperature range.

The next circuit allows ring detection down to a 40-V rms ring signal while providing 60-Hz rejection to about 20-V rms. Zero-crossing filtering can be accomplished either at the input bridge rectifier or at the output. Dependable ring detection demands that the circuit responds only to ring signals, rejecting spurious noise of similar amplitude, such as dialing transients. The configuration shown relies on the fact that ring signals are composed of continuous frequency bursts, whereas dialing transients are much lower in repetition rate. The dc bridge-filter combination at the H11L input has a time constant; it cannot react to widely spaced dialing transients, but will detect the presence of relatively long duration bursts, causing the H11L to activate the downstream interconnect circuits at a precisely defined threshold.

WIRELESS TELEPHONE EAVESDROPPER

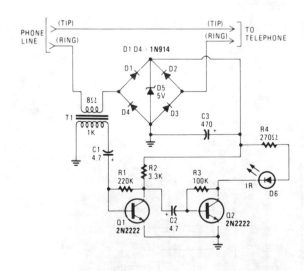

HANDS-ON ELECTRONICS   *Fig. 102-25*

The IR transmitter connects to a telephone circuit, and transmits both sides of all telephone conversations to any line-of-sight location, within 40 feet. No power is taken from the central office, as long as all phones remain on-hook. The current flows through the phone and back to the central office, thereby keying their equipment. We tap into the telephone line by connecting the IR transmitter circuit in series with either the tip or ring. When the telephone is off-hook, current will flow through the diode bridge polarity protector and supply the power for the IR transmitter. The phone's audio information is taken off the line by transformer T1. The 1000-Ω winding of the transformer connects to a two-stage transistor audio amplifier/modulator. A 2000-Ω potentiometer could be added to the input of the two-stage amplifier to control the modulation level, and another potentiometer could be added in place of R3 to adjust the IR's idle current.

TELEPHONE AMPLIFIER

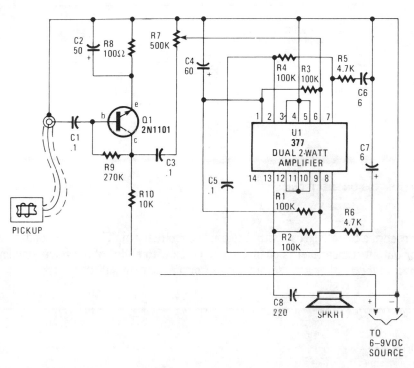

TAB BOOKS

Fig. 102-26

Audio from the telephone is inductively coupled to the base of Q1, which is used as a preamp. The preamp provides a gain of about 75 dB, to boost the input signal from about 4 mV to about 300 mV pk-pk. If you use a higher gain transistor, increase the value of R9 to produce a Q point, measured from minus to the collector of Q1, of one-half the supply voltage. The Q1 output signal is coupled through C3 to R7, which serves as a volume or drive-level control, to U1, a dual, 2 w amplifier connected in cascade. Pins 1 through 7 serve as a driver for the final amplifier, pins 8 through 13. Compensation and balance is accomplished by components R1 through R6 and C4, C6, and C7. Pins 3 through 5, and 10 through 12 should be tied to the negative supply rail.

TELEPHONE TAP

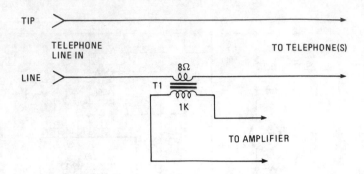

TIP

TELEPHONE LINE IN

LINE

T1

8Ω

1K

TO TELEPHONE(S)

TO AMPLIFIER

Fig. 102-27

Amplify or record a telephone call with the simple circuit shown. The 8-Ω secondary winding of a miniature transistor output transformer is connected in series with either of the telephone lines. The 1000-Ω primary winding can feed either a cassette recorder or an audio amplifier.

103

Temperature Controls

The sources of the following circuits are contained in the Sources section beginning on page 782. The figure number contained in the box of each circuit correlates to the sources entry in the Sources section.

ZERO-POINT SWITCHING TEMPERATURE CONTROL

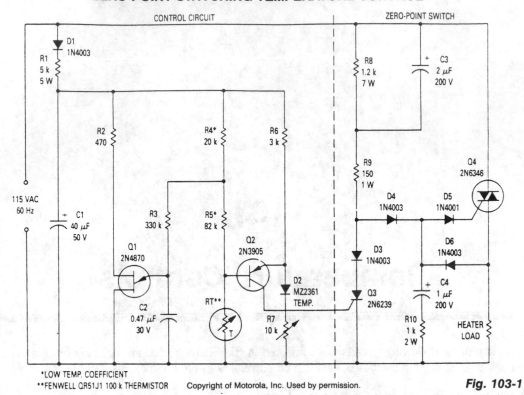

*LOW TEMP. COEFFICIENT
**FENWELL QR51J1 100 k THERMISTOR   Copyright of Motorola, Inc. Used by permission.

Fig. 103-1

This modulated triac zero-point switching circuit controls heater loads operating from 115 Vac. Circuit operation is best described by splitting the circuit into two parts. The circuit at right is the zero-point switch; to the left is the proportional control for the zero-point switch.

SERVO-SENSED HEATER PROTECTOR

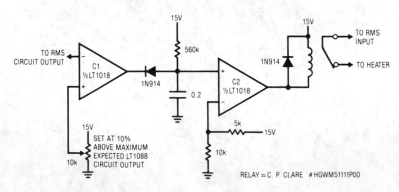

LINEAR TECHNOLOGY TECH.   *Fig. 103-2*

SERVO-SENSED HEATER PROTECTOR (*Cont.*)

This circuit responds quickly enough to prevent damage from most overloads. C1's input is connected to the output of the LT1088 servo circuit. If the LT1088 circuit's output exceeds the threshold at C1's other input, C1 trips, discharging the 2-μF capacitor. This causes C2's output to become low, energize the relay, and break the heater circuit. The 560-KΩ resistor provides a long recharge for the capacitor, preventing chattering action. This arrangement's speed of response is limited by the rms circuit's slew rate, about 0.2 V/ms. For reasonable overloads, the LT1088's temperature increases about 1°C/ms. A 10-V LT1088 output step takes 50 ms, causing a temperature rise of about 50°C.

TEMPERATURE CONTROLLER

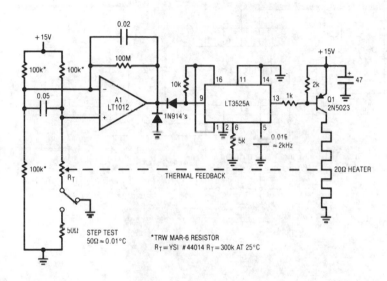

Fig. 103-3

When power is applied, the thermistor, a negative tc device, is at a high value. A1 saturates positive. This forces the LT3525A switching regulator's output low, biasing Q1. As the heater warms, the thermistor's value decreases. When its inputs finally balance, A1 comes out of saturation and the LT3525A pulse-width modulates the heater via Q1, completing a feedback path. A1 provides gain and the LT3525A is highly efficient. The 2-kHz, pulse-width modulated heater power is much faster than the thermal loop's response, and the oven sees an even, continuous heat flow.

PROPORTIONAL TEMPERATURE CONTROLLER

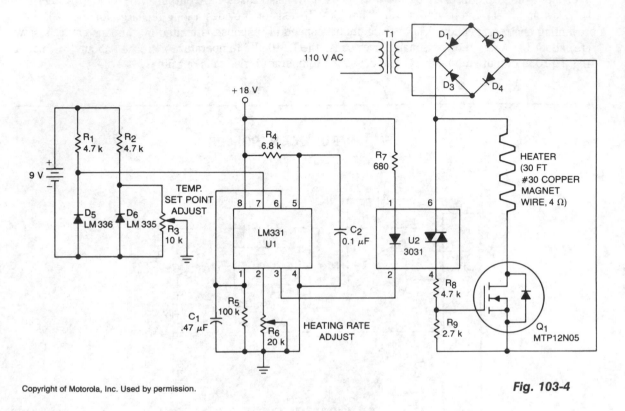

Copyright of Motorola, Inc. Used by permission.

Fig. 103-4

This temperature controller operates as a *pulse snatching* device, which allows it to run at its own speed and turn on at the zero crossing of the line frequency. Zero crossing turn-on reduces the generation of line noise transients. TMOS Power FET, Q1, is used to turn on a heater.

Temperature sensor D6 provides a dc voltage proportional to temperature that is applied to voltage-to-frequency converter U1. Output from U1 is a pulse train proportional to temperature offset that is applied to the input of triac optoisolator U2. The anode supply for the triac is a 28 V pk-pk, full-wave rectified sine wave. The optoisolator ORs the pulse train from U1 with the zero crossing of U2's anode supply, supplying a gate turn on signal for Q1. Therefore, TMOS power FET Q1 can only turn the heater on at the zero crossing of the applied sine wave. The maximum temperature, limited by the sensor and the insulation of the wire, is 130°C for the components shown.

PIEZOELECTRIC FAN-BASED TEMPERATURE CONTROLLER

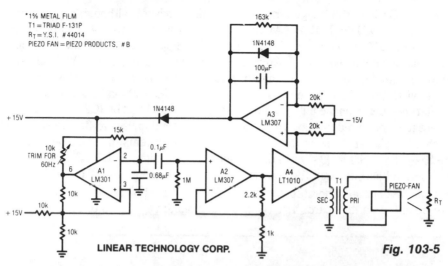

LINEAR TECHNOLOGY CORP.

Fig. 103-5

The fan employed is one of the new electrostatic type which is very reliable, because it contains no wearing parts. These devices require high-voltage drive. When power is applied, the thermistor, located in the fan's exhaust stream, is at a high value. This value unbalances the A3 amplifier driven bridge. A1 receives no power and the fan does not run. As the instrument enclosure warms, the thermistor value decreases until A3 begins to oscillate. A2 provides isolation and gain, and A4 drives the transformer to generate high voltage for the fan. In this fashion, the loop acts to maintain a stable instrument temperature by controlling the fan's exhaust rate. The 100-μF time constant across the error amplifier pins is typical of such configurations. Fast time constants will produce audibly annoying *hunting* in the servo. Optimal values for this time constant and gain depend upon the thermal and airflow characteristics of the enclosure being controlled.

ELECTRONIC HEAT SNIFFER

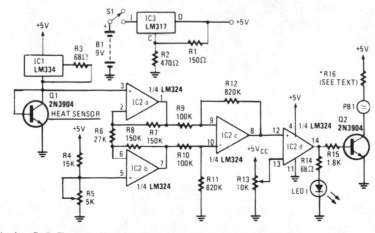

Fig. 103-6

ELECTRONIC HEAT SNIFFER (*Cont.*)

Sensing element Q1 is a 2N3904 general-purpose npn transistor, although any general-purpose npn unit in a TO-92 style case will do. IC1, an LM334, supplies Q1 with a constant current that is independent of temperature. An LM324 quad op amp, IC2, forms a high input-impedance differential amplifier (IC2a, IC2b, and IC2c) with a gain of about 99. IC2d is used as a voltage comparator. When Q1 senses a rise or fall in temperature, the base-to-emitter voltage decreases. That decrease in voltage causes the input to IC2a at pin 3 to deviate from the reference voltage that's fed to IC2b at pin 5, which is set by potentiometers R5. The difference between the input and the reference is amplified by IC2c. That amplified voltage is fed to IC2d where it is compared to a control voltage set by potentiometer R13. The setting of R13 determines the threshold and is set at a point that's equal to the ambient temperature. The output of IC2d at pin 14 is fed to the base of transistor Q2. When the output of IC2d is high, LED1 lights and Q2 turns on. With Q2 turned on, a ground path through the transistor is provided for buzzer PB1.

The circuit can be built on perforated construction board using point-to-point wiring. All components, except Q1, are mounted on the board. Transistor Q1 is mounted at the tip of the heat-sensing probe.

104

Temperature Sensors

The sources of the following circuits are contained in the Sources section beginning on page 782. The figure number contained in the box of each circuit correlates to the sources entry in the Sources section.

Thermocouple Multiplex System
0 – 63°C Temperature Sensor
Isolated Temperature Sensor

THERMOCOUPLE MULTIPLEX SYSTEM

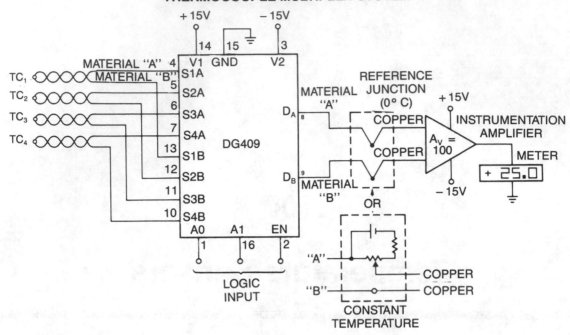

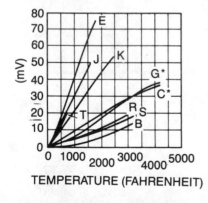

Output Voltage vs. Temperature of Several Common Thermocouples

TEMPERATURE (FAHRENHEIT)

ANSI SYMBOL

T	Copper vs Constantan
E	Chromel vs Constantan
J	Iron vs Constantan
K	Chromel vs Alumel
G*	Tungsten vs Tungsten 26% Rhenium
C*	Tungsten 5% Rhenium vs Tungsten 26% Rhenium
R	Platinum vs Platinum 13% Rhodium
S	Platinum vs Platinum 10% Rhodium
B	Platinum 6% Rhodium vs Platinum 30% Rhodium

*Not ANSI Symbol

Used with permission of Omega Engineering, Inc., Stamford, Conn., 06907

SILICONIX

Fig. 104-1

    To decouple the sensors from the meter amplifier, either a reference junction at 0°C or a bucking voltage set at room temperature may be used. The latter method is simpler, but is sensitive to changes in ambient temperature. The table above shows the output voltage vs temperature of several common types of thermocouples.

O – 63°C TEMPERATURE SENSOR

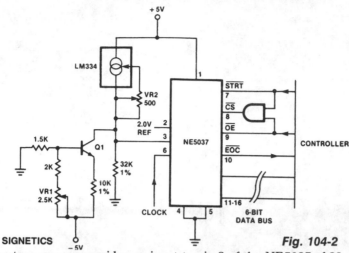

Fig. 104-2

The temperature sensor provides an input to pin 3 of the NE5037 of 32 mV/°C. This 32 mV is the value of one LSB for the NE5037. The LM334 is a three-terminal temperature sensor and provides a current of 1 μA for each degree Kelvin. The 32-KΩ resistor provides the 32 mV for each microamp through it, while the transistor bleeds off 273 μA of the temperature sensor (LM334) current. This bleeding lowers the reading by 273 K, thus converting from Kelvin to Celsius. To read temperature, conversion is started by sending a momentary low signal to pin 7 of the NE5037. When pin 10 of the NE5037 becomes low, conversion is complete and a low is applied to pin 9 of the NE5037 to read data on pins 11 and through 16. Note that this temperature data is in straight binary format. The controller can be a microprocessor in a temperature control application, or discrete circuitry in a simple temperature reporting application.

ISOLATED TEMPERATURE SENSOR

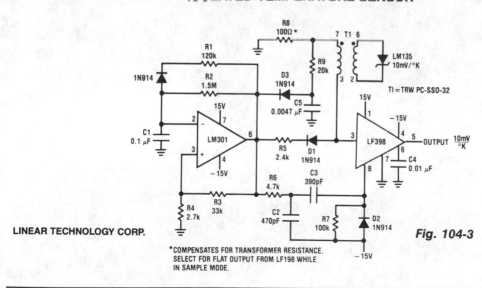

LINEAR TECHNOLOGY CORP.

*COMPENSATES FOR TRANSFORMER RESISTANCE.
SELECT FOR FLAT OUTPUT FROM LF198 WHILE
IN SAMPLE MODE.

Fig. 104-3

631

105

Temperature-to-Time Converters

The sources of the following circuits are contained in the Sources section beginning on page 782. The figure number contained in the box of each circuit correlates to the sources entry in the Sources section.

Two Simple Temperature-to-Time Converters

TWO SIMPLE TEMPERATURE-TO-TIME CONVERTERS

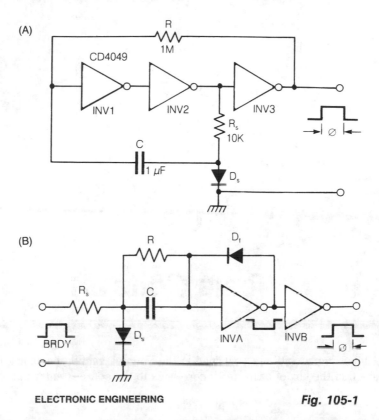

ELECTRONIC ENGINEERING   **Fig. 105-1**

Both of these converters use CMOS inverters. Figure 105-1A shows a free-running circuit having both the pulse duration and pulse pause dependent on temperature of the diode D_S. It can be used where a synchronization between the converter and something else is not required.

Figure 105-1B shows a one shot circuit that produces a pulse with its duration dependent of temperature of diode D_S. The additional diode D_f should have inverse current low enough to not influence the discharging process in the network rc when the INVA output is low. A silicon component or a GaAsP LED can be used.

The converter is intended for a digital system producing a RADY pulse which disappears after the conversion process is ended. The pulse duration is approximately:

$$= 2RC \frac{V_D}{V_{DD}}$$

where V_D is the sensor diode forward voltage and V_{DD} is the supply voltage of the CMOS chip.

Resistance R must be much higher than R_S. A 0.1-μF capacitor can be applied in parallel with D_S, if necessary, to repulse stray pickup and noise in a long cable. The circuits described can be used with a temperature sensitive resistor instead of the diode D_S.

106

Tesla Coils

The sources of the following circuits are contained in the Sources section beginning on page 782. The figure number contained in the box of each circuit correlates to the sources entry in the Sources section.

SIMPLE TESLA COIL

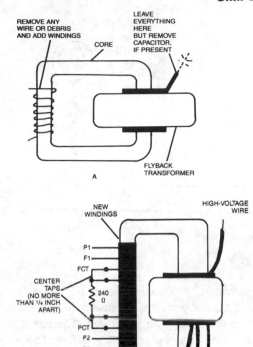

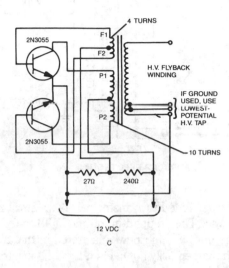

GERNSBACK PUBLICATIONS INC.

Fig. 106-1

The Tesla coil described here can generate 25,000 V. So, even though the output current is low, **be very careful!** The main component is a flyback transformer from a discarded TV.

A new primary winding is needed. Begin by winding 5 turns of #18 wire on the core. Then, twist a loop in the wire, and finish by winding five more turns. Wrap with electrical tape, but leave the loop exposed.

A four-turn winding must be wound over the ten-turn winding that you've just finished. That is done the same way. First wind two turns of #18 wire, then make a loop, and finish by winding two more turns. Again, wrap the new winding with electrical tape, leaving the loop exposed.

When the windings are finished, the two loops shouldn't be more than 1/4-inch apart, but take care that they do not touch. Connect a 240-Ω resistor between the two loops. The modified transformer now should look like the one shown. Connect the transformer as shown. The 27-Ω resistor and two transistors should be mounted on a heatsink and must be insulated from it.

The output of the high-voltage winding should begin to oscillate as soon as the circuit is connected to a 12-Vdc power supply. If it does not, reverse the connections to the base leads of the transistors. In normal operation, you should be able to draw 1-inch sparks from the high-voltage lead using an insulated screwdriver.

TESLA COIL

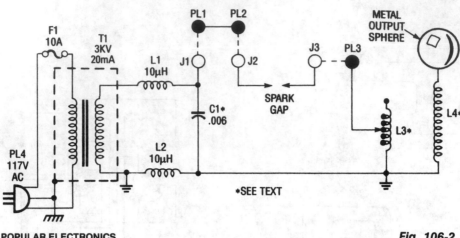

Fig. 106-2

Power is fed to transformer T1, a small neon-sign transformer, which steps the voltage up to about 3000 Vac. The stepped-up output of T1 is fed through L1 and L2 and across C1, causing the capacitor to charge until enough power is stored in the unit to produce an arc across the spark gap. The spark gap, which momentarily connects C1 and L3 in parallel, determines the amount of current transferred between C1 and L3.

The arcing across the spark gap sends a series of high-voltage pulses through L3, giving a sort of oscillated effect. The energy fed through L3 is transferred to L4 via the magnetic coupling between the two coils. Because of the turn ratio that exists between L3 and L4, an even higher voltage is produced across L4. Coil L4 steps up the voltage, which collects on the top-capacitance sphere. There, it causes an avalanche breakdown of the surrounding air, giving off a luminous discharge.

The rotary spark gap is a simple add-on circuit for the Tesla Coil, consisting of a variable dc power supply and a small, 5000-rpm, dc motor. The circuit allows you to vary the output of the Tesla coil by adjusting the rotating speed of the motor. A rotary gap is far more efficient than a stationary gap, because the stationary gap could cut-out and require readjustment.

107

Thermometer Circuits

The sources of the following circuits are contained in the Sources section beginning on page 782. The figure number contained in the box of each circuit correlates to the sources entry in the Sources section.

Differential Thermometer
Temperature-Reporting Digital Thermometer
Electronic Thermometer
Temperature Measuring Add-On for DMM
    Digital Voltmeter

Implantable Ingestible Electronic
    Thermometer
Simple Linear Thermometer
Thermometer Adapter

DIFFERENTIAL THERMOMETER

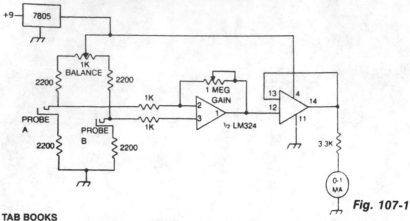

Fig. 107-1

TAB BOOKS

The differential thermometer uses two probes and shows the temperature difference between them, rather than the exact temperature. The thermometer uses a conventional meter as an indicator, and it covers a total range of 20° – 10°low to 10°high.

TEMPERATURE-REPORTING DIGITAL THERMOMETER

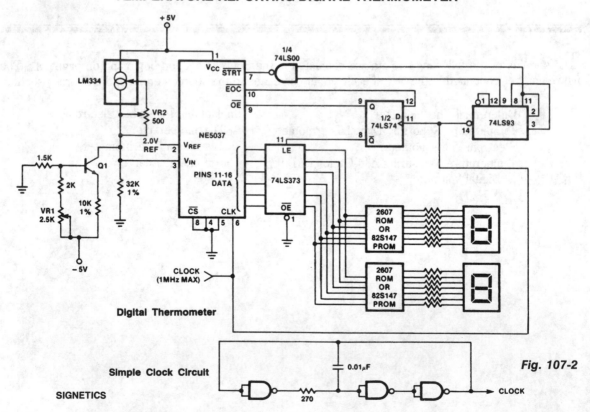

Fig. 107-2

Digital Thermometer

Simple Clock Circuit

SIGNETICS

The ROMs or PROMs must have the correct code for converting the data from the NE5037—used as address for the ROMs or PROMs—to the appropriate segment driver codes. The displayed amount could easily be converted to degrees Fahrenheit, °F, by the controller of (0 – 63°temperature sensor) or through the (P) ROMs. When doing this, a third (hundreds) digit (P)ROM and display will be needed for displaying temperatures above 99°F. An expensive clock can be made from NAND gates or inverters as shown.

ELECTRONIC THERMOMETER

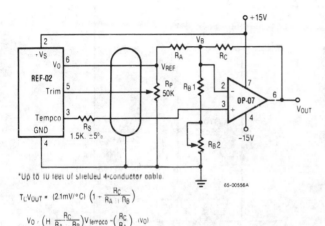

	Resistor Values		
TCV$_{OUT}$ Slope(s)	10mV/°C	100mV/°C	10mV/°F
Temperature Range	–55°C to +125°C	–55°C to +125°C	–65°F to +257°F
Output Voltage Range	–0.55V to +1.25V	–5.5V to +12.5V	–0.67V to +2.57V
Zero Scale	0V at 0°C	0V at 0°C	0V at 0°F
R$_A$ (±1% Resistor)	9.09KΩ	15KΩ	8.25KΩ
R$_{B1}$ (±1% Resistor)	1.5KΩ	1.82KΩ	1.0KΩ
R$_{B2}$ (Potentiometer)	200Ω	500Ω	200Ω
R$_C$ (±1% Resistor)	5.11KΩ	84.5KΩ	7.5KΩ

*Up to 10 feet of shielded 4-conductor cable.

65-00556A

$$T_L V_{OUT} = (2.1mV/°C)\left(1 + \frac{R_C}{R_A \parallel R_B}\right)$$

$$V_O = \left(H \frac{R_C}{R_A} \frac{R_C}{R_B}\right)V_{tempco} - \left(\frac{R_C}{R_A}\right)(V_O)$$

Fig. 107-3

Reprinted with permission from Raytheon Co., Semiconductor Division.

This circuit uses the +5 V reference output and the op amp to level shift and amplify the 2.1 mV/°C Tempco output into a voltage signal dependent on the ambient temperature. Different scaling can be obtained by selecting appropriate resistors from the table giving output slopes calibrated in degrees Celsius or degrees Fahrenheit. To calibrate, first measure the voltage on the Tempco pin, V_{TEMPCO}, and the ambient room temperature, T_A in °C. Put those values into the following equation:

$$\frac{V_{TEMPCO} \text{ (in mV)}}{(S)(T_A + 273)}$$

Where S = Scale factor for your circuit selected from the table in mV. Then turn the circuit power off, short V_{OUT} at pin 6 of the REF-02 to ground, and while applying exactly 100.00 mV to the op amp output, adjust R_{B2} to that $V_B = (X)(100 \text{ mV})$. Now remove the short and the 100-mV source, reapply circuit power and adjust R_P so that the op-amp output voltage equals $(T_A)(S)$. The system is now exactly calibrated.

TEMPERATURE MEASURING ADD-ON FOR DMM DIGITAL VOLTMETER

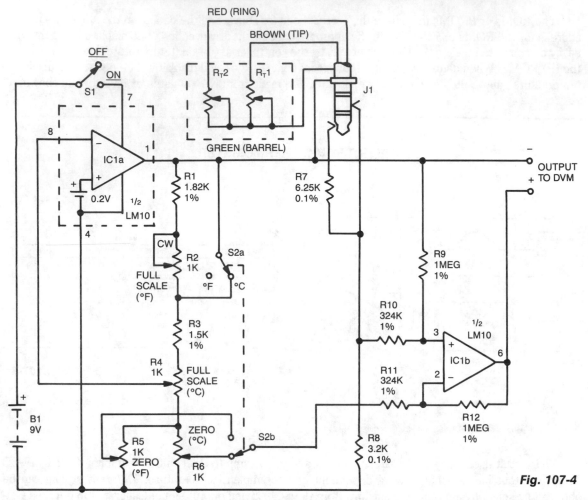

Fig. 107-4

The DVM-to-temperature adapter is built around a single IC, National's LM10. That micropower IC contains a stable 0.2 V reference, a reference amplifier and a general-purpose op amp. The circuit is designed for a linear temperature range of 0 to 100°C (32 to 212°F). The 0.2-V reference and reference amplifier provide a stable, fixed-excitation voltage to the Wheatstone bridge. The voltage is determined by a feedback network consisting of R1 through R6. Switch S2a configures the feedback to increase the voltage from 0.6 V on the Celsius range to 1.08 V on the Fahrenheit range. These differences compensate for the fact that one degree Fahrenheit produces a smaller resistance change than does one degree Celsius.

Resistors R1 through R16 also form the fixed leg of the Wheatstone bridge, nulling the bridge output at zero degrees. Since 0°C is different from 0°F, S2b is used to select the appropriate offset.

The LM10's op amp, along with R9 through R12, form a differential amplifier that boosts the bridge output to 10 mV per degree. Since a single supply is used, and since the output must be able to swing both positive and negative, the output is referenced to the bridge supply voltage, rather than to the common supply.

IMPLANTABLE INGESTIBLE ELECTRONIC THERMOMETER

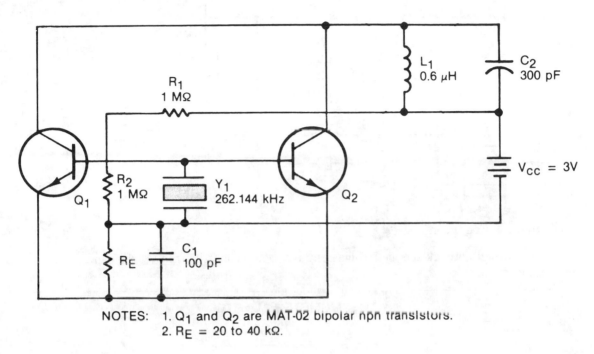

NOTES: 1. Q_1 and Q_2 are MAT-02 bipolar npn transistors.
        2. R_E = 20 to 40 kΩ.

NASA

Fig. 107-5

This oscillator circuit includes a quartz crystal that has a nominal resonant frequency of 262,144 Hz and is cut in the orientation that gives a large linear coefficient of frequency variation with the temperature. In this type of circuit, the oscillation frequency is controlled primarily by the crystal—as long as the gain-bandwidth product is at least four times the frequency. In this case, the chosen component values yield a gain-bandwidth product of 1 MHz. Inductor L1 can be made very small: 100 to 200 turns with a diameter of 0.18 in. (4.8 mm) and a length of 0.5 in. (12.7 mm). Although the figure shows two transistors in parallel, one could be used to reduce power consumption or three could be used to boost the output. The general oscillator circuit can be used to measure temperatures from −10 to +140°C. A unit made for use in the human body from about 30 to 40°C operates at 262,144 ±50 Hz with a frequency stability of 0.1 Hz and a temperature coefficient of 9 Hz/°C.

SIMPLE LINEAR THERMOMETER

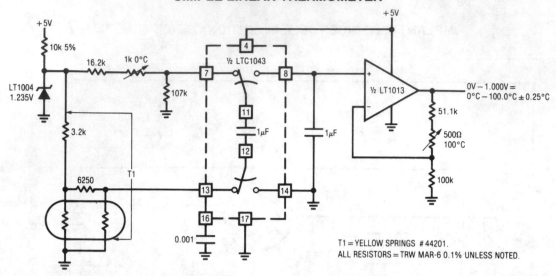

LINEAR TECHNOLOGY CORP.

Fig. 107-6

The thermistor network specified eliminates the need for a linearity trim—at the expense of accuracy and operational range.

THERMOMETER ADAPTER

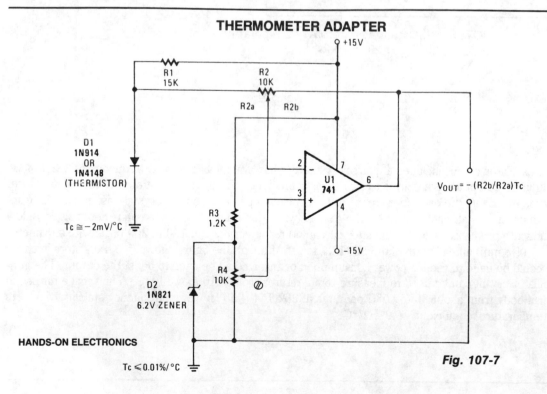

HANDS-ON ELECTRONICS

Fig. 107-7

THERMOMETER ADAPTER (*Cont.*)

A simple op amp and silicon diode are the heart of the temperature-to-voltage converter that will permit you to use an ordinary voltmeter—either analog or digital—to measure temperature. User adjustments make it possible for a reading of either 10 mV or 100 mV to represent 1°F or C.

Temperature sensor D1 is a 1N4148 silicon diode. It has a temperature coefficient of −2 mV/°C. U1, a 741 op amp, is connected as a differential amplifier. A voltage divider consisting of R3 and Zener diode D2 provides a 6.2 V reference voltage. D2 is shunted by potentiometer R4, so that the offset can be adjusted to align the output voltage with either the Celsius or Fahrenheit scale, as desired.

Gain control R2 is adjusted so the output of the op amp is in the scale or voltage range of the meter being used. R4, the offset adjust control, is then adjusted so the output voltage represents either degrees F or C. The thermometer adapter can be calibrated by adjusting R4 while the probe sensor is at a known temperature.

108

Tilt Meter

The sources of the following circuits are contained in the Sources section beginning on page 782. The figure number contained in the box of each circuit correlates to the sources entry in the Sources section.

Digitizer

DIGITIZER

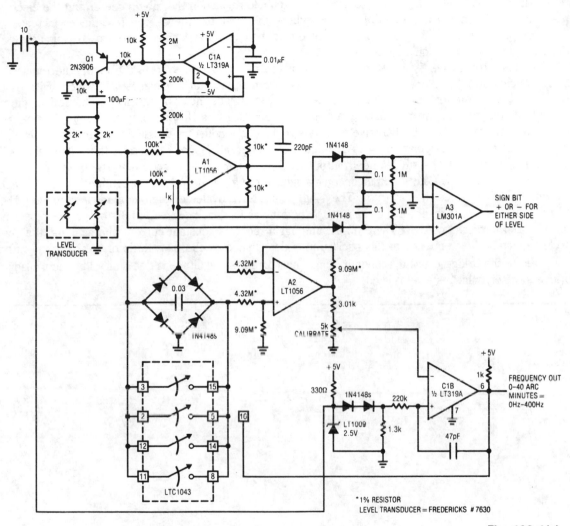

Fig. 108-1(a)

*1% RESISTOR
LEVEL TRANSDUCER = FREDERICKS # 7630

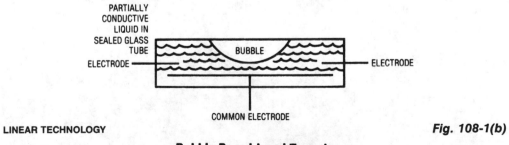

Fig. 108-1(b)

LINEAR TECHNOLOGY

Bubble Based Level Transducer

If the tube is level with respect to gravity, the bubble resides in the tube's center and the electrode resistances to common are identical. As the tube shifts away from level, the resistances increase and decrease proportionally. Transducers of this type must be excited with an ac waveform to avoid damage to the partially conductive liquid inside the tube.

The level transducer is configured with a pair of 2-KΩ resistors to form a bridge. The required ac bridge excitation is developed at C1A, configured as a multivibrator. C1 biases Q1, which switches the LT1009's 2.5-V potential through the 100-μF capacitor to provide the ac bridge drive. The bridge differential output ac signal is converted to a current by A1, operating as a Howland current pump. This current, whose polarity reverses as bridge drive polarity switches, is rectified by the diode bridge. Thus, the 0.03-μF capacitor receives unipolar charge. A2, running at a differential gain of 2, senses the voltage across the capacitor and presents its single-ended output to C1B. When the voltage across the 0.03-μF capacitor becomes high enough, C1B's output becomes high, turning on the paralleled sections of the LTC1043 switch. This discharges the capacitor. The 47-pF capacitor provides enough ac feedback around C1B to allow a complete zero reset for the capacitor. When the ac feedback ceases, C1B's output decreases and the LTC1043 switch goes off. The 0.03-μF unit again receives constant current charging and the entire cycle repeats. The frequency of this oscillation is determined by the magnitude of the constant current delivered to the bridge-capacitor configuration. This current's magnitude is determined by the transducer bridge's offset, which is level related.

109

Time-Delay Circuits

The sources of the following circuits are contained in the Sources section beginning on page 782. The figure number contained in the box of each circuit correlates to the sources entry in the Sources section.

Electronic Time Delay
Timing Threshold and Load Driver
Simple Time Delay

ELECTRONIC TIME DELAY

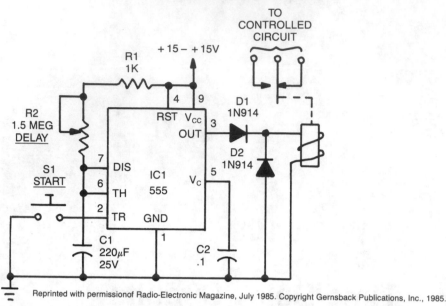

TO CONTROLLED CIRCUIT

Fig. 109-1

The time delay, T, in seconds is: $T = 1.1 \times C1 \times (R1 + R2)$. The resistances are in megohms and capacitances in microfarads. The sum of R1 and R2 should not be less than 1000 Ω nor higher than 20 MΩ. Pressing S1 starts the timing cycle. A low-going pulse, instead of S1 can also be used to initiate the timing cycle. With the values shown and allowing for the tolerances of the 200-μF capacitor, the delay will range from 4 minutes and 50 seconds to 7 minutes and 26 seconds. The output terminal, pin 3, of 555, is normally low and switches high during the timing cycle. The output can either sink or source currents up to 200 mA. Therefore, a load such as a relay coil can be connected between pin 3 and V_{CC} or between pin 3 and ground, depending on circuit requirements. When the relay is connected between pin 3 and ground, it is normally de-energized so it is energized only during the timing cycle. Connecting the relay to ground will save power and allow the IC to run cool.

TIMING THRESHOLD AND LOAD DRIVER

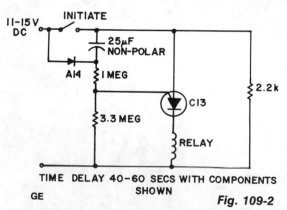

TIME DELAY 40–60 SECS WITH COMPONENTS SHOWN

GE   **Fig. 109-2**

Power is applied to the circuit with the initiate switch open. The 25-μF capacitor charges through the A14, or equivalent, diode and 2.2-KΩ resistor to full supply voltage. When the initiate switch is closed, the low side of the capacitor is suddenly raised to +12 V. This raises the diode side of the capacitor to approximately +24 V. The capacitor immediately begins discharging through the series-connected 1 and 3.3-MΩ resistors. Eventually, the C13 gate becomes forward biased, the device turns, and it applies power to the relay. The delay is virtually independent of supply voltage.

SIMPLE TIME DELAY

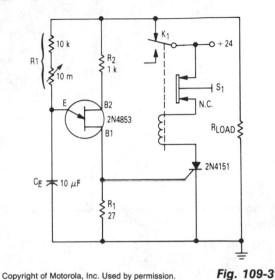

Fig. 109-3

After the first cycle, the relay will normally be energized. When normally closed pushbutton S1 is activated, the SCR turns off, the relay is de-energized, and power is applied to the relaxation oscillator and the load. After a time delay varying from less than a second to approximately 2.5 minutes, as determined by the setting of the 10-MΩ potentiometer, the unijuction will fire and turn on the SCR. The relay will energize until power is removed from the oscillator and the load, and will stay energized until button S1 is pushed again. The UJT trigger output from base 1 directly drives the gate of the SCR. However, where isolation between the UJT trigger or any other type of trigger and the thyristor power circuit is required, then a simple pulse transformer, interfacing the two elements, will suffice.

110

Timers

The sources of the following circuits are contained in the Sources section beginning on page 782. The figure number contained in the box of each circuit correlates to the sources entry in the Sources section.

Sequential Timer
CMOS Precision Programmable
    Laboratory Timer
Long-Time Timer
One-Shot Timer
Three-Minute Timer

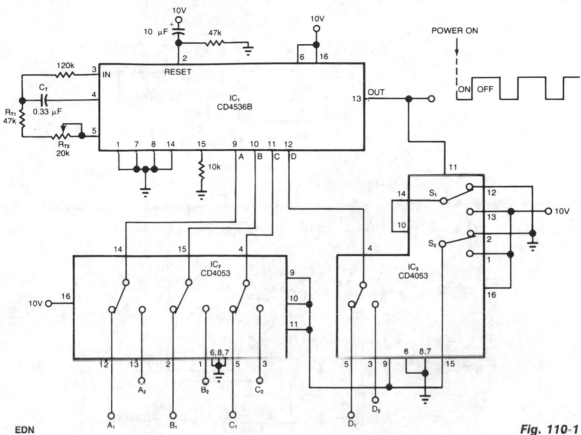

SEQUENTIAL TIMER

EDN

Fig. 110-1

The timer circuit shown gives independent control of the output's on and off intervals, which can range from 0.055 seconds to 30 minutes, relatively unaffected by power-line transients. IC1 is a CMOS programmable-timer chip that includes 24 ripple-binary counter stages; the first eight are bypassed when logic 1 is applied to pin 6. Then, a 4-bit input code at pins A, B, C, and D connects one of the 16 remaining stages to the output at pin 13. The chip includes an oscillator whose timing components are C_T, R_{T1}, and R_{T2}. For this example, you adjust R_{T2} for an internal period T_{IN} of 54.9 ms (18.2 Hz). Then, the output on or off interval is: $T_{OUT} = T_{IN}2^{N-1}$, where N is the number of counter stages in the internal divider chain (See Fig. 110-3). IC2 and IC3 are CMOS triple-spdt analog switches that connect one BCD code ($A1 - D1$) for the on interval and another ($A2 - D2$) for the off interval. You can apply the codes using manual toggle switches or programmable latches. When power is first applied, the switches are in the positions shown, which applies $A1 - D1$ to IC1 and generates the on interval. When the output changes state, all the switches change position and initiate the off interval by applying $A2 - D2$ to IC1. The cycle then repeats. To eliminate race conditions, switches S1 and S2 of IC3 operate in sequence before the remaining four switches operate in parallel. To start the output sequence with an off instead of an on interval, connect a power-on-set signal at pin 1 instead of the power-on-reset signal at pin 2.

CMOS PRECISION PROGRAMMABLE LABORATORY TIMER

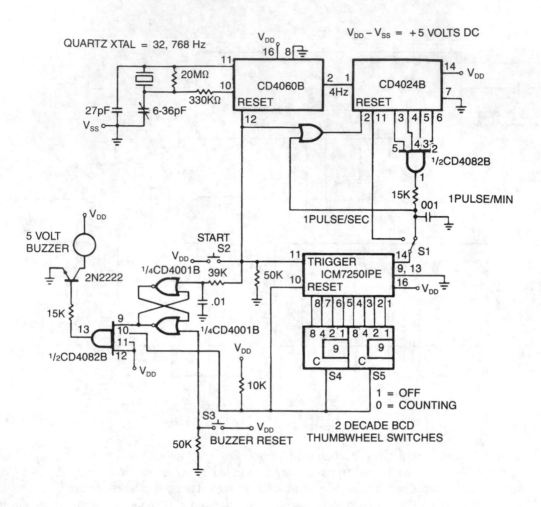

INTERSIL

Fig. 110-2

The time base is first selected with S1 set for seconds or minutes, then units 0–99 are selected on the two thumbwheel switches S4 and S5. Finally, switch S2 is depressed to start the timer. Simultaneously, the quartz crystal-controlled divider circuits are reset, the ICM7250 is triggered and counting begins. The ICM7250 counts until the preprogrammed value is reached, then, the value is reset, pin 10 of the CD4082B is enabled, and the buzzer is turned on. Pressing S3 turns the buzzer off.

LONG-TIME TIMER

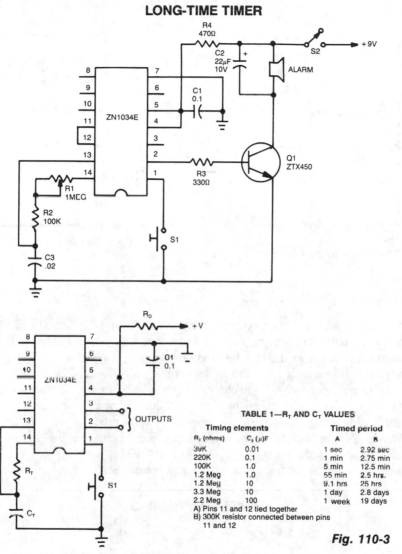

TABLE 1—R_T AND C_T VALUES

Timing elements		Timed period	
R_T (ohms)	C_T (μF)	A	B
39K	0.01	1 sec	2.92 sec
220K	0.1	1 min	2.75 min
100K	1.0	5 min	12.5 min
1.2 Meg	1.0	55 min	2.5 hrs.
1.2 Meg	10	9.1 hrs	25 hrs
3.3 Meg	10	1 day	2.8 days
2.2 Meg	100	1 week	19 days

A) Pins 11 and 12 tied together
B) 300K resistor connected between pins 11 and 12

Fig. 110-3

Reprinted with permission from Radio-Electronics Magazine May 1987. Copyright Gernsback Publications , Inc., 1987.

When used as a stand-alone device, ZN1034E from Ferranti can provide timed intervals ranging from 1 second to 19 days, although the rc time constant is only 220 seconds. The ZN1034E includes an internal voltage regulator, an oscillator, and a 12-stage binary counter. The total delay time provided by the counter is 4095 times the oscillator period. The control logic times-out after 4095 cycles of the oscillator, and delivers high and low output pulses at pins 2 and 3. The output at pin 3 is normally high and decreases at the end of the timed interval. The complementary output at pin 2 is normally low and becomes high at the end of the timed interval. The timing period is initiated by momentarily grounding pin 1. Timing resistor R_T consists of two resistors, R1 and R2, in series. Because R1 has a fixed value of 100 KΩ, the total range of R_T is 100 K to 1.1 MΩ.

ONE-SHOT TIMER

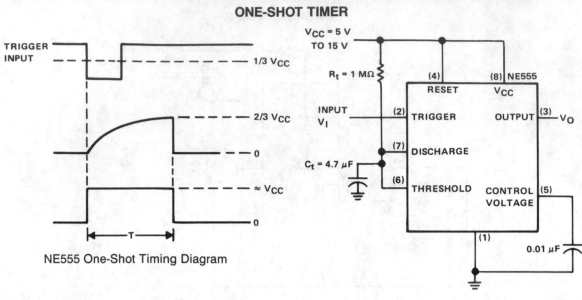

NE555 One-Shot Timing Diagram

Reprinted by permission of Texas Instruments.

Fig. 110-4

This simple circuit consists of only two timing components R_T and C_T, the NE555, and bypass capacitor C2. While not essential for operation, C2 is recommended for noise immunity. During standby, the trigger input terminal is held higher than $1/3\ V_{CC}$ and the output is low. When a trigger pulse appears with a level less than $1/3\ V_{CC}$, the timer is triggered and the timing cycle starts. The output rises to a high level near V_{CC}, and at the same time, C_T begins to charge toward V_{CC}. When the C_T voltage crosses $2/3\ V_{CC}$, the timing period ends with the output falling to zero, and the circuit is ready for another input trigger. Because of the internal latching mechanism, the timer will always time out when triggered, regardless of any subsequent noise, such as bounce, on the trigger input. For this reason, the circuit can also be used as a bounceless switch by using a shorter rc time constant. A 100-KΩ resistor for R_T and a 1-μF capacitor for C_T would give a clean, 0.1 s output pulse when used as a bounceless switch.

THREE-MINUTE TIMER

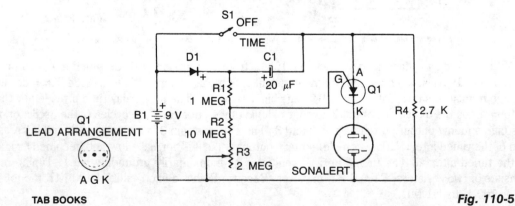

TAB BOOKS

Fig. 110-5

THREE-MINUTE TIMER (*Cont.*)

When S1 is off, C1 charges to within 0.5 V of the battery voltage through diode D1 and resistor R4. When S1 is closed, the anode of the PUT rises to the positive supply voltage. The PUT does not conduct, because battery voltage appears in series with the charge stored on C1, which raises the gate of the PUT to a level positive with respect to the anode. The timer relies on the discharge of capacitor C1 through resistors R1, R2, R3, and R4. Once C1 is at zero volts, the PUT will turn on battery voltage to the Sonalert and cause it to sound.

111

Tone Control Circuits

The sources of the following circuits are contained in the Sources section beginning on page 782. The figure number contained in the box of each circuit correlates to the sources entry in the Sources section.

IC Preamplifier/Tone Control
Ten-Band Octave Equalizer
Three-Band Active Tone Control
Wien-Bridge Filters
Rumble/Scratch Filter

IC PREAMPLIFIER/TONE CONTROL

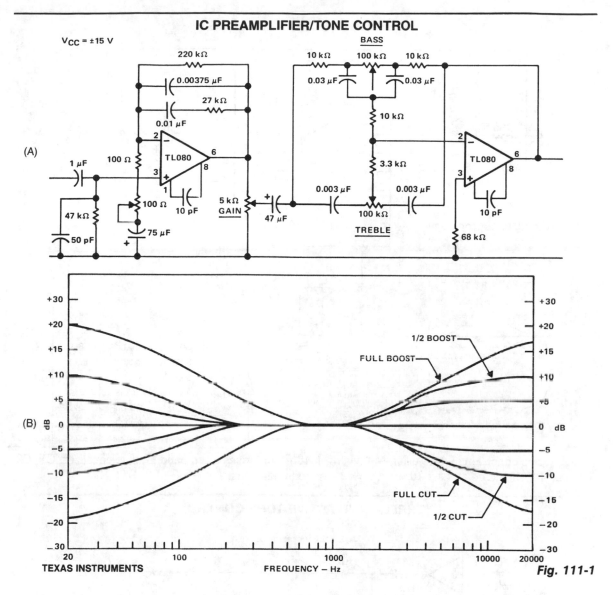

Fig. 111-1

The circuit is a form of the so-called "Americanized" version of the Baxandall negative-feedback tone control. At very low frequencies, the reactance of the capacitor is large enough that they might be considered open circuits, and the gain is controlled by the bass potentiometer. At low to middle frequencies, the reactance of the 0.03-μF capacitors decreases at the rate of 6 dB/octave, and is in parallel with the 200-KΩ potentiometer; so the effective impedance is reduced correspondingly, thereby reducing the gain. This process continues until the 10-KΩ resistors, which are in series with the bass pot, become dominant and the gain levels off at unity. The action of treble circuit is smaller and becomes effective when the reactance of the 0.003-μF capacitors becomes minimal. This complete tone control is in the negative feedback loop of the TL080. Figure B shows the bass and treble tone control response.

TEN-BAND OCTAVE EQUALIZER

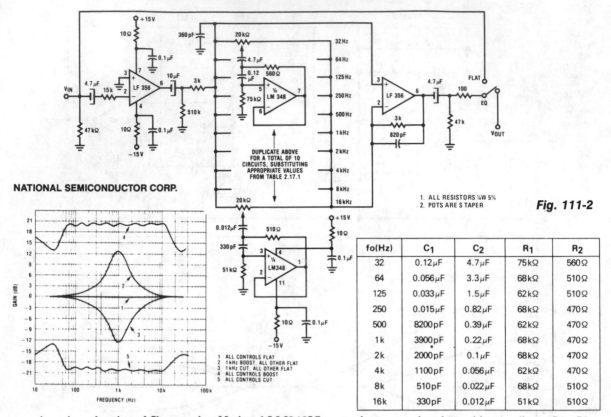

NATIONAL SEMICONDUCTOR CORP.

1. ALL RESISTORS ¼W 5%
2. POTS ARE S TAPER

Fig. 111-2

fo(Hz)	C₁	C₂	R₁	R₂
32	$0.12\,\mu F$	$4.7\,\mu F$	$75\,k\Omega$	$560\,\Omega$
64	$0.056\,\mu F$	$3.3\,\mu F$	$68\,k\Omega$	$510\,\Omega$
125	$0.033\,\mu F$	$1.5\,\mu F$	$62\,k\Omega$	$510\,\Omega$
250	$0.015\,\mu F$	$0.82\,\mu F$	$68\,k\Omega$	$470\,\Omega$
500	$8200\,pF$	$0.39\,\mu F$	$62\,k\Omega$	$470\,\Omega$
1k	$3900\,pF$	$0.22\,\mu F$	$68\,k\Omega$	$470\,\Omega$
2k	$2000\,pF$	$0.1\,\mu F$	$68\,k\Omega$	$470\,\Omega$
4k	$1100\,pF$	$0.056\,\mu F$	$62\,k\Omega$	$470\,\Omega$
8k	$510\,pF$	$0.022\,\mu F$	$68\,k\Omega$	$510\,\Omega$
16k	$330\,pF$	$0.012\,\mu F$	$51\,k\Omega$	$510\,\Omega$

A series of active rf filters using National LM348IC comprises a ten-band graphic equalizer. C1, C2, R1, and R2 should be at least 10% with 5% preferred tolerances.

THREE-BAND ACTIVE TONE CONTROL

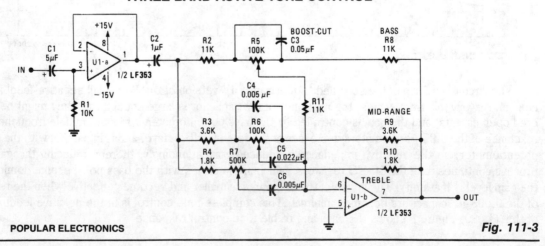

POPULAR ELECTRONICS

Fig. 111-3

WIEN-BRIDGE FILTER

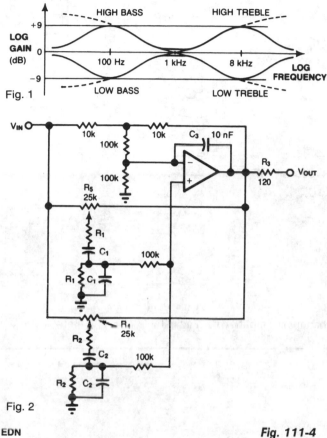

Fig. 1

Fig. 2

EDN

Fig. 111-4

Most audio tone controls affect midband gain, and they often create booming or hissing sounds when activated. You can avoid these problems by using a dual Wien-bridge filter to provide independent control of the treble and bass frequencies.

Experiments with equalizers indicate that the optimum center frequencies are about 100 Hz and 8 kHz. Using the relation $f = (2\pi RC)^{-1}$, set the Fig. 1 values accordingly:

$$100 \text{ Hz: } R1 = 15 \text{ K}\Omega; C1 = 0.1 \ \mu\text{F}$$

$$8 \text{ kHz: } R2 = 16 \text{ K}\Omega; C2 = 1.3 \text{ nF}$$

R3 and C3 provide stability. You obtain a ±9 dB variation of treble and bass by adjusting potentiometers R4 and R5, respectively. The filter's frequency response is shown in Fig. 2.

RUMBLE/SCRATCH FILTER

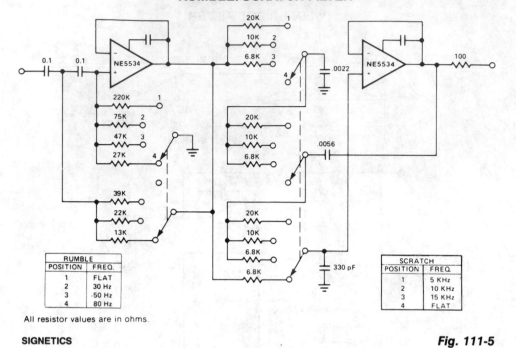

RUMBLE	
POSITION	FREQ.
1	FLAT
2	30 Hz
3	·50 Hz
4	80 Hz

All resistor values are in ohms.

SCRATCH	
POSITION	FREQ.
1	5 KHz
2	10 KHz
3	15 KHz
4	FLAT

SIGNETICS

Fig. 111-5

This is a variable bandpass amplifier with adjustable low- and high-frequency cutoffs.

112

Touch-Switch Circuits

The sources of the following circuits are contained in the Sources section beginning on page 782. The figure number contained in the box of each circuit correlates to the sources entry in the Sources section.

NEGATIVE-TRIGGERED TOUCH CIRCUIT

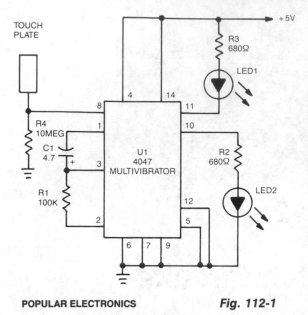

POPULAR ELECTRONICS   **Fig. 112-1**

The 4047 is configured as a monostable multivibrator circuit or one shot that is set up to trigger on a negative-transition of the signal applied to its pin 6 input. The multivibrator's on time is determined by the values of R1 and C1. Although R1 is shown to be a 100-K unit, its value can be anything between 10 K and 1 MΩ. Capacitor C1 can be a nonpolarized capacitor with any practical value above 100 pF. By making R4's value extremely high, the circuit can be used as a touch-triggered one-shot multivibrator. If the value of R4 is reduced to a much lower value, such as 10 KΩ, the circuit can be triggered with a negative pulse through 0.1-μF capacitor connected to pin 6. With a 100-KΩ resistor for R1, and a 4.7-μF electrolytic capacitor for C1, the circuit's on time is about 0.6 second. When R1 is increased to 470 KΩ, the on time of the circuit is increased to over 6 seconds.

POSITIVE-TRIGGERED TOUCH CIRCUIT

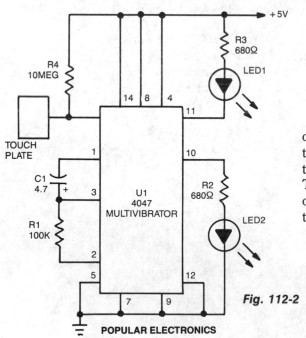

Fig. 112-2

POPULAR ELECTRONICS

LED1 and LED2 indicators turn on and remain on, each time the circuit is triggered. During the timing cycle, U1's Q output at pin 10 becomes positive when the $\overline{Q}$ output at pin 11 becomes negative. The two LEDs can be removed and the Q and $\overline{Q}$ outputs at pins 10 and 11, respectively, can be used to trigger some other circuit.

DIGITAL TOUCH ON/OFF SWITCH

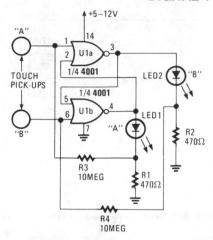

Only one LED can be on when the circuit is at rest. Which LED is illuminated is determined by the touch pick-up that last had human contact. Pickup terminal A controls the on condition of LED1, and terminal B controls the on condition of LED2. A 4001 quad two-input NOR gate is connected in an anti-bounce latching circuit that is activated by touching a pickup.

HANDS-ON ELECTRONICS   *Fig. 112-3*

TWO-TERMINAL TOUCH SWITCH

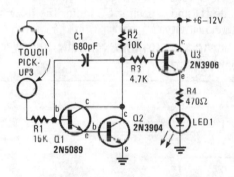

This circuit requires the bridging of two circuits to activate the electronic switch. That circuit does not require a 60-Hz field to operate and can be battery or ac powered. The two-pickup terminals can be made from most any clean metal; they should be about the size of a penny. The input circuitry of the two-terminal touch switch is a high-gain Darlington amplifier that multiplies the small bridging current to a value of sufficient magnitude to turn on Q3, supplying power to LED1. If a quick on and off switching time is desired, the value of C1 should be very small; if a long on-time period is required, the value of C1 can be increased.

HANDS-ON ELECTRONICS   *Fig. 112-4*

TOUCH ON/OFF ELECTRONIC SWITCH

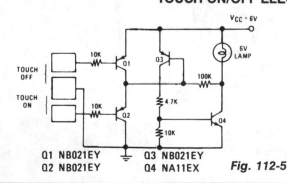

Transistors Q1 and Q2 control latch Q3 and Q4 to switch on the lamp. A high resistance from touching the electrode biases Q1 or Q2 on, setting or resetting the latch.

NATIONAL SEMICONDUCTOR CORP.

Q1 NB021EY   Q3 NB021EY
Q2 NB021EY   Q4 NA11EX   *Fig. 112-5*

LINE-HUM TOUCH SWITCH

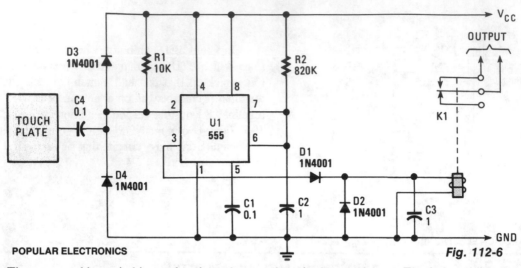

POPULAR ELECTRONICS

Fig. 112-6

The monostable period is set for about 1 second, as is the usual case. The induced line hum comes through C2, providing a continuous string of trigger pulses. The output becomes low for about 10 ms per second as the monostable times out and then retriggers. Diode D1 and capacitor C3 buffer the relay so it doesn't chatter on those 10-ms pulses. Resistor R2 sets the sensitivity.

The relay energizes when the plate is touched and de-energizes, up to one second after the finger is removed. The delay is a function of when the monostable last retriggered.

TOUCH SWITCH

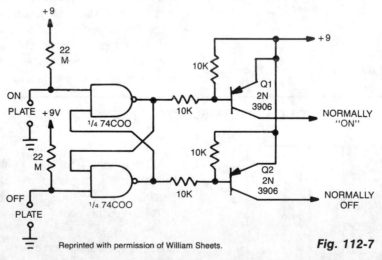

Reprinted with permission of William Sheets.

Fig. 112-7

When the plate is touched, the gate input becomes low, changing the state of the latch. Q1 and Q2 give alternate N-on—N-off outputs.

TOUCH SWITCH

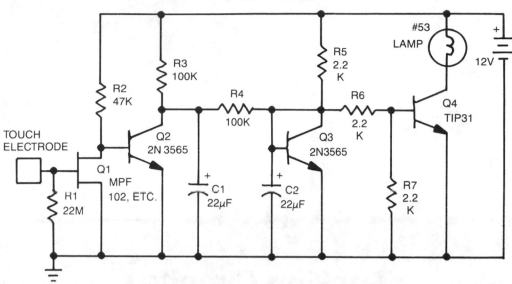

Reprinted with permission of William Sheets.

Fig. 112-8

This touch-actuated switch stays on as long as you keep your finger on the touch plate. R1 sets the input impedance to a high 22 MΩ. Q1 picks up stray signals coupled through your body to the touch plate and amplifies them to turn on Q2, which turns on lamp drivers Q3 and Q4. Lamp I1 is any small 12-V lamp, such as a No. 53—12 V 120 mA. R4 and C1 add a small amount of hysteresis (delay) to keep the light from constantly flickering. A relay can be used for I1.

TOUCH SWITCH

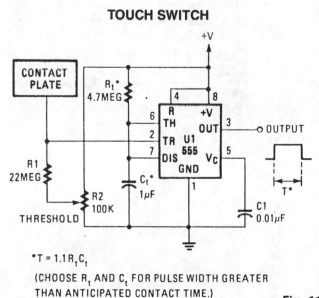

$^*T = 1.1R_t C_t$

(CHOOSE R_t AND C_t FOR PULSE WIDTH GREATER THAN ANTICIPATED CONTACT TIME.)

HANDS-ON ELECTRONICS

Fig. 112-9

113

Tracking Circuits

The sources of the following circuits are contained in the Sources section beginning on page 782. The figure number contained in the box of each circuit correlates to the sources entry in the Sources section.

TRACK AND HOLD

LINEAR TECHNOLOGY CORP.   *2N2369 EMITTER BASE JUNCTION

Fig. 113-1

The 5-MHz track and hold shown here has a 400-kHz power bandwidth driving ±10 V. A buffered input follower drives the hold capacitor, C4, through Q1, a low resistance FET switch. The positive hold command is supplied by TTL logic, with Q3 level shifting to the switch driver, Q2. The output is buffered by A3. When the gate is driven to V– for hold, it pulls the charge out of the hold capacitor. A compensating charge is put into the hold capacitor through C3. The step into hold is made independent of the input level with R7, and adjusted to zero with R10.

Since internal dissipation can be quite high when driving fast signals into a capacitive load, using a buffer in a power package is recommended. Raising the buffer quiescent current to 40 mA with R3 improves frequency response.

POSITIVE AND NEGATIVE VOLTAGE REFERENCE TRACKER

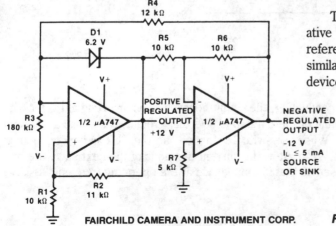

This reference uses an op amp to derive a negative output voltage that tracks with the positive reference voltage. A μA747 dual op amp, or any similar device such as an LM1458 or two μA741 devices, can be used.

$$\text{Positive Output} = V_{D1} \times \frac{R1 + R2}{R2}$$

$$\text{Negative Output} = -\text{Positive Output} \times \frac{R6}{R5}$$

FAIRCHILD CAMERA AND INSTRUMENT CORP.   *Fig. 113-2*

667

SIGNAL TRACK AND HOLD

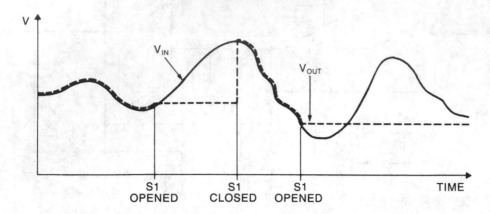

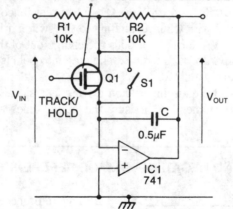

ELECTRONICS TODAY INTERNATIONAL

Fig. 113-3

When the switch is closed or the FET is conducting, the circuit behaves as an inverting amplifier with a gain of R2/R1. Since as the inverting terminal of the op amp is a virtual ground, the capacitor is kept charged to the output voltage by the op amp. When the switch is opened and the FET is nonconducting, the voltage at the output is held constant by the capacitor, the current demands of the next stage are met by the op amp. The value of C should be chosen so that its impedance at the operating frequency is large compared to R1 and R2.

114

Transducer Amplifiers

The sources of the following circuits are contained in the Sources section beginning on page 782. The figure number contained in the box of each circuit correlates to the sources entry in the Sources section.

Differential-to-Single-Ended Voltage
    Amplifier
Equalized Preamp for Magnetic Phono
    Cartridges
Photodiode Amplifier

Tape Playback Amplifier
NAB Record Preamplifier
Magnetic Phono Preamplifier
Two-Pole NAB-Type Preamp
Flat-Response Tape Amplifier

DIFFERENTIAL-TO-SINGLE-ENDED VOLTAGE AMPLIFIER

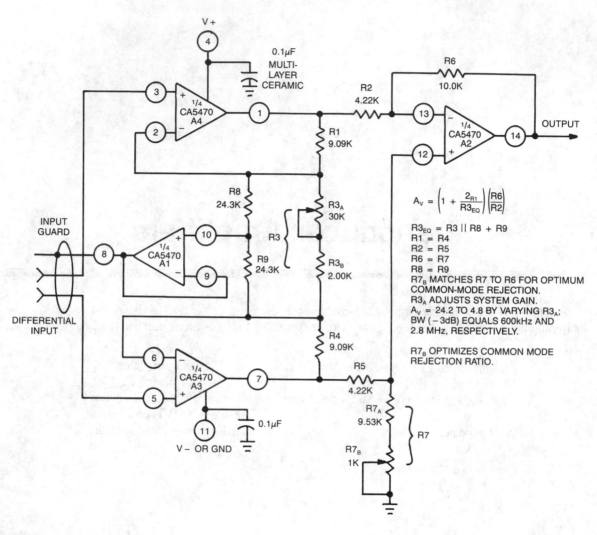

$$A_V = \left(1 + \frac{2_{R1}}{R3_{EQ}}\right)\left|\frac{R6}{R2}\right|$$

$R3_{EQ} = R3 \parallel R8 + R9$
$R1 = R4$
$R2 = R5$
$R6 = R7$
$R8 = R9$
$R7_B$ MATCHES R7 TO R6 FOR OPTIMUM
COMMON-MODE REJECTION.
$R3_A$ ADJUSTS SYSTEM GAIN.
A_V = 24.2 TO 4.8 BY VARYING $R3_A$;
BW (−3dB) EQUALS 600kHz AND
2.8 MHz, RESPECTIVELY.

$R7_B$ OPTIMIZES COMMON MODE
REJECTION RATIO.

GE/RCA

Fig. 114-1

This circuit uses a CA5470 quad microprocessor BiMOS-E op amp. Amplifiers A1 and A2 are employed as a cross-coupled differential input and differential output preamp stage and A3 provides input guard-banding. Amplifier A4 converts the differential outputs of A1 and A2 to a single-ended output.

EQUALIZED PREAMP FOR MAGNETIC PHONOGRAPH CARTRIDGES

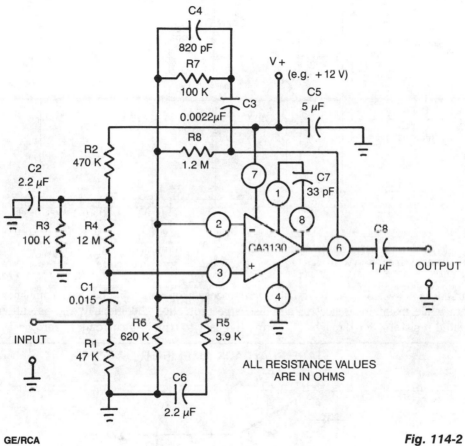

GE/RCA

Fig. 114-2

This circuit uses a CA3130 BiMOS op amp. Amplifier is *equalized* to RIAA playback frequency-response specifications. The circuit is useful as preamplifier following a magnetic tapehead.

PHOTODIODE AMPLIFIER

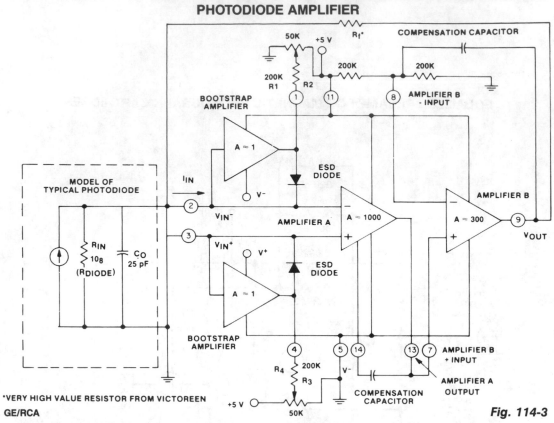

*VERY HIGH VALUE RESISTOR FROM VICTOREEN

GE/RCA

Fig. 114-3

This circuit uses a CA5422 dual BiMOS microprocessor op amp. The bootstrap amplifiers minimize bias currents while maintaining electrostatic discharge protection. Additionally, the potentiometers and their associated resistors, R1 through R4, permit the user to trim bias currents to zero.

TAPE PLAYBACK AMPLIFIER

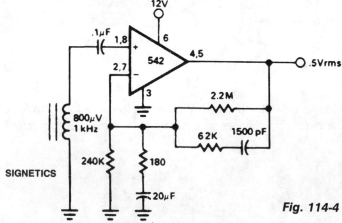

SIGNETICS

Fig. 114-4

NAB RECORD PREAMPLIFIER

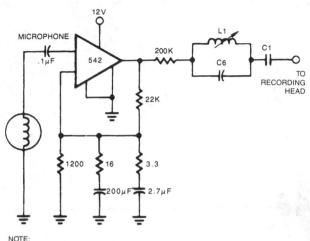

NOTE:
All resistor values are in Ω.

SIGNETICS

Fig. 114-5

TWO-POLE NAB-TYPE PREAMP

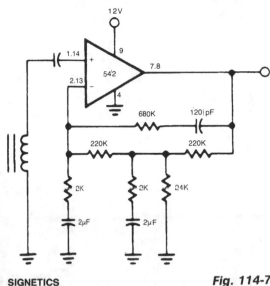

SIGNETICS

Fig. 114-7

MAGNETIC PHONO PREAMPLIFIER

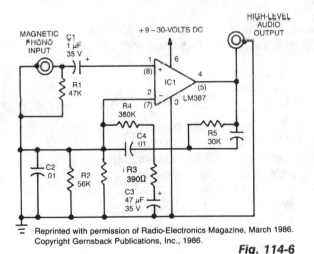

Fig. 114-6

This simple stereo amplifier uses a National LM387IC. The pin numbers in parentheses are for one channel, and those not in parentheses are for the other channel. The supply voltage can be +9 to +30 Vdc at about 10 mA. The output voltage swing is about $V_{CC} - 2$ V pk-pk. The preamp should be able to deliver at least 5 V.

FLAT-RESPONSE TAPE AMPLIFIER

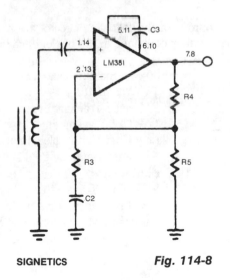

SIGNETICS

Fig. 114-8

115

Transmitters

The sources of the following circuits are contained in the Sources section beginning on page 782. The figure number contained in the box of each circuit correlates to the sources entry in the Sources section.

80-M AMATEUR RADIO TRANSMITTER

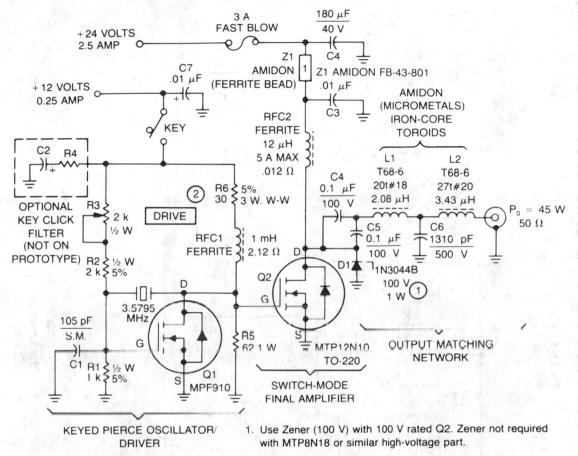

Copyright of Motorola, Inc. Used by permission.

Fig. 115-1

1. Use Zener (100 V) with 100 V rated Q2. Zener not required with MTP8N18 or similar high-voltage part.

2. Adjust DRIVE for minimum oscillation delay on keying.

This transmitter consists of a keyed crystal oscillator/driver and a high efficiency final, each with a TMOS Power FET as the active element. The total parts cost less than $20, and no special construction skills or circuit boards are required.

The Pierce oscillator is unique because the high C_{RSS} of the final amplifier power FET, $700-1200$ pF, is used as part of the capacitive feedback network. In fact, the oscillator will not work without Q2 installed. The MPF910 is a good choice for this circuit because the transistor is capable of driving the final amplifier in a switching mode, while still retaining enough gain for oscillation. To minimize cost, a readily-available color burst TV crystal is used as the frequency-determining element for Q1.

An unusual 84% output efficiency is possible with this transmitter. Such high efficiency is achieved because of the TMOS power FET's characteristics, along with modification of the usual algorithm for determining output matching.

TV TRANSMITTER

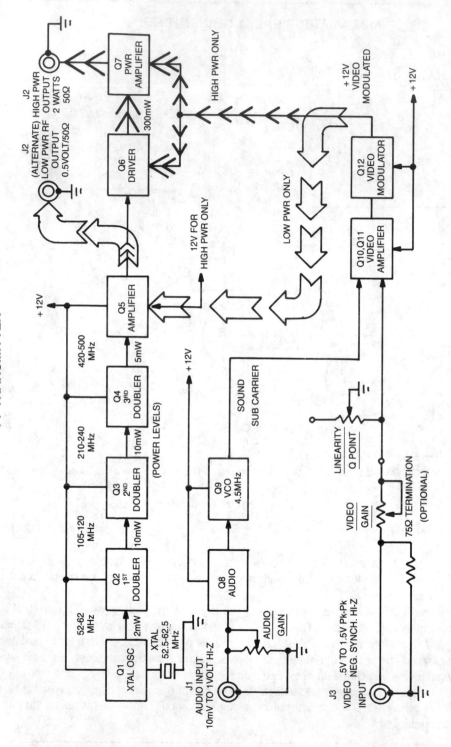

This transmitter is capable of two levels of rf power. For low-power wireless video, like in a house or office, where simultaneous monitoring of program material is desirable without cumbersome hookups, 1 – 30 mW is available. For longer ranges up to several miles, as in amateur (ham) TV, security, and surveillance purposes, 2 W into a 50-Ω load is available.

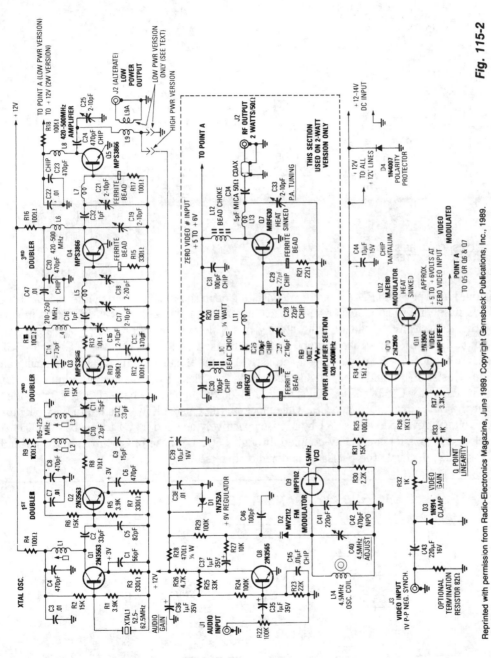

Fig. 115-2

Reprinted with permission from Radio-Electronics Magazine, June 1989. Copyright Gernsback Publications, Inc., 1989.

The video-link transmitter accepts color and B, W video, and audio inputs from VCRs, camcorders, small TV cameras, and microphones. The unit runs on a nominal 12 Vdc and draws 100 mA in the low-power version, or 500 mA in the 2-W version. The kit is available from North Country Radio, P.O. Box 53, Wykagyl Station, NY 10804.

FM VOICE TRANSMITTER

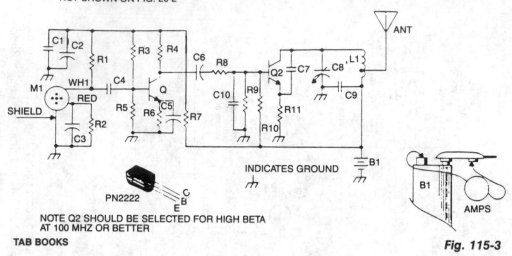

C1, C2 AND R7-DO NOT USE UNLESS
FEED BACK INSTABILITY OCCURS
NOT SHOWN ON FIG. 20-2

PN2222

NOTE Q2 SHOULD BE SELECTED FOR HIGH BETA
AT 100 MHZ OR BETTER

TAB BOOKS

INDICATES GROUND

Fig. 115-3

This is a sensitive, mini-powered FM transmitter consisting of an rf oscillator section interfaced with a high-sensitivity wide passband audio amplifier and capacitance microphone with built-in FET that modulates the base of the rf oscillator transistor. The setting of C8 determines the desired operating frequency—in the standard FM broadcast band, tuned to favor the high end up to 110 MHz. Capacitor C7 supplies the necessary feedback voltage developed across R11 in the emitter circuit of Q2, sustaining an oscillating condition. Resistors R9 and R10 provide the necessary bias of the base-emitter junction for proper operation, and capacitor C10 bypasses any rf to ground fed through to the base circuit. C9 provides an rf return path for the tank circuit of L1 and C8, while blocking the dc supply voltage fed to the collector of Q2. The speech voltage developed across R1 by M1 is capacitively coupled by C4 to the base of Q1. A signal voltage developed across R4 is capacity-coupled through C6 to the base of Q2 through R8. R7 and R8, along with C1 and C2, decouple the oscillator and audio circuits.

1-W CW TRANSMITTER

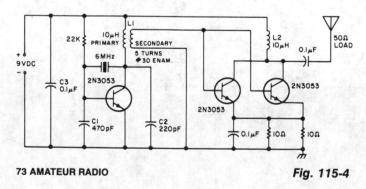

73 AMATEUR RADIO

Fig. 115-4

1-W CW TRANSMITTER (*Cont.*)

This is a little transmitter that could be put into a plastic Easter egg. It delivers approximately 1 W of measured rf output into a 50-Ω dummy load, and creates no heating problems with the circuit. The crystal is a series fundamental type, and the power source provides 9 V with a 2-A supply. The transmitter can operate at another frequency, but C1 and C2 might have to be changed for the circuit to work properly. The secondary of L1 is wound over the center of a 10-μH coil, with five turns of #30 enameled wire.

LOW-COST HALF-DUPLEX INFORMATION TRANSMISSION LINK

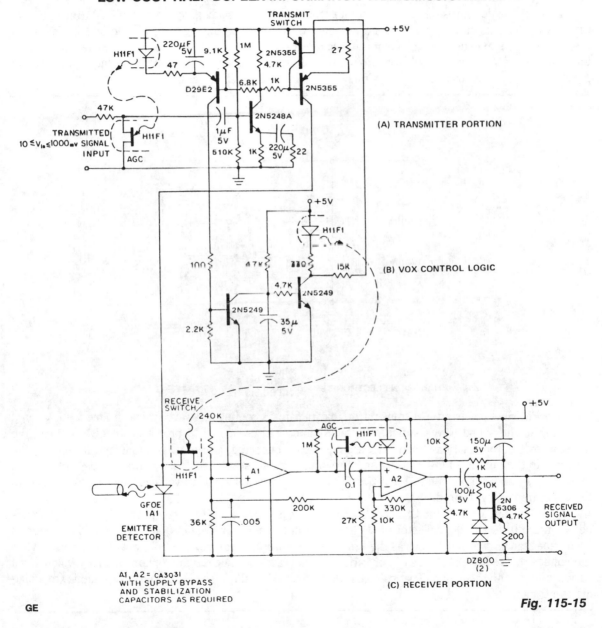

A1, A2 = CA3031
WITH SUPPLY BYPASS
AND STABILIZATION
CAPACITORS AS REQUIRED

(C) RECEIVER PORTION

GE

Fig. 115-15

679

LOW-COST HALF-DUPLEX INFORMATION TRANSMISSION LINK *(Cont.)*

In a half-duplex system, information can flow in both directions, but only one direction at any given time. The conventional method of building a half-duplex link requires a separate emitter and detector, connected with directional couplers, at each end of the fiber. The GFOE1A series of infrared emitting diodes are highly efficient, long-lived emitters, which are also sensitive to the 940 nm infrared they produce. Biased as a photodiode, they exhibit a sensitivity of about 30 nA per μW irradiation at 940 nm. In a suitable bias and switching logic network, they form the basis for a half-duplex information link. A half-duplex link, illustrating the emitter-detector operation of the GFOE1A1, is shown.

This schematic represents a full, general purpose system, including: approximately 50-dB compliance range with 1-V rms output, passive receive, voice-activated switching logic, 100 Hz to 50 kHz frequency response, and inexpensive components and hardware. The system is simple, inexpensive, and can be upgraded to provide more capability through use of higher gain bandwidth amplifier stages. Conversely, performance and cost can be lowered simply by removing undesired features.

FM SNOOPER

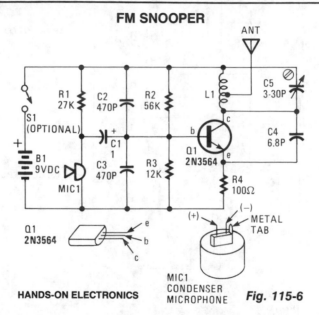

HANDS-ON ELECTRONICS   *Fig. 115-6*

The FM Snooper is an FM transmitter that radiates a continuous wave whose frequency is altered according to the sound waves striking the microphone. An ordinary FM broadcast receiver detects the transmitter's output carrier. When 9-V battery, B1, is connected, a brief surge of current flows from the collector to the emitter of Q1, causing an alternating current, shock oscillation in the resonant LC circuit, to flow back and forth between L1 and C5. So, by varying the value of C5, you can tune the oscillations to the exact frequency desired.

Although tuning capacitor C5 accounts for the major part of the tuning capacitance, the capacitance between the base and the collector of Q1 has a small, but noticeable, effect on the oscillation frequency. That capacitance, which is known as the *junction capacitance*, is not a fixed value, but instead varies when the voltage on the base of the transistor varies. Sound waves striking the microphone induce a voltage that varies in time with the sound. That voltage is applied via C1 to the base of Q1, thereby frequency modulating the transmitter.

VHF TONE TRANSMITTER

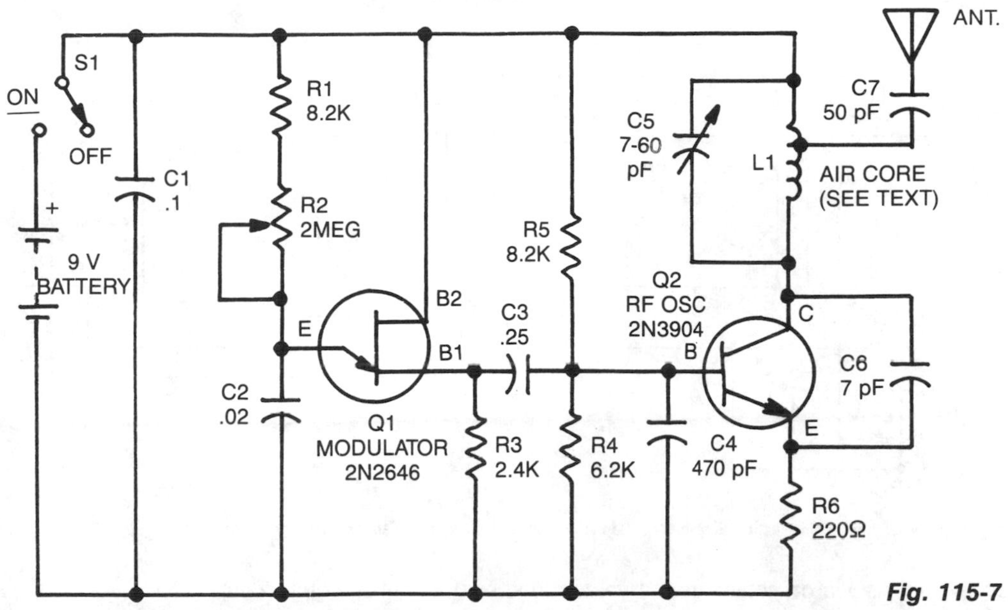

Fig. 115-7

The range of this transmitter is about 50 feet with a short whip antenna. The tone generator, is made up of a unijunction transistor, Q1, and R1, R2, R3, and C2. Transistor Q1 pulses on and off at a rate determined by the time constant of R1 and R2, together with the capacitance of C2 and the B1-emitter junction of Q1. Trimmer potentiometer R2 determines the frequency of the tone generated and allows a range of approximately 100 Hz to over 5 kHz.

Transistor Q2 is the rf oscillator. Its frequency is set by tuned circuits consisting of L1, C5, C6, and the interelectrode capacitance of Q2. The values shown will give a tuning range of about 55 to 108 MHz. Capacitor C6 provides positive feedback from the emitter to the collector of Q2, for oscillation. The audio tone generated by Q1 is applied to the base of Q2, causing the collector current to vary at the frequency of the tone, yielding an amplitude-modulated (AM) signal. This, in turn, varies Q2's collector-to-emitter capacitance, which makes up part of the tuned circuit, and causes the output frequency to vary similarly, producing a frequency-modulated (FM) signal, as well. The rf signal is coupled to the antenna through capacitor C7.

Coil L1 consists of five turns of #18 bare wire, close-wound on a piece of 1/4-inch wooden dowel. The length of the winding is about 1/4 inch. One end of capacitor C7 is soldered to the coil, one turn away from the 9-V supply end, and the other capacitor end is connected to the antenna. To adjust the vertical height and linearity of a TV set, place the tone transmitter near the set and use R2 to select the number of horizontal bars to be displayed. Once the picture is steady and the bars are sharp, adjust the set's vertical controls, so that all the bars are of the same height and are evenly spaced.

The fact that both AM and FM signals are generated makes it possible to use this circuit to check almost any receiver within the transmitter's frequency range. A TV set's sound section (discriminator) will reject the AM portion of the signal, but its video section will respond to it. Similarly, the TV sound section and FM receivers will respond to the FM signal produced.

LOW-FREQUENCY TRANSMITTER

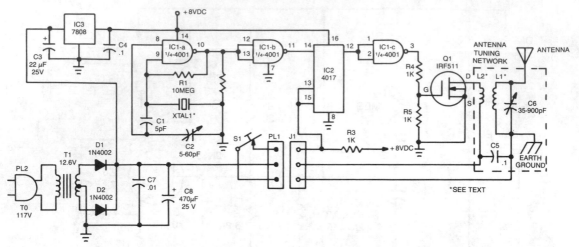

Fig. 115-8

The crystal oscillator, which uses two sections of IC1, a 4001 quad 2-input NOR gate, is a standard and reliable design. The oscillator's 1.85-MHz, square-wave output feeds IC2, a 4017 divide-by-10 counter. The count enable and reset terminals, pins 13 and 15, are normally held high by resistor R3, and the counter is activated by bringing those pins low by closing telegraph key S1—an arrangement that guarantees that the final state of IC2 pin 12 is always high. The high on IC2 pin 12 is inverted by a third section of the 4001, IC1c, to prevent dc current flow through power amplifier Q1 during key-up periods.

WIRELESS FM MICROPHONE

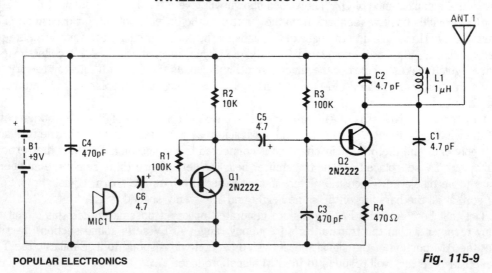

POPULAR ELECTRONICS

Fig. 115-9

WIRELESS FM MICROPHONE (*Cont.*)

Transistor Q1 acts as an amplifier for condenser microphone MIC1. The output of Q1 is applied to the base of transistor Q2 through a 4.7-μF capacitor. C2 and L1 form an LC tank circuit, which is used to set the frequency at which the transmitter operates. Coil L1 is a variable inductor, centered a bit below 1 μH, that is used to adjust the modulating frequency of the circuit. Capacitors C1 and C2 are 4.7 pF units. A lower value can be used to raise the circuit's operating frequency. The microphone and Q1 provide a varying voltage at the base of Q2, with the output of Q2 applied to the LC tank circuit. That causes a modulating action in the tank circuit that, when applied to the antenna, a short piece of wire 6- to 8-inches long, will provide a good, clear FM signal somewhere in the range of 88 to 95 MHz with a range of about 100 feet.

BEACON TRANSMITTER

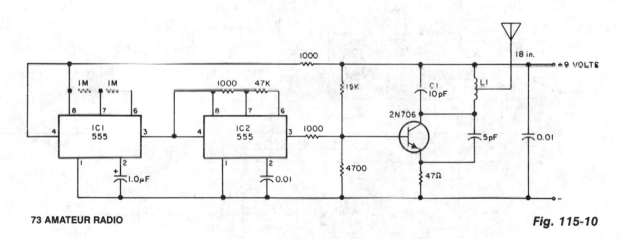

73 AMATEUR RADIO

Fig. 115-10

This transmitter can be used for transmitter hunts, for remote key finding, or for radio telemetry in model rockets. It can be tuned to the two meter band or other VHF bands by charging C1 and L1. L1 is four turns of #20 enameled wire airwound, 0.25 inch in diameter (use a drill bit), 0.2 inch long, center-tapped. The antenna can be 18 inches of any type of wire. IC2 functions as an audio oscillator that is turned on and off by IC1 about once per second. The range of the transmitter is several hundred yards.

40-M CW TRANSCEIVER

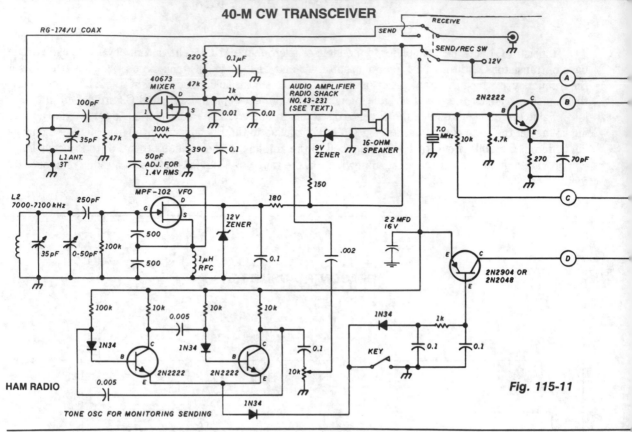

Fig. 115-11

HAM RADIO

VHF MODULATOR

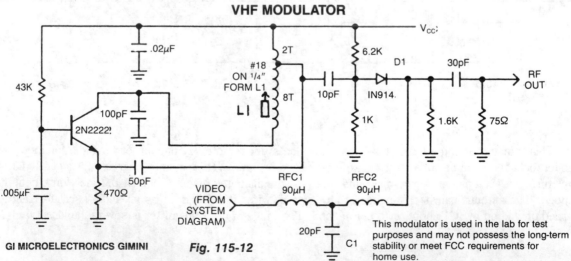

GI MICROELECTRONICS GIMINI   **Fig. 115-12**

This modulator is used in the lab for test purposes and may not possess the long-term stability or meet FCC requirements for home use.

This circuit uses an oscillator (2N2222) and a diode D1 as a nonlinear mixer. The frequency is set by a slug in L1. RFC1, C1, and RFC2 form a low-pass filter to pass video and block rf from the video source.

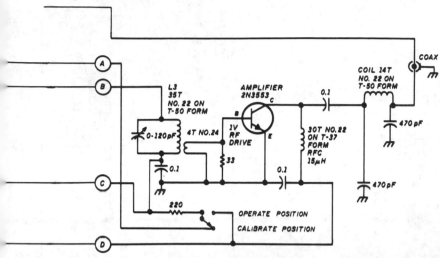

The unit consists of a direct conversion receiver and 1-W transmitter. The direct conversion receiver VFO is tuned just off frequency from the incoming signal. This difference in frequency produces a clean, strong, and solid audio tone signal. Detect the resonant frequency of the transmitter VFO by using the GDO as a field-strength meter.

Because of the large capacitance in the Colpitts VFO, the tuning coil will have fewer turns than the mixer coil. Use the capacitance shown for the VFO gate to ground and to the coil. It will effect the frequency and output. You'll need 1.4-V rms on pin 2 of the mixer to get a good signal from the VFO. The 1000-Ω resistor and 0.01-μF capacitors act as an rf filter from the mixer output.

WIRELESS FM MICROPHONE

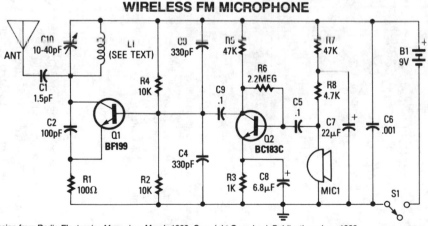

Fig. 115-13

Adjustable capacitor C10, and coil L1 form a tank circuit that, in combination with Q1, C2, and R1, oscillates at a frequency on the FM band. The center frequency is set by adjusting C10. An electret microphone, M1, picks up an audio signal that is amplified by transistor Q2. The audio signal is coupled via C9 to Q1, which frequency modulates the tank circuit. The signal is then radiated from the antenna. The circuit can operate from 9 – 12 Vdc.

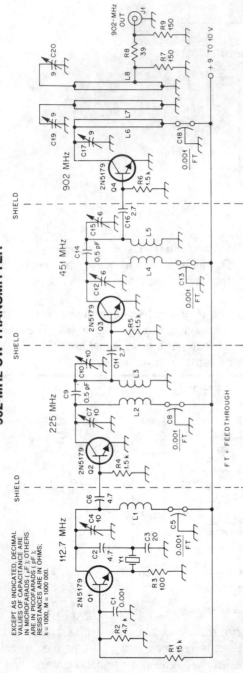

902-MHz CW TRANSMITTER

Fig 2—Schematic diagram of the exciter. Resistors are ¼-W carbon composition. Capacitors are 50-V epoxy-coated ceramic types unless otherwise noted.

EXCEPT AS INDICATED, DECIMAL VALUES OF CAPACITANCE ARE IN MICROFARADS (μF); OTHERS ARE IN PICOFARADS (pF); RESISTANCES ARE IN OHMS; k = 1000, M = 1000 000.

C4, C7, C10—1.5-15 pF miniature air-variable capacitor (Trim-tronics 10-1120-25015-000 or equiv).

C5, C8, C13, C18—470- to 1000-pF ceramic feedthrough capacitor, solder-in type preferred.

C9, C14—0.5-pF "gimmick" capacitor (see text).

C12, C15—1-6 pF miniature air-variable capacitor (Trim-tronics 10-1120-25006-000 or equiv).

C17, C19, C20—0.6-9 pF ceramic piston trimmer capacitor (Voltronics EQT9 or equiv).

J1—Chassis-mount female BNC connector (UG-1094 or equiv).

L1— 5t no. 22 tinned wire, 0.228-in ID (no. 1 drill), spaced 1 wire dia.

L2, L3—4t no. 18 tinned wire, ¼-in ID, spaced 1 wire dia.

L4, L5—2t no. 18 tinned wire, ¼-in ID,

spaced 1 wire dia.

L6, L7, L8—Inductor made from copper strap, 1-in long × 1/8-in wide. See text and Fig 3 for details.

Q1-Q4—2N5179 transistor.

Y1—Fifth-overtone crystal, 80.545 MHz, or seventh-overtone crystal, 112.763 MHz, HC-25 holder, series resonant, 0.005% (avail from JAN Crystals, 2400 Crystal Dr, Ft Meyers, FL 33906 tel 800-237-3063).

Fig. 115-14

QST

The oscillator, Q1, is a standard overtone circuit. A fifth-overtone crystal, 80.545 MHz, is operated on the seventh overtone, 112.763 MHz. C6 couples the output of the oscillator to Q2, which operates as a doubler to 225.5 MHz. A double-tuned circuit using C7, L2, L3, C10 is used in the collector of Q2 to reduce the level of the 112-MHz oscillator signal. The output of Q2 is capacitively coupled at C11 to the base of Q3. The double-tuned circuit in the collector of Q3 with C12, L4, L5, C15, is tuned to 451 MHz. A small capacitance, 2.7 pF, couples the 451-MHz signal to the base of another 2N5179, Q4, which doubles the signal to 902 MHz. The output of the 902-MHz doubler has a triple-tuned circuit using C17, L6, C19, L7, C20, L8 in its collector.

A ONE-TRANSISTOR FM TRANSMITTER

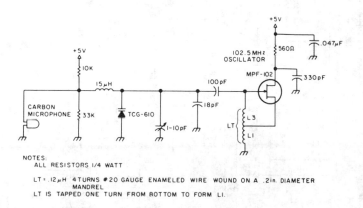

NOTES:
ALL RESISTORS 1/4 WATT

LT = .12 μH 4 TURNS #20 GAUGE ENAMELED WIRE WOUND ON A .2 in. DIAMETER MANDREL
LT IS TAPPED ONE TURN FROM BOTTOM TO FORM L1.

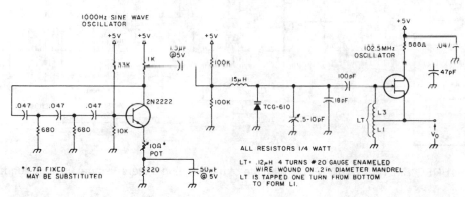

Fig. 115-15

The 2N2222 circuitry is a three-element, phase-shift oscillator circuit, designed to yield a 1,000-Hz sine wave. The 1,000-Hz sine wave is then applied to the TCG-610 varactor diode, 6 pF at 4 V, which changes the tank capacitance, thus varying the rf oscillator frequency at a 1,000-Hz rate. The 1,000-Ω potentiometer in the collector circuit can be adjusted to enable the desired frequency modulation level.

The Hartley rf oscillator, designed around a readily available MPF-102 JFET, has an output that should be relatively stable if it is enclosed in a metal box, thus minimizing changes in tank capacitance. The completed transmitter has a range of 30 feet when not enclosed—without an antenna.

FM MULTIPLEX TRANSMITTER

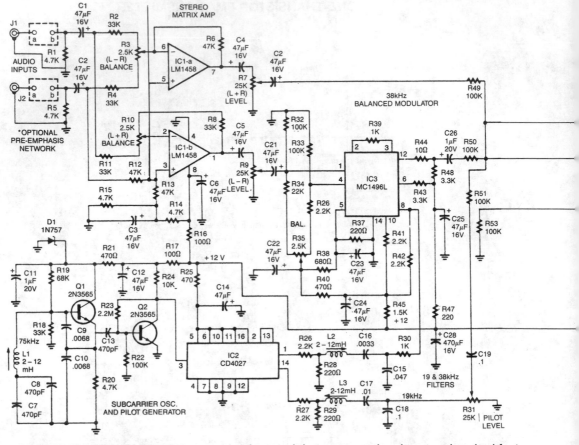

THIS STEREO FM TRANSMITTER is capable of transmitting a stereo signal up to a hundred feet.

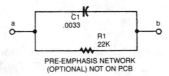

PRE-EMPHASIS NETWORK
(OPTIONAL) NOT ON PCB

THIS PRE-EMPHASIS NETWORK can
be added to the audio inputs of the MPX
transmitter, if necessary.

Reprinted with permission from Radio-Electronics Magazine, March 1988. Copyright Gernsback Publications, Inc., 1988.

RADIO-ELECTRONICS

Fig. 115-16

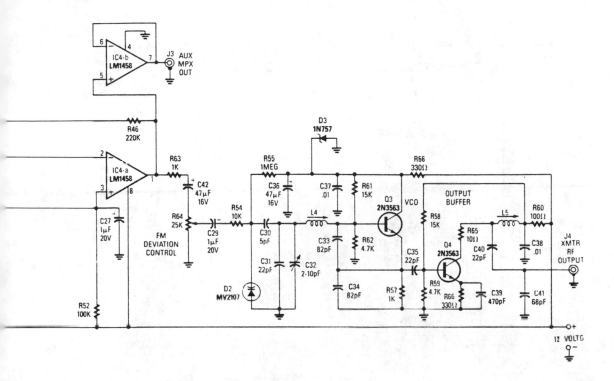

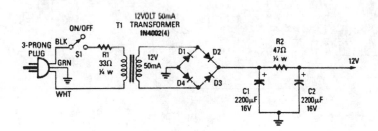

THIS POWER SUPPLY can be used if
you do not want to power the transmitter
with batteries.

This transmitter has a range of up to 100 feet. It generates a complete multiplex stereo signal and is useful for cordless headphone applications in which an inexpensive socket stereo receiver can be used. It can also be used as an FM multiplex generator for receiver alignments. The kit is available from North Country Radio, P.O. Box 53, Wykagyl Station, NY 10804.

QRP CW TRANSCEIVER

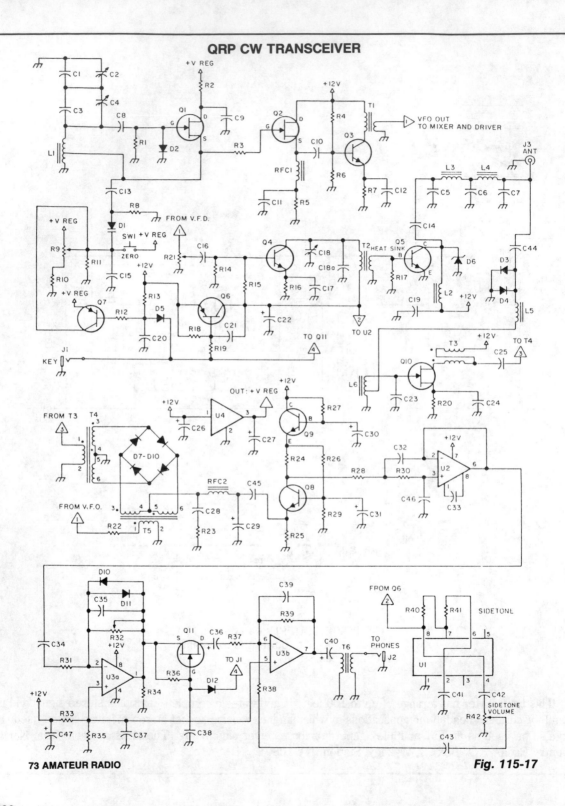

QRP CW TRANSCEIVER (*Cont.*)

This is a 3-W, single-circuit board, VFO-controlled CW transceiver for 40 or 30 meters, featuring a direct-conversion receiver with audio filtering, Receiver Incremental Tuning (RIT), and speaker level audio volume. The transmit frequency is generated by Q1 and its associated components in the VFO. The buffer, Q2, isolates the oscillator from the other circuitry to help keep the VFO stable. Q3 builds up the signal to a more usable level. The driver, Q4, amplifies the signal. The final, Q5, amplifies it to the 3-W level.

Key the transmitter by turning the power to the driver on and off, using Q6 as a switching transistor. Select the frequency by varying the tuning capacitor, C2. The VFO frequency feeds into the diode-ring mixer, and is mixed with the incoming 7- or 10-MHz signal. The *difference*, or *produce*, is the audio frequency. The post-mixer circuitry amplifies the audio signal to speaker level: Q8 preamplifies the signal a little, U2 is an audio filter that attenuates the audio signals above about 700 Hz, and U3 amplifies the signal from the audio filter to listening level.

WIRELESS FM MICROPHONE

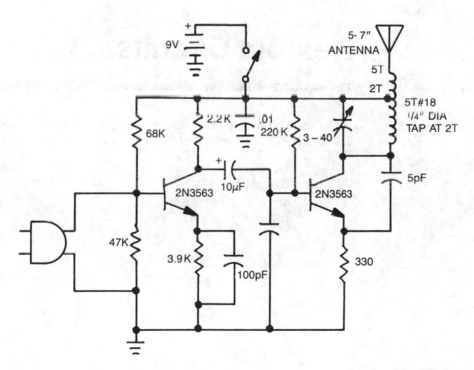

Reprinted with permission of William Sheets.

Fig. 115-18

Use standard rf wiring precautions. The best speech clarity is obtained by using an electret microphone. For music reproduction, substitute a dynamic mike element.

116

Tremolo Circuits

The sources of the following circuits are contained in the Sources section beginning on page 782. The figure number contained in the box of each circuit correlates to the sources entry in the Sources section.

Tremolo
Tremolo
Electronic Tremolo
Tremolo

TREMOLO

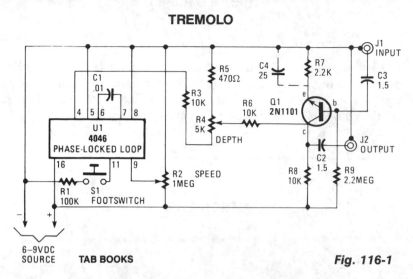

TAB BOOKS

6–9VDC
SOURCE

Fig. 116-1

The VCO of U1, a phased-locked loop, produces a low-frequency square-wave signal from 5 Hz to 2500 Hz, which is controlled by varying the voltage to pin 9 via R2. The frequency threshold is set by R1 and C1. Increasing the value of R1 increases the frequency, and decreasing the value of C1 decreases the threshold frequency. Transistor Q1 is operated as an amplifier with the Q point at 4 V, using a transistor with a voltage gain (beta) of 100. The gain without tremolo is about 2 V. With bypass capacitor C4 in the circuit, the gain is set at 33 dB. The instrument drive cannot be greater than 30 mV with C4 in the circuit (no tremolo), to avoid distortion. Without C4, the drive can be 1.5 V maximum with no distortion. U1's output is coupled to the collector of transistor Q1 via depth control R4. The square-wave signal pulls the audio at a frequency determined by R2. The oscillator is activated by foot switch S1.

TREMOLO

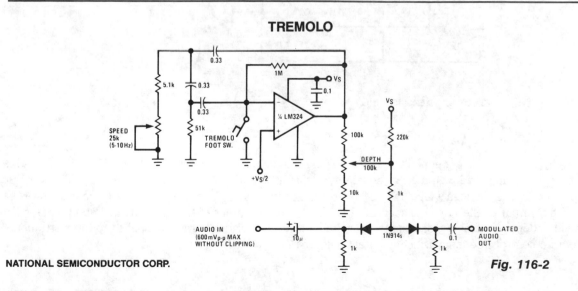

NATIONAL SEMICONDUCTOR CORP.

Fig. 116-2

A phase shift oscillator using the LM324 operates at an adjustable rate, 5 – 10 Hz, set by the speed pot. A portion of the oscillator output is taken from the depth pot and used to modulate the on resistance of two 1N914 diodes operating as voltage controlled attenuators. Care must be taken to restrict the incoming signal level to less than 0.6 V pk-pk, or undesirable clipping will occur.

ELECTRONIC TREMOLO

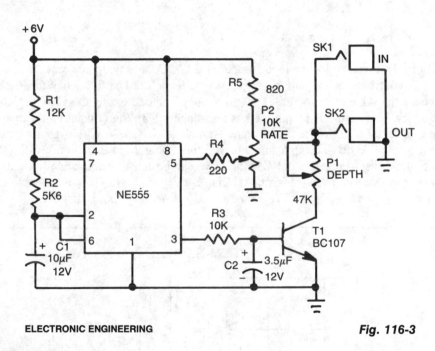

ELECTRONIC ENGINEERING

Fig. 116-3

The tremolo effect is generated by a repeating volume change at a rate usually between 1 and 15 Hz. The timer produces a low frequency square wave that is smoothed by a simple rc integrator. This varying signal modulates the signal input from the instrument. Transistor T1 is used as a voltage controlled resistor. The output of the circuit is connected in parallel to the output of the instrument. Potentiometer P1 provides depth control by adjusting the amplitude of the modulating waveform applied to the instrument. The rate control frequency is set by potentiometer P2.

TREMOLO

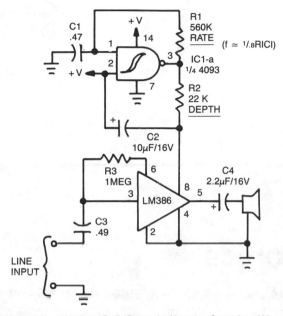

R1
560K
RATE $(f \simeq 1/.8R1C1)$

IC1-a
1/4 4093

R2
22 K
DEPTH

C2
10µF/16V

R3
1MEG

LM386

C4
2.2µF/16V

C3
.49

C1
.47

LINE
INPUT

Fig. 116-4

This simple circuit can color the sound coming from your audio system. Clocking for the circuit is provided by an oscillator built from one quarter of a 4093 quad NAND Schmitt trigger. With the component values shown, it will run at about 5 Hz. The clock frequency is fed to the gain control, pin 8, of an LM386 amplifier. Tremolo is produced by varying the amplifier gain. A trimmer potentiometer can be put in series with R1, to easily experiment with different rates. To experiment, make R1 about 100-KΩ and use a 1-MΩ trimmer. That allows frequencies from about 2 to 20 Hz to pass. Resistor R2 is the depth control. It controls the degree of tremolo. To adjust, put a trimmer in series with R2. Make R2 a 5-KΩ unit and use a 50-KΩ trimmer. Since the tremolo clock uses the gain-control pin of the amplifier, change the value of capacitor C4 in order to change the gain of the amplifier. Make C4 larger to increase the gain or smaller to decrease it. But, don't go any lower than 0.1 µF because you'll be cutting into the bottom-end frequency response.

117

Ultrasonics

The sources of the following circuits are contained in the Sources section beginning on page 782. The figure number contained in the box of each circuit correlates to the sources entry in the Sources section.

RANGING SYSTEM

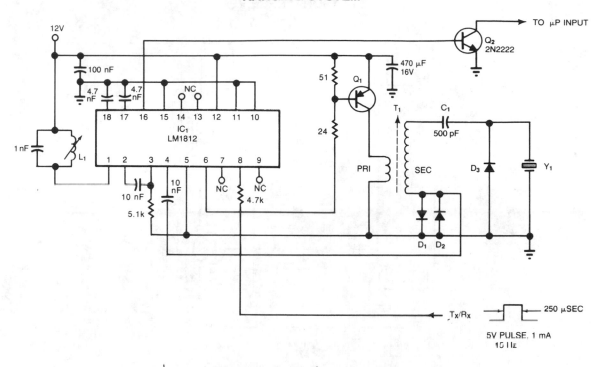

PART	PARTS LIST DESCRIPTION	MANUFACTURER
L₁	15.8 mH ADJUSTABLE #CLN-2A900HM	TOKO AMERICA (SKOKIE, IL)
T₁	PRIMARY 8 TURNS #34* SECONDARY 110 TURNS #30* POTCORE RM8P-A630-3B7 BOBBIN RM8 PCB1-4 CLIPS 991-393-00	FERROXCUBE (SAUGERTIES, NY)
X₁	TRANSDUCER	POLAROID (CAMBRIDGE, MA)
Q₁	D45C6 I_c = 5A MIN; V_CBO = 40V MIN	NATIONAL SEMICONDUCTOR (SANTA CLARA, CA)

*IF MACHINE WOUND, SLIGHTLY LARGER
WIRE SIZES MAY BE USED.

EDN   *Fig. 117-1*

Combine an electrostatic transducer with an ultrasonic transceiver IC to build a ranging system that senses objects at distances from 4 inches to more than 30 feet. Transducer Y1's broadband characteristic simplifies tuning. The secondary of T1 resonates with the 500-pF capacitor C1 at a frequency between 50 and 60 kHz. You tune L1 to this frequency by using an oscilloscope to note the maximum echo sensitivity at pin 1. Step-up transformer T1 provides 150-V bias for the transducer.

ULTRASONIC RECEIVER

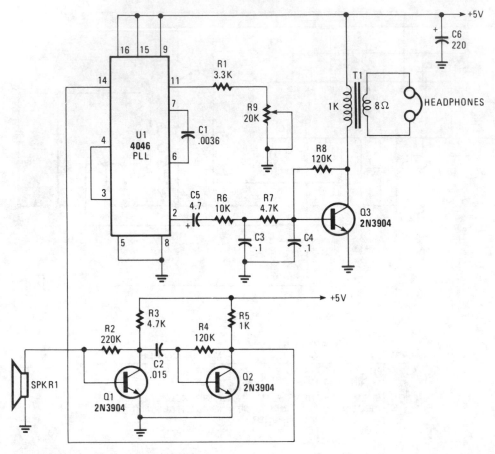

POPULAR ELECTRONICS

Fig. 117-2

The 4046 PPL is used as the heart of a tunable ultrasonic receiver that can be used to locate unheard ultrasonic sounds. The receiver might also be used, along with a simple ultrasonic generator, to send and receive Morse code. The incoming ultrasonic signal is picked up by piezo speaker SPKR1, and amplified by transistors Q1 and Q2. The output is fed to the phase comparator input of U1 at pin 14. The chip's interval VCO is tuned by turning potentiometer R9.

If a 20-kHz signal is picked up by SPKR1 and the VCO is tuned to produce a 19-kHz signal the difference output at pin 2 will be 1 kHz. That 1-kHz signal is amplified by Q3 and coupled through T1 to a pair of headphones. If the received frequency increases to 22 kHz, a 3-kHz tone is heard in the headphones. With the values given in the parts list for C1, R1, and R9, the VCO can be tuned from 12 to well over 42 kHz, which should cover just about anything the piezo sensor can respond to.

ULTRASONIC PEST-REPELLER

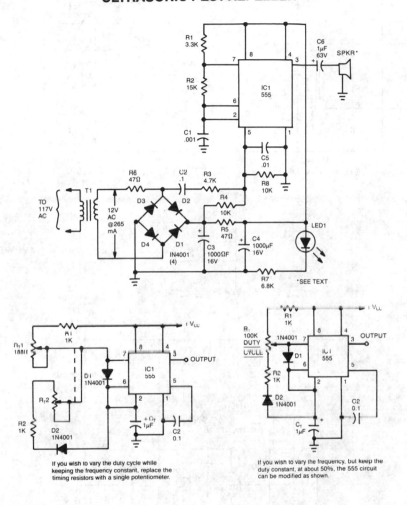

Fig. 117-3

This circuit is a 555 timer IC connected as a square-wave generator. Its base frequency is approximately 45 kHz, as determined by the values of R1, R2, and C1. The 45-kHz *carrier* is frequency modulated by a modified trapizoidal voltage waveform applied to pin 5 of the 555 timer. That modulating voltage is developed by a network consisting of C2, R3, and R4 connected across one leg of the bridge rectifier. The sweep is approximately 20 kHz on each side of the base frequency. The speaker is a 2-inch piezoelectric tweeter.

20-kHz ARC WELDING INVERTER

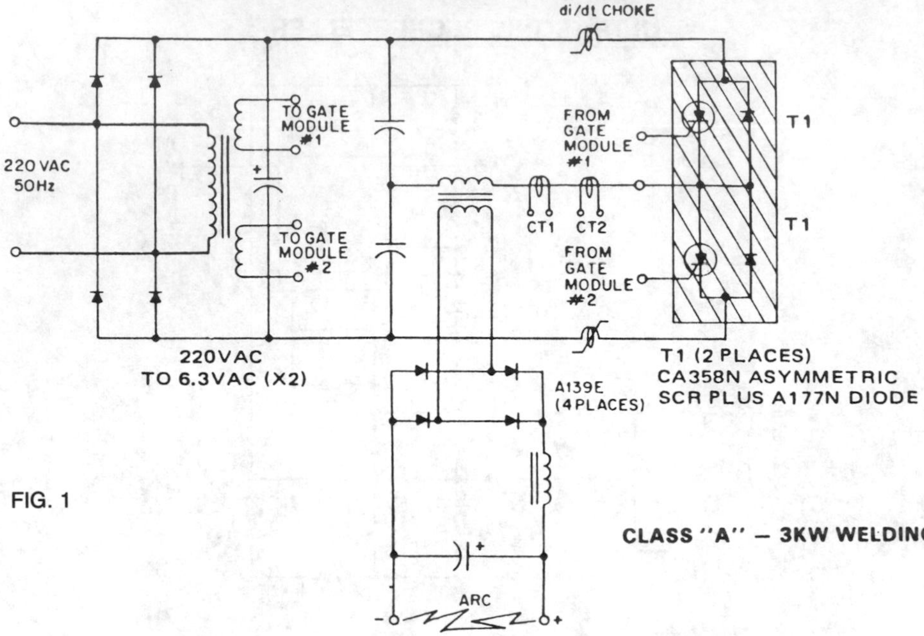

FIG. 1

CLASS "A" — 3KW WELDING INVERTER

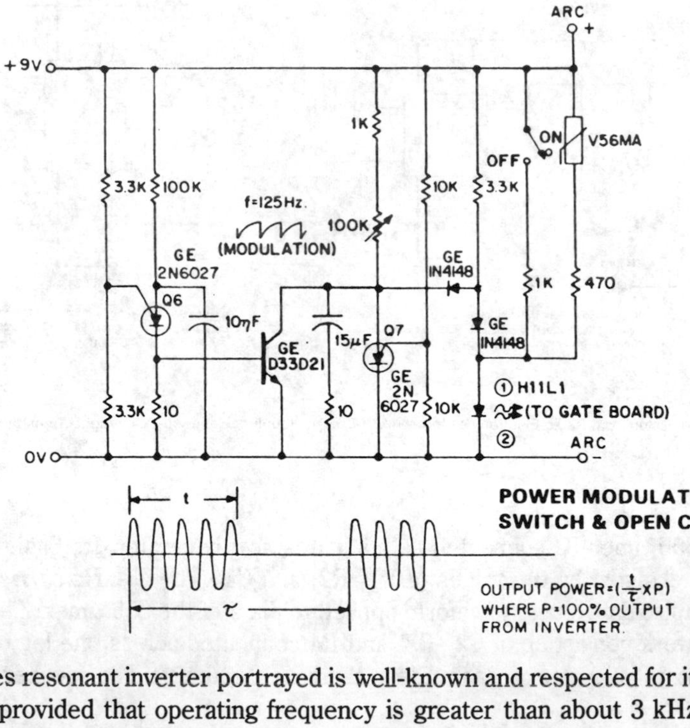

FIG. 2

POWER MODULATOR (WITH ON-OFF SWITCH & OPEN CIRCUIT PROTECTION)

$$\text{OUTPUT POWER} = (\frac{t}{\tau} \times P)$$
WHERE P = 100% OUTPUT FROM INVERTER.

The Class A series resonant inverter portrayed is well-known and respected for its high efficiency, low cost, and small size, provided that operating frequency is greater than about 3 kHz. The disadvantages are, at least in high power versions, the difficulty in effecting smooth RFI-free output voltage modulation

20-kHz ARC WELDING INVERTER (Cont.)

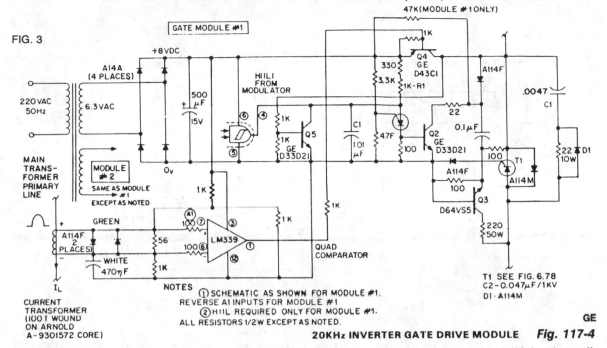

FIG. 3

NOTES
① SCHEMATIC AS SHOWN FOR MODULE #1.
REVERSE A1 INPUTS FOR MODULE #1
② H11L REQUIRED ONLY FOR MODULE #1.
ALL RESISTORS 1/2W EXCEPT AS NOTED.

CURRENT
TRANSFORMER
(100 T WOUND
ON ARNOLD
A-9301572 CORE)

T1 SEE FIG. 6.78
C2-0.047μF/1KV
D1-A114M

GE

20KHz INVERTER GATE DRIVE MODULE   **Fig. 117-4**

without significant added complexity, and a natural tendency to *run away* under no-load (high *Q*) conditions.

The 20 kHz control circuit (see Fig. 2) overcomes these shortcomings by feeding back into the asymmetrical thyristor trigger pulse generators (see Fig. 3) signals that simultaneously shut the inverter down, when its output voltage exceeds a preset threshold, then time-ratio modulates the output. This feedback is accomplished with full galvanic isolation between input and output thanks to an H11L opto-Schmitt coupler. The fundamental 20-kHz gate firing pulses are generated by a PUT relaxation oscillator Q1. The pulses are then amplified by transistors Q2 and Q3. The 20-kHz sinusoidal load current flowing in the primary of the output transformer is then detected by current transformer CT1, with op amp A1 converting the sine wave into a square wave, whose transitions coincide with the load current zero points.

Consequently, each time the output current changes, phase A1 also changes state and, via transistor Q4, either connects the thyristor gate to a −8 Vdc supply for minimum *gate assisted* turn-off time and highest reapplied *dV/dt* capability or disables this supply to prepare the thyristor for subsequent firing.

Modulation intelligence is coupled into this same H11L through two additional PUTs, Q6 and Q7, Q6 oscillates at a fixed 1.25 kHz, which establishes the modulation frequency. The duty cycle is determined by a second oscillator, Q7, whose conduction state, on or off, establishes or removes current from the H11L diode. With a fundamental inverter frequency of 20 kHz and a modulation frequency of 1.25 kHz, the resultant time ratio-controlled power output is given by:

$$P_{\mathrm{OUT}} = \left(P_M \times \frac{t}{\tau} \right)$$

where P_M = 100% continuous output power. Minimum power is one cycle of 20 kHz (50 μs) in the 1.25-kHz modulation frame (800 μs), that is, 6.25% P_M.

ULTRASONIC TRANSCEIVER

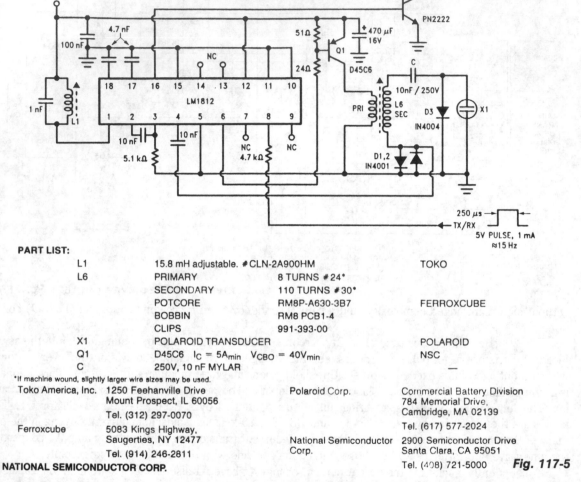

PART LIST:

L1	15.8 mH adjustable. #CLN-2A900HM		TOKO
L6	PRIMARY	8 TURNS #24*	
	SECONDARY	110 TURNS #30*	
	POTCORE	RM8P-A630-3B7	FERROXCUBE
	BOBBIN	RM8 PCB1-4	
	CLIPS	991-393-00	
X1	POLAROID TRANSDUCER		POLAROID
Q1	D45C6 $I_C = 5A_{min}$ $V_{CBO} = 40V_{min}$		NSC
C	250V, 10 nF MYLAR		—

*If machine wound, slightly larger wire sizes may be used.

Toko America, Inc. 1250 Feehanville Drive
Mount Prospect, IL 60056
Tel. (312) 297-0070

Ferroxcube 5083 Kings Highway,
Saugerties, NY 12477
Tel. (914) 246-2811

Polaroid Corp.

National Semiconductor
Corp.

Commercial Battery Division
784 Memorial Drive,
Cambridge, MA 02139
Tel. (617) 577-2024

2900 Semiconductor Drive
Santa Clara, CA 95051
Tel. (408) 721-5000

NATIONAL SEMICONDUCTOR CORP.

Fig. 117-5

The LM1812 is a complete ultrasonic transceiver on a chip designed for use in a variety of pulse-echo ranging applications. The chip operates by transmitting a burst of oscillations with a transducer, then using the same transducer to listen for a return echo. If an echo of sufficient amplitude is received, the LM1812 detector puts out a pulse of approximately the same width as the original burst. The closer the reflecting object, the earlier the return echo. Echos could be received immediately after the initial burst was transmitted, except for the fact that the transducer *rings*.

When transmitting, the transducer is excited with several hundred volts peak to peak, and it operates in a *loudspeaker* mode. Then, when the LM1812 stops transmitting and begins to receive, the transducer continues to vibrate or ring, even though excitation has stopped. The transducer acts as a microphone and produces an ac signal initially the same amplitude as the transmit pulse. This signal dies away as is governed by the transducer's damping factor, but as long as detectable ringing remains, the LM1812's detector will be held on, masking any return echos.

SONAR TRANSDUCER/SWITCH

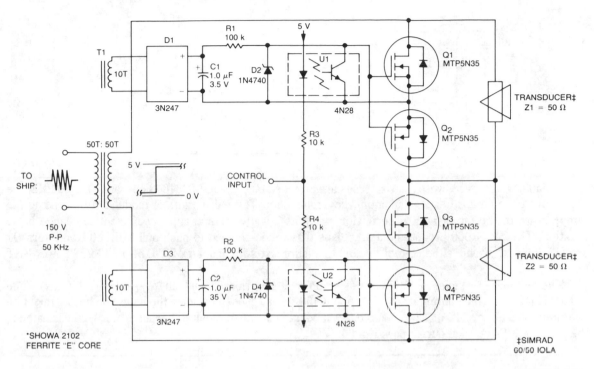

Copyright of Motorola, Inc. Used by permission.

Fig. 117-6

This submersible sonar positioning apparatus generally consists of dual-opposed ultrasonic transducers, alternately excited, with return signals processed and displayed for observation and measurement. Typical transmitter frequencies range from 50 to 200 kHz and pulse widths can be varied from 0.3 to 5 ms, depending on depth and resolution requirements.

The input to the transducer/switch is transformer T1 which provides isolation and impedance matching. The turn ratio of the secondary windings depends on the peak-to-peak amplitude of the transmitter output into the specified load. The transmitted pulse that appears on the secondary winding charges capacitors C1 and C2 through bridge rectifiers D1 and D3. Zener diodes D2 and D4 limit the TMOS gate bias to 12 V; R1 and R2 limit the discharge current from C1 and C2.

The square-wave control input is applied to opto-isolators U1 and U2 through resistors R3 and R4. If the control input is 0 V, U1 is activated; when it changes to +5 V, U2 is activated. When U1 is activated, it saturates and reduces the gate bias to zero, turning Q1 and Q2 off. Q3 and Q4 remain on, effectively shunting transducer Z2. When U2 is activated, it saturates and reduces the bias to zero, turning Q3 and Q4 off. Q1 and Q2 remain on, effectively shunting transducer Z1.

120-кHz 500-W INDUCTION HEATER

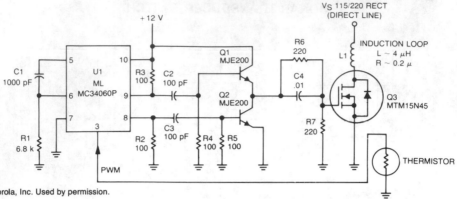

Fig. 117-7

Variable width pulses with fast rise times are provided by U1, and MC34060 operating at 120 kHz, the optimum frequency for heating aluminum alloy containers. The pulse width is modulated by sensing the temperature of the target with a thermistor, using its negative temperature coefficient to change pulse duration. The MC34060 produces output pulses that are ac-coupled to push-pull MJE200 transistors Q1 and Q2. This IC provides the current needed to ensure fast switching for MTM15N45 TMOS power FET Q3.

The estimated efficiency is 80%, based on switching losses and an R_{ON} of 0.4 Ω (max). The MTM15N45, with maximum ratings of 15 A and 450 V, was chosen because the induction heater might be operated from either 115 or 220 V sources. A modest heatsink is required because 100 W is dissipated in the power FETs at a full output power of 500 W.

ULTRASONIC TRANSCEIVER

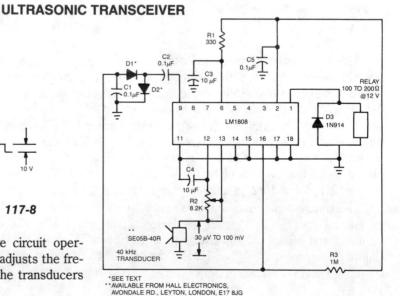

ELECTRONIC DESIGN   *Fig. 117-8*

This ultrasonic transmit/receive circuit operates at 40 kHz. Control resistor R5 adjusts the frequency for best performance with the transducers used.

ULTRASONIC RECEIVER

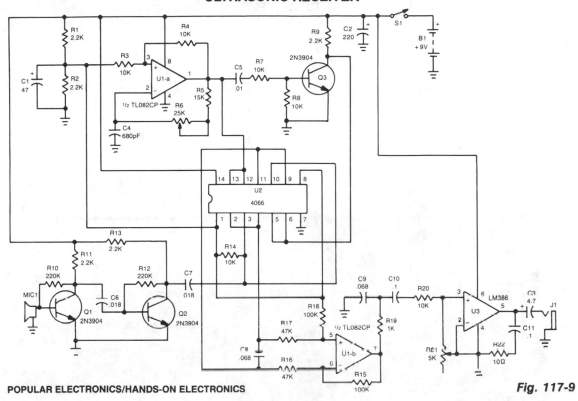

POPULAR ELECTRONICS/HANDS-ON ELECTRONICS

Fig. 117-9

The piezo speaker, MIC1, picks up the incoming ultrasonic signal and feeds it to the base of Q1. The two-transistor booster amplifier, Q1 and Q2, raises the signal to a level that is sufficient to drive one input of this most unusual mixer circuit.

Integrated circuit U2, a quad bilateral switch, functions as an extremely clean balanced-mixer circuit for the superheterodyne receiver. Integrated circuit U1a, 1/2 of a dual op amp, is connected in a variable-frequency square-wave oscillator circuit. Resistors R5, R6, and capacitor C4 determine the frequency and tuning range of the oscillator.

The oscillator's square-wave output is fed along two paths. In one path, the output of U1a is input to pins 12 and 13 of U2. In the other path, the signal is fed to the base of Q3, which is configured as an inverter. The inverter outputs a signal that is 180° out-of-phase with the input signal. The inverted output of Q3 is then fed to U2 at pins 5 and 6. There, the two input signals, the ultrasonic input from MIC1 and the oscillator output, are mixed. The mixing of the ultrasonic input and the square-wave signal produces an audible product that is fed to the input of a differential amplifier, U1b, the second half of the dual op amp, which has a voltage gain of two. The output of U1b at pin 7 is filtered by R19 and C9 to remove the high-frequency content of the mixed signal.

Only the difference frequency is important; the sum frequency, the incoming ultrasonic signal added to the oscillator frequency, is too high for the human ear to hear. The sum frequency is removed by R19 and C9 to produce a clean output signal to feed power-amplifier U3. Resistor R21 functions as the circuit's volume control.

ULTRASONIC-PULSED PEST CONTROLLER

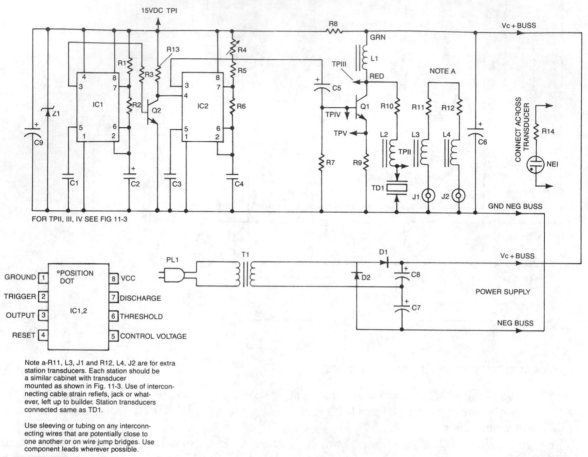

Note a-R11, L3, J1 and R12, L4, J2 are for extra
station transducers. Each station should be
a similar cabinet with transducer
mounted as shown in Fig. 11-3. Use of interconnecting cable strain refiefs, jack or whatever, left up to builder. Station transducers
connected same as TD1.

Use sleeving or tubing on any interconnecting wires that are potentially close to
one another or on wire jump bridges. Use
component leads wherever possible.

TAB BOOKS

Fig. 117-10

IC2 forms a stable oscillator whose frequency and pulse width is determined by the values of R4, R5, R6, and C4. R4 is made adjustable for precise frequency setting. The output of IC2 is pin 3, which is capacitively coupled to the base of Q1. L1 acts as a high-impedance choke to the signal, while allowing the collector of Q1 to be dc-biased. Q1 amplifies the positive pulses from IC2 and step drives the series resonant combination of L2 and TD1. Resistor R10 serves to broaden the response of this resonant circuit. L2 and the inherent capacity of the transducer, TD1, forms a resonant circuit at around 23 kHz. It is usually found that most rodents are bothered when the signal is pulsed with the off exceeding the on time. This timing is accomplished via timer IC1 and timer inverter Q2. IC1 is free running and its periods are determined by R1, R2, and C2 to be approximately two seconds off and two seconds on. The periods are inverted via Q2 and used to gate pin 4 of IC2, the frequency oscillator, turning it on for two seconds and off for three seconds. The power supply is a conventional voltage doubler with a zener regulator for the oscillator voltages.

ULTRASONIC PEST CONTROLLER

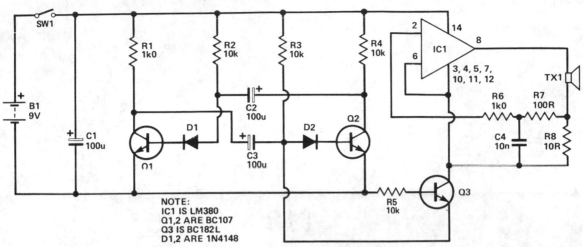

ELECTRONICS TODAY INTERNATIONAL

Fig. 117-11

This circuit consists of two basic parts: an oscillator tuned to 40 kHz, and a voltage doubler with pulse generator. The pulses are about 10 ms long and occur 2 – 3 per s to reduce battery drain and increase the annoyance factor for a cat, dog, hedgehog, etc. The voltage doubling action increases the available output power for any given battery voltage.

118

Video Amplifiers

The sources of the following circuits are contained in the Sources section beginning on page 782. The figure number contained in the box of each circuit correlates to the sources entry in the Sources section.

RGB VIDEO AMPLIFIER

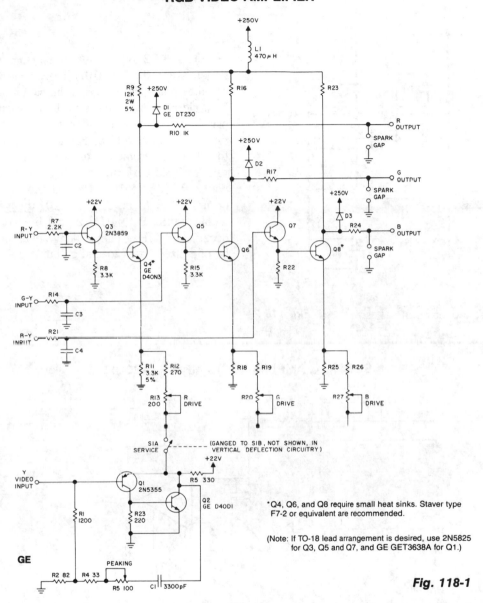

*Q4, Q6, and Q8 require small heat sinks. Staver type F7-2 or equivalent are recommended.

(Note: If TO-18 lead arrangement is desired, use 2N5825 for Q3, Q5 and Q7, and GE GET3638A for Q1.)

Fig. 118-1

Transistors Q1 and Q2 and their associated components provide: a low-impedance output with the necessary power to drive the output stages, give increased gain to high frequencies, and peaking the video for enhanced transient response. Emitter followers Q3, Q5, and Q7 provides low-impedance drive to output stages, Q4, Q6, and Q8. The output stages, with the color difference signals applied to their bases and the luminance signals to their emitters, perform matrixing. The matrixing results in composite output information, to the picture tube, which contains both luminance and chroma information.

VIDEO LINE DRIVING AMPLIFIER

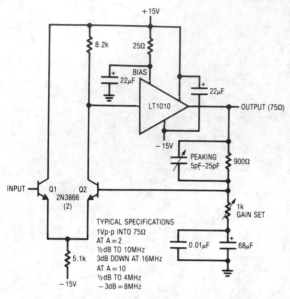

Q1 and Q2 form a differential stage which single-ends into the LT1010. The capacitively terminated feedback divider gives the circuit a dc gain of 1, while allowing ac gains up to 10. Using a 20-Ω bias resistor, the circuit delivers 1 V pk-pk into a typical 75-Ω video load. For applications sensitive to NTSC requirements, dropping the bias resistor value will aid performance. At $A = 2$, the gain is within 0.5 dB to 10 MHz and the -3 dB point occurs at 16 MHz. At $A = 10$, the gain is flat, within ± 0.5 dB to 4 MHz, and the -3 dB point occurs at 8 MHz. The peaking adjustment should be optimized under loaded output conditions.

LINEAR TECHNOLOGY CORP. **Fig. 118-2**

SUMMING AMPLIFIER/CLAMPING CIRCUIT

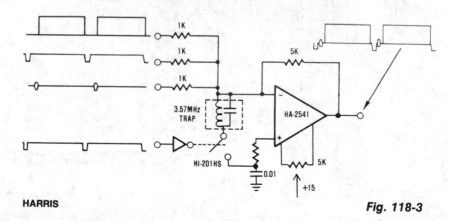

HARRIS **Fig. 118-3**

This circuit is a traditional summing amplifier configuration with the addition of the dc clamping circuit. The operation is quite simple; each component—synchronization, color burst, picture information, etc.— of the composite video signal is applied to its own input terminal of the amplifier. These signals combine algebraically and form the composite signal at the output. The clamping circuit, if used, restores the 0-V reference of the composite signal.

DC GAIN-CONTROLLED VIDEO AMPLIFIER

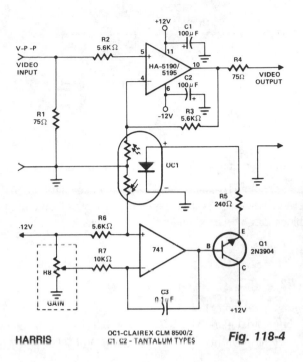

OC1-CLAIREX CLM 8500/2
C1, C2 - TANTALUM TYPES

HARRIS

Fig. 118-4

respectively, series stabilization resistor R2, and power supply bypass capacitors C1 and C2. The circuit differs from standard designs in that the gain control network includes a photoresistor, part of OC1. The optocoupler/isolator OC1 contains two matched photoresistors, both activated by a common LED. The effective resistances offered by these devices are inversely proportional to the light emitted by the LED. One photoresistor is part, with R3, of the HA-5190/5195 gain network, while the other forms a voltage-divider with R6 to control the bias applied to the integrator noninverting terminal.

In operation, the dc voltage supplied by gain control R8 is applied to the integrator inverting input terminal through input resistor R7. Depending on the relative magnitude of the control voltage, the integrator output will either charge or discharge C3. This change in output, amplified by Q1, controls the current supplied to the OC1 LED through series limiting resistor R5. The action continues until the voltage applied to the integrator noninverting input by the R6—photoresistor gain network is changing, adjusting the op amp stage gain. As the control voltage at R8 is readjusted, the OC1 photo-resistances track these changes, automatically readjusting the op amp in accordance with the new control voltage setting.

This amplifier employs a cascaded op amp integrator and transistor buffer, Q1, to drive the gain control element. Except for a simple modification, the HA-5190/5195 stage is connected as a conventional noninverting op amp, and includes input and output impedance matching resistors R1 and R4,

---

75-Ω VIDEO PULSE AMPLIFIER

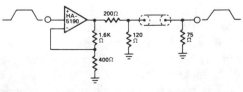

HARRIS

Fig. 118-5

HA-5190 can drive the 75-Ω coaxial cable with signals up to 2.5 V pk-pk without the need for current boosting. In this circuit, the overall gain is approximately unity because of the impedance matching network.

VIDEO GAIN BLOCK

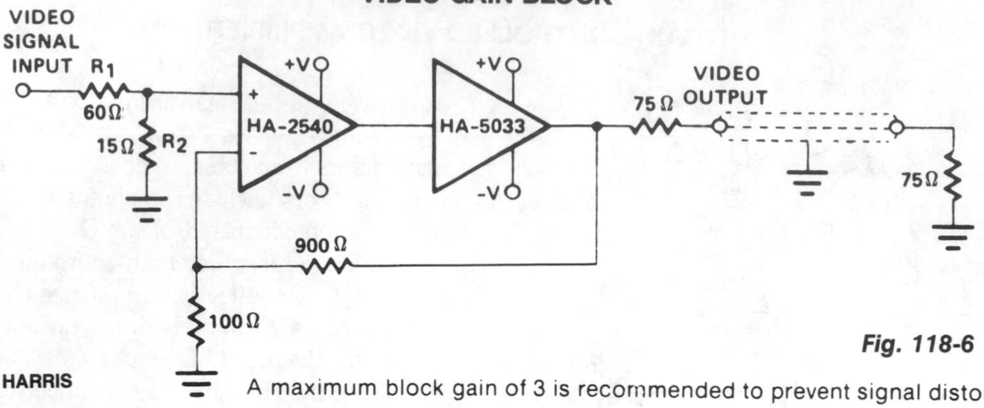

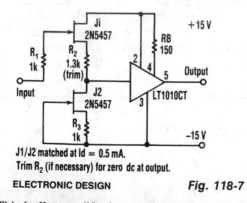

Fig. 118-6

HARRIS

A maximum block gain of 3 is recommended to prevent signal distortion.

This configuration utilizes the wide bandwidth and speed of HA-2540, plus the output capability of HA-5033. Stabilization circuitry is avoided by operating HA-2540 at a closed loop gain of 10, while maintaining an overall block gain of unity. However, gain of the block can be varied using the equation:

$$\frac{V_{OUT}}{V_{IN}} = 5 \frac{R2}{(R1 + R2)}$$

$$\text{where } R1 + R2 = 75 \ \Omega$$

A maximum block gain of 3 is recommended to prevent signal distortion.

This circuit was tested for differential phase and differential gain using a Tektronix 520A vector scope and a Tektronix 146 video signal generator. Both differential phase and differential gain were too small to be measured.

LOW-DISTORTION VIDEO BUFFER

J1/J2 matched at Id = 0.5 mA.
Trim R₂ (if necessary) for zero dc at output.

ELECTRONIC DESIGN

Fig. 118-7

This buffer amplifier's overall harmonic distortion is a low 0.01% or less at 3-V rms output into a 500-Ω load with no overall feedback. The LT1010CT offers a 100 V/μs slew rate, a 20 MHz video bandwidth, and 100 mA of output. A pair of JFETs, J1 and J2 are preselected for a nominal match at the bias level of the linearized source-follower input stage, at about 0.5 mA. The source-bias resistor, R2, of J1 is somewhat larger than R3 so that it can drop a larger voltage and cancel the LT1010CT's offset. J1 and J2 provide an untrimmed dc offset of ±50 mV or less. Swapping J1 and J2 or trimming the R2 value can give a finer match.

The circuit's overall harmonic distortion is low: 0.01% or less at 3-V rms output into a 500-Ω load with no overall feedback. The circuit's response to a ±5 V, 10 kHz square-wave input, band-limited to 1 μs, has no overshoot. If needed, setting bias resistor R_B lower can accommodate even steeper input-signal slopes and drive lower impedance loads with high linearity. The main trade-off for both objectives is more power dissipation. A secondary trade-off is the need for retrimming the source-bias resistor, R2.

119

Video Circuits

The sources of the following circuits are contained in the Sources section beginning on page 782. The figure number contained in the box of each circuit correlates to the sources entry in the Sources section.

RGB-COMPOSITE CONVERTER

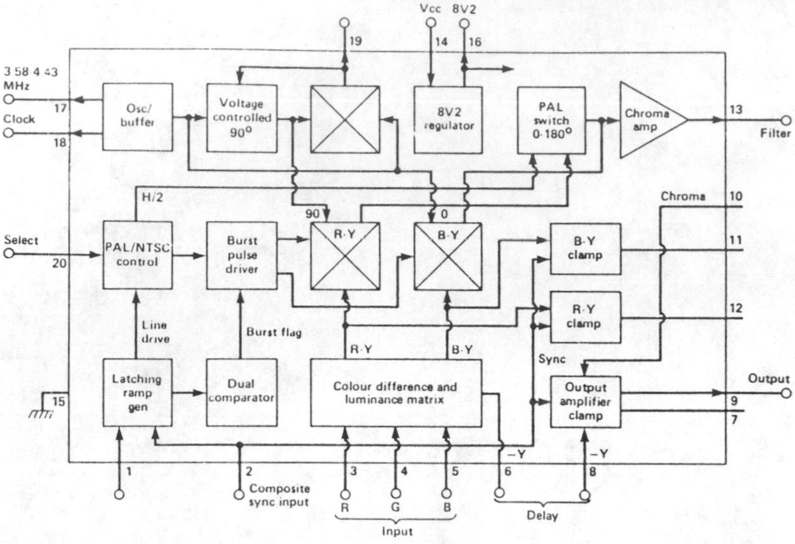

BLOCK DIAGRAM OF THE MC1377P ENCODER IC.

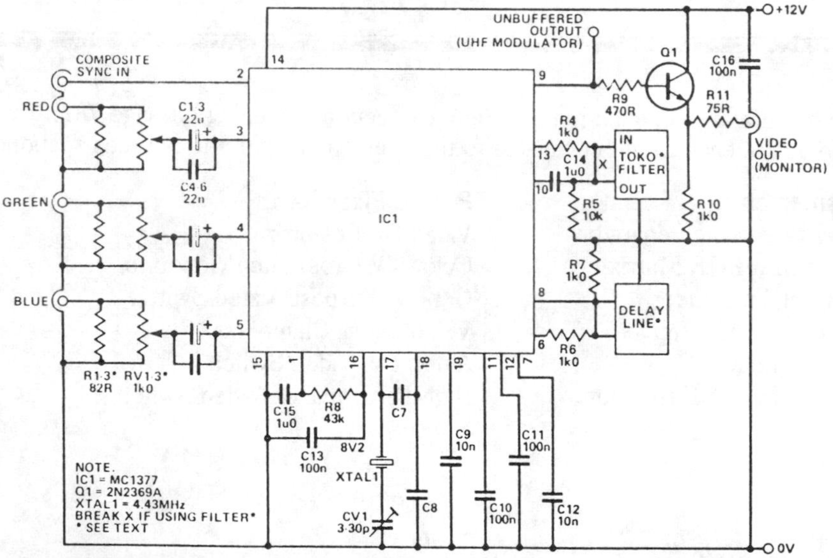

(a) 100% GREEN INPUT (PIN 4)

(b) 100% RED INPUT (PIN 3)

(c) 100% BLUE INPUT (PIN 5)

(d) COMPOSITE OUTPUT (PIN 9)

(e) SYNC INPUT (PIN 2)

(f) CHROMA OUTPUT (PIN 13)

(g) CHROMA INPUT (PIN 10)

(h) LUMINANCE OUTPUT (PIN 6)

(i) LUMINANCE INPUT (PIN 8)

CIRCUIT DIAGRAM OF THE CONVERTER.

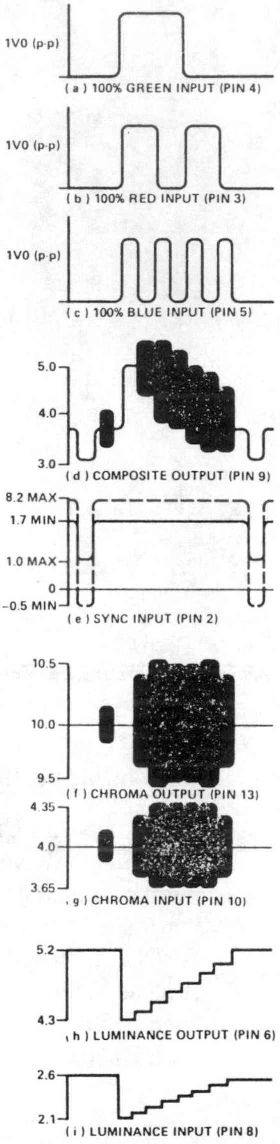

NOTE.
IC1 = MC1377
Q1 = 2N2369A
XTAL1 = 4.43MHz
BREAK X IF USING FILTER*
* SEE TEXT

The signals that should appear at the test points around the chip.

ELECTRONICS DIGEST

Fig. 119-1

The incoming RGB inputs are terminated with resistors R1, R2, and R3 and potentiometers RV1, RV2, and RV3. These provide input impedances of approximately 75 Ω. The presets should be adjusted to provide a maximum input of 1 V pk-pk into the MC1377. The inputs are ac-coupled into the encoder; the large value capacitor is required for the 60 Hz field component.

RGB-COMPOSITE CONVERTER *(Cont.)*

The Colpitts oscillator for the color burst is formed around pins 17 and 18. About 0.5 V pk-pk should appear on pin 17 and 0.25 V rms into pin 18 with the oscillator components removed. The incoming composite sync signal at pin 2 should be negative-going. The device will accept CMOS and TTL directly. If it is necessary to ac-couple the sync, then a pull-up to 8.2 V is required—a regulated 8.2 V is provided on pin 16.

From the composite sync input, the MC1377 generates a ramp which it uses to provide the burst gate pulse. The slope of this ramp can be varied by a potentiometer on pin 1. However, a preset value, shown as 43 KΩ, is usually sufficient. The chrominance filter should be fitted between pins 13 and 10. If the filter is not used, a compensatory potential divider should be fitted (both are shown). We used a prealigned Toko bandpass filter centered on 4.43 MHz. If the chroma filter is fitted, the delay through it, 400 ns, has to be compensated for by a luminance delay line between pins 6 and 8. This line is shorted out if the filter is not fitted. The composite video output from the IC is buffered to provide a low-impedance drive for a monitor, or it can be applied directly to a UHF modulator commonly used in computers.

SINGLE-SUPPLY WIDE-RANGE SYNC SEPARATOR

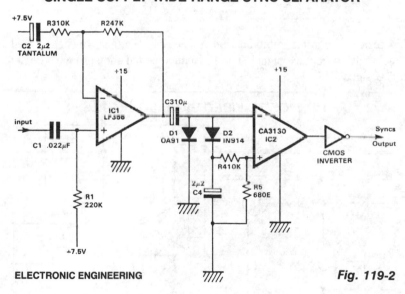

ELECTRONIC ENGINEERING

Fig. 119-2

This circuit extracts the sync pulses from a video signal over a wide range of amplitudes and operates a single +15 V supply. IC1 buffers and amplifies the incoming signal and applies it via C3 to the peak detector, consisting of D2 and C4. It is also applied to one input of a comparator, IC2. The other input of IC2 is set at a voltage corresponding to about 0.065 of the peak video amplitude, by the divider R4/R5. The trigger points of IC2 are set near the bottom of the sync pulses which help prevent spurious noise. These resistors also leak across C4, so they must be chosen as a compromise between excessive ripple and speed of response to falling signal levels. The IC2 output swings between 0 and 15 V and is conveniently CMOS compatible, but further buffering is advisable, hence the CMOS inverter. Maximum input amplitude is set by saturating IC1's output. The minimum acceptable level is set by the forward voltage drop of the dc restoring clamp D1, which should be either a germanium (as shown) or a Schottky diode.

CHROMA DEMODULATOR WITH RGB MATRIX

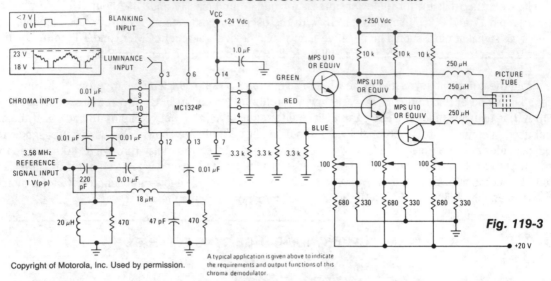

A typical application is given above to indicate the requirements and output functions of this chroma demodulator.

Fig. 119-3

The MC1324 provides chroma demodulation recovering recooling the R, G, and B signals to drive video amps for each color difference signal. The luminance signal and chrominance signal are matrixed to get the R, G, and B signals.

COMPOSITE-VIDEO SIGNAL TEXT ADDER

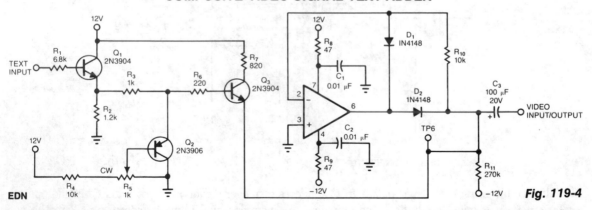

EDN

Fig. 119-4

This circuit shows a simple way to add text information to a composite-video signal that might be floating at some indeterminate dc level. The text generator and composite-video source must have the same sync signal. The video-input and -output signals share the same terminal. C3 couples the video signal to the output of a rectifier circuit that is based on a subvideo-speed op amp. A faster op amp would clamp on individual sync pulses rather than on the video waveform's average value, as is desired. R11 serves as a pulldown resistor and feedback resistor R10 ensures that TP6 remains at ground level. Emitter follower Q1 buffers the text signal, and R5 serves as a gain control. A simple clamp circuit, Q2, is sufficient for regulating amplitude, because the text signal contains no gray-scale levels. Q3 couples the text signal into the op-amp clamp circuit.

PAL/NTSC DECODER WITH RGB INPUTS

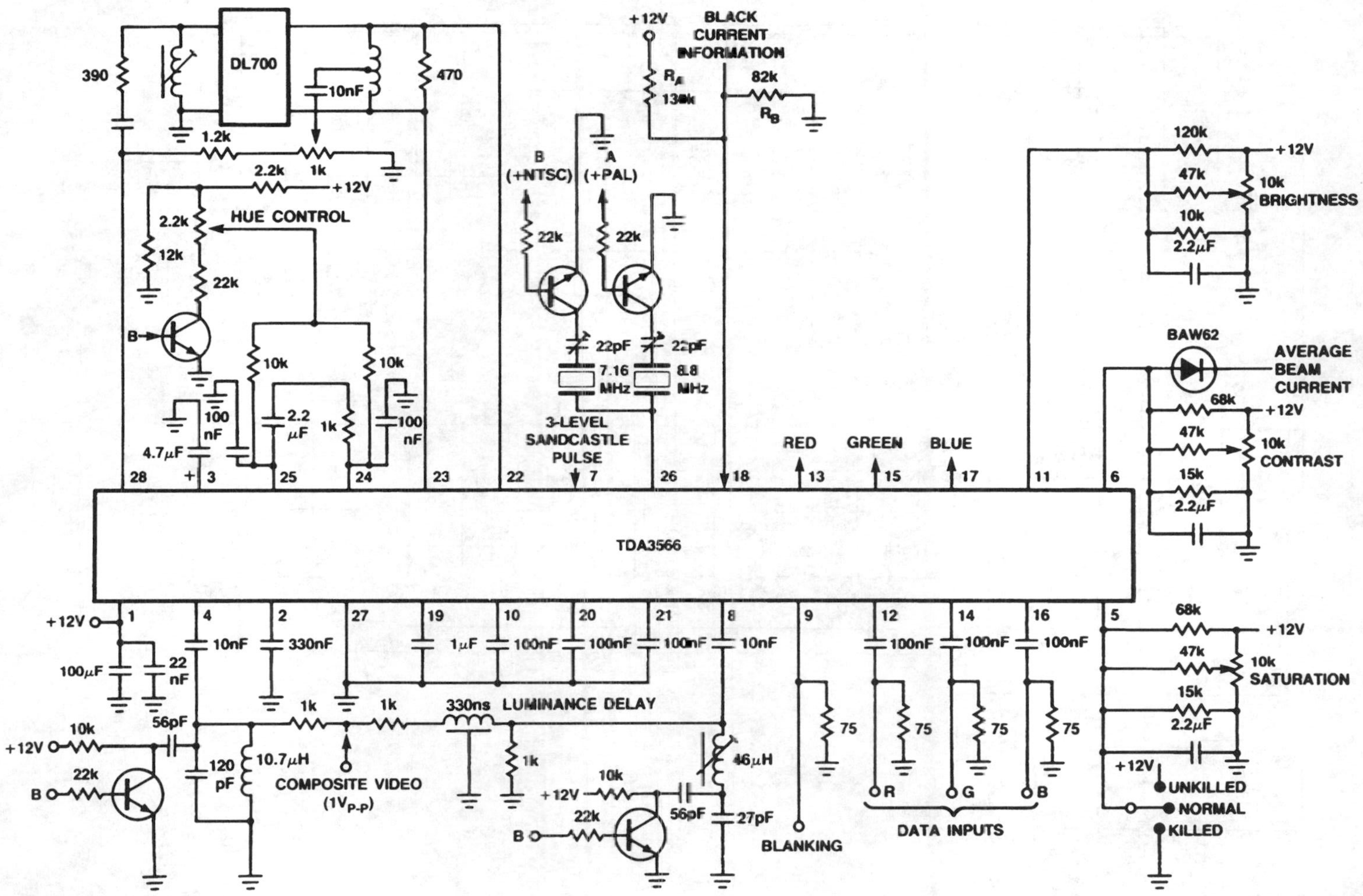

SIGNETICS

Fig. 119-5

This circuit shows the TDA3566 for a PAL/NTSC Decoder.

WIRELESS VIDEO CAMERA LINK

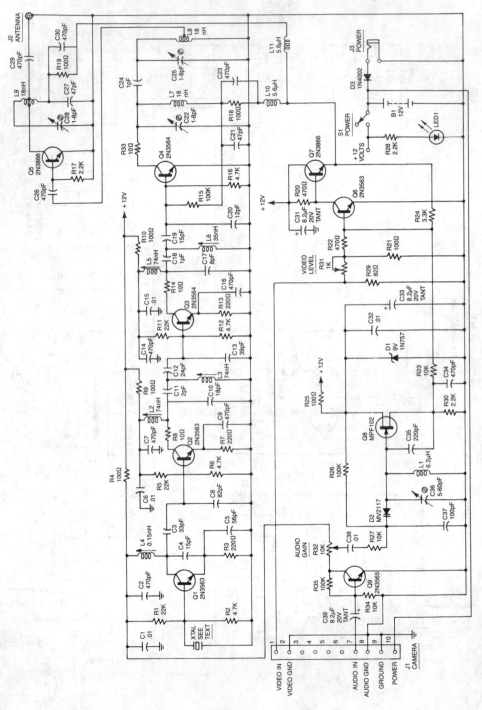

Fig. 119-6

Reprinted with permission from Radio-Electronics Magazine, February 1986. Copyright Gernsback Publications, Inc., 1986.

WIRELESS VIDEO CAMERA LINK *(Cont.)*

This high-performance video-camera link transmits signals from your video camera to your VCR, or from your VCR to TVs throughout your home. The first stage of the rf chain is a crystal-controlled oscillator, Q1, with a frequency of 60 to 65 MHz, which is one-eighth of the final output frequency. The oscillator produces a signal of about +6 dBm (4 mW) that drives three stages of frequency doublers. The combined action of those doublers multiplies the input frequency by eight for a final output frequency of (nominally) 500 MHz. Double-tuned circuits are used between each stage to help reduce spurious outputs that might cause unwanted interference. The video input signal from your VCR, video camera, etc. drives a video modulator, Q6 and Q7, that adds the video signal to the +12 V line supplying power to the final doubler, Q4, and the output amplifier, Q5. That method of modulation is similar to the way a conventional AM-radio transmitter is modulated. The video modulator has a nominal bandwidth of five MHz. The audio input is applied to Q8, which operates as a VCO running at a nominal frequency of 4.5 MHz to produce the modulated sound carrier. For simplicity, Q8 is a free-running oscillator, since the ±25 kHz frequency deviation that is required would be very difficult to produce at that frequency with a crystal-controlled oscillator. Besides, most TV sound systems will accept a ±10 kHz error in the sound-carrier frequency without producing undue distortion, and that greatly simplifies the circuitry required. The kit is available from North Country Radio, P.O. Box 53, Wykagyl Station, NY 10804.

VIDEO SWITCH WITH VERY HIGH OFF ISOLATION

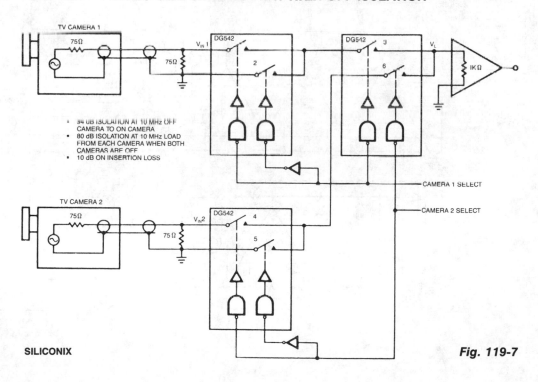

SILICONIX

Fig. 119-7

VIDEO PALETTE

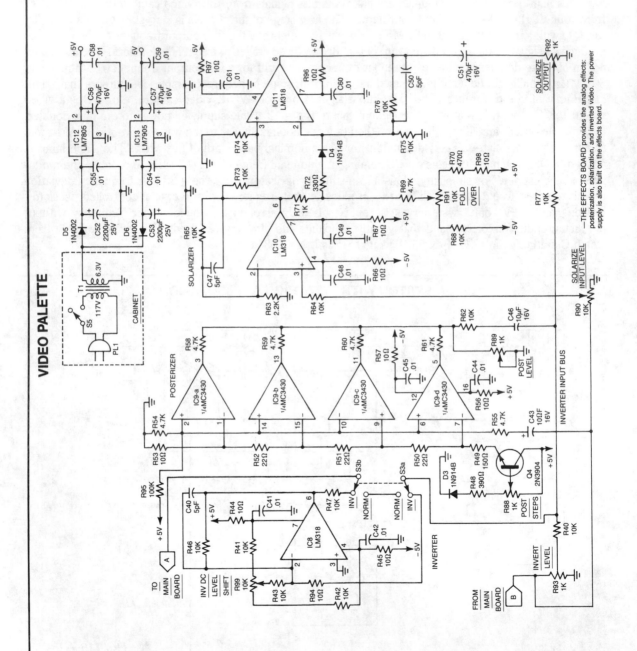

—THE EFFECTS BOARD provides the analog effects: posterization, solarization, and inverted video. The power supply is also built on the effects board.

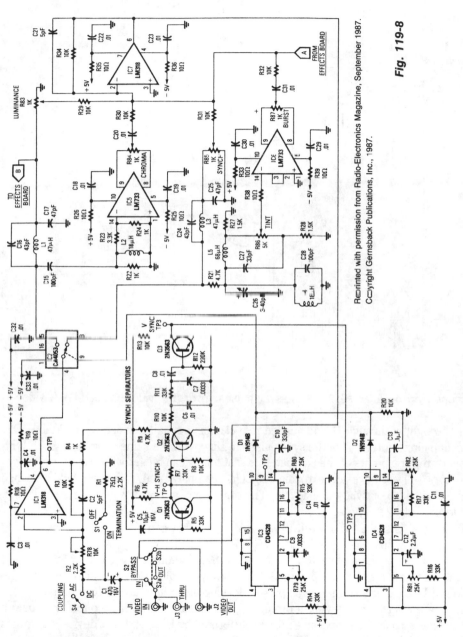

Fig. 119-8

Reprinted with permission from Radio-Electronics Magazine, September 1987.
Copyright Gernsback Publications, Inc., 1987.

This system consists of two parts. The main board dissects the video signal and provides independent level control for burst, chroma, luminance, and sync signals, as well as phase and polarity. The video signal is reassembled in the output in corrected or modified format as required by the user. The effects board produces luminance inversion or can generate discrete luminance steps (posterization) or a nonlinear gray scale (solarization) to achieve simulation of photographic effects commonly seen in various special-effect photographic processes. The kit is available from North Country Radio, P.O. Box 53, Wykagyl Station, NY 10804.

PICTURE FIXER/INVERTER

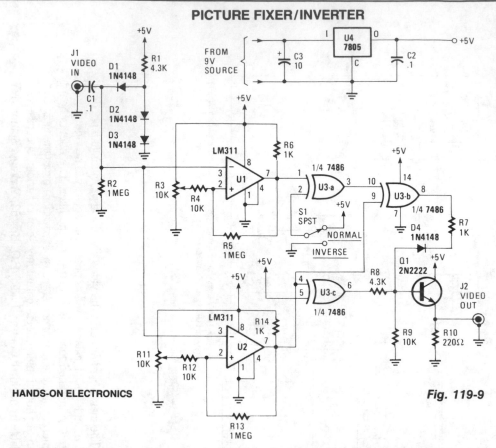

HANDS-ON ELECTRONICS

Fig. 119-9

The circuit will accept a video signal, separate the sync pulses, invert the video, and add new video to the old sync pulses.

The video signal is brought in through J1 and applied to a clamping circuit consisting of C1, D1, D2, D3, R1, and R2. The clamp circuit forces all of the sync pulses to align with the same dc voltage level. With the video voltage clamped, the trip points of the comparators that follow can be set with trimmer resistors R3 and R11. The resistors will not have to be readjusted. One comparator, U1, is adjusted to change states with a change in either video or sync-pulse levels. The other comparator, U2, is adjusted to trip on changes of sync-pulse levels only.

The output of U1 now consists of a logic level, 0 to +5 V, signal that contains both sync pulses and video. The composite signal is coupled to an EXCLUSIVE-OR gate, U3a, where it is either inverted or not inverted, depending upon the position of switch NORM/REV S1. The output, at pin 3 of U3a, is next sent to U3b. There the composite signal is combined with the sync-pulse only signal from U2. The EXCLUSIVE-OR action of U3b cancels out the sync pulses, leaving only video at the IC's output.

Since the sync pulses are inverted as they pass through U2, they must be inverted once more before being combined with the video signal. That final inversion is performed by U3c, and that device's output is combined with that of U3b via D4, R7, R8, and R9. The newly combined signal is buffered by emitter-follower Q1, and sent to the outside world via J2. The circuit can be powered by a 9- to 12-V wall-mount power supply. The supply voltage is regulated down to 5 V by U4.

VIDEO DC RESTORER

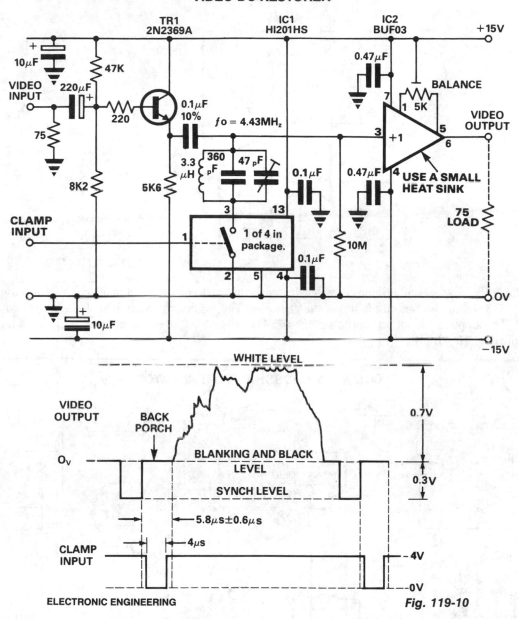

ELECTRONIC ENGINEERING

Fig. 119-10

The main requirement for efficient dc restoration is to provide a short time-constant during the clamp period, with a long time-constant during the active line time. The switch within the Harris HI201HS has an on resistance of 30 Ω and the PMI buffer, BUF03, has an input resistance of 50 MΩ. The tuned circuit presents a high impedance at the 4.43-MHz color subcarrier frequency so that the color-burst signal is retained if the video signal contains this information.

COLOR VIDEO AMPLIFIER

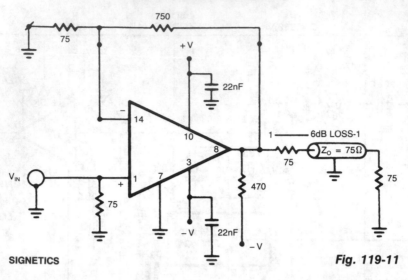

SIGNETICS

Fig. 119-11

The NE5539 wideband op amp is easily adapted for use as a color video amplifier. The gain varies less than 0.5% from the bottom to the top of the staircase. The maximum differential phase is approximately +0.1°. The amplifier circuit was optimized for a 75-Ω input and output termination impedance for a gain of approximately 10 (20 dB).

COLOR TV CROSSHATCH GENERATOR

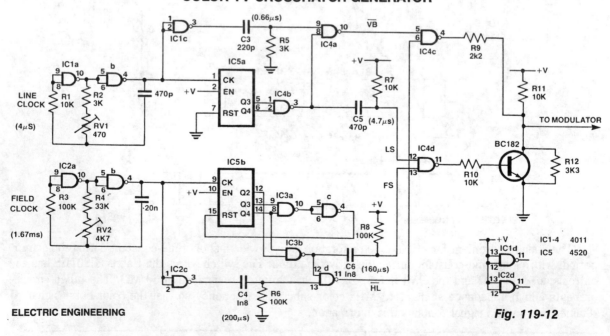

ELECTRIC ENGINEERING

Fig. 119-12

COLOR TV CROSSHATCH GENERATOR (*Cont.*)

This circuit provides a simple, low-cost crosshatch generator for convergence and geometry adjustments on color TVs. The generator is driven by two clocks, one for the horizontal drive, IC1ab, and one for the vertical drive, IC2ab. The clock outputs are applied to the two binary counters contained in IC5 which generate the line and field sync pulses and respective blanking periods. Line clock pulses, buffered by IC1c, are differentiated by C3/R5 to produce the vertical bars. These bars are gated by IC4a which suppresses the bars during the line blanking period produced by the coincidence of Q3, Q4 outputs of IC5a, detected by IC4b. This output is also differentiated by C5/R7 to produce the line sync pulse, *LS*. A similar process is used to generate horizontal lines and the field sync, except that in order to give the correct aspect ratio, the count of IC5b is reset at 12, coincidence of Q3 and Q4. In coincidence of Q2, Q4 is used to generate the sync pulse *FS* and the blanking period. The line and field sync pulses, *LS* and *FS*, are combined in IC4d.

GENERAL PURPOSE VIDEO SWITCH

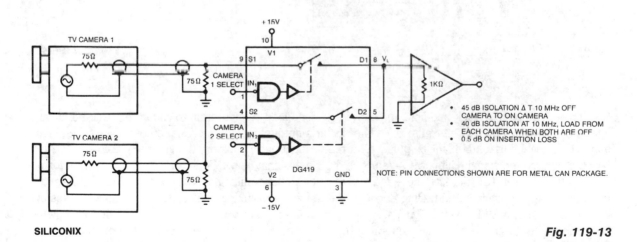

SILICONIX

Fig. 119-13

The circuit shown provides 40-dB isolation at 6 MHz and is good for general purpose video switching.

VIDEO SIGNAL CLAMP

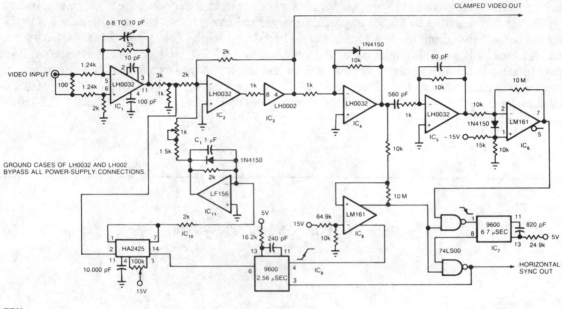

EDN

Fig. 119-14

The circuit uses a track-and-hold amplifier in a closed-loop configuration to clamp the back-porch voltage of a standard video waveform to 0 V. The circuit's outputs include a clamped composite-video signal and a TTL-level horizontal-blanking pulse. Differential input buffer IC1 and the summing amplifier IC2 isolate the input video signal. Clipper IC4 removes the video signal, leaving only the synchronization information. Differentiator IC5 detects the edges of the horizontal blanking pulses and produces pulses that correspond to the leading and trailing edges of the horizontal blanking pulses. IC6 clips these pulses and converts them to a TTL level. IC7 uses these clipped pulses to generate a TTL-level window that, when combined with the horizontal pulse generated by IC8, forms a TTL representation of the original horizontal pulse. This representation is synchronized to the input waveform. IC9 uses the trailing edge of this reconstructed waveform to generate the track pulse for track-and-hold amplifier IC10. IC11 filters IC10's dc output and, after gain adjustment, feeds it back to IC2's summing node.

AUTOMATIC VIDEO SWITCH

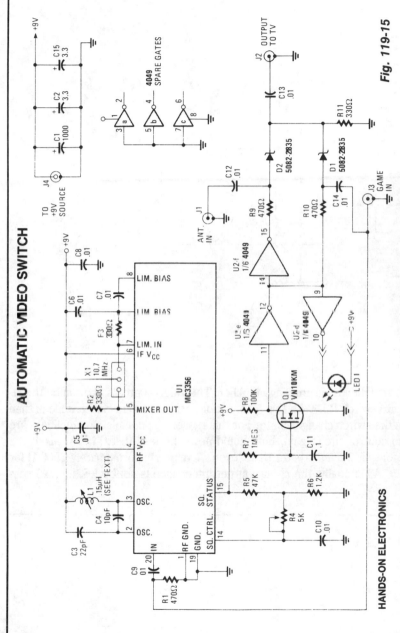

Fig. 119-15

HANDS-ON ELECTRONICS

Turn on a game, computer, videodisc player, or whatever, and its output takes priority over the antenna-derived signal—the antenna is disconnected and the alternate source is fed to the monitor. When the alternate video source is no longer detected, the switch automatically reconnects the antenna. When the rf carrier is detected, a high logic output appears at pin 15 of U1—squelch status. The signal is then buffered to VMOS FET Q1. This FET can drive CMOS inverter/buffer U2, can be an *open-drain* output to drive a relay, or can convert to a 5 V logic level. When pin 15 of U1 becomes high, Q1 turns on, pulling pin 11 of U2e low; in turn, pin 12 becomes high. The output of U2e is fed to U2f, forcing its output, at pin 15, low. When the signal at U2 pin 12 is high, D1 is biased on, allowing the signal at J3 to flow through C14, D1, and out to J2 through C13. When the signal at J3 is removed, U1 pin 15 decreases. That decrease causes U2 pin 12 to decrease and U2 pin 15 to increase. When the signal at U2 pin 15 is high, it biases D2 on, allowing the signals at J1 to flow through C12, D2, and out to J2 through C13.

727

HIGH-PERFORMANCE VIDEO SWITCH

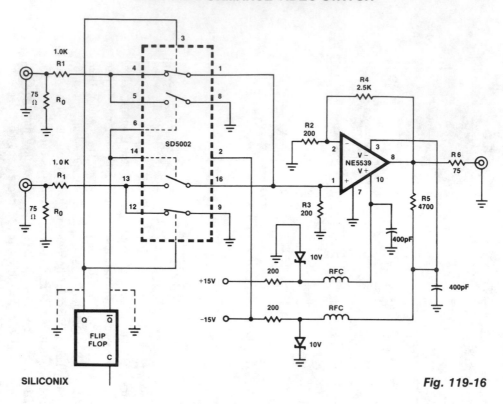

SILICONIX

Fig. 119-16

This figure shows a one-of-two switch with a summing amplifier. The video source's line can be terminated either externally or internally to switch R0. With this termination resistor, a load change of less than 1 Ω will be *seen* by the source when the switch changes state. For this reason, input isolation amplifiers are not necessary. R4 can be varied to control circuit gain, but should never be less than 1400 Ω since the NE5539 is internally compensated for gain values greater than seven. A value of approximately 2500 Ω for R4 will set circuit gain to near unity. Additionally, the circuit output impedance is set by R6, and R5 sets the output dc offset to near zero.

120

Voice Circuits

The sources of the following circuits are contained in the Sources section beginning on page 782. The figure number contained in the box of each circuit correlates to the sources entry in the Sources section.

Ac Line-Voltage Announcer
Dialed Phone Number Vocalizer
Computer Speech Synthesizer
Allophone Generator
Electronic Voice Substitute

AC LINE-VOLTAGE ANNOUNCER

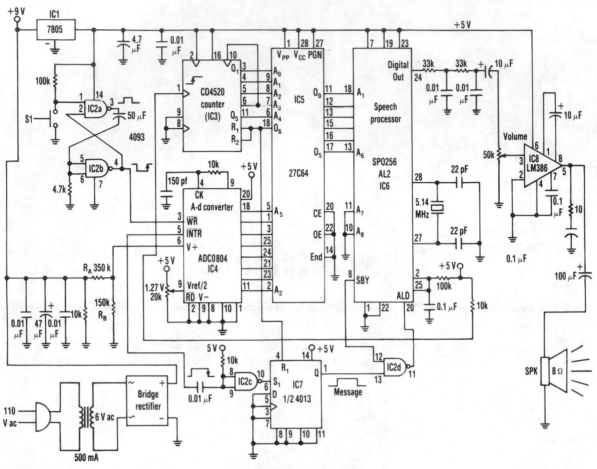

Fig. 120-1

The range of this ac-voltage monitor is 100 to 140 Vac, with a resolution of 1 V. The speech processor interprets an 8-bit binary input code from an analog-to-digital converter. The processor's pulse-code-modulated output then passes through a filter and an amplifier before driving the circuit's speaker to vocalize the corresponding number. Each time switch S1 is pressed, the speech-processor program enunciates the monitored voltage readings from 100 to 140 V, depending on the code at the input of a 27C64 EPROM.

The voltage-monitoring circuit consists of a bridge rectifier, filter capacitors, and a 10-KΩ load resistor. A divider, R_A and R_B, limits the input voltage to a maximum 2.55 V. The a/d converter, IC4, then sends the voltage reading to the 27C64 EPROM, IC5. Pressing S1 sends a negative transient pulse to the write, WR, input of the a/d converter, IC4, which initiates a 100-μs conversion process.

DIALED PHONE NUMBER VOCALIZER

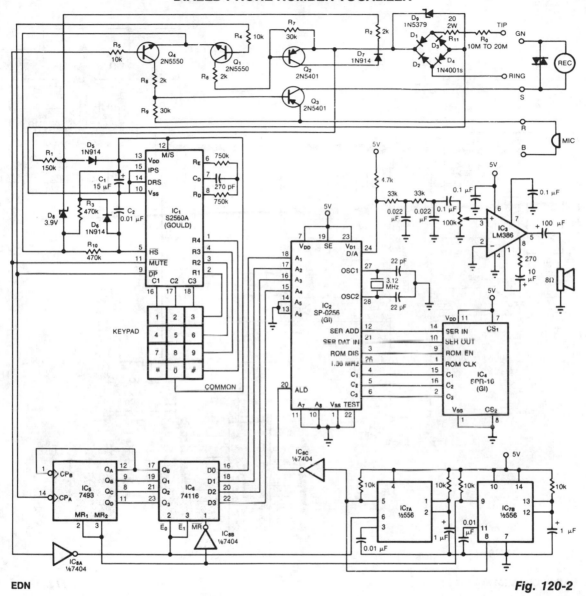

EDN

Fig. 120-2

By vocalizing the numbers and symbols of its keypad, the phone provides an audible confirmation that is useful to the blind. The serial-interface, 2 K-byte × 8-bit ROM (IC4) stores programmed sequences of instructions that are executed by the speech-processor chip IC2—manufactured by the General Instrument Corp. When you depress a key, tone-dialing chip IC1 issues the corresponding number of pulses at its DP output. Counter IC5 totals the pulses, and IC6 latches the resulting 4-bit digital word. This word, converted to serial format by IC2, becomes an address that selects a block of memory within IC4.

COMPUTER SPEECH SYNTHESIZER

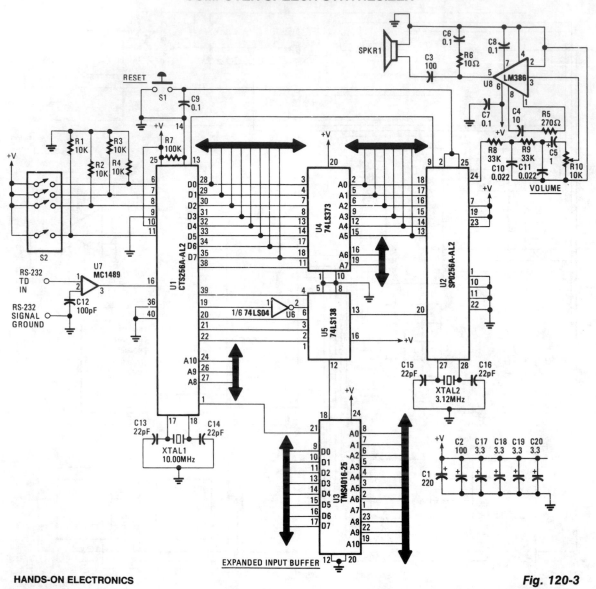

Fig. 120-3

This text-to-speech converter is built around the SPO256-AL2 speech processor and the CT6256-AL2 text-to-speech converter chips—manufactured by General Instruments. The circuit is set up to receive standard ASCII code from virtually any microcomputer or dumb terminal that is equipped with an RS-232 port—such as a serial-printer or modem port. If a microcomputer is used, the synthesizer can be activated from a terminal-emulator of any communications program, or from any programming language such as BASIC.

COMPUTER SPEECH SYNTHESIZER (*Cont.*)

The serial input from the RS-232 port enters the circuit through U7, the MC1489 RS-232 receiver chip, and is converted from an RS-232 level to a TTL-level signal. The CTS256-AL2 chip, U1, then converts the ASCII characters into allophone codes and sends those codes to U3, the TMS4016 external-RAM chip. The codes are then transferred to the SPO256-AL2, U2, through the 74LS373 octal latch, U4. Finally, the SPO256-AL2 sends out an audio signal to the LM386 audio amplifier, U8, through some high-pass filtering, and on to the speaker. The 74LS138d, U5, and the 74LSO4, U6, provide control logic.

ALLOPHONE GENERATOR

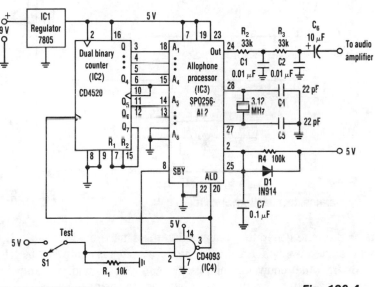

ELECTRONIC DESIGN

Fig. 120-4

The circuit, a general-purpose system with many uses, vocalizes 59 allophones contained in the speech processor. After filtering and amplification, its pulse-code-modulated output can drive an 8-Ω speaker. The processor's address pins, A1 to A6, define 64 speech-entry points.

Closing the test switch to the NAND gate lowers its output, thereby loading an address and triggering the ALD input for an allophone cycle. The CD4520 dual binary counter, IC2, counts from 0 to 63 in binary code until its Q7 output resets it on the number 64 count. To generate a phrase, just add an EPROM between IC2 and IC3 that contains a program for a predetermined sequence of allophones.

ELECTRONIC VOICE SUBSTITUTE

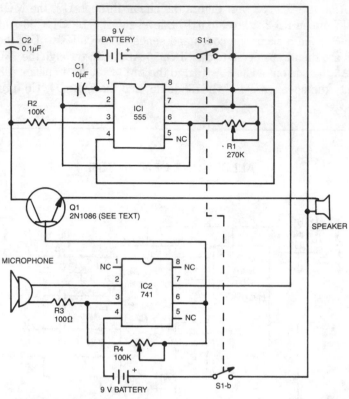

GERNSBACK PUBLICATIONS INC.

Fig. 120-5

The 555 acts as the tone generator configured in the astable mode. Its pin 3, square-wave output is transformed into a triangle wave by R1 and C2. The *voice's* pitch is controlled by R1. Transistor Q1 can be 2N1086, 2N1091, or any other equivalent npn germanium type. Sounds are amplified by the 741, and the IC's output drives the transistor to saturation. When the transistor is in the saturated state, the triangle wave is able to reach the speaker, and your new *voice* can be heard.

121

Voltage-Controlled Oscillators

The sources of the following circuits are contained in the Sources section beginning on page 782. The figure number contained in the box of each circuit correlates to the sources entry in the Sources section.

BALANCED TMOS VCO

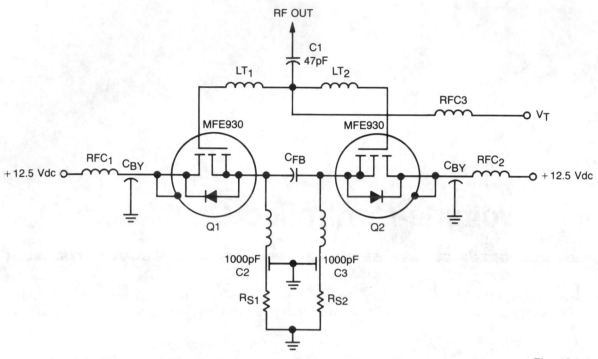

RF OUT

C1
47pF

LT$_1$ LT$_2$

RFC3

V$_T$

MFE930   MFE930

RFC$_1$ C$_{BY}$ C$_{FB}$ C$_{BY}$ RFC$_2$

+ 12.5 Vdc   + 12.5 Vdc

Q1   Q2

1000pF   1000pF
C2   C3

R$_{S1}$ R$_{S2}$

Fig. 121-1

This TMOS VCO operates in push-pull to produce 4 W at 70 MHz. It consists of two MFE930 TMOS devices in a balanced VCO that generally provide better linearity than the single-ended types. Varactors are not used because the design takes advantage of the large change in *Miller* capacitance, C_{RSS}, that is available in TMOS gate structures.

In the balanced VCO, the fundamental (f_O) and/or twice the fundamental ($2f_O$) can be coupled from the circuit at separate nodes. This makes the balanced oscillator very useful in phase-locked loops. The fundamental:

$$f_O = \frac{1}{2} (L_F C_{RSS})^{-1/2}$$

where:

$$L_F = 0.68 \ \mu H$$

WAVEFORM GENERATOR/STABLE VCO

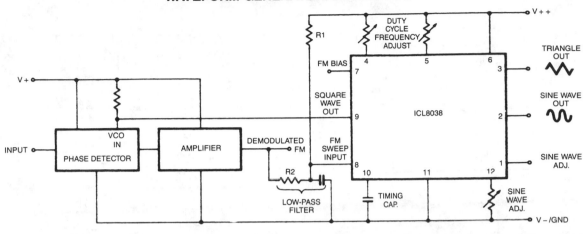

INTERSIL

Fig. 121-2

In this circuit, a waveform generator is used as a stable VCO in a Phase-Locked Loop (PLL).

VARIABLE-CAPACITANCE DIODE-SPARKED VCO

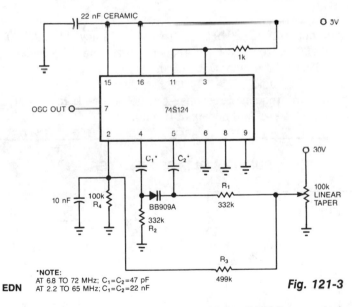

*NOTE:
AT 6.8 TO 72 MHz; $C_1 = C_2 = 47$ pF
AT 2.2 TO 65 MHz; $C_1 = C_2 = 22$ nF

EDN

Fig. 121-3

You can transform a 741S124 multivibrator into a wideband VCO by replacing it conventional fixed capacitor with a variable-capacitance diode. The only disadvantage of this scheme is the 30-V biasing voltage that the diode requires. Capacitors C1 and C2 couple the Philips BB909A variable-capacitance diode to the 74S124. R1 and R2 are large enough to isolate ground and control voltages from the timing capacitors. Resistors R3 and R4 form a voltage divider for the 74S124's control input.

LOGARITHMIC SWEEP VCO

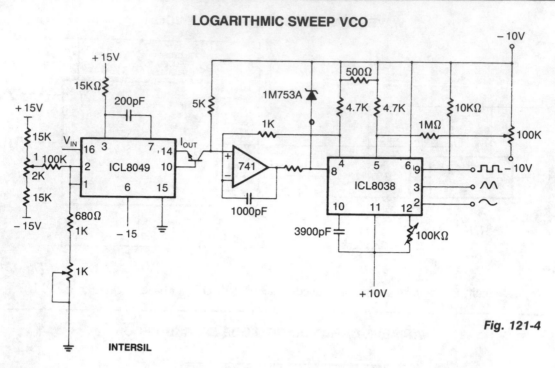

Fig. 121-4

INTERSIL

This circuit uses the output of the ICL8049 to control the frequency of the ICL3038 waveform generator; the 741 op amp is used to linearize the voltage-frequency response. The input voltage to the 8049 can be, for example, from the horizontal sweep signal of an oscilloscope; the output of the 8038 will then sweep logarithmically across the audio range. By feeding this to the equipment being measured and detecting the output, a standard frequency response can be obtained. If the output is fed through an ICL8048 before being displayed, a standard bode plot results.

SUPPLY VOLTAGE SPLITTER

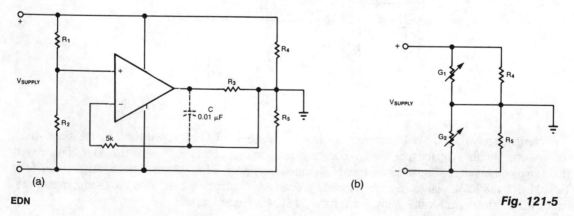

(a)

(b)

EDN

Fig. 121-5

SUPPLY VOLTAGE SPLITTER (*Cont.*)

This simple circuit can convert a single supply voltage, such as a battery, into a bipolar supply. Sense resistors R1 and R2 establish relative magnitudes for the resulting positive and negative voltages. Their rail-to-rail value, of course, equals V_{SUPPLY}. R4 and R5 represent the load impedances. For example, equal-value sense resistors produce $1/2$ V_{SUPPLY} across each of the load resistors, R4 and R5. The op amp maintains these equal voltages by sinking or sourcing current through R3; the op amp's action is equivalent to that of variable conductances G1 and G2 in shunt with each load resistor. Choose a value for R3 so that the largest voltage across it, the greatest load-current mismatch, won't exceed the op amp's output-voltage capability for the application. You can add a buffer amplifier at the op amp's output to provide greater load currents. If you need bypass capacitors across the load resistors as well, connect a capacitor (dashed lines) to ensure that the amplifier remains stable.

3 – 5 V REGULATED OUTPUT CONVERTER

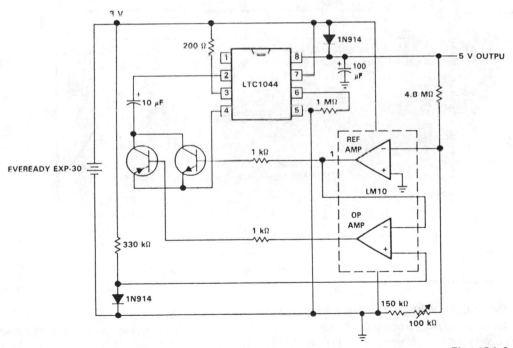

Fig. 121-6

VCO

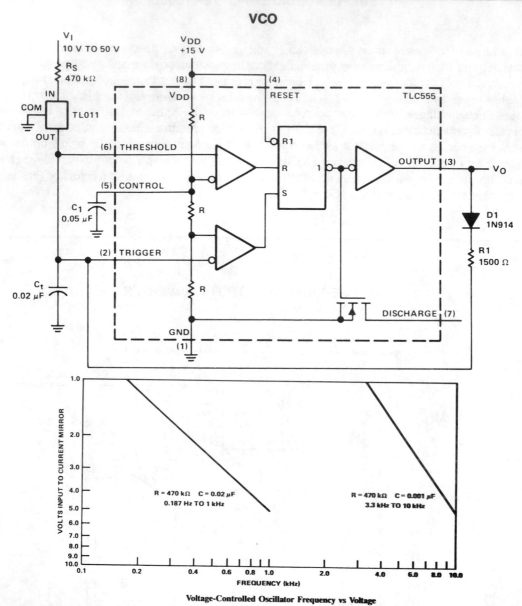

Voltage-Controlled Oscillator Frequency vs Voltage

Fig. 121-7

At startup, the voltage in the trigger input at pin 2 is less than the trigger level voltage, $1/3\ V_{DD}$, causing the timer to be triggered via pin 2. The output of the timer at pin 3 becomes high, allowing capacitor C_t to charge very rapidly through diode D1 and resistor R1.

When capacitor C_t charges to the upper threshold voltage $2/3\ V_{DD}$, the flip-flop is reset, the output at pin 3 decreases, and capacitor C_t discharges through the current mirror, TLO11. When the voltage at pin 2 reaches $1/3\ V_{DD}$, the lower threshold or trigger level, the timer triggers again and the cycle is repeated.

SIMPLE VCO

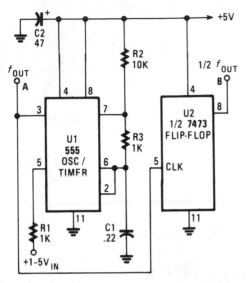

POPULAR ELECTRONICS   *Fig. 121-8*

The output frequency of the VCO, U1, varies inversely with the input voltage. With a 1-V input, the oscillator output frequency is about 1500 Hz; with a 5-V input, the output frequency drops to around 300 Hz. The output frequency range of U1 can be altered by varying the values of C1, R2, and R3. Increasing the value of any those three components will lower the oscillator frequency, and decreasing any of those values will raise the frequency. Output-waveform symmetry suffers since the frequency varies from one extreme to the other. At the highest frequency, the waveform is almost equally divided. But when the frequency drops, the output of the circuit turns into a narrow pulse. If a symmetrical waveform is required, add the second IC, U2, half of a 7473P dual TTL J-K flip-flop, to the oscillator circuit. The signal frequency output by U2 is $1/2$ of the input.

122

Voltage Converters

The sources of the following circuits are contained in the Sources section beginning on page 782. The figure number contained in the box of each circuit correlates to the sources entry in the Sources section.

UNIPOLAR-TO-DUAL SUPPLY CONVERTER

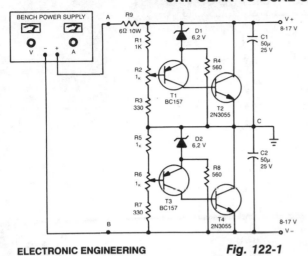

The outputs in this circuit are independently variable and can be loaded unsymmetrically. The output voltage remains constant, irrespective of load and changes. By varying potentiometers R2 or R6, the output voltages can be conveniently set. Outputs can be varied between 8 and 17 V, so that the standard ± 9, ± 12, and ± 15 V settings can be made. This converter is designed for a maximum load current of 1 A and the output impedance of both supplies of 0.35 Ω. This circuit is not protected against shortcircuits, but uses the protection provided by the dc input source. This circuit is ideal for biassing operation amplifier circuits.

ELECTRONIC ENGINEERING   **Fig. 122-1**

EFFICIENT SUPPLY SPLITTER

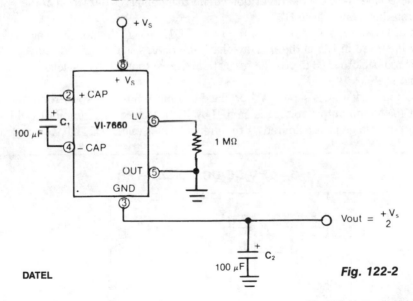

DATEL   **Fig. 122-2**

In this application, the VI-7660 is connected as a voltage splitter. Note that the *normal* output pin is connected to ground and the *normal* ground pin is used as the output. The switches that allow the charge pumping are bidirectional; therefore, charge transfer can be performed in reverse. The 1-MΩ resistor is used to avoid start-up problems by forcing the internal regulator on. An application for this circuit would be driving low-voltage, ± 7.5 Vdc, circuits from ± 15 Vdc supplies, or low-voltage logic from 9 to 12 V batteries.

HIGH-EFFICIENCY FLYBACK VOLTAGE CONVERTER

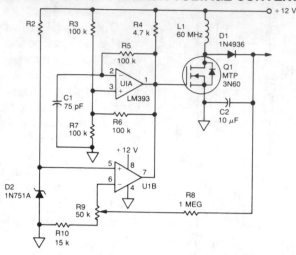

Fig. 122-3

U1 is a dual voltage comparator with open collector outputs. The A side is an oscillator operating at 100 kHz, and the B side is part of the regulation circuit that compares a fraction of the output voltage to a reference generated by zener diode D2.

The output of U1A is applied directly to the gate of Q1. During the positive half-cycle of the Q1 gate voltage, energy is stored in L1; in the negative half, the energy is discharged into C2. A portion of the output voltage is fed back to U1B to provide regulation. The output voltage is adjustable by changing feedback potentiometer R9.

Using the component values shown will produce a nominal 300-V output from a 12-V source. However, the circuit maximum output voltage is limited by R10; a lower value for R10 yields a higher output voltage. The output voltage is also limited by the breakdown of values Q1, L1, D1, and C2.

3 – 25 V DC-DC CONVERTER

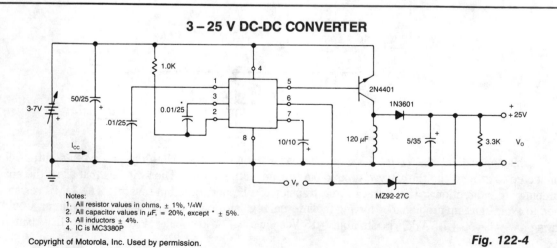

Notes:
1. All resistor values in ohms, ± 1%, 1/4W
2. All capacitor values in µF, = 20%, except * ± 5%.
3. All inductors ± 4%.
4. IC is MC3380P

Fig. 122-4

REGULATED 15-V_{OUT} 6-V DRIVEN CONVERTER

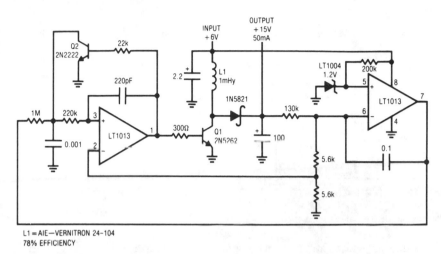

L1 = AIE—VERNITRON 24–104
78% EFFICIENCY

LINEAR TECHNOLOGY CORP.

Fig. 122-5

This converter delivers up to 50 mA from a 6-V battery with 78% efficiency. This flyback converter functions by feedback controlling the frequency of inductive flyback events. The inductor's output, rectified and filtered to dc, biases the feedback loop to establish a stable output. If the converter's output is below the loop setpoint, A2's inputs unbalance and current is fed through the 1-MΩ resistor at A1. This ramps the 1000-pF value positive. When this ramp exceeds the 0.5-V potential at A1's positive input, the amplifier switches high. Q2 turns on, discharging the capacitor to ground. Simultaneously, regenerative feedback through the 200-pF value causes a positive-going pulse at A1's positive input, sustaining A1's positive output. Q1 comes on, allowing inductor, L1, current to flow. When A1's feedback pulse decays, its output becomes low, turning off Q1. Q1's collector is pulled high by the inductor's flyback and the energy is stored in the 100-μF capacitor. The capacitor's voltage, which is the circuit output, is sampled by A2 to close a loop around A1/Q1. This loop forces A1 to oscillate at whatever frequency is required to maintain the 15-V output.

In-phase transformer windings for the drain and gate of TMOS power FET Q1 cause the circuit to oscillate. Oscillation starts when the feedback coupling capacitor, C1, is charged from the supply line via a large resistance; R2 and R3 limit the collector current to Q2. During *pump-up*, the on time is terminated by Q2, which senses the ramped source current of Q1. C1 is charged on alternate half-cycles by Q2 and forward-biased by zener D2.

When the regulated level is reached, forward bias is applied to Q2, terminating the on time earlier at a lower peak current. When this occurs, the frequency increases in inverse proportion to current, but the energy per cycle decreases in proportion to current squared. Therefore, the total power coupled through the transformer to the secondary is decreased.

1.5-W OFFLINE CONVERTER

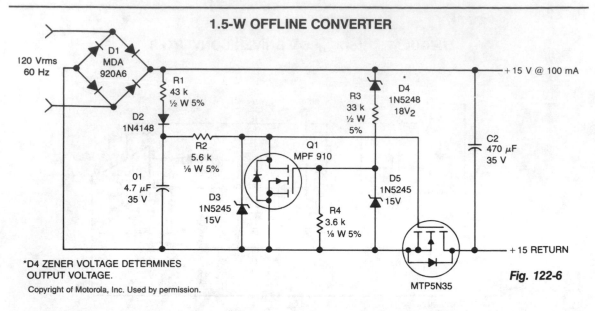

*D4 ZENER VOLTAGE DETERMINES
OUTPUT VOLTAGE.

Copyright of Motorola, Inc. Used by permission.

Fig. 122-6

This nonisolated, unregulated, minimum component converter fills the void between low-power zener regulation and the higher power use of a 60-Hz input transformer. It is intended for use wherever a nonisolated supply can be used safely.

The circuit operates by conducting only during the low-voltage portion of the rectified sine wave. R1 and D2 charge C1 to approximately 20 V, which is maintained by Q1. This voltage is applied to the gate of Q2, turning it on. When the rectified output voltage exceeds the zener voltage of D4, Q1 turns on, shunting the gate of Q2 to ground, turning it off.

DUAL OUTPUT ±12 OR ±15 V DC-DC CONVERTER

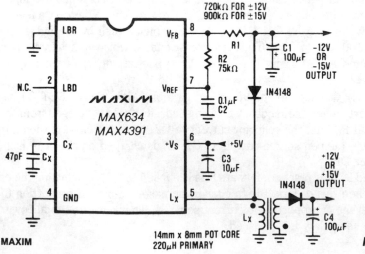

MAXIM

14mm x 8mm POT CORE
220μH PRIMARY

Fig. 122-7

DUAL OUTPUT ±12 OR ±15 V DC-DC CONVERTER (*Cont.*)

The buck-boost configuration of the MAX634 is well suited for dual output dc-dc converters. Only a second winding on the inductor is needed. Typically, this second winding is bifilar—primary and secondary are wound simultaneously using two wires in parallel. The inductor core is usually a toroid or a pot core. The negative output voltage is fully regulated by the MAX634. The positive voltage is semiregulated, and will vary slightly with load changes on either the positive or negative outputs.

12-TO-16 V CONVERTER

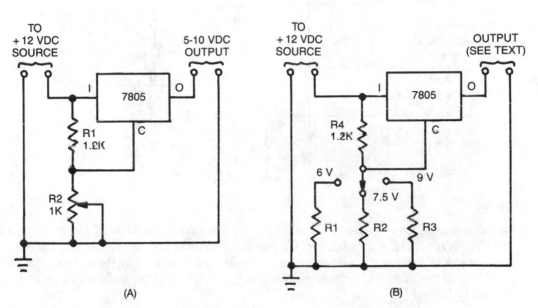

Fig. 122-8

Many devices operate from a car's 12-V electrical system. Some require 12 V; others require some lesser voltage. An automobile battery's output can vary from 12 to 13.8 V under normal circumstances. The load requirements of the device might vary. This circuit maintains a constant voltage regardless of how those factors change. Simple circuit, A, uses a 7805 voltage regulator. In addition to a constant output, this IC provides overload and short-circuit protection. That unit is a 5-V, 1-A regulator, but when placed in circuit B, it can provide other voltages as well. When the arm of potentiometer R1 is moved toward ground, the output varies from 5 to about 10 V.

SELF-OSCILLATING FLYBACK-SWITCHING CONVERTER

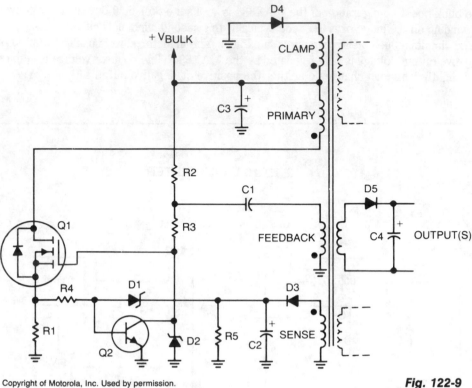

Fig. 122-9

Regulation is provided by taking the rectified output of the sense winding and applying it as a bias to the base of Q2 via zener D1. The collector of Q2 then removes drive from the gate of Q1. Therefore, if the output voltage should increase, Q2 removes the drive to Q1 earlier, shortening the on time, and the output voltage will remain the same. Dc outputs are obtained by merely rectifying and filtering secondary windings, as done by D5 and C4.

123

Voltage-to-Frequency Converters

The sources of the following circuits are contained in the Sources section beginning on page 782. The figure number contained in the box of each circuit correlates to the sources entry in the Sources section.

1 Hz-TO-30 MHz VOLTAGE-TO-FREQUENCY CONVERTER

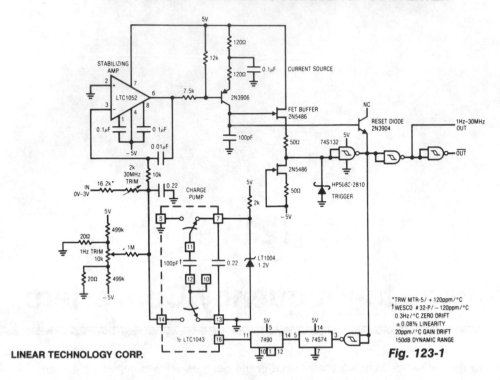

LINEAR TECHNOLOGY CORP.

Fig. 123-1

Circuit has a 1 Hz-to-30 MHz output, 150-dB dynamic range, for a 0 to 5 V input. It maintains 0.08% linearity over its entire 7^{1}/3 decade range with a full-scale drift of about 20 ppm/°C. To get the additional bandwidth, the fast JFET buffer drives the Schottky TTL Schmitt trigger. The Schottky diode prevents the Schmitt trigger from ever seeing negative voltage at its input. The Schmitt's input voltage hysteresis provides the limits which the oscillator runs between. The 30-MHz, full-scale output is much faster than the LTC1043 can accept, so the digital divider stages are used to reduce the feedback frequency signal by a factor of 20. The remaining Schmitt sections furnish complementary outputs.

DIFFERENTIAL-INPUT VOLTAGE-TO-FREQUENCY CONVERTER

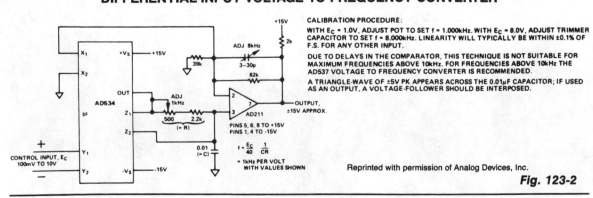

CALIBRATION PROCEDURE:

WITH E_C = 1.0V, ADJUST POT TO SET f = 1.000kHz. WITH E_C = 8.0V, ADJUST TRIMMER CAPACITOR TO SET f = 8.000kHz. LINEARITY WILL TYPICALLY BE WITHIN ±0.1% OF F.S. FOR ANY OTHER INPUT.

DUE TO DELAYS IN THE COMPARATOR, THIS TECHNIQUE IS NOT SUITABLE FOR MAXIMUM FREQUENCIES ABOVE 10kHz. FOR FREQUENCIES ABOVE 10kHz THE AD537 VOLTAGE TO FREQUENCY CONVERTER IS RECOMMENDED.

A TRIANGLE-WAVE OF ±5V PK APPEARS ACROSS THE 0.01μF CAPACITOR; IF USED AS AN OUTPUT, A VOLTAGE-FOLLOWER SHOULD BE INTERPOSED.

Reprinted with permission of Analog Devices, Inc.

Fig. 123-2

750

LOW-COST VOLTAGE-TO-FREQUENCY CONVERTER

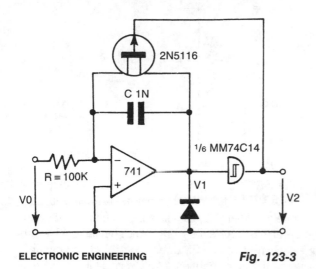

Fig. 123-3

The 741 op amp integrator signal is fed into the Schmitt trigger input of an inverter. When the signal reaches the magnitude of the positive-going threshold voltage, the output of the inverter is switched to zero. The inverter output controls the FET switch directly. For a gate voltage of zero, the FET channel turns on to low resistance and the capacitor is discharged. The discharge current depends on the on resistance of the FET. When the capacitor C1 is discharged to the negative-going threshold voltage level of the inverter, the inverter output is switched to ± 12 V. This switch causes the FET channel to be switched off, and the discharging process is switched into a charging process again. Using the components shown, an output frequency of about 10 kHz with 0.1% linearity can be obtained.

WIDE-RANGE VOLTAGE-TO-FREQUENCY CONVERTER

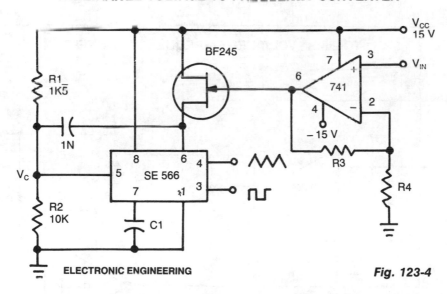

Fig. 123-4

This circuit is based upon the change of frequency of the function generator with the input voltage V_{IN}. Generally, the frequency depends upon the capacitance and resistor connected to pin 6. This resistor is replaced by the FET. The frequency range is adjustable by changing the input voltage, V_{IN}; the converter will give a range of 10 Hz to 1 MHz.

WIDE-RANGE VOLTAGE-TO-FREQUENCY CONVERTER

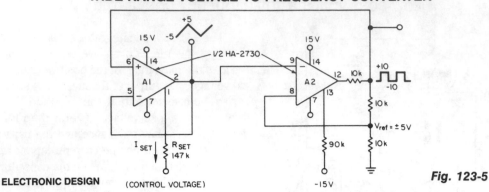

Fig. 123-5

This circuit uses a programmable op amp such as the HA2730—a two-amplifier monolithic chip with independent programming ports for each amplifier—whose slew rate and other parameters vary linearly with a so-called *set current*. The converter circuit uses one amplifier, A1, as a slewing amplifier and other, A2, as a comparator function. The control voltage V_C, determines A1's slew rate. And, because A1's output voltage swing is constant, the modulation of its set current results in direct control of the circuit's frequency. A1's internal compensation capacitor acts as the timing component. An internal bipolar current source, whose current magnitude is directly proportional to the set current of pin 1, then determines the charge-discharge rate. A conversion nonlinearity of $\pm 0.03\%$ of full scale over 3 decades and $\pm 1.5\%$ of full scale over 4.3 decades of frequency is possible. The frequency range is adjustable by a change in the resistance, R.

5 kHz-TO-2 MHz VOLTAGE-TO-FREQUENCY CONVERTER

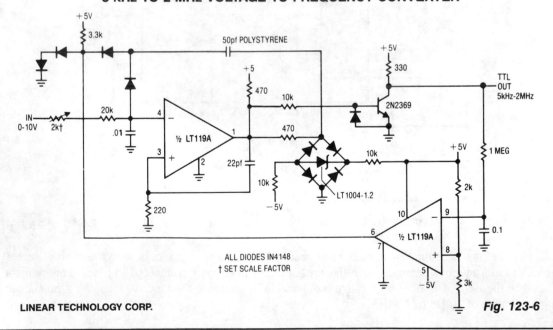

Fig. 123-6

PRESERVED INPUT VOLTAGE-TO-FREQUENCY CONVERTER

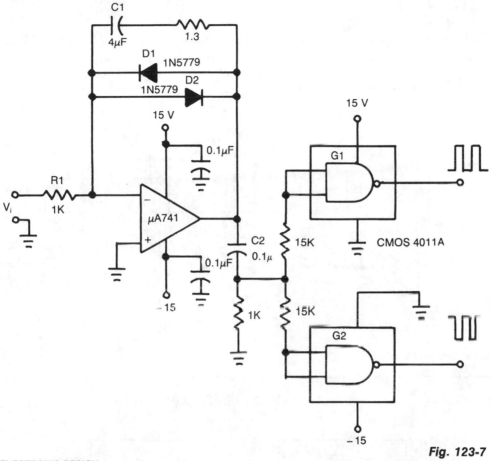

Fig. 123-7

The input voltage, V1, causes C1 to charge and produce a ramp voltage at the output of the 741 op amp. Diodes D1 and D2 are four-layer devices. When the voltage across C1 reaches the breakover voltage of either diode, the diode conducts to discharge C1 rapidly and the op amp output goes abruptly to zero. This rapid discharge action applies a narrow pulse to G1 and G2. Positive discharge pulses produced by a positive V1 are coupled to the output only through G1, while negative pulses are coupled only through G2.

Because of the forward breakover current of diodes D1 and D2, the circuit won't operate below a minimum input voltage. An increase of R1 increases this minimum voltage and reduces the circuit's dynamic range. The minimum input voltage with R1 at 1 KΩ is in the range of 10 to 50 mV. This input dead zone, when input signal V1 is near zero is desirable in applications that require a signal to exceed a certain level before an output is generated.

1 Hz-TO-10 MHz VOLTAGE-TO FREQUENCY CONVERTER

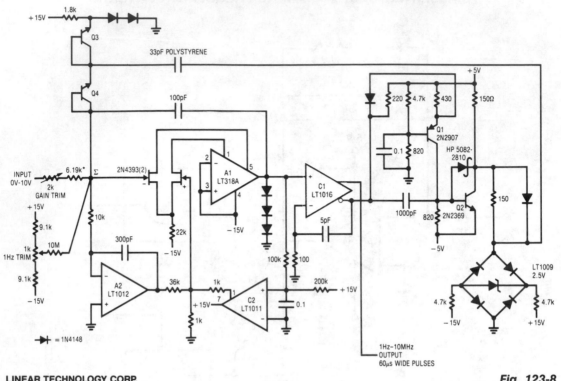

LINEAR TECHNOLOGY CORP.

Fig. 123-8

VOLTAGE-TO-FREQUENCY CONVERTER

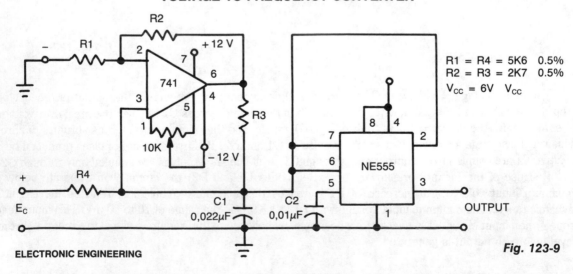

R1 = R4 = 5K6   0.5%
R2 = R3 = 2K7   0.5%
V_{CC} = 6V V_{CC}

ELECTRONIC ENGINEERING

Fig. 123-9

This circuit can accept positive or negative or differential control voltages. The output frequency is zero when the control voltage is zero. The 741 op amp forms a current source controlled by the voltage E_C to charge the timing capacitor C1 linearly. NE555 is connected in the astable mode, so that the capacitor charges and discharges between $1/3$ V_{CC} and $2/3$ V_{CC}. The offset is adjusted by the 10-K potentiometer so that the frequency is zero when the input is zero. For the component values shown: $f \approx 4.2$ E_C kHz. If two dc voltages are applied to the ends of R1 and R4, the output frequency will be proportional to the difference between the two voltages.

1 HZ-TO-1.25 MHZ VOLTAGE-TO-FREQUENCY CONVERTER

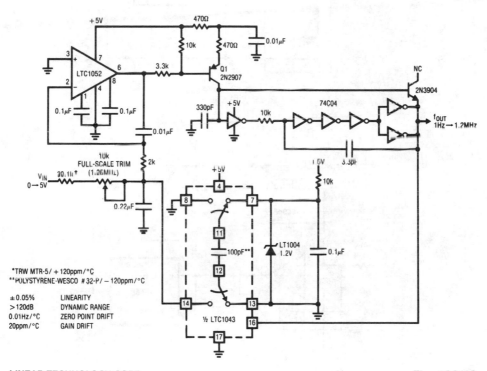

LINEAR TECHNOLOGY CORP.

Fig. 123-10

This stabilized voltage-to-frequency converter features 1 Hz – 1.25 MHz operation, 0.05% linearity, and a temperature coefficient of typically 20 ppm/°C. This circuit runs from a single 5-V supply. The converter uses a charge feedback scheme to allow the LTC1052 to close a loop around the entire circuit, instead of simply controlling the offset. This approach enhances linearity and stability, but introduces the loop's settling time into the overall voltage-to-frequency step-response characteristic.

ACCURATE VOLTAGE-TO-FREQUENCY CONVERTER

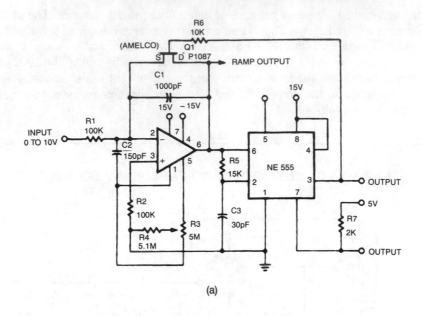

(a)

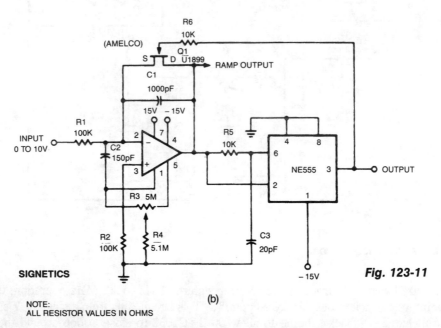

(b)

SIGNETICS

NOTE:
ALL RESISTOR VALUES IN OHMS

Fig. 123-11

This linear voltage-to-frequency converter, a, achieves good linearity over 0 to −10 V. Its mirror image, b, provides the same linearity over 0 to +10 V, but it is not DTL/TTL compatible.

VOLTAGE-TO-FREQUENCY CONVERTER

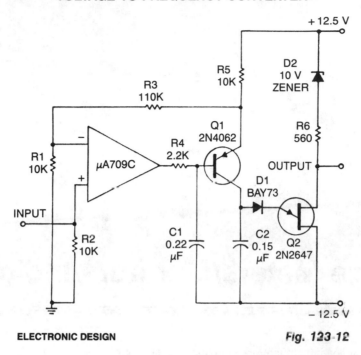

ELECTRONIC DESIGN

Fig. 123-12

This circuit consists of a UJT oscillator in which the timing charge capacitor C2 is linearly dependent on the input signal voltage. The charging current is set by the voltage across resistor R5, which is accurately controlled by the amplifier.

124

Voltage Meters/Monitors/Indicators

The sources of the following circuits are contained in the Sources section beginning on page 782. The figure number contained in the box of each circuit correlates to the sources entry in the Sources section.

Voltage-Level Indicator
$4^{1}/_{2}$-Digit DVM
Full-Scale Four-Decade $3^{1}/_{2}$-Digit DVM
Over/Under Voltage Monitor
High Input Resistance Dc Voltmeter
Dc Voltmeter
Voltage Freezer
Multiplexed Common-Cathode
   LED-Display ADC
Ac Voltmeter
FET Voltmeter
Sensitive Rf Voltmeter

Voltage Monitor
Audio Millivoltmeter
High-Input Resistance Voltmeter
Frequency Counter
Audio Millivoltmeter
Low-Voltage Indicator
FET Voltmeter
Simplified Voltage-Level Sensor
Peak Program Detector
Wide-Range AC Voltmeter
Visible Voltage Indicator

VOLTAGE-LEVEL INDICATOR

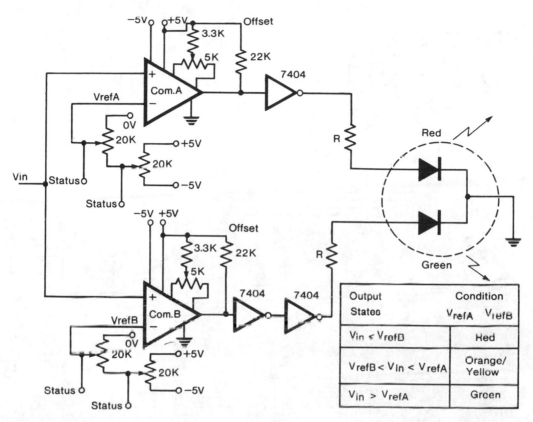

Output States	Condition	
	V_{refA}	V_{refB}
$V_{in} < V_{refB}$	Red	
$V_{refB} < V_{in} < V_{refA}$	Orange/ Yellow	
$V_{in} > V_{refA}$	Green	

ELECTRONIC ENGINEERING

Fig. 124-1

A tricolor LED, acts as the visual indicator of the voltage level. The voltage to be measured is connected to the two comparators in parallel. The first 20-KΩ trimmer defines a voltage between ±5 V and this becomes the full-scale value of the reference voltage. The second trimmer is a fine adjustment to give any reference voltage between 0 V and the full-scale voltage. Thus, it is possible to select both positive and negative reference voltages. During the initialization procedure, a voltage, equal to the reference voltage of each comparator, is connected to the input terminal, and the offset balance potentiometer is adjusted to give a reading between the high and low output voltage levels. The inverter following comp A ensures that, whatever the input voltage, at least one diode is lit. The two inverters following comp B leave the voltage largely unchanged, but provide the current necessary to illuminate the diode. The value of the resistance should be chosen so that the current through any single diode does not exceed the specified limit, usually 30 mA. The LED contains a red and a green diode with a common cathode. When both diodes are lit, a third color, orange, is emitted. With $V_{ref A}$ greater than $V_{ref B}$, the output states given in the diagram apply.

4¹/₂-DIGIT DVM

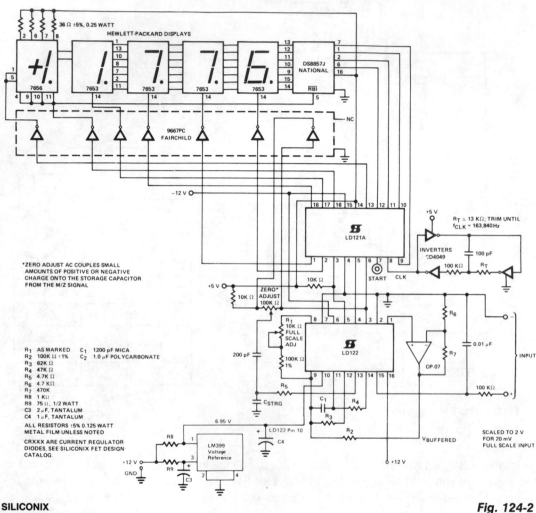

SILICONIX

Fig. 124-2

- 1-μV resolution
- Overrange blinking
- 0 – 19.999 mV input voltages
- Zero adjust-to-null offset introduced by PC board leakage and the comparator.

FULL-SCALE FOUR-DECADE 3¹/₂-DIGIT DVM

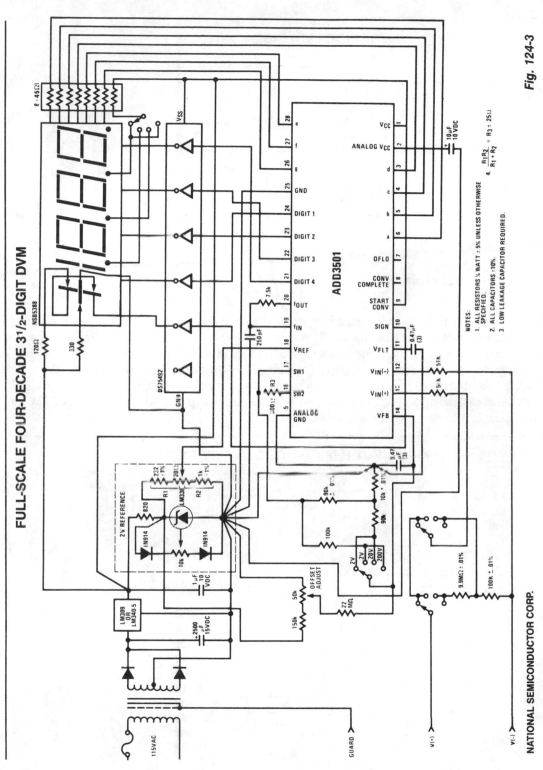

Fig. 124-3

NATIONAL SEMICONDUCTOR CORP.

This DVM circuit uses a National ADD3501 DVM chip and an LM336 reference IC to create a simple DVM with relatively few components. When making a single range panel meter, the range switching components can be left out, as required.

OVER/UNDER VOLTAGE MONITOR

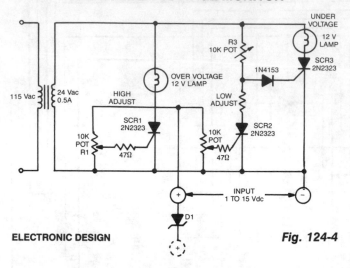

ELECTRONIC DESIGN

Fig. 124-4

Any potential from 1 to 15 V can be monitored with this circuit. Two lamps alert any undesirable variation. The voltage differential from lamp turn-on to turn-off is about 0.2 V at any setting. High and low set points are independent of each other. The SCRs used in the circuit should be the sensitive gate type. R3 must be experimentally determined for the particular series of SCRs used. This is done by adjusting R3 to the point where the undervoltage lamp turns on when no signal is present at the SCR2 gate. Any 15-V segment can be monitored by putting the zener diode, D1, in series with the positive input lead. The low set-point voltage will then be the zener voltage plus 0.8 V.

HIGH INPUT RESISTANCE DC VOLTMETER

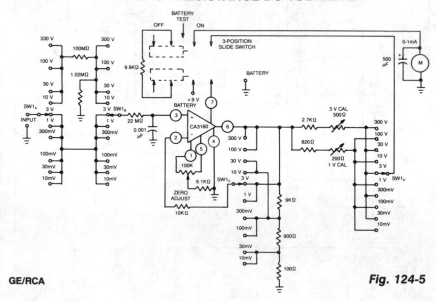

GE/RCA

Fig. 124-5

HIGH INPUT RESISTANCE DC VOLTMETER (*Cont.*)

This voltmeter exploits a number of the CA3160 BiMOS op amp's useful characteristics. The available voltage ranges from 10 mV to 300 V. Powered by a single 8.4-V mercury battery, this circuit, with zero input, consumes approximately 500 μA. Thus, at full-scale input, the total supply current will increase by 1000 μA.

DC VOLTMETER

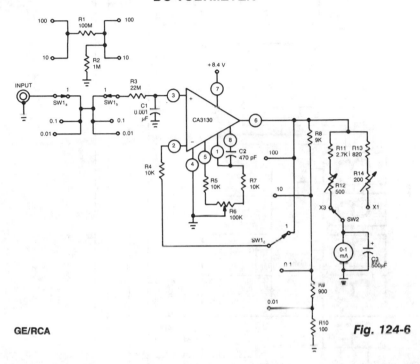

GE/RCA

Fig. 124-6

This dc voltmeter, with high input resistance, uses a CA3130 BiMOS op amp and measures voltages from 10 mV to 300 V. Resistors R12 and R14 are used individually to calibrate the meter for full-scale deflection. Potentiometer R6 is used to null the op amp and meter on the 10-mV range by shorting the input terminals, then adjusting R6 for the first indication of upscale meter deflection.

VOLTAGE FREEZER

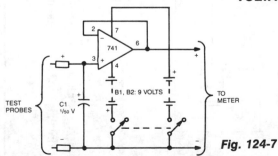

Fig. 124-7

This circuit reads and stores voltages, thus freezing the meter reading even after the probes are removed. The op amp is configured as a unity-gain voltage follower, with C1 situated at the input to store the voltage. For better performance, use an LF13741 or a TL081 op amp in place of the 741. These two are JFET devices and offer a much higher input impedance than the 741.

Reprinted with permission of Radio-Electronics Magazine, November 1982.

MULTIPLEXED COMMON-CATHODE LED-DISPLAY ADC

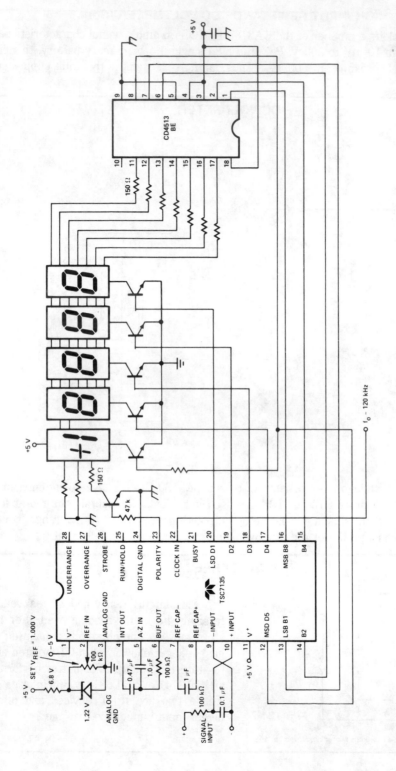

Fig. 124-8

Copyright Teledyne Industries, Inc.

Here, a Teledyne TSC7135 DVM chip is used to drive a multiplexed 5-digit display. A CD4513BE CMOS IC, for common cathode drive, is used as a segment driver selected by pins 17–20 of the DVM chip. The transistors can be any suitable npn type such as 2N3904, etc.

AC VOLTMETER

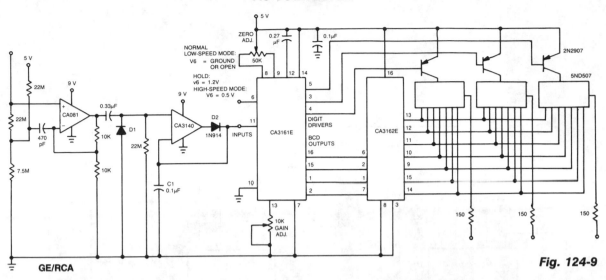

Fig. 124-9

GE/RCA

CA081 and CA3140 BiMOS op amp offer minimal loading on the circuits being measured. The wide bandwidth and high slew rate of the CA081 allow the meter to operate up to 0.5 MHz.

FET VOLTMETER

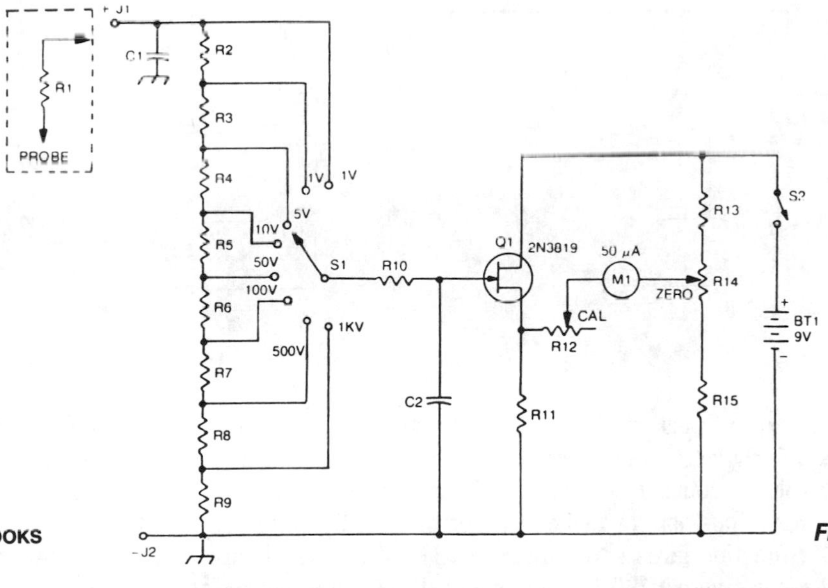

TAB BOOKS

Fig. 124-10

A 2N3819 FET provides a *solid-state VOM*. The 2N3819 acts as a *cathode follower* in a VOM. The bias offset (meter null) is obtained with R14 and R12 sets full-scale calibration. R2 through R9 should total about 10 MΩ. R10 is a protective resistor, and C2 provides ac bypassing to limit rf and noise pickup.

SENSITIVE RF VOLTMETER

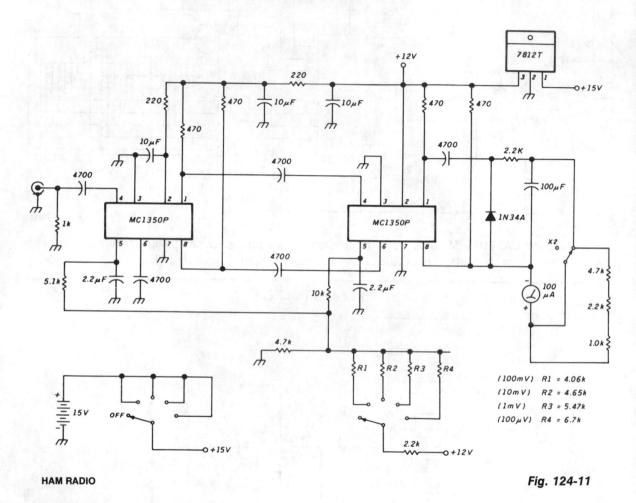

Fig. 124-11

This schematic shows a peak-reading diode voltmeter driven by two stages of amplification. A 100-μF capacitor provides a fairly large time constant, which results in satisfactory meter damping. The limited differential output voltage coupled with an overdamped meter prevents most *needle pinning* when you select an incorrect range position, or make other errors. An SPST toggle switch selects additional series resistance. This X2 function gives some more overlap of the sensitivity ranges. The resistance values shown are correct for use with a 100-μA meter with 1500-Ω internal resistance.

VOLTAGE MONITOR

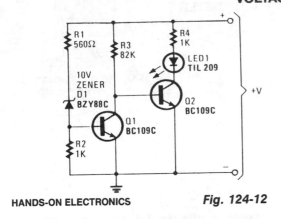

HANDS-ON ELECTRONICS

Fig. 124-12

If the battery voltage exceeds about 11 V, current flows through R1, D1, and R2. The voltage produced as a result of current flow through R2 is sufficient to bias transistor Q1 into conduction. That places the collector voltage of Q1 virtually at ground. Therefore, Q2, driven from the collector of Q1, is cut off, LED1 and current-limiting resistor R4 are connected in the collector circuit of Q2. With Q2 in the cut-off state, the LED does not light. Should Q1's base voltage drop below approximately 0.6 V, Q1 turns off, biasing Q2 on and illuminating LED1 to indicate that the battery voltage has fallen below the 11 V threshold level.

AUDIO MILLIVOLTMETER

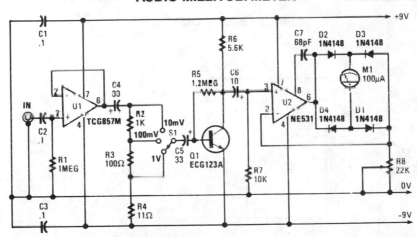

POPULAR ELECTRONICS/HANDS-ON ELECTRONICS

Fig. 124-13

Capacitor C4 couples the output of U1 to a simple attenuator, which is used to provide a loss of 0 dB, 20 dB, or 40 dB, depending on the setting of range switch S1. The circuits sensitivity is 10-V rms for full-scale deflection, so the attenuator gives additional ranges of 100-mV and 1-V rms. The attenuator output is connected through capacitor C5 to common-emitter amplifier Q1, which has a high-voltage gain of 40 dB.

To get linear scaling on the meter, we have to use an active-rectifier circuit built around U2. That IC is connected so that its noninverting input is biased to the 0-V bus via R7. Capacitor C6 couples the output of Q1 to the noninverting input of U2; C7 is the compensation capacitor for U2.

The voltage gain of U2 is set by the difference in resistance between the output and the inverting input, and between the inverting input and the ground bus. One resistance is made up of the diode-bridge rectifier D1 through D4, the other by resistor R8. This circuit has a nearly flat frequency response to about 200 kHz.

HIGH-INPUT RESISTANCE VOLTMETER

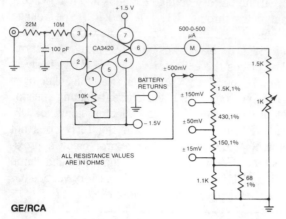

A resistance of 1,000,000 MΩ takes advantage of the high input impedance of the CA3420 BiMOS op amp. Only two 1.5-V AA-type penlite batteries are required for use. Full-scale deflection is ±500 mV, ±150 mV, and ±15 mV.

GE/RCA

Fig. 124-14

FREQUENCY COUNTER

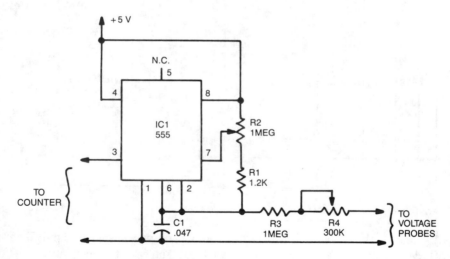

Fig. 124-15

The output frequency from IC pin 3 is determined by the voltage input to pin 6. A standard frequency counter can be used to measure voltages directly over a limited range from 0 to 5 V. In this circuit, the 555 is wired as an astable multivibrator. Resistor R2 determines the output frequency when the input to the circuit (the voltage measured by the voltage probes) is zero. R4 is a scaling resistor that adjusts the output frequency so that a change in the input voltage of 1 V will result in a change in the output frequency of 10 Hz. That will happen when the combined resistance of R3 and R4 is 1.2 MΩ. To calibrate short the voltage probes together, adjust R2 until the reading on the frequency counter changes to 00 Hz. Then, use the voltage probes to measure an accurate 5-V source and adjust R4 until the frequency counter reads 50 Hz.

AUDIO MILLIVOLTMETER

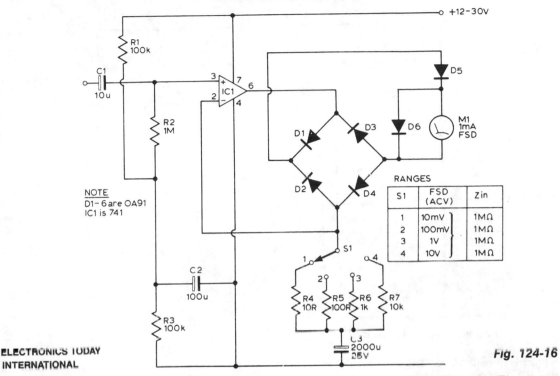

NOTE
D1–6 are OA91
IC1 is 741

RANGES

S1	FSD (ACV)	Zin
1	10mV	1MΩ
2	100mV	1MΩ
3	1V	1MΩ
4	10V	1MΩ

ELECTRONICS TODAY
INTERNATIONAL

Fig. 124-16

This circuit has a flat response from 8 Hz to 50 kHz at −3 db on the 10-mV range. The upper limit remains the same on the less sensitive ranges, but the lower frequency limit covers under 1 Hz.

LOW-VOLTAGE INDICATOR

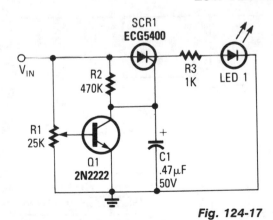

Fig. 124-17

Input terminal V_{IN} is connected to the +V line of the circuit that the indicator is to monitor, and the grounds of both circuits are connected together. The position of potentiometer R1's wiper determines Q1's base voltage. As long as the transistor gets enough bias voltage to remain on, the low voltage at the collector will keep the SCR from firing. As the battery voltage starts to fall, the transistor's base voltage will fall as well. When Q1 turns off (V_{IN} drops), the collector voltage increases. That voltage provides enough gate drive to turn on the SCR, which turns on the LED. The LED could also be a buzzer or almost any other type of warning device.

Reprinted with permission of Radio-Electronics Magazine, January 1986. Copyright Gernsback Publications, Inc., 1986.

FET VOLTMETER

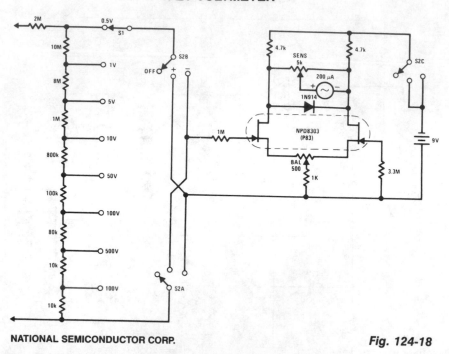

Fig. 124-18

This FETVM replaces the function of the VTVM and rids the instrument of the usual line cord. In addition, FET drift rates are far superior to vacuum tube circuits, allowing a 0.5 V full-scale range which is impractical with most vacuum tubes. The low leakage, low noise NPD8303 is ideal for this application.

SIMPLIFIED VOLTAGE-LEVEL SENSOR

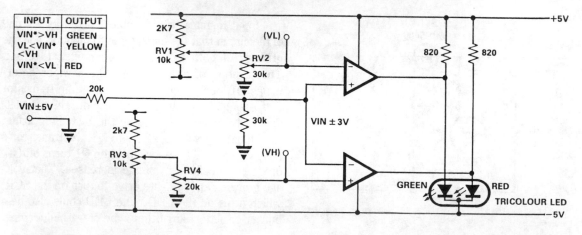

INPUT	OUTPUT
VIN*>VH	GREEN
VL<VIN*<VH	YELLOW
VIN*<VL	RED

Fig. 124-19

SIMPLIFIED VOLTAGE-LEVEL SENSOR (*Cont.*)

This circuit uses only one IC, either 1, LM393 dual comparator or $^1/_2$, LM339 quad comparator. RV1 and RV3 set the full scale reference voltage, and RV2 and RV4 set the switching thresholds to a value between 0 V and the full-scale reference. The change in input voltage needed to fully switch the output state is less than 0.05 mV (typical).

An alternative is:

INPUT	OUTPUT
V_{IN}, V_H	red
V_H, V_{IN}, V_L	yellow
V_{IN} V_L	green

PEAK PROGRAM DETECTOR

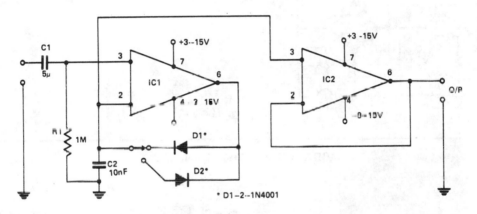

ELECTRONICS TODAY INTERNATIONAL

Fig. 124-20

This circuit will allow a multimeter to display the positive or negative peaks of an incoming signal. A 741, IC1, is used in the noninverting mode with R1 defining the input impedance. D1 or D2 will conduct on a positive or negative peak, charging C2 until the inverting input is at the same dc level as the incoming peak. This level will maintain the voltage until a higher peak is detected, then this will be stored by C2. Another 741, IC2, prevents loading by the multimeter. Connected in the noninverting mode as a unity gain buffer, output impedance is less than 1 Ω. This circuit has a useful frequency response from 10 Hz to 100 kHz at ±1 dB. High linearity is ensured by placing the diodes in the feedback loop of IC1, effectively compensating for the 0.6 V bias that these components require.

WIDE-RANGE AC VOLTMETER

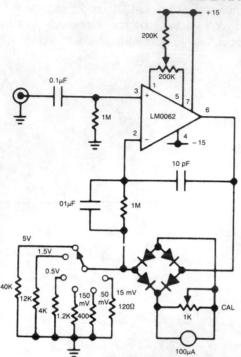

NATIONAL SEMICONDUCTOR CORP.

In this circuit, a diode bridge is used as a meter rectifier. The offset voltage is compensated for by the op amp, since the bridge is in the feedback network.

Fig. 124-21

VISABLE VOLTAGE INDICATOR

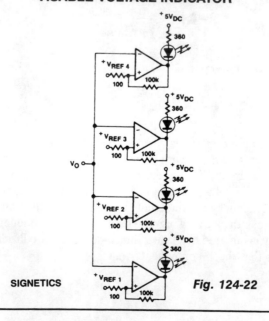

SIGNETICS

Fig. 124-22

125

Voltage References

The sources of the following circuits are contained in the Sources section beginning on page 782. The figure number contained in the box of each circuit correlates to the sources entry in the Sources section.

Bipolar Reference Source
Expanded-Scale Analog Meter
Digitally Controlled Voltage Reference

BIPOLAR REFERENCE SOURCE

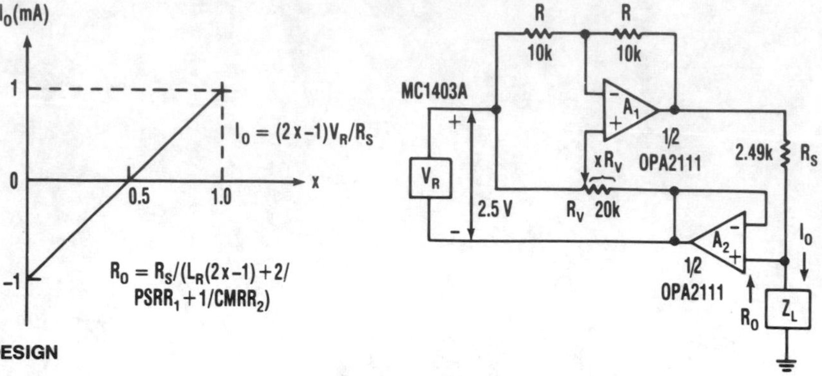

$$I_0 = (2x-1)V_R/R_S$$

$$R_0 = R_S/(L_R(2x-1)+2/PSRR_1+1/CMRR_2)$$

Fig. 125-1

This current source has continuous control of the magnitude and polarity of its amplifier gain and needs only one voltage reference. The circuit includes reference V_R, voltage-amplifier circuit A1 with gain-setting resistor R_S, and bootstrap-follower amplifier A2. The bootstrapping converts the circuit to a current source and allows the load to be grounded. Any voltage developed across load Z_L feeds back to the reference and voltage amplifier, making their functions immune to that voltage. Then the current-source circuitry floats, instead of the load.

The voltage reference is connected to both the inverting and noninverting inputs of A1; this provides a balanced combination of positive and negative gain. The inverting connection has equal feedback resistors, R, for a gain of -1, and the noninverting connection varies according to the fractional setting, X, of potentiometer R_V. X controls the noninverting gain and adjusting it counters the effect of some of the inverting gain. The value of X is the portion of R_V's resistance from the noninverting input of A1 to the temporarily grounded output of A2. Between potentiometer extremes, the current varies with $X \pm 1$ mA.

EXPANDED-SCALE ANALOG METER

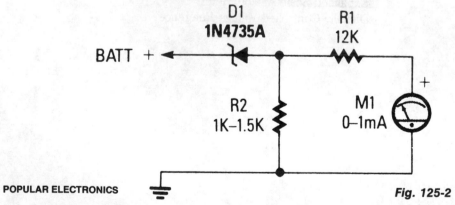

Fig. 125-2

The circuit consists of 0–1 mA meter M1, 6.2-V zener diode D1, and 12-KΩ, 1% resistor R1. R2 is included in the circuit as a load resistor for the zener diode. The value of R2 isn't critical; use a value of 1000 to 1500 Ω. The meter reads from 6 to 18 volts, which is perfect for checking a car's charging system.

DIGITALLY-CONTROLLED VOLTAGE REFERENCE

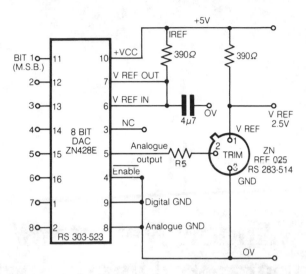

ELECTRONIC ENGINEERING

Fig. 125-3

This circuit shows a simple method of achieving a voltage reference which can be trimmed using an 8-bit DAC with an on-chip voltage reference. The analogue output from the DAC drives the trim pin of the ZNREF025 via a resistor, R_S. When $R_S = 0$, this circuit will produce a trim range exceeding $\pm 5\%$ of the nominal reference voltage. When R_S is greater than zero, the trim range is reduced. It was found that after dividing the trim range by two, the needed value for R_S was approximately 1 MΩ.

The reference voltage of the ZNREF025 can be set to an accuracy determined by the trim range of the device itself and the accuracy of the DAC. Increasing R_S reduces the percentage trim range and hence increases the effective voltage resolution. Other voltage references from the ZNREF series can be used with this circuit if other voltages are required. This voltage reference can be used to set the value of V_{REF} to a much tighter tolerance, than the data sheet specification of $\pm 1\%$, in a much wider range of operating conditions. Applications could include any system with automatic self-calibration of instrumentation, such as in electronic weighing scales.

126

Window Detectors/
Comparators/Discriminators

The sources of the following circuits are contained in the Sources section beginning on page 782. The figure number contained in the box of each circuit correlates to the sources entry in the Sources section.

DIGITAL FREQUENCY WINDOW

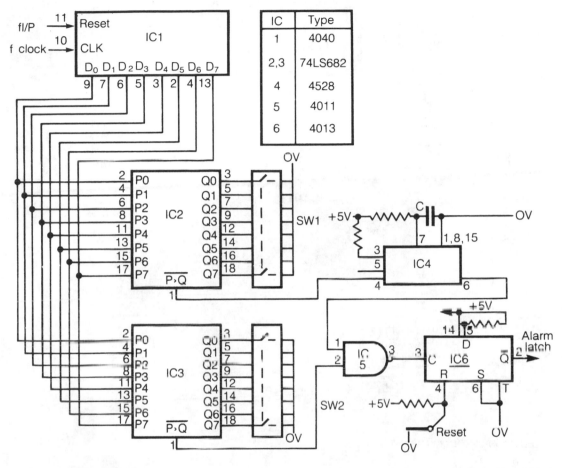

IC	Type
1	4040
2,3	74LS682
4	4528
5	4011
6	4013

ELECTRONIC ENGINEERING

Fig. 126-1

This circuit detects frequency variation above or below preset limits. IC1 is a binary counter clocked at F_{CLK}. The outputs are compared with switch preset values by IC2 and IC3. The input signal, which must be a positive-going pulse, is used to reset IC1. The *P greater than Q* output of the comparators is at logic O for input frequencies below the preset values. Above the preset count, a pulse train is output.

IC2, detects a low input by supplying the pulse train to a retriggerable monostable, IC4. When the input frequency falls below the preset value in SW1, the monostable is no longer triggered and its output falls to logic O. IC3 detects the frequency high state SW2, and outputs directly when this occurs. The outputs from both comparators can then be latched as shown, using IC5 and IC6. The clock frequency is related to input and switch values: switch value = $F_{CLK/input}$. The time constant of IC4 is not critical, but must obviously exceed the maximum input pulse period.

WINDOW DETECTOR

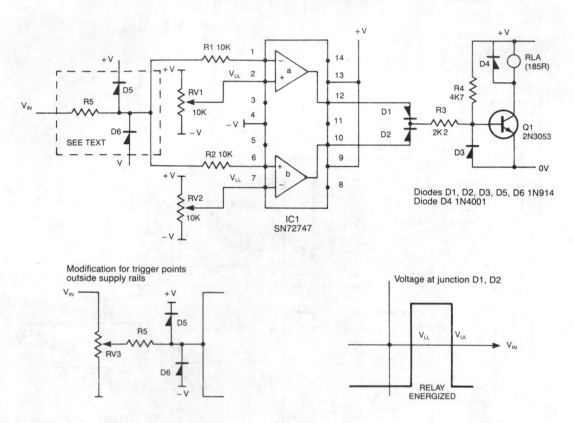

Diodes D1, D2, D3, D5, D6 1N914
Diode D4 1N4001

IC1
SN72747

Modification for trigger points
outside supply rails

Voltage at junction D1, D2

RELAY
ENERGIZED

ELECTRONICS TODAY INTERNATIONAL

Fig. 126-2

This circuit de-energizes a normally energized relay if the input voltage goes above or below two individually set voltages. The transistor driving the relay is normally turned on by R4, so the relay is normally energized. If the cathode of D1 or D2 is taken negative, Q1 will turn off and the relay will de-energize. The IC is a 72747 dual op amp used without feedback, so the full gain of about 100dB is available. The amplifier output will thus swing from full positive to full negative for a few mV change at the input. The relay is therefore only energized if V_{IN} is between V_{UL} and V_{LL}. The two limits can be set anywhere between the supply rails, but obviously V_{UL} must be more positive than V_{LL}. If V_{IN} can go outside the supply rails, D5, D6, and R5 should be added to prevent damage to IC1. If V_{UL} and V_{LL} are required to be outside the supply rails, V_{IN} can be reduced by RV3. The supplies can be any value, providing that the voltage across them is not more than 30 V.

WINDOW DETECTOR

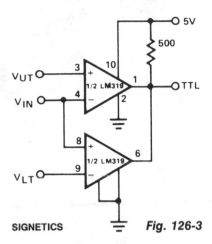

$V_{OUT} = 5V$ for $V_{LT} < V_{IN} < V_{UT}$

$V_{OUT} = 0$ for $V_{IN} < V_{LT}$ or $V_{IN} > V_{UT}$

SIGNETICS

Fig. 126-3

WINDOW DETECTOR

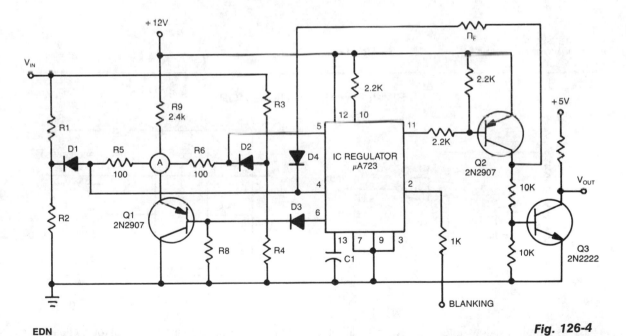

EDN

Fig. 126-4

The detector circuit compares the output voltage of two separate voltage dividers with a fixed reference voltage. The resultant absolute error signal is amplified and converted to a logic signal that is TTL compatible.

WINDOW DETECTOR

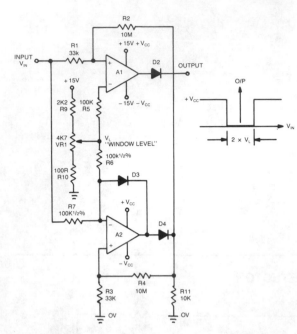

This novel window detector uses only two op amps. The width of the window can be changed by the 4.7-KΩ potentiometer.

ELECTRONIC ENGINEERING   **Fig. 126-5**

SIMPLE WINDOW DETECTOR

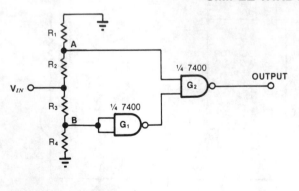

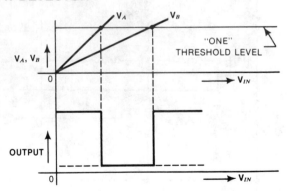

EDN   **Fig. 126-6**

This simple window detector uses only half of a 7400 quad NAND gate plus four resistors, chosen so that the voltage at point A exceeds the voltage at point B for any input voltage. With no input applied or when V_{IN} is at ground, the output of gate G1 is one; hence G2's output is also one. As the input voltage increases, V_A rises faster than V_B. When V_A reaches an acceptable one level, the circuit's output drops to zero. As the input continues to increase, V_B rises to an acceptable level, changing the output of G2 to one.

MULTIPLE APERTURE WINDOW DISCRIMINATOR

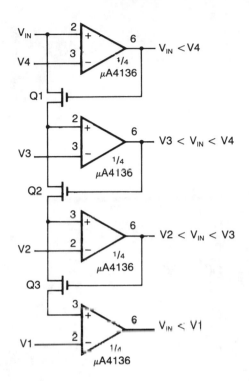

The circuit shown here uses μA4136 comparators and FETs Q1 through Q3.

FAIRCHILD CAMERA AND INSTRUMENT CORP.

Fig. 126-7

Sources Index

Fig. 3-14. Harris, Analog Product Data Book, 1988, p. 10-110.

Fig. 3-15. Linear Technology Corp., Linear Applications Handbook, 1987, p. AN18-3.

Fig. 3-16. Linear Technology Corp., Linear Applications Handbook, 1987, p. AN21-5.

Fig. 3-17. Harris, Analog Product Data Book, 1988, p. 10-150.

Fig. 3-18. Reprinted from EDN, 9/1/88, © 1989 Cahners Publishing Co., a division of Reed Publishing USA.

Fig. 3-19. Intersil, Applications Handbook, 1988, p. 3-181.

Fig. 3-20. Maxim, Maxim Advantage, p. 44.

Fig. 3-21. Electronic Engineering, 9/88/ p. 28.

Chapter 4

Fig. 4-1. Siliconix, Integrated Circuits Data Book, 1988, p. 6-148.

Fig. 4-2. Harris, Analog Product Data Book, 1988, p. 10-48.

Fig. 4-3. Signetics, 1987 Linear Data Manual Vol 2: Industrial, 2/87, p. 5-311.

Fig. 4-4. Maxim, Seminar Applications Book, 1988/89, p. 38.

Chapter 5

Fig. 5-1. Reprinted from EDN, 4/77, © 1989 Cahners Publishing Co., a division of Reed Publishing USA.

Chapter 6

Fig. 6-1. Harris, Analog Product Data Book, 1988, p. 10-173.

Fig. 6-2. Harris, Analog Product Data Book, 1988, p. 10-13.

Fig. 6-3. Signetics, RF Communications Handbook, 1989, p. 2-14 and 2-15.

Chapter 7

Fig. 7-1. Harris, Analog Product Data Book, 1988, p. 10-108.

Fig. 7-2. Signetics, Analog Data Manual, 1983, p. 10-20.

Fig. 7-3. ZeTeX (formerly Ferranti), Technical Handbook Super E-Line Transistors, 1987, p. SE-153.

Fig. 7-4. Harris, Analog Product Data Book, 1988, p. 10-161.

Fig. 7-5. Hands-On Electronics, Summer 1984, p. 74

Fig. 7-6. Radio-Electronics, 8/88, p. 33.

Fig. 7-7. Signetics, RF Communications Handbook, 1989, p. 1-61 and 1-62.

Fig. 7-8. Popular Electronics, 7/89, p 26.

Fig. 7-9. Popular Electronics, Fact Card No. 110.

Fig. 7-10. Harris, Analog Product Data Book, 1988, p. 10-174.

Fig. 7-11. Harris, Analog Product Data Book, 1988, p. 10-174.

Fig. 7-12. Hands-On Electronics, 5/87, p. 96.

Fig. 7-13. QST, 1/89, p. 20.

Fig. 7-14. Hands-On Electronics/Popular Electronics, 11/88, p. 39.

Chapter 8

Fig. 8-1. Gernsback Publications Inc., 42 New Ideas, 1984, p. 9.

Fig. 8-2. ZeTeX (formerly Ferranti), Technical Handbook Super E-Line Transistors, 1987, p. SE-158.

Fig. 8-3. Popular Electronics, 8/89, p. 22.

Fig. 8-4. Gernsback Publications Inc. 42 New Ideas, 1985, p. 28.

Fig. 8-5. Reprinted with permission from Electronic Design. Copyright 1989, Penton Publishing.

Fig. 8-6. ZeTeX, (formerly Ferranti), Technical Handbook Super E-Line Transistors, 1987, p. SE-155.

Fig. 8-7. Radio Electronics, 7/85, p. 55.

Fig. 8-8. Motorola, Motorola TMOS Power FET Design Ideas, 1985, p 3.

Fig. 8-9. Reprinted from EDN, 8/81, © 1989 Cahners Publishing Co., a division of Reed Publishing USA.

Fig. 8-10. Intersil, Component Data Catalog, 1987, p. 2-108.

Fig. 8-11. Signetics, 1987 Linear Data Manual Vol. 2: Industrial, 2/87, p. 7-67.

Fig. 8-12. Hands-On Electronics, 9-10/86, p. 27.

Fig. 8-13. TAB Books, The Giant Book of Easy-to-Build Electronic Projects, 1982, p. 186.

Fig. 8-14. TAB Books, 101 Sound, Light, and Power IC Projects, 1986, p. 139.

Fig. 8-15. Hands-On Electronics, 5/87, p. 95.

Fig. 16. Hands-On Electronics/Popular Electronics, 12/88, p. 25.

Fig. 8-17. Popular Electronics/Hands-On Electronics, 3/89, p. 36.

Fig. 8-18. Radio-Electronics, 4/87, p. 67.

Chapter 9

Fig. 9-1. Motorola, Motorola Thyristor Device Data, Series A 1985, p. 1-6-54.

Fig. 9-2. Reprinted from EDN, 7/21/88, © 1989 Cahners Publishing Co., a division of Reed Publishing USA.

Fig. 9-3. Linear Technology Corp., Linear Databook, 1986, p. 4-15.

Fig. 9-4. Radio-Electronics, 7/86, p. 67.

Fig. 9-5. Motorola, Motorola Thyristor Device Data, Series A 1985, p. 1-6-46.

Fig. 9-6. ZeTeX (formerly Ferranti), Technical Handbook Super E-Line Transistors, 1987, p. SE-164.

Fig. 9-7. Popular Electronics, 7/89, p. 81.

Fig. 9-8. Gernsback Publications, Inc., 42 New Ideas, 1984, p. 9.

Chapter 10

Fig. 10-1. Reprinted from EDN, 7/21/88, © 1989 Cahners Publishing Co., a division of Reed Publishing USA.

Fig. 10-2. Popular Electronics/Hands-On Electronics, 5/89, p. 25.

Fig. 10-3. Motorola, Motorola TMOS Power FET Design Ideas, 1985, p. 5.

Fig. 10-4. Maxim, 1986 Power Supply Circuits, p. 26.

Fig. 10-5. Hands-On Electronics, Spring 1985, p. 49.

Fig. 10-6. National Semiconductor Corp., 1984 Linear Supplement Databook, p. S2-7.

Fig. 10-7. Reprinted from EDN, 9/5/85, © 1989 Cahners Publishing Co., a division of Reed Publishing USA.

Fig. 10-8. Motorola, Motorola TMOS Power FET Design Ideas, 1985, p. 6.

Fig. 10-9. Maxim, 1986 Power Supply Circuits, p. 121.

Fig. 10-10. Reprinted from EDN, 1/8/87, © 1989 Cahners Publishing Co., a division of Reed Publishing USA.

Fig. 10-11. Linear Technology, Application Note 8, p. 2.

Fig. 10-12. GE, Optoelectronics, Third Edition, Ch. 6, p. 148.

Chapter 11

Fig. 11-1. Harris, Analog Product Data Book, 1988, p. 10-183.

Fig. 11-2. Reprinted with permission from Electronic Design. Copyright 1989, Penton Publishing.

Fig. 11-3. Harris, Analog Product Data Book, 1988, p. 10-106.

Fig. 11-4. Signetics, 1987 Linear Data Manual Vol 2: Industrial, 2/87, p. 4-78.

Fig. 11-5. Linear Technology, 1986 Linear Databook, p. 2-45.

Chapter 12

Fig. 12-1. Reprinted from EDN, 7/10/86, © 1989 Cahners Publishing Co., a division of Reed Publishing USA.

Fig. 12-2. Ham Radio, 7/86, p. 88.

Chapter 13

Fig. 13-1. Intersil, Applications Handbook, 1988, p. 3-138.

Fig. 13-2. Hands-On Electronics, Fall 1984, p. 68

Chapter 14

Fig. 14-1. Radio-Electronics, 1/89, p. 55.

Fig. 14-2. Radio-Electronics, 2/89, p. 55

Fig. 14-3. Radio-Electronics, 2/89, p. 55

Fig. 14-4. Signetics, 1987 Linear Data Manual Vol. 1: Communications, 2/87, p. 5-36.

Chapter 15

Fig. 15-1. Signetics, 1987 Linear Data Manual Vol. 2: Industrial, 12/2/86, p. 6-78.

Fig. 15-2. Harris, Analog Product Data Book, 1988, p. 2-106.

Fig. 15-3. TAB Books, 44 Electronics Projects for the Darkroom, p. 101.

Fig. 15-4. Reprinted from EDN, 10/78, © 1989 Cahners Publishing Co,. a division of Reed Publishing USA.

Chapter 16

Fig. 16-1. Texas Instruments, Linear and Interface Circuits Applications, 1985, Vol. 1, p. 3-46 and 3-47.

Fig. 16-2. Linear Technology, 1986 Liner Databook, p. 2-44.

Fig. 16-3. Reprinted from EDN, 4/75, © 1989 Cahners Publishing Co., a division of Reed Publishing USA.

Fig. 16-4. Reprinted from EDN, 4/79, © 1989 Cahners Publishing Co., a division of Reed Publishing USA.

Fig. 16-5. National Semiconductor Corp., 1984 Linear Supplement Databook, p. S1-42.

Fig. 16-6. Linear Technology, 1986 Linear Databook, p. 2-29.

Fig. 16-7. Reprinted from EDN, 2/18/88, © 1989 Cahners Publishing Co., a division of Reed Publishing USA.

Fig. 16-8. Harris, Analog Product Data Book, 1988, p. 10-16.

Chapter 17

Fig. 17-1. Signetics, 1987 Linear Data Manual Vol. 1: Communications, 2/87, p. 4-373 and 4-375.

Fig. 17-2. Signetics, RF Communications Handbook, 1989, p. 2-17 and 2-18.

Fig. 17-3. Signetics, RF Communications Handbook, 1989, p. 2-15 and 2-16.

Fig. 17-4. Signetics, RF Communications Handbook, 1989, p. 2-17 and 2-18.

Chapter 18

Fig. 18-1. Electronic Engineering, Applied Ideas, 4/86, p. 33.

Fig. 18-2. Siliconix, Integrated Circuits Data Book, 1988, p. 6-40 and 6-41.

Fig. 18-3. Hands-On Electronics/Popular Electronics, 11/88, p. 43.

Fig. 18-4. Linear Technology Corp., Linear Databook, 1986, p. 2-98.

Fig. 18-5. Siliconix, Integrated Circuits Data Book, 1988, p. 13-200.

Fig. 18-6. Siliconix, Integrated Circuits Data Book, 1988. p. 5-140.

Fig. 18-7. Reprinted from EDN, 4/28/88, © 1989 Cahners Publishing Co., a division of Reed Publishing USA.

Fig. 18-8. Harris, Analog Product Data Book, 1988, 2-106.

Fig. 18-9. Reprinted from EDN, 10/27/88, © 1989 Cahners Publishing Co., a division of Reed Publishing USA.

Fig. 18-10. Reprinted from EDN, 9/1/88, © 1989 Cahners Publishing Co., a division of Reed Publishing USA.

Fig. 18-11. Reprinted from EDN, 6/22/89, © 1989 Cahners Publishing Co., a division of Reed Publishing USA.

Fig. 18-12. Reprinted with permission from Electronic Design. Copyright 1988, Penton Publishing.

Fig. 18-13. Reprinted from EDN, 9/1/88, © 1989 Cahners Publishing Co., a division of Reed Publishing USA.

Fig. 18-14. Intersil, Component Data Catalog, 1987, p. 13-22.

Fig. 18-15. Electronic Engineering, Applied Ideas, 10/88, p. 37.

Fig. 18-16. Electronic Engineering, Applied Ideas, 12/88, p. 22.

Chapter 19

Fig. 19-1. Teledyne Semiconductor, Data Acquisition IC Handbook, 1985, p. 9-7.

Fig. 19-2. Radio Electronics, 9/89, p. 47.

Fig. 19-3. Signetics, 1987 Linear Data Manual Vol. 2: Industrial, 2/87, p. 7-65.

Fig. 19-4. Maxim, Seminar Applications Book, 1988/89, p. 149.

Fig. 19-5. Ham Radio, 5/89, p. 26.

Fig. 19-6. Popular Electronics, 10/89, p. 42.

Fig. 19-7. Maxim, Seminar Applications Book, 1988/89, p. 83.

Fig. 19-8. Reprinted from EDN, 10/2/86, © 1989 Cahners Publishing Co., a division of Reed Publishing USA.

Fig. 19-9. Linear Technology Corp., Linear Databook, 1986, p. 8-13.

Fig. 19-10. Linear Technology Corp., Linear Databook, 1986, p. 8-43.

Fig. 19-11. Maxim, Seminar Applications Book, 1988/89, p. 78.

Fig. 19-12. Reprinted from EDN, 3/7/85, © 1989 Cahners Publishing Co., a division of Reed Publishing USA.

Fig. 19-13. Reprinted from EDN, 8/17/89, © Cahners Publishing Co., a division of Reed Publishing USA.

Fig. 19-14. Linear Technology Corp., Linear Databook, 1986, p. 2-112.

Fig. 19-15. Reprinted from EDN, 10/1/87, © 1989 Cahners Publishing Co., a division of Reed Publishing USA.

Fig. 19-16. Reprinted from EDN, 3/3/88, © 1989 Cahners Publishing Co., a division of Reed Publishing USA.

Fig. 19-17. Signetics, 1987 Linear Data Manual Vol. 2: Industrial, 2/87, p. 7-62.

Chapter 20

Fig. 20-1. QST, 2/89, p. 21.

Fig. 20-2. Reprinted from EDN, 11/10/88, © 1989 Cahners Publishing Co., a division of Reed Publishing USA.

Fig. 20-3. Hands-On Electronics / Popular Electronics, 1/89, p. 59.

Fig. 20-4. Reprinted from EDN, 3/74, © 1989 Cahners Publishing Co., a division of Reed Publishing USA.

Fig. 20-5. Intersil, Component Data Catalog, 1987, p. 14-49.

Fig. 20-6. Gernsback Publications Inc., 42 New Ideas, 1984, p. 18.

Fig. 20-7. R-E Experimenters Handbook, 1987, p. 151.

Chapter 21

Fig. 21-1. Motorola, MECL System Design Handbook, 1983, p. 226.

Fig. 21-2. Reprinted from EDN, 2/78, © 1989 Cahners

Publishing Co., a division of Reed Publishing USA.

Fig. 21-3. Radio-Electronics, 2/86, p. 46.

Fig. 21-4. Reprinted from EDN, 6/83, © 1989 Cahners Publishing Co., a division of Reed Publishing USA.

Fig. 21-5. Reprinted from EDN, 6/83, © 1989 Cahners Publishing Co., a division of Reed Publishing USA.

Fig. 21-6. Radio-Electronics, 2/87, p. 96.

Fig. 21-7. Motorola, MECL System Design Handbook, 1983, p. 228.

Fig. 21-8. Reprinted from EDN, 5/73, © 1989 Cahners Publishing Co., a division of Reed Publishing USA.

Fig. 21-9. NASA Tech Briefs, 8/89, p. 20.

Fig. 21-10. Linear Technology Corp., Linear Applications Handbook, 1987, p. AN3-14.

Fig. 21-11. RF Design, 3/87, p. 31.

Fig. 21-12. 73 Amateur Radio, 1/89, p. 35.

Fig. 21-13. RF Design, 3/87, p. 31.

Fig. 21-14. RF Design, 3/87, p. 31.

Chapter 22

Fig. 22-1. R-E Experiments Handbook, 1989, p. 23.

Fig. 22-2. Hands-On Electronics / Popular Electronics, 1/89, p. 97.

Fig. 22-3. Reprinted from EDN, 7/21/88, © 1989 Cahners Publishing Co., a division of Reed Publishing USA.

Fig. 22-4. Hands-On Electronics / Popular Electronics, 1/89, p. 84.

Fig. 22-5. Reprinted from EDN, 9/29/88, © 1989 Cahners Publishing Co., a division of Reed Publishing USA.

Chapter 23

Fig. 23-1. Reprinted from EDN, 10/29/87, © 1989 Cahners Publishing Co., a division of Reed Publishing USA.

Fig. 23-2. Reprinted from EDN, 3/4/87, © 1989 Cahners Publishing Co., a division of Reed Publishing USA.

Fig. 23-3. Reprinted from EDN, 10/3/85, © 1989 Cahners Publishing Co., a division of Reed Publishing USA.

Chapter 24

Fig. 24-1. Signetics, 1987 Linear Data Manual Vol. 1: Communications, 2/87, p. 4-303.

Chapter 25

Fig. 25-1. Reprinted from EDN, 3/21/85, © 1989 Cahners Publishing Co., a division of Reed Publishing USA.

Fig. 25-2. Harris, Analog Product Data Book, 1988, p. 2-106.

Fig. 25-3. Teledyne Semiconductor, Data Acquisitions IC Handbook, 1985, p. 15-15.

Fig. 25-4. Siliconix, Integrated Circuits Data Book, 3/85, p.5-8.

Fig. 25-5. Signetics, RF Communications Handbook, 1989, p. 3-22.

Fig. 25-6. Reprinted from EDN, 1/20/79, © 1989 Cahners Publishing Co., a division of Reed Publishing USA.

Fig. 25-7. Gernsback Publications Inc., 42 New Ideas, 1984, p. 6.

Fig. 25-8. National Semiconductor Corp., Linear Databook, 1982, p. 3-97.

Fig. 25-9. Radio-Electronics, 3/89, p. 12.

Fig. 25-10. Harris, Analog Product Data Book, 1988, p. 10-167.

Fig. 25-11. Reprinted from EDN, 9/19/85, © 1989 Cahners Publishing Co., a division of Reed Publishing USA.

Fig. 25-12. Reprinted from EDN, 4/76, © 1989 Cahners Publishing Co., a division of Reed Publishing USA.

Fig. 25-13. Texas Instruments, Linear and Interface Circuits Applications, 1985, Vol. 1, p. 7-11 and 7-12.

Fig. 25-14. Radio Electronics, 4/89, p. 60.

Fig. 25-15. Reprinted from EDN, 12/8/88, © 1989 Cahners Publishing Co., a division of Reed Publishing USA.

Fig. 25-16. Reprinted from EDN, 2/20/76, © 1989 Cahners Publishing Co., a division of Reed Publishing USA.

Fig. 25-17. Reprinted from EDN, 10/17/85, © 1989 Cahners Publishing Co., a division of Reed Publishing USA.

Chapter 26

Fig. 26-1. Siliconix, Integrated Circuits Data Book, 3/85, p. 4-15.

Fig. 26-2. Precision Monolithics Inc., 1981 Full Line Catalog, p. 6-85.

Fig. 26-3. Intersil, Applications Handbook, 1988, p. 2-38.

Fig. 26-4. Intersil, Applications Handbook, 1988, p. 3-183.

Fig. 26-5. Harris, Analog Product Data Book, 1988, p. 10-17.

Fig. 26-6. GE/RCA, BiMOS Operational Amplifiers Circuit Ideas, 1987, p. 12.

Chapter 27

Fig. 27-1. Electronic Engineering, 12/78, p. 31.

Chapter 28

Fig. 28-1. Siliconix, Mospower Applications Handbook, p. 6-161.

Fig. 28-2. Motorola, Motorola Thyristor Device Data, Series A, 1985, p.1-5-27.

Fig. 28-3. Reprinted from EDN, 7/10/86, © 1989 Cahners Publishing Co., a division of Reed Publishing USA.

Fig. 28-4. Siliconix, Integrated Circuits Data Book, 3/85, p. 1-11.

Chapter 29

Fig.29-1. Signetics, Fiber-Optic Communication Data and Applications, 1988, p. 5-3.

Fig. 29-2. Signetics, Fiber-Optic Communication Data and Applications, 1988, p. 3-63 and 3-64.

Fig. 29-3. Signetics, Fiber-Optic Communication Data and Applications, 1988, p. 2-3 and 2-4.

Fig. 29-4. Signetics, 1987 Linear Data Manual Vol. 1: Communications, 2/87, p. 5-56 and 5-58.

Fig. 29-5. Signetics, Fiber-Optic Communication Data and Applications, 1988, p. 3-3.

Fig. 29-6. Signetics, Fiber-Optic Communication Data and Applications, 1988, p. 3-7 and 3-9.

Chapter 30

Fig. 30-1. Ham Radio, 1/85, p. 51.

Fig. 30-2. Hands-On Electronics, 8/87, p. 65.

Chapter 31

Fig. 31-1. Siliconix, Integrated Circuits Data Book, 1988, p. 13-181.

Fig. 31-2. Reprinted from EDN, 9/29/88, © 1989 Cahners Publishing Co., a division of Reed Publishing USA.

Fig. 31-3. Reprinted with permission from Electronic Design. Copyright 1989, Penton Publishing,

Fig. 31-4. Harris, Analog Product Data Book, 1988, p. 10-16.

Fig. 31-5. Harris, Analog Product Data Book, 1988, p. 10-175.

Fig. 31-6. Intersil, Component Data Catalog, 1987, p. 7-45.

Fig. 31-7. Reprinted from EDN, 3/16/89, © 1989 Cahners Publishing Co., a division of Reed Publishing USA.

Fig. 31-8. Raytheon, Linear and Integrated Circuits, 1989, p. 4-189.

Fig. 31-9. Popular Electronics, Fact Card No. 101.

Fig. 31-10. Popular Electronics, Fact Card No. 104.

Fig. 31-11. Popular Electronics, Fact Card No. 59.

Fig. 31-12. Popular Electronics, Fact Card No. 59.

Fig. 31-13. Popular Electronics, Fact Card No. 117.

Fig. 31-14. Popular Electronics, Fact Card No. 117.

Fig. 31-15. Popular Electronics, Fact Card No. 117.

Fig. 31-16. Popular Electronics, Fact Card No. 101.

Chapter 32

Fig. 32-1. Hands-On Electronics, 3/87, p. 97.

Fig. 32-2. Ideas for Design.

Fig. 32-3. Reprinted from EDN, 7/73, © 1989 Cahners Publishing Co., a division of Reed Publishing USA.

Fig. 32-4. Popular Electronics, 9/89, p. 88.

Fig. 32-5. Reprinted from EDN, 7/73, © 1989 Cahners Publishing Co., a division of Reed Publishing USA.

Fig. 32-6. Reprinted from EDN, 7/73, © 1989 Cahners Publishing Co., a division of Reed Publishing USA.

Fig. 32-7. Reprinted from EDN, 7/73, © 1989 Cahners Publishing Co., a division of Reed Publishing USA.

Fig. 32-8. Reprinted from EDN, 7/73, © 1989 Cahners Publishing Co., a division of Reed Publishing USA.

Fig. 32-9. Popular Electronics, Fact Card No. 65.

Fig. 32-10. Texas Instruments, Linear and Interface Circuits Applications, 1985, Vol. 1, p. 7-17.

Fig. 32-11. Reprinted from EDN, 7/73, © 1989 Cahners Publishing Co., a division of Reed Publishing USA.

Fig. 32-12. Hands-On Electronics, 3/87, p. 97.

Fig. 32-13. ZeTeX, (formerly Ferranti), Technical Handbook Super E-Line Transistors, 1987, p. SE-153.

Fig. 32-14. Reprinted from EDN, 7/73, © 1989 Cahners Publishing Co., a division of Reed Publishing USA.

Fig. 32-15. Popular Electronics, 9/89, p. 23.

Fig. 32-16. Reprinted from EDN, 7/73, © 1989 Cahners Publishing Co., a division of Reed Publishing USA.

Fig. 32-17. Reprinted from EDN, 9/78, © 1989 Cahners Publishing Co., a division of Reed Publishing USA.

Fig. 32-18. Popular Electronics/Hands-On Electronics, 5/89, p. 86.

Chapter 33

Fig.33-1. Linear Technology Corp., Linear Applications Handbook, 1987, p. AN5-5.

Chapter 34

34-1. National Semiconductor Corp., Linear Applications Databook, p. 1079.

Fig. 34-2. Radio-Electronics, 6/88, p. 49.

Fig. 34-3. TAB Books, The Build-It Book of Electronic Projects, p. 18.

Fig. 34-4. GE/RCA, BiMOS Operational Amplifiers Circuit Ideas, 1987, p. 23.

Fig. 34-5. Hands-On Electronics, 9–10/86, p. 24.

Fig. 34-6. Reprinted with permission from Electronic Design. Copyright 1989, Penton Publishing.

Fig. 34-7. Reprinted from EDN, 2/21/85, © 1989 Cahners Publishing Co., a division of Reed Publishing USA.

Chapter 35

Fig. 35-1. Signetics, RF Communications Handbook, 1989, p. 1-33.

Fig. 35-2. Signetics, RF Communications Handbook, 1989, p. 1-33.

Fig. 35-3. Intersil, Component Data Catalog, 1987, p. 7-44.

Fig. 35-4. Hands-On Electronics, Fact Card No. 29.

Chapter 36

Fig. 36-1. Electronic Engineering, 9/87, p. 27.

Fig. 36-2. Reprinted from EDN, 5/26/88, © 1989 Cahners Publishing Co., a division of Reed Publishing USA.

Fig. 36-3. Reprinted from EDN, 2/77, © 1989 Cahners Publishing Co., a division of Reed Publishing USA.

Fig. 36-4. Reprinted from EDN, 6/8/89, © 1989 Cahners Publishing Co., a division of Reed Publishing USA.

Fig. 36-5. Signetics, RF Communications Handbook, 1989, p. 4-10 and 4-11.

Fig. 36-6. Reprinted from EDN, 11/24/88, © 1989 Cahners Publishing Co., a division of Reed Publishing USA.

Chapter 37

Fig. 37-1. Reprinted from EDN, 5/83, © 1989 Cahners Publishing Co., a division of Reed Publishing USA.

Chapter 38

Fig. 38-1. Reprinted from EDN, 3/21/85, © 1989 Cahners Publishing Co., a division of Reed Publishing USA.

Fig. 38-2. Reprinted from EDN, 11/78, © 1989 Cahners Publishing Co., a division of Reed Publishing USA.

Fig. 38-3. Reprinted from EDN, 12/13/84, © 1989 Cahners Publishing Co., a division of Reed Publishing USA.

Fig. 38-4. Popular Electronics/Hands-On Electronics, 4/89, p. 23.

Fig. 38-5. GE/RCA, BiMOS Operational Amplifiers Circuit Ideas, 1987, p. 7.

Fig. 38-6. Electronic Engineering, 9/88, p. 34.

Fig. 38-7. Electronic Engineering, 9/86, p. 34.

Fig. 38-8. Reprinted from EDN, 4/14/88, © 1989 Cahners Publishing Co., a division of Reed Publishing USA.

Fig. 38-9. Electronic Engineering, 4/88, p. 33.

Fig. 38-10. Reprinted from EDN, 5/26/88, © 1989 Cahners Publishing Co., a division of Reed Publishing USA.

Fig. 38-11. GE/RCA BiMOS Operational Amplifiers Circuit Ideas, 1987, p. 8.

Fig. 38-12. Reprinted with permission from Electronic Design. Copyright 1989, Penton Publishing.

Fig. 38-13. Signetics, 1987 Linear Data Manual Vol. 1: Communications, 2/87, p. 4-312.

Fig. 38-14. Texas Instruments, Linear and Interface Circuits Applications, 1985, Vol. 1, p. 7-18.

Fig. 38-15. Intersil, 1978 Databook.

Fig. 38-16. Harris, Analog Product Data Book, 1988, p. 10-109.

Fig. 38-17. Siliconix, Integrated Circuits Data Book, 1981, p. 8-51.

Fig. 38-18. Hands-On Electronics/Popular Electronics, 1/89, p. 84.

Fig. 38-19. GE/RCA BiMOS Operational Amplifiers Circuit Ideas, 1987, p. 9.

Fig. 38-20. National Semiconductor Corp., Linear Applications Databook, p. 1118.

Fig. 38-21. Raytheon, Linear and Integrated Circuit, 1984, p. 12-8.

Fig. 38-22. Reprinted from EDN, 6/78, © 1989 Cahners Publishing Co., a division of Reed Publishing USA.

Fig. 38-23. Maxim, Seminar Applications Book, 1988/89, p. 45.

Fig. 38-24. Raytheon, Linear and Integrated Circuits, 1984, p. 12-7.

Fig. 38-25. Harris, Analog Product Data Book, 1988, p. 10-15.

Fig. 38-26. Intersil, Component Data Catalog, 1987, p. 7-104.

Fig. 38-27. Popular Electronics, Fact Card No. 98.

Fig. 38-28. Hands-On Electronics, Fact Card No. 86.

Fig. 38-29. Harris, Analog Product Data Book, 1988, p. 10-168.

Fig. 38-30. Hands-On Electronics, Fact Card No. 86.

Fig. 38-31. Raytheon, Linear and Integrated Circuits,

1989, p. 4-188.

Fig. 38-32. GE/RCA BiMOS Operational Amplifiers Circuit Ideas, 1987, p. 8.

Fig. 38-33. GE/RCA, BiMOS Operational Amplifiers Circuit Ideas, 1987, p. 9.

Fig. 38-34. Hands-On Electronics, Fact Card No. 89.

Fig. 38-35. Signetics, 1987 Linear Data Manual Vol. 1: Communications, 2/87, p. 4-311.

Fig. 38-36. Intersil, Component Data Catalog, 1987, p. 7-44.

Fig. 38-37. Hands-On Electronics, Fact Card No. 89.

Chapter 39

Fig. 39-1. Popular Electronics/Hands-On Electronics, 3/89, p. 24.

Fig. 39-2. Popular Electronics, 6/89, p. 27.

Fig. 39-3. Elektor, 6/78, p. 6-18.

Chapter 40

Fig. 40-1. GE/RCA, BiMOS Operational Amplifiers Circuit Ideas, 1987, p. 27.

Fig. 40-2. Linear Technology Corp., Linear Databook, 1986, p. 2-99.

Fig. 40-3. Reprinted from EDN, 8/75, © 1989 Cahners Publishing Co., a division of Reed Publishing USA.

Fig. 40-4. Reprinted from EDN, 9/73, © 1989 Cahners Publishing Co., a division of Reed Publishing USA.

Fig. 40-5. Linear Technology Design Notes, 3/89, Number 5.

Fig. 40-6. General Instrument Microelectronics, Application Note 1601, p. 3.

Fig. 40-7. Reprinted from EDN, 2/76, © 1989 Cahners Publishing Co., a division of Reed Publishing USA.

Fig. 40-8. Popular Electronics, Fact Card No. 65.

Fig. 40-9. Reprinted from EDN, 10/30/86, © 1989 Cahners Publishing Co., a division of Reed Publishing USA.

Chapter 41

Fig. 41-1. Reprinted from EDN, 7/7/88, © 1989 Cahners Publishing Co., a division of Reed Publishing USA.

Fig. 41-2. Texas Instruments, Linear and Interface Circuits Applications, 1987, p. 12-9.

Fig. 41-3. Reprinted from EDN, 7/11/85, © 1989 Cahners Publishing Co., a division of Reed Publishing USA.

Fig. 41-4. Texas Instruments, Linear and Interface Circuits Applications, 1987, p. 12-8.

Chapter 42

Fig. 42-1. Motorola, RF Data Manual, 1986, p. 6-226.

Fig. 42-2. Motorola, RF Data Manual, 1986, p. 6-85.

Fig. 42-3. 73 Amateur Radio, 3/89, p. 66.

Fig. 42-4. Signetics, RF Communications Handbook, 1989, p. 1-31.

Fig. 42-5. Siliconix, Small-Signal FET Data Book, 1989, p. 4-158, 4-159, and 9-42.

Fig. 42-6. R-E Experiments Handbook, 1989, p. 156.

Fig. 42-7. R-E Experiments Handbook, 1989, p. 33.

Fig. 42-8. Reprinted from EDN, 1/7/88, © 1989 Cahners Publishing Co., a division of Reed Publishing USA.

Chapter 43

Fig. 43-1. NASA, NASA Tech Briefs, p. 55.

Chapter 44

Fig. 44-1. Elektor Electronics, 7–8/87 Supplement, p. 36.

Fig. 44-2. Reprinted from EDN, 5/74, © 1989 Cahners Publishing Co., a division of Reed Publishing USA.

Fig. 44-3. Hands-On Electronics, Fact Card No. 57.

Fig. 44-4. 73 Magazine, 10/83, p. 53.

Fig. 44-5. Practical Wireless, 5/85, p. 37.

Chapter 45

Fig. 45-1. Radio-Electronics, 8/88, p. 37.

Fig. 45-2. GE, optoelectronics, Third Edition, Ch. 6, p. 115.

Fig. 45-3. Popular Electronics/Hands-On Electronics, 4/89, p. 83.

Fig. 45-4. ZeTeX (formerly Ferranti), Technical Handbook Super E-Line Transistors, 1987, p. SE-155.

Fig. 45-5. Popular Electronics/Hands-On Electronics, 4/89, p. 82.

Fig. 45-6. Reprinted from EDN, 11/13/86, © 1989 Cahners Publishing Co., a division of Reed Publishing USA.

Fig. 45-7. Hands-On Electronics, Fact Card No. 83.

Fig. 45-8. Popular Electronics, Fact Card No. 94.

Fig. 45-9. Hands-On Electronics, Fact Card No. 83.

Chapter 46

Fig. 46-1. Linear Technology, Application Note 9, p. 6.

Fig. 46-2. Maxim, Maxim Advantage, p. 45.

Fig. 46-3. Linear Technology Corp., Linear Applications Handbook, 1987, p. AN3-2.

Fig. 46-4. GE/RCA, BiMOS Operational Amplifiers Circuit Ideas, 1987, p. 19.

Fig. 46-5. GE/RCA, BiMOS Operational Amplifiers Circuit Ideas, 1987, p. 15.

Fig. 46-6. Harris, Analog Product Data Book, 1988, p. 10-181.

Fig. 46-7. Maxim, Maxim Advantage, p. 45.

Fig. 46-8. Siliconix, Integrated Circuits Data Book, 1988, p. 5-172.

Chapter 47

Fig. 47-1. Siliconix, Integrated Circuits Data Book, 3/85, p. 10-64.

Fig. 47-2. Linear Technology Corp., Linear Databook, 1986, p. 8-40.

Chapter 48

Fig. 48-1. Fairchild Camera & Instrument Corp., Linear Databook, 1982, p. 4-88.

Fig. 48-2. National Semiconductor Corp., Linear Databook, 1982, p. 10-63.

Fig. 48-3. Hands-On Electronics/Popular Electronics, 1/89, p. 39.

Fig. 48-4. Hands-On Electronics/Popular Electronics, 11/88, p. 39.

Fig. 48-5. Popular Electronics, Fact Card No. 59.

Chapter 49

Fig. 49-1. R-E Experimenters Handbook, 1987, p. 129.

Fig. 49-2. Popular Electronics, 6/89, p. 25.

Fig. 49-3. Maxim, Seminar Applications Book, 1988/89, p. 81.

Fig. 49-4. Motorola, Motorola TMOS Power FET Design Ideas, p. 35.

Fig. 49-5. Linear Technology Corp., Linear Applications Handbook, 1987, p. AN3-14.

Fig. 49-6. Popular Electronics, 10/89, p.26.

Chapter 50

Fig. 50-1. Elektor Electronics, 7 – 8/87 Supplement, p. 55.

Fig. 50-2. Reprinted from EDN, 10/27/88, © Cahners Publishing Co., a division of Reed Publishing USA.

Fig. 50-3. Hands-On Electronics/Popular Electronics, 1/89, p. 27.

Fig. 50-4. Motorola, Motorola TMOS Power FET Design Ideas, 1985, p. 22.

Fig. 50-5. Motorola, Motorola Thyristor Device Data, Series A, 1985, p. 1-6-54.

Fig. 50-6. Hands-On Electronics, Fact Card No. 37.

Fig. 50-7. Motorola, Motorola Thyristor Device Data, Series A, 1985, p. 1-6-48.

Fig. 50-8. Modern Electronics, 11/85, p. 44.

Fig. 50-9. Motorola, Motorola Thyristor Device Data, Series A, 1985, p. 1-6-10.

Fig. 50-10. GE, Optoelectronics, Third Edition, Ch. 6, p. 106.

Fig. 50-11. ZeTeX (formerly Ferranti), Technical Handbook Super E-Line Transistors, 1987, p. SE-158.

Fig. 50-12. Reprinted with permission from Electronic Design. Copyright 1989, Penton Publishing.

Fig. 50-13. GE, Optoelectronics, Third Edition, Ch. 6, p. 110.

Chapter 51

Fig. 51-1. TAB Books, Build Your Own Laser, Phaser, Ion Ray Gun, 1983, p. 13.

Fig. 51-2. Siliconix, Mospower Applications Handbook, p. 6-184.

Chapter 52

Fig. 52-1. GE, Optoelectronics, Third Edition, Ch. 6, p. 109.

Fig. 52-2. GE, Optoelectronics, Third Edition, Ch. 6, p. 112.

Fig. 52-3. Motorola, Motorola TMOS Power FET Design Ideas, 1985, p. 19.

Fig. 52-4. Texas Instruments, Linear and Interface Circuits Applications, 1985, Vol. 1, p. 3-3 and 3-4.

Fig. 52-5. Reprinted from EDN, August 1976, © 1989 Cahners Publishing Co., a division of Reed Publishing USA.

Fig. 52-6. Harris, Analog Product Data Book, 1988, p. 10-182.

Fig. 52-7. GE, Optoelectronics, Third Edition, Ch. 6, p. 109.

Fig. 52-8. GE, Optoelectronics, Third Edition, Ch. 6, p. 107.

Fig. 52-9. GE/RCA, BiMOS Operational Amplifiers Circuit Ideas, 1987, p. 7.

Fig. 52-10. Hands-On Electronics, Fact Card No. 57.

Fig. 52-11. GE, Optoelectronics, Third Edition, Ch. 6, p. 108.

Fig. 52-12. GE, Application Note 200.35, p. 15.

Chapter 53

Fig. 53-1. Hands-On Electronics, 9/87, p. 96.

Fig. 53-2. Popular Electronics, Fact Card No. 113.

Fig. 53.3 Harris, Analog Product Data Book, 1988, p. 10-54.

Chapter 54

Fig. 54-1. Signetics, 1987 Linear Data Manual Vol. 2: Industrial, 2/87, p. 5-350 and 5-352.

Chapter 55

Fig. 55-1. Reprinted with permission from Electronic Design. Copyright 1989, Penton Publishing.

Fig. 55-2. Hands-On Electronics, Fact Card No. 29.

Fig. 55-3. Hands-On Electronics, Fact Card No. 29.

Chapter 56

Fig. 56-1. Reprinted from EDN, 5/2/85, © 1989 Cahners Publishing Co., a division of Reed Publishing USA.

Fig. 56-2. Popular Electronics, 10/89, p. 104.

Fig. 56-3. Hands-On Electronics, Winter 1985, p. 31.

Fig. 56-4. Ham Radio, 4/86, p. 24.

Fig. 56-5. Gernsback Publications Inc., 42 New Ideas, 1984, p. 24.

Fig. 56-6. Ham Radio, 12/88, p. 19.

Fig. 56-7. Texas Instruments, Linear and Interface Circuits Applications, 1987, p. 12-5.

Fig. 56-8. Reprinted from EDN, 3/23/85, © 1989 Cahners Publishing Co., a division of Reed Publishing USA.

Fig. 56-9. 73 Amateur Radio, 6/89, p. 44.

Fig. 56-10. Hands-On Electronics, 2/87, p. 92.

Fig. 56-11. Elektor Electronics, 7–8/87 Supplement, p. 23.

Fig. 56-12. GE/RCA, BiMOS Operational Amplifiers Circuit Ideas, 1987, p. 20.

Fig. 56-13. GE, Optoelectronics, Third Edition, Ch. 6, p. 111.

Fig. 56-14. Hands-On Electronics, 7–8/86, p. 51.

Fig. 56-15. Intersil, Component Data Catalog, 1987, p. 14-70.

Fig. 56-16. Intersil, Component Data Catalog, 1987, p. 14-121.

Fig. 56-17. NASA, NASA Tech Briefs, 1/89, p. 19.

Fig. 56-18. Reprinted from EDN, 5/83, © 1989 Cahners Publishing Co., a division of Reed Publishing USA.

Fig. 56-19. 73 Amateur Radio, 8/88, p. 24.

Fig. 56-20. Linear Technology Corp., Linear Databook, 1986, p. 2-96.

Fig. 56-21. Radio-Electronics, 9/87, p. 32.

Fig. 56-22. Hands-On Electronics/Popular Electronics, 12/88, p. 61.

Fig. 56-23. Siliconix, Integrated Circuits Data Book, 3/85, p. 10-137.

Fig. 56-24. GE/RCA BiMOS Operational Amplifiers Circuit Ideas, 1987, p. 12.

Fig. 56-25. Electronic Engineering, 11/86, p. 34.

Fig. 56-26. Popular Electronics, 10/89, p. 84.

Fig. 56-27. Teledyne Semiconductor, Data Acquisition IC Handbook , 1985, p. 15-11.

Fig. 56-28. GE, Optoelectronics, Third Edition, Ch. 6, p. 112.

Fig. 56-29. Reprinted from EDN, 3/79, © 1989 Cahners Publishing Co., a division of Reed Publishing USA.

Chapter 57

Fig. 57-1. Reprinted from EDN, 3/7/85, © 1989 Cahners Publishing Co., a division of Reed Publishing USA.

Fig. 57-2. Electronic Engineering, Applied Ideas, 5/88, p. 25.

Fig. 57-3. Signetics. 1987 Linear Data Manual Vol. 2: Industrial, 2/87, p. 7-63.

Fig. 57-4. Reprinted from EDN, 9/15/88, © 1989 Cahners Publishing Co., a division of Reed Publishing USA.

Chapter 58

Fig. 58-1. Radio-Electronics, 12/85, p. 38.

Chapter 59

Fig. 59-1. Popular Electronics/Hands-On Electronics, 3/89, p. 82.

Fig. 59-2. GE, Application Note 90.16, p. 25.

Fig. 59-3. Popular Electronics/Hands-On Electronics, 3/89, p. 83.

Fig. 59-4. Popular Electronics/Hands-On Electronics, 4/89, p. 22.

Fig. 59-5. Popular Electronics/Hands-On Electronics, 3/89, p. 82.

Fig. 59-6. Radio-Electronics, 10/88, p. 51.

Fig. 59-7. Radio-Electronics, 4/85, p. 40.

Fig. 59-8. Siliconix, Small-Signal FET Data Book, 1989, p. 6-69 and 6-70.

Fig. 59-9. R-E Experimenter's Handbook, 1989, p. 41.

Fig. 59-10. RF Design, 7/89, p. 53.

Fig. 59-11. GE, Application Note 200.85, p. 16.

Fig. 59-12. GE, Application Note 200.85, p. 19.

Fig. 59-13. Harris, Analog Product Data Book, 1988, p. 10-109.

Fig. 59-14. Linear Technology Corp., Linear Applications Handbook, 1987, p. AN13-24.

Fig. 59-15. Gernsback Publications Inc., 44 New Ideas, 1985, p. 2.

Fig. 59-16. Radio-Electronics, 6/89, p. 42.

Fig. 59-17. Hands-On Electronics, 3/87, p. 26.

Chapter 60

Fig. 61-1. National Semiconductor Corp., Linear Applications Databook, p. 1028.

Fig. 60-2. Popular Electronics, 7/89, p. 23.

Fig. 60-3. Harris, Analog Product Data Book, 1988, p. 10-170.

Fig. 60-4. Popular Electronics, Fact Card No. 107.

Fig. 60-5. Harris, Analog Product Data Book, 1988, p. 10-176.

Chapter 61

Fig. 61-1. Reprinted from EDN, 11/28/85, © 1989 Cahners Publishing Co., a division of Reed Publishing USA.

Fig. 61-2. Signetics, 1987 Linear Data Manual Vol.2: Industrial, 10/10/86, p. 4-260.

Fig. 61-3. Electronic Design, 11/24/88, p. 222.

Fig. 61-4. Reprinted from EDN, 11/28/85, © 1989 Cahners Publishing Co., a division of Reed Publishing USA.

Fig. 61-5. Popular Electronics, 10/89, p. 23.

Fig. 61-6. Signetics, 1987 Linear Data Manual Vol. 2: Industrial, 2/87, p. 7-59.

Fig. 61-7. Signetics, 1987 Linear Data Manual Vol. 2: Industrial, 2/78, p. 7-59.

Fig. 61-8. Texas Instruments, Linear and Interface Circuits Applications, 1985, Vol. 1, p. 4-3.

Fig. 61-9. Signetics, RF Communications Handbook, 1989, p. 4-9.

Chapter 62

Fig. 62-1. Motorola, Motorola TMOS Power FET Design Ideas, 1985, p. 53.

Fig. 62-2. Motorola, Motorola TMOS Power FET Design Ideas, 1985, p. 29.

Fig. 62-3. Reprinted with permission from Electronic Design. Copyright 1988, Penton Publishing.

Fig. 62-4. Reprinted from EDN, 8/20/87, © 1989 Cahners Publishing Co., a division of Reed Publishing USA.

Fig. 62-5. Reprinted from EDN, 3/77, © 1989 Cahners Publishing Co., a division of Reed Publishing USA.

Fig. 62-6. Signetics, 1987 Linear Data Manual Vol.2: Industrial, 2/87, p. 7-63.

Fig. 62-7. Linear Technology Corp., Linear Applications Handbook, 1987, p. AN1-7.

Fig. 62-8. NASA, NASA Tech Briefs, 2/89, p. 26.

Fig. 62-9. Linear Technology Corp., Linear Applications Handbook, 1987, p. AN11-7.

Fig. 62-10. Maxim, Maxim Advantage, p. 47.

Fig. 62-11. Signetics, 1987 Linear Data Manual Vol.2: Industrial, 2/87, p. 8-94 and 8-95.

Fig. 62-12. Fairchild Camera and Instrument Corp., Linear Databook, 1982, p. 4-89.

Fig. 62-13. Texas Instruments, Linear and Interface Circuits Applications, 1987, p. 10-22.

Fig. 62-14. Texas Instruments, Linear and Interface Circuits Applications, 1985, Vol. 1, p. 7-22.

Fig. 62-15. ZeTeX (formerly Ferranti), Technical Handbook Super E-Line Transistors, 1987, p. SE-162.

Fig. 62-16. Linear Technology Corp., Linear Databook, 1986, p. 5-15.

Chapter 63

Fig. 63-1. Siliconix, Integrated Circuits Data Book, 1988, p. 13-203.

Fig. 63-2. Maxim, Seminar Applications Book, 1988/89, p. 61.

Fig. 63-3. Intersil, Applications Handbook, 1988, p. 5-6.

Fig. 63-4. Siliconix, Integrated Circuits Data Book, 1988, p. 13-200.

Fig. 63-5. Harris, Analog Product Data Book, 1988, p. 10-23.

Fig. 63-6. Intersil, Applications Handbook, 1988, p. 5-6.

Fig. 63-7. Harris, Analog Product Data Book, 1988, p. 10-12.

Chapter 64

Fig. 64-1. Signetics, Analog Data Manual, 1983, p. 4-26.

Fig. 64-2. Signetics, Analog Data Manual, 1983, p. 4-32.

Fig. 64-3. Signetics, Analog Data Manual, 1983, p. 4-33.

Chapter 65

Fig. 65-1. Texas Instruments, Linear and Interface Circuits Applications, 1985, Vol. 1, p. 3-10 and 3-11.

Fig. 65-2. National Semiconductor Corp., Linear Applications, p. 1083.

Chapter 66

Fig. 66-1. Siliconix, Integrated Circuits Data Book, 3/85, p. 10-211.

Chapter 67

Fig. 67-1. GE, Optoelectronics, Third Edition, Ch. 6, p. 141.

Fig. 67-2. GE, Optoelectronics, Third Edition, Ch. 6, p. 135.

Fig. 67-3. GE, Optoelectronics, Third Edition, Ch. 6, p. 144.

Fig. 67-4. GE, Optoelectronics, Third Edition, Ch. 6, p. 134.

Fig. 67-5. GE, Optoelectronics, Third Edition, Ch. 6, p. 147.

Fig. 67-6. GE, Optoelectronics, Third Edition, Ch. 6, p. 127.

Fig. 67-7. Reprinted from EDN, 1/82, © 1989 Cahner Publishing Co., a division of Reed Publishing USA.

Fig. 67-8. GE, Optoelectronics, Third Edition, Ch. 6, p. 123.

Fig. 67-9. GE, Optoelectronics, Third Edition, Ch. 6, p. 126.

Fig. 67-10. GE, Optoelectronics, Third Edition, Ch. 6, p. 108.

Fig. 67-11. GE, Optoelectronics, Third Edition, Ch. 6, p. 122.

Fig. 67-12. GE, Optoelectronics, Third Edition, Ch. 6, p. 135.

Fig. 67-13. GE, Optoelectronics, Third Edition, Ch. 6, p. 145.

Fig. 67-14. GE, Optoelectronics, Third Edition, Ch. 6, p. 138.

Fig. 67-15. Reprinted from EDN, 11/10/88, © 1989 Cahner Publishing Co., a division of Reed Publishing USA.

Chapter 68

Fig. 68-1. Reprinted from EDN, 8/7/86, © 1989 Cahner Publishing Co., a division of Reed Publishing USA.

Fig. 68-2. Texas Instruments, Linear and Interface Circuits Applications, 1985, Vol. 1, p. 7-17.

Fig. 68-3. GE, Application Note 90.16, p. 27.

Fig. 68-4. Elektor Electronics, 7 – 8/87 Supplement, p. 36.

Fig. 68-5. Harris, Analog Product Data Book, 1988, p. 10-15.

Fig. 68-6. Elektor Electronics, 7 – 8/87 Supplement, p. 63.

Fig. 68-7. Reprinted from EDN, 1/79, © 1989 Cahner Publishing Co., a division of Reed Publishing USA.

Fig. 68-8. Elektor Electronics, 7 – 8/87 Supplement, p. 59.

Fig. 68-9. Reprinted from EDN, 7/74, © 1989 Cahners Publishing Co., A division of Reed Publishing USA.

Fig. 68-10. Harris, Analog Product Data Book, 1988, p. 10-97.

Fig. 68-11. Reprinted from EDN, 1/77, © 1989 Cahners Publishing Co., a division of Reed Publishing USA.

Fig. 68-12. Radio-Electronics, 8/86, p. 83.

Fig. 68-13. Linear Technology Corp., Linear Applications Handbook, 1987, p. AN20-11.

Fig. 68-14. Fairchild Camera and Instrument, Linear Databook, 1982, p. 4-71.

Fig. 68-15. Reprinted from EDN, 8/76, © 1989 Cahners Publishing Co., a division of Reed Publishing USA.

Fig. 68-16. Reprinted from EDN, 5/11/89, © 1989 Cahners Publishing Co., a division of Reed Publishing USA.

Fig. 68-17. Reprinted from EDN, 4/82, © 1989 Cahners Publishing Co., a division of Reed Publishing USA.

Fig. 68-18. Hands-On Electronics, Fact Card No. 49.

Fig. 68-19. Reprinted from EDN, 2/2/89 © 1989 Cahners Publishing Co., a division of Reed Publishing USA.

Fig. 68-20. Hands-On Electronics/Popular Electronics, 1/89, p. 26.

Fig. 68-21. GE/RCA, BiMOS Operational Amplifier Circuit Ideas, 1987, p. 9.

Fig. 68-22. Reprinted from EDN, 1/78, © 1989 Cahners Publishing Co., a division of Reed Publishing USA.

Fig. 68-23. Harris, Analog Product Data Book, 1988, p. 10-108.

Chapter 69

Fig. 69-1. Siliconix, Integrated Circuits Data Book, 3/85, p. 10-62.

Fig. 69-2. Elektor Electronics, 7-8/87 Supplement, p. 50.

Fig. 69-3. Texas Instruments, Linear and Interface Circuits Applications, 1985, Vol. 1, p. 7-13.

Fig. 69-4. Reprinted from EDN, 3/82, © 1989 Cahners Publishing Co., a division of Reed Publishing USA.

Fig. 69-5. Radio-Electronics, 10/84, p. 32.

Fig. 69-6. 73 Amateur Radio, 8/88, p. 47.

Fig. 69-7. Texas Instruments, Linear and Interface Circuits Applications, 1985, Vol. 1, p. 4-2.

Fig. 69-8. Signetics, 1987 Linear Data Manual Vol. 2: Industrial, 2/87, p. 7-62.

Fig. 69-9. Ham Radio, 12/88, p. 26.

Chapter 70

Fig. 70-1. Harris, Analog Product Data Book, 1988, p. 10-14.

Fig. 70-2. Reprinted from EDN, 8/78, © 1989 Cahners Publishing Co., a division of Reed Publishing USA.

Fig. 70-3. Signetics, 1987 Linear Data Manual Vol. 1: Communications, 2/87, p. 4-67 and 4-68.

Chapter 71

Fig. 71-1. Popular Electronics/Hands-On Electronics, 4/89, p. 26.

Fig. 71-2. Hands-On Electronics, 1/87, p. 101.

Fig. 71-3. Texas Instruments, Linear and Interface Circuits Applications, 1985, Vol. 1, p. 7-14.

Fig. 71-4. Hands-On Electronics, 11/87, p. 32.

Fig. 74-11. Hands-On Electronics, 12/86, p. 92.

Chapter 75

Fig. 75-1. Siliconix, Mospower Applications Handbook, p. 6-62.

Fig. 75-2. 73 Amateur Radio, 4/88, p. 20.

Fig. 75-3. Reprinted from EDN, 12/8/88, © 1989 Cahners Publishing Co., a division of Reed Publishing USA.

Fig. 75-4. Linear Technology Corp., Linear Application Handbook, 1987, p. AN8-4.

Fig. 75-5. Ham Radio, 7/89, p. 20.

Fig. 75-6. Popular Electronics, Fact Card No. 100.

Fig. 75-7. Popular Electronics, Fact Card No. 95.

Chapter 76

Fig. 76-1. Hands-On Electronics, 1-2/86, p. 96.

Fig. 76-2. Maxim, 1986 Power Supply Circuits, p. 120.

Chapter 77

Fig. 77-1. Hands-On Electronics, 12/87, p. 73.

Fig. 77-2. Linear Technology Corp., Linear Applications Handbook, 1987, p. AN9-9.

Fig. 77-3. Microwaves and RF, 3/86, p. 143.

Fig. 77-4. Reprinted from EDN, 6/72, © 1989 Cahners Publishing Co., a division of Reed Publishing USA.

Fig. 77-5. Reprinted from EDN, 7/7/88, © 1989 Cahners Publishing Co., a division of Reed Publishing USA.

Fig. 77-6. Reprinted from EDN, 10/16/86, © 1989 Cahners Publishing Co., a division of Reed Publishing USA.

Fig. 77-7. Reprinted from EDN, 3/73, © 1989 Cahners Publishing Co., a division of Reed Publishing USA.

Fig. 77-8. Harris, Analog Product Data Book, 1988, p. 10-182.

Fig. 77-9. Linear Technology Corp., Application Note 9, p. 9.

Fig. 77-10. Hands-On Electronics, 9/87, p. 96.

Chapter 78

Fig. 78-1. Linear Technology Corp., Linear Applications Handbook, 1987, p. AN3-5.

Fig. 78-2. Siliconix, Integrated Circuits Data Book, 1988, p. 5-128.

Fig. 78-3. Intersil, Applications Handbook, 1988, p. 2-34.

Fig. 78-4. Harris, Analog Product Data Book, 1988, p. 10-13.

Fig. 78-5. Harris, Analog Product Data Book, 1988, p. 10-13.

Fig. 78-6. Harris, Analog Product Data Book, 1988, p. 10-169.

Chapter 79

Fig. 79-1. 73 Magazine, 7/77, p. 35.

Fig. 79-2. Reprinted from EDN, 4/83, © 1989 Cahners Publishing Co., a division of Reed Publishing USA.

Fig. 79-3. TAB Books, 104 Weekend Electronics Projects, p. 70.

Fig. 79-4. Contributed by William Sheets.

Fig. 79-5. Reprinted from EDN, 6/12/86, © 1989 Cahners Publishing Co., a division of Reed Publishing USA.

Fig. 79-6. Motorola, Motorola TMOS Power FET Design Ideas, 1985, p. 16.

Chapter 80

Fig. 80-1. Hands-On Electronics, 7/87, p. 47.

Fig. 80-2. RF Design, 12/86, p. 41.

Fig. 80-3. Popular Electronics/Hands-On Electronics, 5/89, p. 85.

Fig. 80-4. Hands-On Electronics, Fact Card No. 57.

Fig. 80-5. Siliconix, Small-Signal FET Data Book, 1/86, p. 7-28.

Chapter 81

Fig. 81-1. Popular Electronics, 9/89, p. 41.

Fig. 81-2. Motorola, Motorola TMOS Power FET Design Ideas, 1985, p. 15.

Fig. 81-3. Reprinted from EDN, 11/10/88, © Cahners Publishing Co., a division of Reed Publishing USA.

Fig. 81-4. Reprinted from EDN, 5/82, © 1989 Cahners Publishing Co., a division of Reed Publishing USA.

Fig. 81-5. Reprinted from EDN, 11/27/86, © 1989 Cahners Publishing Co., a division of Reed Publishing USA.

Fig. 81-6. Linear Technology Corp., Linear Databook, 1986, p. 5-33.

Chapter 82

Fig. 82-1. Reprinted from EDN, 11/13/86, © 1989 Cahners Publishing Co., a division of Reed Publishing USA.

Fig. 82-2. Harris, Analog Product Data Book, 1988, p. 10-14.

Chapter 83

Fig. 83-1. Signetics, 1987 Linear Data Manual Vol. 1: Communications, 11/14/86, p. 7-33.

Fig. 83-2. Signetics, 1987 Linear Data Manual Vol. 1: Communications, 11/14/86, p. 4-115.

Fig. 83-3. Radio-Electronics, 8/87, p. 39.

Fig. 83-4. Signetics, RF Communications Handbook,

1989, p. 3-80 and 3-81.

Fig. 83-5. Signetics, 1987 Linear Data Manual Vol. 2: Industrial, 2/87, p. 7-66 and 7-67.

Fig. 83-6. RF Design, 3/87, p. 46.

Fig. 83-7. 73 Amateur Radio, 10.86, p. 54.

Fig. 83-8. GE, Optoelectronics, Third Edition, Ch. 6, p. 128.

Fig. 83-9. Hands-On Electronics/Popular Electronics, 1/89, p. 35.

Fig. 83-10. Signetics, 1987 Linear Data Manual Vol. 1: Communications, 2/87, p. 7-25.

Chapter 84

Fig. 84-1. Signetics, 1987 Linear Data Manual Vol. 2: Industrial, 11/6/86, p. 4-137.

Fig. 84-2. GE/RCA, BiMOS Operational Amplifiers Circuit Ideas, 1987, p. 17.

Chapter 85

Fig. 85-1. Signetics, 1987 Linear Data Manual Vol. 2: Industrial, 2/87, p. 7-66 and 7-67.

Fig. 85-2. Hands-On Electronics/Popular Electronics, 12/88, p. 24.

Fig. 85-3. National Semiconductor Corp., Voltage Regulator Handbook, p. 10-59.

Chapter 86

Fig. 86-1. QST, 9/85, p. 41.

Fig. 86-2. Motorola, RF Data Manual, 1986, p. 6-221.

Fig. 86-3. Microwaves and RF, 9/85, p. 191.

Fig. 86-4. Teledyne Semiconductor, Data and Design Manual, 1981, p. 11-178.

Fig. 86-5. Harris, Analog Product Data Book, 1988, p. 10-58.

Fig. 86-6. GE/RCA, BiMOS Operational Amplifiers Circuit Ideas, 1987, p. 20.

Fig. 86-7. Motorola, RF Data Manual, 1986, p. 6-182.

Fig. 86-8. Radio-Electronics, 3/82, p. 59.

Fig. 86-9. Motorola, RF Data Manual, 1986, p. 6-236.

Chapter 87

Fig. 87-1. Intersil, Component Data Catalog, p. 7-96.

Fig. 87-2. Harris, Analog Product Data Book, 1988, p. 10-22.

Fig. 87-3. Harris, Analog Product Data Book, 1988, p. 10-54.

Fig. 87-4. Harris, Analog Product Data Book, 1988, p. 10-22.

Fig. 87-5. Harris, Analog Product Data Book, 1988, p. 10-37.

Fig. 87-6. Siliconix, Integrated Circuits Data Book, 1988, 5-127.

Fig. 87-7. Harris, Analog Product Data Book, 1988, p. 10-14.

Fig. 87-8. Harris, Analog Product Data Book, 1988, p. 10-22.

Fig. 87-9. GE/RCA, BiMOS Operational Amplifier Circuit Ideas, 1987, p. 13.

Chapter 88

Fig. 88-1. Popular Electronics/Hands-On Electronics, 3/89, p. 24.

Fig. 88-2. Popular Electronics/Hands-On Electronics, 12/88, p. 26.

Chapter 89

Fig. 89-1. Linear Technology Corp., Linear Applications Handbook, 1987, p. AN5-7.

Fig. 89-2. GE/RCA, BiMOS Operational Amplifiers Circuit Ideas, 1987, p. 7.

Fig. 89-3. Linear Technology Corp., Linear Databook Supplement, 1988, p. S2-34.

Fig. 89-4. Hands-On Electronics, 9/87, p. 97.

Chapter 90

Fig. 90-1. Popular Electronics, 10/89, p. 84.

Fig. 90-2. National Semiconductor, Linear Databook, 1982, p. 3-289.

Fig. 90-3. Hands-On Electronics, Spring 1985, p. 35.

Fig. 90-4. National Semiconductor Corp., Audio/Radio Handbook, 1980, p. 4-29.

Fig. 90-5. Hands-On Electronics, 5-6/86, p. 86.

Fig. 90-6. Reprinted from EDN, 8/4/88, © 1989 Cahners Publishing Co., a division of Reed Publishing USA.

Fig. 90-7. Everyday Electronics, 5/88, p. 292.

Fig. 90-8. Hands-On Electronics, Winter 1985, p. 72.

Fig. 90-9. Radio-Electronics, 12/81, p. 53.

Fig. 90-10. Reprinted with permission from Electronic Design. Copyright 1980, Penton Publishing.

Fig. 90-11. National Semiconductor Corp., Audio/Radio Handbook, 1980, p. 4-39.

Fig. 90-12. National Semiconductor Corp., Linear Databook, 1982, p. 3-289.

Fig. 90-13. Texas Instruments, Complex Sound Generator Bulletin No. DL-S 12612, p. 15.

Chapter 91

Fig. 91-1. GE, Optoelectronics, Third Edition, Ch. 6, p. 131.

Chapter 92

Fig. 92-1. Reprinted from EDN, 4/17/86, © 1989 Cahners Publishing Co., a division of Reed Publishing USA.

Fig. 92-2. ZeTeX (formerly Ferranti), Technical Handbook Super E-Line Transistors, 1987, p. SE-154.

Fig. 92-3. Reprinted from EDN, 12/8/88, © 1989 Cahners Publishing Co., a division of Reed Publishing USA.

Chapter 93

Fig. 93-1. Gernsback Publications Inc., 42 New Ideas, 1984, p. 4.

Fig. 93-2. Hands-On Electronics, 11/87, p. 92.

Fig. 93-3. Radio-Electronics, 12/81, p. 54.

Fig. 93-4. Radio-Electronics, 12/81, p. 53.

Fig. 93-5. Texas Instruments, Complex Sound Generator Bulletin No. DL-S 12612, p. 11.

Chapter 94

Fig. 94-1. Reprinted from EDN, 8/5/78, © 1989 Cahners Publishing Co., a division of Reed Publishing USA.

Fig. 94-2. Hands-On Electronics, 3/87, p. 26.

Chapter 95

Fig. 95-1. Harris, Analog Product Data Book, 1988, p. 10-97.

Fig. 95-2. Siliconix, Integrated Circuits Data Book, 3/85, p. 5-8.

Chapter 96

Fig. 96-1. Reprinted from EDN, 4/30/87, © 1989 Cahners Publishing Co., a division of Reed Publishing USA.

Fig. 96-2. Reprinted from EDN, 5/28/87, © 1989 Cahners Publishing Co., a division of Reed Publishing USA.

Fig. 96-3. Exar, Telecommunications Databook, 1986, p. 9-24.

Fig. 96-4. Exar, Telecommunications Databook, 1986, p. 9-24.

Chapter 97

Fig. 97-1. Texas Instruments, Linear and Interface Circuits Applications, 1985, Vol. 1, p. 7-28.

Fig. 97-2. Intersil, Component Data Catalog, 1987, p. 7-96.

Chapter 98

Fig. 98-1. Hands-On Electronics, 2/87, p. 65.

Chapter 99

Fig. 99-1. Motorola, TMOS Power FET Design Ideas, 1985, p. 48.

Fig. 99-2. Reprinted from EDN, 2/18/88, Cahner Publishing Co., a division of Reed Publishing USA.

Fig. 99-3. Reprinted from EDN, 5/79, © 1989 Cahner Publishing Co., a division of Reed Publishing USA.

Fig. 99-4. Siliconix, Mospower Applications Handbook, p. 6-185.

Fig. 99-5. Reprinted from EDN, 6/76, © 1989 Cahners Publishing Co., a division of Reed Publishing USA.

Chapter 100

Fig. 100-1. Reprinted from EDN, 3/3/88, © 1989 Cahners Publishing Co., a division of Reed Publishing USA.

Fig. 100-2. Reprinted from EDN, 5/26, © 1989 Cahners Publishing Co., a division of Reed Publishing USA.

Fig. 100-3. Electronic Engineering, Applied Ideas, 11/88, p. 28.

Chapter 101

Fig. 101-1. Hands-On Electronics, 11/87, p. 84.

Fig. 101-2. Popular Electronics/Hands-On Electronics, 5/89, p. 87.

Fig. 101-3. Hands-On Electronics, 11/87, p. 85.

Chapter 102

Fig. 102-1. Exar, Telecommunications Databook, 1986, p. 4-2.

Fig. 102-2. Texas Instruments, Linear and Interface Circuits Applications, 1985, Vol. 1, p. 7-23.

Fig. 102-3. Hands-On Electronics, Winter 1985, p. 49.

Fig. 102-4. Signetics, 1987 Linear Data Manual Vol. 1: Communications, 12/2/86, p. 6-23.

Fig. 102-5. Popular Electronics, 9/89, p. 27.

Fig. 102-6. National Semiconductor Corp., 1984 Linear Supplement Databook, p. S13-15.

Fig. 102-7. Popular Electronics, Fact Card No. 104.

Fig. 102-8. Exar, Telecommunications Databook, 1986, p. 11-61.

Fig. 102-9. Siliconix, Integrated Circuits Data Book, 3/85, p. 7-12.

Fig. 102-10. 73 Magazine, 10.78, p. 78.

Fig. 102-11. National Semiconductor Corp., 1984 Linear Supplement Databook, p. S13-16.

Fig. 102-12. Reprinted from EDN, 8/78, © 1989

Cahners Publishing Co., a division of Reed Publishing. USA.

Fig. 102-13. Siliconix, Integrated Circuits Data Book, 3/85, p. 7-11.

Fig. 102-14. Reprinted from EDN, 10/20/79, © 1989 Cahners Publishing Co., a division of Reed Publishing USA.

Fig. 102-15. Intersil, Databook 1987, p. 7-8.

Fig. 102-16. GE, Optoelectronics, Third Edition, Ch. 6, p. 128.

Fig. 102-17. Reprinted from EDN, 8/79, © 1989 Cahners Publishing Co., a division of Reed Publishing USA.

Fig. 102-18. Radio-Electronics, 10/83, p. 56.

Fig. 102-19. Electronic Engineering, 2/87, p. 40.

Fig. 102-20. Reprinted from EDN, 4/27/89, © 1989 Cahners Publishing Co., a division of Reed Publishing USA.

Fig. 102-21. Hands-On Electronics, 10/87, p. 95.

Fig. 102-22. Hands-On Electronics, 9-10/86, p. 88.

Fig. 102-23. Popular Electronics/Hands-On Electronics, 5/89, p. 86.

Fig. 102-24. GE, Optoelectronics, Third Edition, Ch. 6, p. 124.

Fig. 102-25. Hands-On Electronics, 7-8/86, p. 87.

Fig. 102-26. TAB Books, 101 Sound, Light, and Power IC Projects.

Fig. 102-27. Hands-On Electronics, 9-10/86, p. 105.

Chapter 103

Fig. 103-1. Motorola, Motorola Thyristor Device Data, Series A 1985, p. 1-6-60.

Fig. 103-2. Linear Technology Corp., Application Note 22, p. 10.

Fig. 103-3. Linear Technology Corp., Linear Applications Handbook, 1987, p. AN5-1.

Fig. 103-4. Motorola, Motorola TMOS Power FET Design Ideas, 1985, p. 24.

Fig. 103-5. Linear Technology Corp., Linear Applications Handbook, 1987, p. AN4-7.

Fig. 103-6. Radio-Electronics, 5/85, p. 110.

Chapter 104

Fig. 104-1. Siliconix, Integrated Circuits Data Book, 1988, p. 13-204.

Fig. 104-2. Signetics, 1987 Linear Data Manual Vol. 2: Industrial, 11/14/86, p. 5-58.

Fig. 104-3. Linear Technology Corp., Linear Databook, 1986, p. 8-43.

Chapter 105

Fig. 105-1. Electronic Engineering, Applied Ideas, 11/88, p. 28.

Chapter 106

Fig. 106-1. Gernsback Publications Inc., 42 New Ideas, 1984, p. 18.

Fig. 106-2. Popular Electronics, 8/89, p. 29.

Chapter 107

Fig. 107-1. TAB Books, 44 Electronics Projects for the Darkroom.

Fig. 107-2. Signetics, 1987 Linear Data Manual Vol. 2: Industrial, 11/14/86, p. 5-58.

Fig. 107-3. Raytheon, Linear and Integrated Circuits, 1989, p. 8-16.

Fig. 107-4. R-E Experimenters Handbook, 1987, p. 11.

Fig. 107-5. NASA, NASA Tech Briefs, 10/87, p. 34.

Fig. 107-6. Linear Technology Corp., Linear Applications Handbook, 1987, p. AN3-14.

Fig. 107-7. Hands-On Electronics, 9-10/86, p. 32.

Chapter 108

Fig. 108-1. Linear Technology Corp., Linear Applications Handbook, 1987, p. AN7-12.

Chapter 109

Fig. 109-1. Radio-Electronics, 7/85, p. 16.

Fig. 109-2. GE, Semiconductor Data Handbook, Second Edition, p. 905.

Fig. 109-3. Motorola, Motorola Thyrister Device Data, Series A 1985, p. 1-6-43.

Chapter 110

Fig. 110-1. Reprinted from EDN, 9/13/86, © 1989 Cahners Publishing Co., a division of Reed Publishing USA.

Fig. 110-2. Intersil, Databook 1987, p. 7-101.

Fig. 110-3. Radio-Electronics, 5/87, p. 129.

Fig. 110-4. Texas Instruments, Linear and Interface Circuits Applications, 1985, Vol. 1, p. 7-13.

Fig. 110-5. TAB Books, The Build-It Book of Electronic Projects, p. 32.

Chapter 111

Fig. 111-1. Texas Instruments, Linear and Interface Circuits Applications, 1985, Vol. 1, p. 3-11 and 3-12.

Fig. 111-2. National Semiconductor Corp., Audio/Radio Handbook, 1980, p. 2-61.

Chapter 118

Fig. 118-1. GE, Application Note 90.88, p. 7.

Fig. 118-2. Linear Technology Corp., Linear Applications Handbook, 1987, p. AN4-3.

Fig. 118-3. Harris, Analog Product Data Book, 1988, p. 10-149.

Fig. 118-4. Harris, Analog Product Data Book, 1988, p. 10-58.

Fig. 118-5. Harris, Analog Product Data Book, 1988, p. 10-54.

Fig. 118-6. Harris, Analog Product Data Book, 1988, p. 10-96.

Fig. 118-7. Reprinted with permission from Electronic Design. Copyright 1989, Penton Publishing.

Chapter 119

Fig. 119-1. Electronics Digest, Spring 1988, p. 63.

Fig. 119-2. Electronic Engineering, 5/84, p. 36.

Fig. 119-3. Motorola, Linear Integrated, Circuits, p. 5-37.

Fig. 119-4. Reprinted from EDN, 8/18/88, © 1989 Cahners Publishing Co., a division of Reed Publishing USA.

Fig. 119-5. Signetics, Linear Data Manual Vol. 3: Video, p. 10-58.

Fig. 119-6. Radio-Electronics, 2/86, p. 51.

Fig. 119-7. Siliconix, Integrated Circuits Data Book, 3/85, p. 10-67.

Fig. 119-8. Radio-Electronics, 9/87, p. 41.

Fig. 119-9. Hands-On Electronics, Winter 1985, p. 44.

Fig. 119-10. Electronic Engineering, 8/86, p. 32.

Fig. 119-11. Signetics Linear Data Manual Vol. 3: Video, p. 11-95.

Fig. 119-12. Electronic Engineering, 7/84, p. 27.

Fig. 119-13. Siliconix, Integrated Circuits Data Book, 3/85, p. 10-66.

Fig. 119-14. Reprinted from EDN, 5/16/85, © 1989 Cahners Publishing Co., a division of Reed Publishing USA.

Fig. 119-15. Hands-On Electronics, 7-8/86, p. 78.

Fig. 119-16. Siliconix, Small-Signal FET Data Book, 1/86, p. 7-101.

Chapter 120

Fig. 120-1. Reprinted with permission from Electronic Design. Copyright 1989, Penton Publishing.

Fig. 120-2. Reprinted from EDN, 1/7/88, © 1989 Cahners Publishing Co., a division of Reed Publishing USA.

Fig. 120-3. Hands-On Electronics, 10/88, p. 30.

Fig. 120-4. Reprinted with permission from Electronic Design. Copyright 1988, Penton Publishing.

Fig. 120-5. Gernsback Publications Inc., 42 New Ideas, 1984, p. 16.

Chapter 121

Fig. 121-1. Motorola, Motorola TMOS Power FET Design Ideas, 1985, p. 47.

Fig. 121-2. Intersil, Component Data Catalog, 1987, p. 6-29.

Fig. 121-3. Reprinted from EDN, © 1989 Cahners Publishing Co., a division of Reed Publishing USA.

Fig. 121-4. Intersil, Applications Handbook, 1988, p. 6-6.

Fig. 121-5. Reprinted from EDN, 10/1/87, © 1989 Cahners Publishing Co., a division of Reed Publishing USA.

Fig. 121-6. Texas Instruments, Linear and Interface Circuit Applications, 1985, p. 7-19.

Fig. 121-7. Texas Instruments, Linear Circuits Data Book, 1989, p. 2-73.

Fig. 121-8. Popular Electronics, 10/89, p. 105.

Chapter 122

Fig. 122-1. Electronic Engineering, 11/76, p. 23.

Fig. 122-2. Datel, Data Conversion Components, p. 6-18.

Fig. 122-3. Motorola, Motorola TMOS Power FET Design Ideas, 1985, p. 41.

Fig. 122-4. Motorola, Linear Integrated Circuits, p. 5-145.

Fig. 122-5. Linear Technology Corp., Linear Applications Handbook, 1987, p. AN8-9.

Fig. 122-6. Motorola, Motorola TMOS Power FET Design Ideas, 1985, p. 39.

Fig. 122-7. Maxim, 1986 Power Supply circuits, p. 44.

Fig. 122-8. Radio-Electronics, 4/85, p. 80.

Fig. 122-9. Motorola, Motorola TMOS Power FET Design Ideas, 1985, p. 36.

Chapter 123

Fig. 123-1. Linear Technology Corp., Linear Applications Handbook, 1987, p. AN9-14.

Fig. 123-2. Analog Devices, Data Acquisition Databook, 1982, p. 6-27.

Fig. 123-3. Electronic Engineering, 12/75, p. 11.

Fig. 123-4. Electronic Engineering, 7/76, p. 23.

Fig. 123-5. Reprinted with permission from Electronic Design. Copyright 1975, Penton Publishing.

Fig. 123-6. Linear Technology Corp., Linear Databook, 1986, p. 5-79.

Fig. 123-7. Reprinted with permission from Electronic Design. Copyright 1975, Penton Publishing.

Fig. 123-8. Linear Technology Corp., Linear Databook, 1986, p. 5-33.

Fig. 123-9. Electronic Engineering, 10/77, p. 17.

Fig. 123-10. Linear Technology Corp., Linear Applications Handbook, 1987, p. AN9-13.

Fig. 123-11. Signetics, 1987 Linear Data Manual Vol. 2: Industrial, 2/87, p. 7-63.

Fig. 123-12. Reprinted with permission from Electronic Design. Copyright 1967, Penton Publishing.

Chapter 124

Fig. 124-1. Electronic Engineering, 10/84, p. 41.

Fig. 124-2. Siliconix, Integrated Circuits Data Book, 3/85, p. 3-21.

Fig. 124-3. National Semiconductor Corp., CMOS Databook, 1981, p. 3-41.

Fig. 124-4. Reprinted with permission from Electronic Design. Copyright 1970, Penton Publishing.

Fig. 124-5. GE/RCA, BiMOS Operational Amplifiers Circuit Ideas, 1987, p. 16.

Fig. 124-6. GE/RCA, BiMOS Operational Amplifiers Circuit Ideas, 1987, p. 16.

Fig. 124-7. Radio-Electronics, 11/82, p. 92.

Fig. 124-8. Teledyne, Teledyne Semiconductor Databook, p. 9.

Fig. 124-9. GE/RCA, BiMOS Operational Amplifiers Circuit Ideas, 1987, p. 15.

Fig. 124-10. TAB Books, Third Book of Electronic Projects, p. 37.

Fig. 124-11. Ham Radio, 7/89, p. 62.

Fig. 124-12. Hands-On Electronics, 4/87, p. 93.

Fig. 124-13. Popular Electronics/Hands-On Electronics, 4/89, p. 25.

Fig. 124-14. GE/RCA, BiMOS Operational Amplifiers Circuit Ideas, 1987, p. 11.

Fig. 124-15. Gernsback Publications Inc., 44 New Ideas, 1985, p. 8.

Fig. 124-16. Electronics Today International, 10/78, p. 95.

Fig. 124-17. Radio-Electronics, 1/86, p. 104.

Fig. 124-18. National Semiconductor Corp., Transistor Databook, 1982, p. 7-26.

Fig. 124-19. Electronic Engineering, 9/85, p. 25.

Fig. 124-20. Electronics Today International, 6/76, p. 42.

Fig. 124-21. National Semiconductor Corp., Linear Databook, 1982, p. 362.

Fig. 124-22. Signetics, 1987 Linear Data Manual Vol. 2: Industrial, 11/14/86, p. 5-269.

Chapter 125

Fig. 125-1. Reprinted with permission from Electronic Design. Copyright 1989, Penton Publishing.

Fig. 125-2. Popular Electronics, 6/89, p. 22.

Fig. 125-3. Electronic Engineering, Applied Ideas, 3/89, p. 32.

Chapter 126

Fig. 126-1. Electronic Engineering, 7/88, p. 27.

Fig. 126-2. Electronics Today International, 6/76, p. 40.

Fig. 126-3. Signetics, Analog Data Manual, 1982, p. 8-10.

Fig. 126-4. Reprinted from EDN, 5/73, © 1989 Cahners Publishing Co., a division of Reed Publishing USA.

Fig. 126-5. Electronic Engineering, 1/83, p. 31.

Fig. 126-6. Reprinted from EDN, 8/20/78, © 1989 Cahners Publishing Co., a division of Reed Publishing USA.

Fig. 126-7. Fairchild Camera and Instrument Corp., Linear Databook, 1982, p. 4-180.

Index

Numbers preceded by a ''I,'' ''II,'' ''III,'' or ''IV'' are from *Encyclopedia of Electronic Circuits* Vol. I, II, III, or IV, respectively.

A

absolute-value amplifier, I-31
absolute-value circuit,I-37, IV-274
absolute-value full wave rectifier, II-528
absolute-value Norton amplifier, III-11
ac bridge circuit, II-81
ac flasher, III-196
ac linear coupler, analog, II-412
ac motor
  control for, II-375
  three-phase driver, II-383
  two-phase driver, II-382
ac sequential flasher, II-238
ac switcher, high-voltage optically
  coupled, III-408
ac timer, .2 to 10 seconds, adjustable,
  II-681
ac-coupled amplifiers, dynamic, III-17
ac/dc indicator, IV-214
ac-to-dc converter, I-165
  fixed power supplies, IV-395
  full-wave, IV-120
  high-impedance precision rectifier, I-
   164
acid rain monitor, II-245, III-361

acoustic-sound receiver/transmitter, IV-
  311
active antennas, III-1-2, IV-1-4
  basic designs, IV-3
  wideband rod, IV-4
  with gain, IV-2
active clamp-limiting amplifiers, III-15
active crossover networks, I-172-173
active filters (*see also* filter circuits)
  band reject, II-401
  bandpass, II-221, II-223, III-190
  digitally tuned low power, II-218
  five pole, I-279
  high-pass, second-order, I-297
  low-pass, digitally selected break
   frequency, II-216
  low-power, digitally selectable center
   frequency, III-186
  low-power, digitally tuned, I-279
  programmable, III-185
  RC, up to 150 kHz, I-294
  state-variable, III-189
  ten-band graphic equalizer using, II-
   684
  three-amplifier, I-289
  tunable, I-294

universal, II-214
  variable bandwidth bandpass, I-286
active integrator, inverting buffer, II-
  299
adapters
  dc transceiver, hand-held, III-461
  program, second-audio, III-142
  traveller's shaver, I-495
adder, III-327
AGC, II-17
AGC amplifiers
  AGC system for CA3028 IF amplifier,
   IV-458
  rf, wideband adjustable, III-545
  squelch control, III-33
  wideband, III-15
air conditioner, auto, smart clutch for,
  III-46
air flow detector, I-235, II-242, III-364
air flow meter (*see* anemometer)
air-pressure change detector, IV-144
air-motion detector, III-364
airplane propeller sound effect, II-592
alarms (*see also* detectors; indicators;
  monitors;sensors; sirens), III-3-9,
  IV-84-87

buzzers (*cont.*)
   continuous tone 2kHZ, I-11
   gated 2kHz, I-12

C

cable bootstrapping, I-34
cable tester, III-539
calibrated circuit, DVM auto, I-714
calibrated tachometer, III-598
calibration standard, precision, I-406
calibrators
   crystal, 100 kHz, I-185
   electrolytic-capacitor reforming
      circuit, IV-276
   ESR measurer, IV-279
   oscilloscope, II-433, III-436
   portable, I-644
   square-wave, 5 V, I-423
   tester, IV-265
   wave-shaping circuits, high slew
      rates, IV-650
cameras (*see* photography-related
     circuits; television-related circuits;
     video circuits)
canceller, central image, III-358
capacitance buffers
   low-input, III-498
   low-input, stabilized, III-502
capacitance meters, I-400, II-91-94,
     III-75-77
   A/D, three-and-a-half digit, III-76
   capacitance-to-voltage, II-92
   digital, II-94
capacitance multiplier, I-416, II-200
capacitance relay, I-130
capacitance switched light, I-132
capacitance-to-pulse width converter,
     II-126
capacitance-to-voltage meter, II-92
capacitor discharge
   high-voltage generator, III-485
   ignition system, II-103
capacity tester, battery, III-66
car port, automatic light controller, II-
     308
cars (*see* automotive circuits)
carrier-current circuits, III-78-82, IV-
     91-93
   AM receiver, III-81
   audio transmitter, III-79
   data receiver, IV-93
   data transmitter, IV-92
   FM receiver, III-80
   intercom, I-146
   power-line modem, III-82
   receiver, I-143
   receiver, single transistor, I-145
   receiver, IC, I-146

remote control, I-146
   transmitter, I-144
   transmitter, integrated circuit, I-145
carrier-operated relay (COR), IV-461
carrier system receiver, I-141
carrier transmitter with on/off 200kHz
     line, I-142
cascaded amplifier, III-13
cassette bias oscillator, II-426
cassette interface, telephone, III-618
cassette-recorders (*see* tape-recorder
     circuits)
centigrade thermometer, I-655, II-648,
     II-662
central image canceller, III-358
charge pool power supply, III-469
charge pumps
   positive input/negative output, I-418,
     III-360
   regulated for fixed power supplies,
     IV-396
chargers (*see* battery charger)
chase circuit, I-326, III-197
Chebyshev filters (*see also* filter cir-
     cuits)
   bandpass, fourth-order, III-191
   fifth-order multiple feedback low-
     pass, II-219
   high-pass, fourth-order, III-191
chime circuit, low-cost, II-33
chopper amplifier, I-350, II-7, III-12
checkers
   buzz box continuity and coil, I-551
   car battery condition, I-108
   crystal, I-178, I-186
   zener diode, I-406
chroma demodulator with RGB matrix,
     III-716
chug-chug sound generator, III-576
circuit breakers (*see also* protection
     circuits)
   12ns, II-97
   ac, III-512
   high-speed electronic, II-96
   trip circuit, IV-423
circuit protection (*see* protection cir-
     cuits)
clamp-on-current probe compensator,
     II-501
clamp-limiting amplifiers, active, III-15
clamping circuits
   video signal, III-726
   video summing amplifier and, III-710
class-D power amplifier, III-453
clippers, II-394, IV-648
   audio-powered noise, II-396
   audio clipper/limiter, IV-355
   zener-design, fast and symmetrical,
     IV-329

clock circuits, II-100-102, III-83-85
   60Hz clock pulse generator, II-102
   adjustable TTL, I-614
   comparator, I-156
   crystal oscillators, micropower
     design, IV-122
   digital, with alarm, III-84
   gas discharge displays, 12-hour, I-253
   oscillator/clock generator, III-85
   phase lock, 20-Mhz to Nubus, III-
     105
   run-down clock for games, IV-205
   sensor touch switch and clock, IV-
     591
   single op amp, III-85
   source, clock source, I-729
   three-phase from reference, II-101
   TTL, wide-frequency, III-85
   Z80 computer, II-121
clock generators
   oscillator, I-615, III-85
   precision, I-193
   pulse generator, 60 Hz, II-102
clock radio, I-542
   AM/FM, I-543
CMOS circuits
   555 astable true rail to rail square
     wave generator, II-596
   9-bit, III-167
   coupler, optical, III-414
   crystal oscillator, III-134
   data acquisition system, II-117
   flasher, III-199
   inverter, linear amplifier from, II-11
   mixer, I-57
   optical coupler, III-414
   oscillator, III-429, III-430
   short-pulse generator, III-523
   timer, programmable, precision, III-
     652
   touch switch, I-137
   universal logic probe, III-499
coaxial cable, five-transistor pulse
     booster, II-191
Cockcroft-Walton cascaded voltage
     doubler, IV-635
code-practice oscillator, I-15, I-20, I-
     22, II-428-431, IV-373, IV-375, IV-
     376
coil drivers, current-limiting, III-173
coin flipper circuit, III-244
color amplifier, video, III-724
color-bar generator, IV-614
color organ, II-583, II-584
color video amplifier, I-34
Colpitts crystal oscillator, I-194, I-572,
     II-147
   1-to-20 MHz, IV-123
   frequency checker, IV-301

control circuits (*cont.*)
 full-wave SCR, I-375
 heater, I-639
 hi-fi tone, high-Z input, I-676
 high-power, sensitive contacts for, I-371
 LED brightness, I-250
 light-level, I-380
 light-level, 860 W limited-range low-cost, I-376
 light-level, brightness, low-loss, I-377
 liquid level, I-388
 model train and/or car, I-453, I-455
 motor controllers (*see* motor control circuits)
 on/off, I-665
 phase control, hysteresis-free, I-373
 power tool torque, I-458
 sensitive contact, high power, I-371
 servo system, III-384
 single-setpoint temperature, I-641
 speed control (*see* speed controllers)
 switching, III-383
 temperature, I-641-643
 temperature-sensitive heater, I-640
 three-phase power-factor, II-388
 tone control (*see* tone controls)
 voltage-control, pulse generator and, III-524
 water-level sensing, I-389
 windshield wiper, I-105
conversion and converters, I-503, II-123-132, III-109-122, IV-110-120
 3-5 V regulated output, III-739
 4-18 MHz, III-114
 4-to-20-mA current loop, IV-111
 5V-to-isolated 5V at 20MA, III-474
 5V/0.5A buck, I-494
 9-to-5 V converter, IV-119
 12 V- to 9-, 7.5-, or 6-V, I-508
 12-to-16 V, III-747
 +50V feed forward switch mode, I-495
 +50 V push-pull switched mode, I-494
 100 MHz, II-130
 100 V/10.25 A switch mode, I-501
 ac-to-dc, I-165
 ac-to-dc, high-impedance precision rectifier, I-164
 analog-to-digital (*see* analog-to-digital conversion)
 ATV rf receiver/converter, IV-420 MHz, low-noise, IV-496, IV-497
 BCD-to-analog, I-160
 BCD-to-parallel, multiplexed, I-169
 buck/boost, III-113
 calculator-to-stopwatch, I-153

capacitance-to-pulse width, II-126
current-to-frequency, IV-113
current-to-frequency, wide-range, I-164
current-to-voltage, I-162, I-165
current-to-voltage, grounded bias and sensor, II-126
current-to-voltage, photodiode, II-128
dc-dc, 3-25 V, III-744, IV-118
dc-to-dc, +3-to-+5 V battery, IV-119
dc-to-dc, 1-to-5 V, IV-119
dc-to-dc, bipolar, no inductor, II-132
dc-to-dc, fixed 3- to 15-V supplies, IV-400
dc-to-dc, isolated +15V., III-115
dc-to-dc, push-pull, 400 V, 60 W, I-210
dc-to-dc, regulating, I-210, I-211, II-125, III-121
dc-to-dc, step up-step down, III-118
digital-to-analog (*see* digital-to-analog conversion)
fixed power supply, III-470
flyback, I-211
flyback, self oscillating, I-170, II-128
flyback, voltage, high-efficiency, III-744
frequency, I-159
frequency-to-voltage (*see* frequency-to-voltage conversion)
high-to-low impedance, I-41
intermittent converter, power-saving design, IV-112
light intensity-to-frequency, I-167
logarithmic, fast-action, I-169
low-frequency, III-111
ohms-to-volts, I-168
oscilloscope, I-471
period-to-voltage, IV-115
pico-ampere, 70 voltage with gain, I-170
PIN photodiode-to-frequency, III-120
polarity, I-166
positive-to-negative, III-112, III-113
peak-to-peak, ac-dc, precision, II-127
pulse height-to-width, III-119
pulse train-to-sinusoid, III-122
pulse width-to-voltage, III-117
radio beacon converter, IV-495
rectangular-to-triangular waveform, IV-116-117
regulated 15-Vout 6-V driven, III-745
resistance-to-voltage, I-161-162
RGB-composite video signals, III-714
RMS-to-dc, II-129, I-167
RMS-to-dc, 50-MHz thermal, III-117
RGB-to-NTSC, IV-611

sawtooth wave converter, IV-114
shortwave, III-114
simple LF, I-546
sine-to-square wave, I-170, IV-120
square-to-sine wave, III-118
square-to-triangle wave, TTL, II-125
temperature-to-frequency, I-168
temperature-to-time, III-632-633
triangle-to-sine wave, II-127
TTL-to-MOS logic, II-125, I-170
two-wire to four-wire audio, II-14
unipolar-to-dual voltage supply, III-743
video, a/d and d/a, IV-610-611
video, RGB-to-NTSC, IV-611
VLF, I-547
VLF, rf converter, IV-497
voltage ratio-to-frequency, III-116
voltage, III-742-748, III-742
voltage, negative voltage, uP-controlled, IV-117
voltage, offline, 1.5-W, III-746
voltage-to-current, I-166, II-124, III-110, IV-118
voltage-to-current, power, I-163
voltage-to-current, zero IB error, III-120
voltage-to-frequency (*see* voltage-to-frequency conversion)
voltage-to-pulse duration, II-124
WWV-to-SW rf converter, IV-499
coprocessor socket debugger, III-104
countdown timer, II-680
counters, I-133-139, III-123-130
 analog circuit, II-137
 attendance, II-138
 binary, II-135
 divide-by-N, CMOS programmable, I-257
 divide-by-*n*, 1+ GHz, IV-155
 divide-by-odd-number, IV-153
 frequency, III-340, III-768, IV-300
 frequency, 1.2 GHz, III-129
 frequency, 10-MHz, III-126
 frequency, 100 MHz, periodic, II-136
 frequency, low-cost, III-124
 frequency, preamp, III-128
 frequency, tachometer and, I-310
 geiger, I-536-537
 microfarad counter, IV-275
 odd-number divider and, III-217
 preamplifier, oscilloscope, III-438
 precision frequency, I-253
 programmable, low-power wide-range, III-126
 ring, 20 kHz, II-135
 ring, incandescent lamp, I-301
 ring, low cost, I-301
 ring, low-power pulse circuit, IV-437

VCO driver, op-amp design, IV-362
drop-voltage recovery for long-line
     systems, IV-328
drum sound effect, II-591
dual-tone decoding, II-620
dual-tracking regulator, III-462
duplex line amplifier, III-616
duty-cycle detector, IV-144
duty-cycle meter, IV-275
duty-cycle monitor, III-329
duty-cycle multivibrator, 50-percent,
     III-584
duty-cycle oscillators
     50-percent, III-426
     variable, fixed-frequency, III-422
dwell meters
     breaker point, I-102
     digital, III-45

E

eavesdropper, telephone, wireless, III-
     620
echo effect, analog delay line, IV-21
edge detector, I-266, III-157
EEPROM pulse generator, 5V-pow-
     ered, III-99
EKG simulator, three-chip, III-350
elapsed-time timer, II-680
electric-fence charger, II-202
electric-vehicle battery saver, III-67
electrolytic-capacitor reforming circuit,
     IV-276
electrometer, IV-277
electrometer amplifier, overload pro-
     tected, II-155
electronic dice, IV-207
electronic locks, II-194-197, IV-161-
     163
     combination, I-583, II-196
     digital entry lock, IV-162
     keyless design, IV-163
     three-dial combination, II-195
electronic music, III-360
electronic roulette, II-276, IV-205
electronic ship siren, II-576
electronic switch, push on/off, II-359
electronic theremin, II-655
electronic thermometer, II-660
electronic wake-up call, II-324
electrostatic detector, III-337
emergency lantern/flasher, I-308
emergency light, I-378, IV-250
emissions analyzer, automotive
     exhaust, II-51
emulators, II-198-200
     capacitance multiplier, II-200
     JFET ac coupled integrator, II-200
     resistor multiplier, II-199
     simulated inductor, II-199

encoders
     decoder and, III-14
     telephone handset tone dial, I-634,
          III-613
     tone, I-67, I-629
     tone, two-wire, II-364
engine tachometer, I-94
enlarger timer, II-446, III-445
envelope detectors, III-155
     AM signals, IV-142
     low-level diodes, IV-141
envelope generator/modulator, musical,
     IV-22
EPROM, Vpp generator for, II-114
equalizers, I-671, IV-18
     ten-band, graphic, active filter in, II-
          684
     ten-band, octave, III-658
equipment-on reminder, I-121
exhaust emissions analyzer, II-51
expanded-scale meters
     analog, III-774
     dot or bar, II-186
expander circuits (see compressor/
     expander circuits)
extended-play circuit, tape-recorders,
     III-600
extractor, square-wave pulse, III-584

F

555 timer
     astable, low duty cycle, II-267
     beep transformer, III-566
     integrator to multiply, II-669
     RC audio oscillator from, II-567
     square wave generator using, II-595
fader, audio fader, IV-17
fail-safe semiconductor alarm, III-6
fans
     infrared heat-controlled fan, IV-226
     speed controller, automatic, III-382
Fahrenheit thermometer, I-658
fault monitor, single-supply, III-495
fax/telephone switch, remote-con-
     trolled, IV-552-553
feedback oscillator, I-67
fence charger, II-201-203
     battery-powered, II-202
     electric, II-202
     solid-state, II-203
FET circuits
     dual-trace scope switch, II-432
     input amplifier, II-7
     probe, III-501
     voltmeter, III-765, III-770
fiber optics, II-204-207, III-176-181
     driver, LED, 50-Mb/s, III-178
     interface for, II-207
     link, I-268, I-269, I-270, III-179

motor control, dc, II-206
     receiver, 10 MHz, II-205
     receiver, 50-Mb/s, III-181
     receiver, digital, III-178
     receiver, high-sensitivity, 30nw, I-
          270
     receiver, low-cost, 100-M baud rate,
          III-180
     receiver, low-sensitivity, 300nW, I-
          271
     receiver, very-high sensitivity, low
          speed, 3nW, I-269
     repeater, I-270
     speed control, II-206
     transmitter, III-177
field disturbance sensor/alarm, II-507
field-strength meters, II-208-212, III-
     182-183, IV-164-166
     1.5-150 MHz, I-275
     adjustable sensitivity indicator, I-274
     high-sensitivity, II-211
     LF or HF, II-212
     microwave, low-cost, I-273
     rf sniffer, II-210
     sensitive, I-274, III-183
     signal-strength meter, IV-166
     transmission indicator, II-211
     tuned, I-276
     UHF fields, IV-165
     untuned, I-276
filter circuits, II-213-224, III-184-192,
     IV-167-177
     active (see active filters)
     antialiasing/sync-compensation, IV-
          173
     audio, biquad, III-185
     audio, tunable, IV-169
     bandpass (see bandpass filters)
     band-reject, active, II-401
     biquad, I-292-293
     biquad, audio, III-185
     biquad, RC active bandpass, I-285
     bridge filter, twin-T, programmable,
          II-221
     Butterworth, high-pass, fourth-order,
          I-280
     Chebyshev (see Chebyshev filters)
     CW, razor-sharp, II-219
     full wave rectifier and averaging, I-
          229
     high-pass (see high-pass filters)
     low-pass (see low-pass filters)
     networks of, I-291
     noise, dynamic, III-190
     noisy signals, III-188
     notch (see notch filters)
     programmable, twin-T bridge, II-221
     rejection, I-283
     ripple suppressor, IV-175

probes (*cont.*)
  clamp-on-current compensator, II-501
  CMOS logic, I-523
  FET, III-501
  general purpose rf detector, II-500
  ground-noise, battery-powered, III-500
  logic probes (*see* logic probes)
  microvolt, II-499
  optical light probe, IV-369
  pH, I-399, III-501
  prescaler, 650 MHz amplifying, II-502
  rf, I-523, III-498, III-502, IV-433
  single injector-tracer, II-500
  test, 4-220V, III-499
  three-in-one test set: logic probe, signal tracer, injector, IV-429
  tone, digital IC testing, II-504
  universal test probe, IV-431
process control interface, I-30
processor, CW signal, I-18
product detector, I-223
programmable amplifiers, II-334, III-504-508
  differential-input, programmable gain, III-507
  inverting, programmable-gain, III-505
  noninverting, programmable-gain, III-505
  precision, digital control, III-506
  precision, digitally programmable, III-506
  programmable-gain, selectable input, I-32
  variable-gain, wide-range digital control, III-506
projectors (*see* photography-related circuits)
proportional temperature controller, III-626
protection circuits, II-95-99, III-509-513
  12ns circuit breaker, II-97
  automatic power down, II-98
  circuit breaker, ac, III-512
  circuit breaker, electronic, high-speed, II-96
  compressor protector, IV-351
  crowbars, electronic, II-99, III-510
  heater protector, servo-sensed, III-624
  line protectors, computer I/O, 3 uP, IV-101
  line dropout detector, II-98
  line-voltage monitor, III-511
  low-voltage power disconnector, II-97
  overvoltage, II-96, IV-389

overvoltage, fast, III-513
overvoltage, logic, I-517
polarity-protection relay for power supplies, IV-427
power-down, II-98
power-failure alarm, III-511
power-line connections monitor, ac, III-510
power supply, II-497, I-518
reset-protection for computers, IV-100
proximity sensors, I-135-136, I-344, II-505-507, III-514-518, IV-341-346
  alarm for, II-506
  capacitive, III-515
  field disturbance sensor/alarm, II-507
  infrared-reflection switch, IV-345
  relay-output, IV-345
  SCR alarm, III-517
  self-biased, changing field, I-135
  switch, III-517
  UHF movement detector, III-516
pseudorandom sequencer, III-301
pulse circuits, IV-435-440
  amplitude discriminator, III-356
  coincidence detector, II-178
  counter, ring counter, low-power, IV-437
  delay, dual-edge trigger, III-147
  detector, missing-pulse, III-159
  divider, non-integer programmable, III-226, II-511
  extractor, square-wave, III-584
  generator, 555-circuit, IV-439
  generator, delayed-pulse generator, IV-440
  generator, free-running, IV-438
  generator, logic troubleshooting applications, IV-436
  generator, transistorized design, IV-437
  height-to-width converters, III-119
  oscillator, fast, low duty-cycle, IV-439
  oscillator, start-stop, stable design, IV-438
  pulse train-to-sinusoid converters, III-122
  sequence detector, II-172
  stretcher, IV-440
  stretcher, negative pulse stretcher, IV-436
  stretcher, positive pulse stretcher, IV-438
pulse generators, II-508-511
  2-ohm, III-231
  300-V, III-521
  astable multivibrator, II-510
  clock, 60Hz, II-102
  CMOS short-pulse, III-523
  delayed, II-509

EEPROM, 5V-powered, III-99
interrupting pulse-generation, I-357
logic, III-520
programmable, I-529
sawtooth-wave generator and, III-241
single, II-175
two-phase pulse, I-532
unijunction transistor design, I-530
very low duty-cycle, III-521
voltage-controller and, III-524
wide-ranging, III-522
pulse supply, high-voltage power supplies, IV-412
pulse-dialing telephone, III-610
pulse-position modulator, III-375
pulse-width-to-voltage converters, III-117
pulse-width modulators (PWM), IV-326
  brightness controller, III-307
  control, microprocessor selected, II-116
  modulator, III-376
  motor speed control, II-376, III-389
  multiplier circuit, II-264, III-214
  out-of-bounds detector, III-158
  proportional-controller circuit, II-21
  servo amplifier, III-379
  speed control/energy-recovering brake, III-380
  very short, measurement circuit, III-336
pulse/tone dialer, single-chip, III-603
pulsers, laser diode, III-311
pump circuits
  controller, single chip, II-247
  positive input/negative output charge, I-418
push switch, on/off, electronic, II-359
push-pull power supply, 400V/60W, II-473
pushbutton power control switch, IV-388
PUT battery chargers, III-54
PUT long-duration timer, II-675
pyrometer, optical, I-654

Q

Q-multipliers
  audio, II-20
  transistorized, I-566
QRP CW transmitter, III-690
QRP SWR bridge, III-336
quad op amp, simultaneous waveform generator using, II-259
quadrature oscillators, III-428
  square-wave generator, III-585
quartz crystal oscillator, two-gate, III-136

quick-deactivating battery sensor, III-61

R

race-car motor/crash sound generator, III-578
radar detectors, II-518-520, IV-441-442
  one-chip, II-519
radiation detectors, II-512-517
  alarm, II-4
  micropower, II-513
  monitor, wideband, I-535
  photomultiplier output-gating circuit, II-516
  pocket-sized Geiger counter, II-514
radiation-hardened 125A linear regulator, II-468
radio
  AM car-radio to short-wave radio converter, IV-500
  AM demodulator, II-160
  AM radio, power amplifier, I-77
  AM radio, receivers, III-81, III-529, III-535
  AM/FM, clock radio, I-543
  AM/FM, squelch circuit, II-547, III-1
  amateur radio, III-260, III-534, III-675
  automotive, receiver for, II-525
  clock, I-542
  direction finder, radio signals, IV-148-149
  FM (see FM transmissions)
  portable-radio 3 V fixed power supplies, IV-397
  radio beacon converter, IV-495
  receiver, AM radio, IV-455
  receiver, old-time design, IV-453
  receiver, reflex radio receiver, IV-452
  receiver, short-wave receiver, IV-454
  receiver, TRF radio receiver, IV-452
radio beacon converter, IV-495
radio-control circuits
  audio oscillator, II-567, III-555
  motor speed controller, I-576
  phase sequence reversal by, II-438
  oscillator, emitter-coupled, II-266
  receiver/decoder, I-574
  single-SCR design, II-361
radioactivity (see radiation detectors)
rain warning beeper, II-244, IV-189
RAM, non-volatile CMOS, stand-by power supply, II-477
ramp generators, I-540, II-521-523, III-525-527, IV-443-447
  accurate, III-526
  integrator, initial condition reset, III-527

linear, II-270
  variable reset level, II-267
  voltage-controlled, II-523
ranging system, ultrasonic, III-697
reaction timer, IV-204
read-head pre-amplifier, automotive circuits, III-44
readback system, disc/tape phase modulated, I-89
readout, rf current, I-22
receiver audio circuit, IV-31
receivers and receiving circuits (see also transceivers; transmitters), II-524-526, III-528-535, IV-448-461
  50kHz FM optical transmitter, I-361
  acoustic-sound receiver, IV-311
  AGC system for CA3028 IF amplifier, IV-458
  AM, III-529, IV-455
  AM, carrier-current circuit, III-81
  AM, integrated, III-535
  analog, I-545
  ATV rf receiver/converter, 420 MHz, low-noise, IV-496, IV-497
  car radio, capacitive diode tuning/ electronic MW/LW switching, II-525
  carrier current, I-143, I-146
  carrier current, single transistor, I-145
  carrier system, I-141
  carrier-operated relay (COR), IV-461
  CMOS line, I-546
  data receiver/message demuxer, three-wire design, IV-130
  fiber optic, 10 MHz, II-205
  fiber optic, 50-Mb/s, III-181
  fiber optic, digital, III-178
  fiber optic, high-sensitivity, 30nW, I-270
  fiber optic, low-cost, 100 M baud rate, III-180
  fiber optic, low-sensitivity, 300nW, I-271
  fiber optic, very high-sensitivity, low speed 3nW, I-269
  FM, carrier-current circuit, III-80
  FM, MPX/SCA, III-530
  FM, narrow-band, III-532
  FM, tuner, III-529
  FM, zero center indicator, I-338
  FSK data, III-533
  ham-band, III-534
  IF amplifier, IV-459
  IF amplifier, preamp, 30 MHz, IV-460
  IF amplifier/receiver, IV-459
  infrared, I-342, II-292, III-274, IV-220-221

laser, IV-368
LF receiver, IV-451
line-type, digital data, III-534
line-type, low-cost, III-532
  monitor for, II-526
  optical, I-364, II-418
  optical light receiver, IV-367, IV-368
  PLL/BC, II-526
  pulse-frequency modulated, IV-453
  radio control, decoder and, I-574
  radio receiver, AM, IV-455
  radio receiver, old-time design, IV-453
  radio receiver, reflex, IV-452
  radio receiver, TRF, IV-452
  regenerative receiver, one-transistor design, IV-449
  RS-232 to CMOS, III-102
  short-wave receiver, IV-454
  signal-reception alarm, III-270
  superheterodyne receiver, 3.5-to-10 MHz, IV-450-451
  tracer, III-357
  transceiver/mixer, HF, IV-457
  ultrasonic, III-698, III-705
  zero center indicator for FM, I-338
recording amplifier, I-90
recording devices (see tape-recorder circuits)
rectangular to triangular waveform converter, IV-116-117
rectifiers, II-527-528, III-536-537
  absolute value, ideal full wave, II-528
  averaging filter, I-229
  bridge rectifier, fixed power supplies, IV-398
  broadband ac active, IV-271
  diodeless, precision, III-537
  full-wave, I-234, III-537, IV-328, IV-650
  half wave, I-230, II-528, IV-325
  half-wave, fast, I-228
  high-impedance precision, for ac/dc converter, I-164
  inverter/rectifier, programmable op-amp design, IV-364
  low forward-drop, III-471
  precision, I-422
  synchronous, phase detector-selector/balanced modulator, III-441
redial, electronic telephone set with, III-606
reference voltages, I-695, III-773-775
  ± 10V, I-696
  ± 3V, I-696
  ± 5V, I-696
  0- to 20 V power, I-694, I-699
  amplifier, I-36
  bipolar output, precision, I-698

W

Other Bestsellers of Related Interest

ENCYCLOPEDIA OF ELECTRONIC CIRCUITS
Vol. 1—Rudolf F. Graf

". . . schematics that encompass virtually the entire spectrum of electronics technology . . . This is a well worthwhile book to have handy." —**Modern Electronics**

Discover hundreds of the most versatile electronic and integrated circuit designs, all available at the turn of a page. You'll find circuit diagrams and schematics for a wide variety of practical applications. Many entries also include clear, concise explanations of the circuit configurations and functions. 768 pages, 1,762 illustrations. Book No. 1938, $29.95 paperback, $60.00 hardcover

Volume 2, 740 pages, 1,762 illustrations. Book No. 3138, $32.95 paperback, $60.00 hardcover

Volume 3, 832 pages, 1,050 illustrations. Book No. 3348, $32.95 paperback, $60.00 hardcover

ELECTRONIC DATABOOK—4th Edition
—Rudolf F. Graf

If it's electronic, it's here—detailed and comprehensive! Use this book to broaden your electronics information base. Revised and expanded to include all up-to-date information, this fourth edition makes any electronic job easier and less time-consuming. You'll find information that will aid in the design of local area networks, computer interfacing structure, and more! 528 pages, 132 illustrations. Book No. 2958, $24.95 paperback only

UNDERSTANDING ELECTRONICS—3rd Edition
—R. H. Warring, Edited by G. Randy Slone

Design and build your own circuits with the classic reference that's now more complete than ever. Revised with state-of-the-art information on all the modern advances in electronics, you'll find thorough coverage of the basics of electronics, and everything from AC and DC power to the developing new fields of photoelectronics and digital computing. 230 pages, 188 illustrations. Book No. 3044, $12.95 paperback only

MASTERING ELECTRONICS MATH
—2nd Edition—R. Jesse Phagan

A self-paced text for hobbyists and a practical toolbox reference for technicians, this book guides you through the practical calculations needed to design and troubleshoot circuits and electronic components. Clear explanations and sample problems illustrate each concept, including how each is used in common electronics applications. If you want to gain a strong understanding of electronics math and stay on top of your profession, this book will be a valuable tool for you. 344 pages, 270 illustrations. Book No. 3589, $17.95 paperback, $27.95 hardcover

BOB GROSSBLATT'S GUIDE TO CREATIVE CIRCUIT DESIGN—Robert Grossblatt

Radio Electronics' popular columnist Robert Grossblatt brings his unique circuit design philosophy and style to this hands-on guide. Emphasizing the importance of scientific method over technical knowledge, he walks you through the circuit design process—from brainwork to paperwork to boardwork—and suggests ways for making your bench time as efficient as possible. 248 pages, 129 illustrations. Book No. 3610, $17.95 paperback, $28.95 hardcover

INTERNATIONAL ENCYCLOPEDIA OF INTEGRATED CIRCUITS—2nd Edition
—Stan Gibilisco

The most thorough coverage of foreign and domestic integrated circuits is available today in this giant resource. Seven separate sections detail thousands of ICs and their applications, including all relevant information, charts, and tables. This second edition of a unique, all-in-one reference tells what each IC is, what it does, how it does it, and what its relationship is to other ICs and their applications. 1,168 pages, 4,605 illustrations. Book No. 3802, $84.95 hardcover only

Prices Subject to Change Without Notice.

Look for These and Other TAB Books at Your Local Bookstore

To Order Call Toll Free 1-800-822-8158
(In PA, AK, and Canada call 717-794-2191)

or write to TAB Books, Blue Ridge Summit, PA 17294-0840.

Title	Product No.	Quantity	Price

☐ Check or money order made payable to TAB Books

Charge my ☐ VISA ☐ MasterCard ☐ American Express

Acct. No. _____ Exp. _____

Signature: _____

Name: _____

Address: _____

City: _____

State: _____ Zip: _____

Subtotal $ _____

Postage and Handling
($3.00 in U.S., $5.00 outside U.S.) $ _____

Add applicable state and local
sales tax $ _____

TOTAL $ _____

TAB Books catalog free with purchase; otherwise send $1.00 in check or money order and receive $1.00 credit on your next purchase.

Orders outside U.S. must pay with international money in U.S. dollars

TAB Guarantee: If for any reason you are not satisfied with the book(s) you order, simply return it (them) within 15 days and receive a full refund. BC